Principles of Pharmacology

Principles of Pharmacology reviews the fundamentals of pharmacology with a focus on the various body systems. Drug classifications, names, sources, and development serve as essential foundations. Furthermore, this volume uniquely discusses that understanding pharmacokinetics, pharmacodynamics, pharmacogenomics, and drug toxicity is crucial before addressing medications, clinical indications, mechanisms of action, adverse effects, and contraindications. Readers will encounter clinical case studies at the end of each chapter, accompanied by questions that provide real-world scenarios. Helpful appendices address high-alert medications, dosage calculations, and answer keys.

Key Features:

- This book highlights both basic pharmacology, including pharmacodynamics and pharmacokinetics, and clinical pharmacology.
- Each chapter will cover indications, mechanisms of action, adverse effects, and contraindications.
- This book begins with a detailed introduction to the origins of pharmacology and drug effects on various body systems, which are crucial to today's pharmacology environment.
- Pharmacology is a multidisciplinary science by its nature. Hence, disciplinary areas include chemistry, biology, and neuroscience.

This book benefits health providers without in-depth training in pharmacology, such as pharmacy technicians, medical assistants, dental students, medical students, nurse practitioners, physician assistants, surgical technicians, and physical therapists.

Principles of Pharmacology

Jahangir Moini MD, MPH
Professor of Science and Health (retired)
Eastern Florida State College, Palm Bay, FL.

CRC Press
Taylor & Francis Group
Boca Raton London New York

CRC Press is an imprint of the
Taylor & Francis Group, an **informa** business

First edition published 2026
by CRC Press
2385 NW Executive Center Drive, Suite 320, Boca Raton FL 33431

and by CRC Press
4 Park Square, Milton Park, Abingdon, Oxon, OX14 4RN

CRC Press is an imprint of Taylor & Francis Group, LLC

© 2026 Taylor & Francis Group, LLC

ISBN: 9781032610795 (hbk)
ISBN: 9781032609478 (pbk)
ISBN: 9781003461913 (ebk)

DOI: 10.1201/9781003461913

Typeset in Times
by codeMantra

Dedication

To my precious grandchildren, Artemis, Anastasia, Anabelle, and Laila; to my wonderful wife, Hengameh; and to my beautiful daughters, Mahkameh and Morvarid.

Contents

PART I Fundamentals of Pharmacology

PART II   Drugs Affecting the Nervous System

PART III *Drugs Affecting the Cardiovascular System*

About the Author

Dr. Moini was an assistant professor in the Department of Preventive Medicine at Tehran University Medical School for nine years. For 18 years, he was the Director of Epidemiology for the Brevard County Health Department. For 15 years, he was the Director of Science and Health for Everest University in Melbourne, FL. He was also a Professor of Science and Health at Everest University for 24 years. He was a Professor of Science and Health at Eastern Florida State College for six years but is now retired. He is an international author of 62 books. His *Anatomy and Physiology for Healthcare Professionals* was translated and released in Japan and South Korea in 2020. His *Complications of Diabetes Mellitus* was translated into Spanish and released in 2023.

Preface

The author welcomes you to this first edition of *Principles of Pharmacology*. He aims to provide thorough and up-to-date essential information on pharmacology and drug effects on various body systems. This book also includes a discussion of special topics. It is intended for all pharmacy technicians, medical students, nurses, nurse practitioners, and physician assistants. It can be translated into many different languages and will be a valuable resource worldwide.

This book is organized into three parts. In Part I, the chapters discuss general principles of pharmacology. Part II focuses on drug effects on various body systems. Part III covers drugs affecting multiple systems.

The book's nine appendices focus on the top 100 drugs by prescription, drug-food and drug-drug interactions, new drugs and formulations, the ISMP's list of high-alert medications, a drug identification guide, prototype drugs and indications, case studies (including practical scenarios with critical thinking questions), dosage calculations, and an answer key that corresponds with the chapter questions and case studies.

The features and content most beneficial to readers include real-world case studies with critical thinking questions. Four-color printing helps illustrate the non-text components and makes the e-book version more engaging. Each chapter's figures, graphs, and tables help illuminate the text material to engage readers and give them additional insights and understanding. There is also a glossary and an index.

Acknowledgments

The author appreciates the contributions of everyone who assisted in creating this book, including the Taylor & Francis team, Hilary Lafoe, Senior Editor, Adam Mlodzinski, Editorial Assistant, Gabrielle Perez, Project Editor; Morvarid Moini Kihm, DMD, MPH, and Greg Vadimsky.

Part I

Fundamentals of Pharmacology

1 Drug Classifications, Drug Names, and Drug Sources

LEARNING OBJECTIVES

After studying this chapter, readers should be able to:

1. List the father of medicine and the father of botany.
2. Describe the concepts of pharmacology.
3. Explain the importance of the Orange Book.
4. Describe the pharmacoepidemiology of the drug development process.
5. Review drug classifications and illicit drugs.
6. Compare the three different types of drug names.
7. List the various sources from which drugs are derived.
8. Summarize the biogenetically engineered sources.

OVERVIEW

Pharmacology is a broad science that studies the origin, nature, chemistry, effects, and uses of drugs. The study of a drug's biological effects on patients, as used in medical treatment, is called *clinical pharmacology*. This focuses on the rate at which the body absorbs a drug, its distribution, excretion, and potential toxicity or adverse effects. Clinical pharmacology studies drug effects in patients and healthy volunteers. Drugs are administered in higher quantities than ever before in history. Approximately 4 billion prescriptions are written every year in the United States alone. Nearly 50% of all Americans take at least one prescription drug regularly. Substances used for therapeutic purposes are drugs or medicines, biologics, or alternative therapies. Drug names include chemical, generic, and brand or trade names. Drug sources include plants, animals, minerals, synthetic or semisynthetic sources, and genetic engineering. With the evolution of pharmacology, the healthcare industry plays a more crucial role than ever in ensuring the safe and effective use of medications.

THE HISTORY OF PHARMACOLOGY

In early history, disease prevention was linked to beliefs in superstitions, evil spirits, and magical powers. The first sources of drugs ever used came from plants. Herbal medicine has been practiced in nearly every culture throughout time. Among the first medicinal agents were alcohol and opium, which are still used today. Specific agents considered poisons, such as *curare*, still have limited therapeutic use. The first documented pharmacological writings come from Egypt and are over 3,000 years old. The Chinese recorded a 40-volume collection of plant remedies in approximately 2,700 B.C. At about 2,600 B.C., the earliest pharmacy record was created in Babylonia, Mesopotamia. In China (about 2,000 B.C.), Shen Nung is believed to have written the first *Pen T'sao*, a native herbal documentation of 365 drugs. The Egyptian *Ebers Papyrus* (circa 1,500 B.C.) listed over 700 remedies for treating various conditions.

In the 4th century B.C., Hippocrates declared in Greece that natural laws governed the understanding of health and disease. Human dissections began for early medical practitioners to understand how individual organs functioned. Hippocrates was a philosopher, physician, and pharmacist, today called "the father of medicine." He is believed to have originated the *Oath of Hippocrates* (or "Hippocratic Oath"), which is still used today.

At about 300 B.C., Theophrastus (the "father of botany") wrote about classifications of plants based on their various parts. Between 120 and 63 B.C., Mithridates (the "father of **toxicology**") studied the adverse effects of plants. In the Greco-Roman era, the Romans widely adopted many Greek medical discoveries. The Romans organized this medical (and pharmaceutical) knowledge, converting theory into rules of medical practice.

In the 1st century A.D., Dioscorides developed a list of 600 plants and their classifications (by substance) called *De Materia Medica*. He helped transition the Greek system of knowledge to the Roman system of science. *De Materia Medica* was widely used as the most significant source of pharmaceutical knowledge until the 16th century.

The Greek physician Galen created De *Methodo Medendi (On the Art of Healing)* between A.D. 129 and 210. Galen criticized other physicians who did not prepare their remedies. After the end of the Roman Empire (approximately A.D. 476), significant advances in medicine and pharmacy occurred in the Arab countries. The first formularies, dosage forms, and pharmacy shops appeared. Specific dosage forms included syrups and confections mixed with sugar and honey but containing medicinal agents. The first pharmacy shop appeared in Baghdad (approximately A.D. 762). In 1190, the first designated hospital pharmacy was opened in Marrakech, Morocco. Later, in Salerno, Italy, what may have been Europe's first university was established, which significantly contributed to pharmacy and medicine.

The first pharmacy guilds developed around the 12th century. Between 1231 and 1240, the Magna Carta of Pharmacy was established, separating pharmacy from medicine. Between 1493 and 1541, the Renaissance considerably grew these and many other fields. Paracelsus in Switzerland was the first scientist to advocate using single drugs (instead of complex mixtures of medicines) to treat

DOI: 10.1201/9781003461913-2

FIGURE 1.1   Paracelsus. (Source: Wikimedia Commons (public domain) https://upload.wikimedia.org/wikipedia/commons/thumb/4/4a/Paracelsus.jpg/640px-Paracelsus.jpg.)

individual diseases (see Figure 1.1). He was listed as the *father of pharmacology* because he believed all substances were potential poisons based on their dosage.

In the 17th century, the English physiologist William Harvey first explained the beneficial and harmful effects of drugs. At this time, the first recorded reference to the word *pharmacology* occurred. He also demonstrated intravenous drug administration for the first time. However, a conflict arose between physicians and pharmacists during this period. Many physicians did not believe that pharmacists had sufficient medical knowledge and that they diagnosed and prescribed without being qualified. Rules for practice were tightened in many countries, with Germany using governmental controls for pharmacy regulation.

Some countries appointed *Royal Apothecaries* to provide pharmacy and other services to royal families. Pharmacopeias were developed to enable governments to protect public health by standardizing medicines.

In the newly developed United States of America, pharmacy began in the 18th century. Benjamin Franklin established the first hospital in 1751, with Jonathan Roberts as its first pharmacist. By 1789, the College of Philadelphia employed a materia medica and pharmacy professor. During this century, there were four different sources from which medications could be purchased: the "dispensing doctor," the apothecary shop, the general store, and the wholesale druggist.

In the 19th century, it was first demonstrated that specific drugs worked in certain body sites. Ehrlich discovered antibiotics, followed by other scientists discovering *insulin*. The Philadelphia College of Pharmacy was founded in 1821, and William Proctor was elected (in 1846) to the

chair of the Professorship of Pharmacy there. He introduced controls into the practice of pharmacy. Several periods of development in manufacturing pharmacy evolved during this century. These included the formative period (1867–1874), botanical research period (1875–1882), standardization period (1882–1894), organic chemical synthesis (1883, ongoing), and the biological period (1895, ongoing).

Thousands of new drugs were developed in the 20th century, saving millions of lives. In 1901, industrialized pharmacies used hormones, which utilized large-scale manufacturing techniques. Similarly, this type of manufacturing began with vitamins in 1909 and antibiotics in 1940. Industrialization brought standardization, biologically prepared products, complex chemical synthesis, and increased use of parenteral medications. Retail pharmacists came to accept mass-produced pharmacy products over time. The pharmaceutical industry created new "needs," which also helped retail pharmacies. Retail pharmacies never lost their usefulness with the general public, even though the activity of actually preparing many medications became industrialized. Table 1.1 lists significant events in the history of pharmacology.

As the Patient Care Era of pharmacy evolved, there was an increased focus on research to develop new medicines. Pharmaceutical research became more rational and targeted. Better drugs were designed to meet specific patient needs, utilizing computers for more intensive study and research. However, multiple available drug therapies led to adverse drug reactions, interactions, and varying degrees of outcomes. The roles of pharmacists became more focused on patient advising and education to ensure the maximum effects of medications with minimum harm. This *drug review* (or *drug monitoring*) was more patient-focused as part of the concept of *pharmaceutical care*, which evolved in the late 1980s. Pharmacists now dispense drugs and work with patients to ensure the best possible outcomes.

Biotechnology and **genetic engineering** promise more success in patient care in the future. Modifying the patient's genetic makeup may cure or partially cure some diseases. Recombinant DNA technology is used to produce certain new medications. These new areas of drug development offer future cures for currently incurable diseases.

YOU SHOULD REMEMBER

To understand pharmacology, healthcare professionals must thoroughly grasp anatomy, physiology, chemistry, microbiology, immunology, pathophysiology, and genetics.

Check Your Knowledge

1. Where did the first pharmacological writings originate?
2. Who is considered the "father of pharmacology," and what is the reason behind this title?

TABLE 1.1

Pharmacology History – Events

Time Period	Location	Event
3,000+ B.C.	Egypt	First pharmacological writings
2,700 B.C.	China	Plant remedies recorded
1,500 B.C.	Egypt	*Ebers Papyrus*
4th century B.C.	Greece	Hippocrates' natural laws
1st century B.C.	Greece	Dioscorides' *De Materia Medica*
16th century A.D.	Switzerland	Paracelsus advocated single drugs instead of complex mixtures of drugs.
17th century A.D.	England	William Harvey explained the beneficial and harmful effects of drugs and demonstrated intravenous administration for the first time; this is the first use of the term *pharmacology*.
18th century A.D.	England	Edward Jenner developed the first vaccine (for smallpox).
19th century A.D.	Germany	Paul Ehrlich discovered antibiotics.
19th century	Various	Morphine, strychnine, quinine, and nicotine were created.
1827	Germany	First commercial sale of morphine
1847	Estonia	The first department of pharmacology was established, which led to the first official recognizing of pharmacology as a distinct discipline.
1879	Russia	First vaccine developed for cholera by Waldemar Haffkine
1881	France	First vaccine for anthrax, developed by Louis Pasteur
1885	France	First vaccine for rabies developed by Louis Pasteur and Emile Roux
1890	Germany	Emil von Behring develops tetanus and diphtheria vaccines
1890	USA	John Jacob Abel found the first pharmacology department in the United States at the University of Michigan
1895	Poland/France	Pierre and Marie Curie invented the technique of making X-rays
1897	England	Sir Almoth Edward Wright developed typhoid vaccine
1898	Poland/France	Pierre and Marie Curie discovered radium
1899	Germany	Felix Hoffman developed aspirin
1922	Romania	Insulin first used to treat diabetes
1926	USA	First vaccine developed for whooping cough by Louis Sauer
1927	France	First vaccine developed for tuberculosis by Albert Calmette and Camille Guerin
1928	England	Sir Alexander Fleming discovered penicillin
1945	USA; England	Benadryl developed by George Rieveschi; first vaccine developed for influenza
1948	USA	Cortisone discovered by Edward Kendall and Philip Hench
1952	USA	Jonas Salk developed the first vaccine for polio
1964	USA	First vaccine developed for measles by John Franklin Enders
1967	USA	First vaccine developed for mumps by Maurice Hilleman
1970	USA	First vaccine developed for rubella by the Wistar Institute
1974	Japan	First vaccine developed for chicken pox by Michiaki Takahashi
1977	USA	First vaccine developed for pneumonia
1980	USA	Patent awarded for recombinant DNA to Stanley Cohen and Herbert Boyer
1981	Chile	First vaccine developed for hepatitis B by Pablo Valenzuela
2006	USA	Merck & Company marketed the first vaccine for human papillomavirus

3. Who developed the first vaccine in history, and what was the name of the vaccine?
4. In which country was insulin first used in the treatment of diabetes?
5. Who discovered penicillin, and in what year?

THE CONCEPTS OF PHARMACOLOGY

A drug is a chemical substance that affects a living organism by changing its structures or functions. Nearly all chemicals can be considered drugs because they can have some impact on a living organism. *Pharmacology* studies the history, sources, properties, and effects on living organisms.

The term is derived from two Greek words: pharmacy (meaning "medicine") and *logos* (meaning "study"). More than 11,000 brand-name, generic, and combination drugs are currently available. These are listed in the Food and Drug Administration (FDA) document titled the *Approved Drug Products with Therapeutic Equivalence Evaluations*, which is informally called the "**Orange Book**."

The Orange Book **details** the strength of every generic drug approved for distribution in the United States. Each product is rated based on its therapeutic equivalence to the "innovator drug." State pharmacy boards and managed care organizations utilize this as a reference to decide which generic products can be substituted for innovative

TABLE 1.2

Therapeutic Drug Classification

Therapeutic Classification	Indication
Antihypertensives	Lowering blood pressure
Antianginals	Angina blood clotting regulation
Antidysrhythmics	Restoring normal heart rhythm
Antihyperlipidemics	Lowering blood cholesterol
Anticoagulants	Blood clotting regulation

TABLE 1.3

Pharmacologic Drug Classifications

Classification	Mechanism of Action
Adrenergic antagonist (or blocker)	Blocks physiological reactions to stress
Angiotensin-converting enzyme inhibitor	Blocks hormonal activity
Calcium channel blocker	Blocks heart calcium channels
Diuretic	Lowers plasma volume
Vasodilator	Dilates peripheral blood vessels

drugs. The two primary drug classifications are *therapeutic* and *pharmacologic*. Therapeutic classifications and indications for specific diseases are presented in Table 1.2. The pharmacologic drug classification and mechanism of action are outlined in Table 1.3.

However, it would be best if you understood that most drugs have multiple classifications. An example is the drug epinephrine. It is classified as an autonomic nervous system agent, a vasoconstrictor, an adrenergic agonist, a bronchodilator, a **sympathomimetic**, an agent for **anaphylaxis**, an antiglaucoma agent, an ocular mydriatic, a **catecholamine**, and a topical hemostatic. This indicates that it falls under both pharmacological and therapeutic classifications. You must understand all of the sorts of drugs.

The study of pharmacology is more complicated because drugs may elicit various responses in patients due to age, body mass, health status, genetics, and gender.

The therapeutic objective of a drug is to deliver maximum benefit with minimal harm to the patient. The primary concern when administering medication is the strength of the patient's response. A drug's efficacy is determined by its concentration at the sites of action. All medicines can be toxic under certain circumstances.

Check Your Knowledge

1. Why is the Orange Book important in pharmacology?
2. Why is the study of pharmacology more complicated?

PHARMACOEPIDEMIOLOGY

Pharmacoepidemiology studies the determinants of both intended and unintended effects of drugs, vaccines,

biologics, medical procedures, and medical devices. It also studies utilization, safety signal detection, comparative effectiveness, cost–benefit, and benefit–risk analyses. Pharmacoepidemiology combines the principles of pharmacology and epidemiology to study the use and effects of drugs in large populations. In the future, pharmacoepidemiology can be improved by studying drugs in more real-world settings. Close monitoring of drug effects can help avoid future toxicities; closer monitoring of drugs after they enter the marketplace could have prevented this delay. Pharmacoepidemiology is a vital component of the drug development process that will become increasingly important in the years to come.

DRUG CLASSIFICATIONS

Drugs are usually classified by various formulations designed to maximize their effectiveness. Standard classifications of dosage forms include tablets, capsules, troches, suppositories, solutions, suspensions, emulsions, topical dosage forms, implants, and parenteral products. Of these forms, drugs are also classified as prescription (**legend**), nonprescription or over-the-counter (OTC), investigational, and illicit (illegal) drugs. Besides drugs or medications, other substances may be used therapeutically, including biologics and *alternative therapies*. Biologics are agents naturally produced in animal cells, by the body, or by microorganisms. They include hormones, monoclonal antibodies, interferons, natural blood products and components, and vaccines. Alternative therapies, known as *complementary therapies*, include herbs, natural plant extracts, minerals, vitamins, dietary supplements, and techniques that some consider "unconventional." Alternative therapies also include acupuncture, biofeedback, hypnosis, and massage.

PRESCRIPTION DRUGS

Prescription drugs can only be prescribed by legally authorized health practitioners for intended uses by appropriate patients. A *prescription* is typically the method for ordering medication or other therapy for a patient. It details the drug's name, dosage regimen, and the patient for whom it is written. Pre-printed prescription forms are commonly used. The components of a prescription include the following:

- Patient's name, address, age, or birth date
- The date the prescriber wrote the prescription
- The Rx symbol
- The prescribed medication's name and dosage strength
- Dispensing instructions for the pharmacist
- Directions for the patient (known as the Signa)
- Refill or specialized labeling instructions, or both
- The prescriber's signature, address, and telephone number

Some parts of the prescription may be written in Latin instead of English in the United States. In hospitals or other institutional settings, medication prescriptions are usually written on forms called "Physician's Order Sheets." These forms typically create duplicate copies by applying pressure to the top sheet, allowing documents to be sent to the pharmacy, the records department, or other facility areas.

Medications must be kept away from unauthorized individuals wherever they are stored. Many require environmental controls to ensure a stable temperature and humidity. Medications are commonly stored in various ways, including at room temperature, in refrigerators, and in freezers. Expiration dates on medication labels indicate how long medications are considered to be at full potency and, therefore, can be administered. When a month and year are shown, the enclosed drug expires on the last day of that month in the given year. Beyond that, the manufacturer cannot guarantee the medication's full potency and stability. When drugs are stored in ways besides what is listed on their labels, their potency may be affected.

Controlled substances are agents identified by the government as capable of causing physical or psychological dependence, or both. The Controlled Substances Act of 1970 establishes five classifications of controlled substances. These substances are subject to stricter controls than other prescription medications.

NONPRESCRIPTION DRUGS

Nonprescription drugs can be bought and used without a prescription, as determined by the FDA. These drugs, often referred to as *OTC* drugs, are deemed safe for use as long as their administration directions are followed. Since 1972, OTC drugs have been classified into three categories, with three classifications established:

- **Category I**: safe and effective for therapeutic uses claimed for them
- **Category II**: not safe and effective
- **Category III**: additional data required to establish safety and **efficacy**. An effective drug elicits the responses for which it is administered. All new drugs must be proven effective before being released for marketing.

Many formerly prescription-only drugs are now available in various forms as OTC products. Since the 1970s, over 60 prescription medications have become OTC medications, and many more are considering the switch from prescription to OTC. However, consumers should remember that although OTC medications do not require prescriptions, they can still cause toxicity. They can interact with numerous prescription drugs and other substances, and certain medical conditions exclude the use of specific OTC drugs. Efforts to standardize their labels are ongoing, aimed at making them more informative yet easy to understand.

The labels are titled "Drug Facts," which list active ingredients first, followed by appropriate uses, warnings, directions, and inactive ingredients. Healthcare professionals must determine which OTC drugs patients should avoid based on their conditions, requiring continuous evaluation. All patients must have a complete medication profile documented, including both OTC and prescription drugs.

Americans spend approximately $20 billion yearly on OTC drugs, and 40% take at least one OTC drug every 2 days. These drugs account for 60% of all doses administered in the United States. Initial therapy usually begins with an OTC drug before a patient sees a healthcare professional to obtain a prescription. In the average home medicine cabinet, there are 24 OTC drug products. Patients must self-evaluate adverse effects that may develop because of a medication. They must avoid masking symptoms such as cough, fever, or pain, which a severe underlying disorder may cause.

Since herbal supplements and vitamins do not require FDA approval for safety and efficacy, they may cause additional problems in infants, children, and older adults. This is because of differences in their absorption, biotransformation, and elimination rates. Infants and children have immature levels of these functions. Older adults experience physiological changes because of aging, and these changes occur.

ILLICIT DRUGS

Illicit drugs are used and distributed illegally. They include drugs that are not legal to be sold in the United States or that can be legally sold under certain circumstances but have been manufactured unlawfully, diverted, or stolen from normal distribution channels. Usually, they are used to alter moods or feelings and not for medical purposes. For more discussion on substance abuse, see Chapter 13.

Check Your Knowledge

1. What are the differences between prescription and nonprescription drugs?
2. Which government agency approves a prescription drug for general distribution?

DRUG NAMES

During the earliest stages of a drug's development, it is usually known only by its **chemical name**. This name is systematically derived and identifies the drug's chemical structure and composition. **Chemical names** are often complex and contain hyphens between the individual terms. Therefore, a code designation may be chosen as a temporary name, which is generally discarded once the drug reaches the market. The code designation is often the only identifier used when drugs are investigated for future use. Only a few chemical names, such as calcium gluconate, lithium carbonate, and sodium chloride, are easy to remember. Sometimes, a drug's name and classification are based

TABLE 1.4

Common Drug Name Example

Brand Name	Generic Name	Chemical Name
Nexium	Esomeprazole magnesium	(S)-5-methoxy-2-[(4-methoxy-3,5-dimethylpyridin-2-yl) methylsulfinyl]-1H-benzo(d)imidazole
Lipitor	Atorvastatin calcium	[R-(R*, R*)]-2-(4-fluorophenyl)-ß, δ-dihydroxy-5-(1-methylethyl)-3-phenyl-4-[(phenylamino) carbonyl]-1Hpyrrole-1-heptanoic acid, calcium salt (2:1) trihydrate
Plavix	Clopidogrel bisulfate	Methyl (+)-(S)-α-(2-chlorophenyl)-6,7-dihydrothieno[3,2-C]pyridine-5(4H)acetate sulfate (1:1)
Advair Diskus	Fluticasone propionate	S-(fluoromethyl) 6α, 9-difluoro-11β, 17-dihydroxy-16α-methyl-3oxoandrosta-1,4-diene-17β-carbothioate, 17-propionate
OxyContin	Oxycodone HCl	4, 5α-epoxy-14-hydroxy-3-methoxy-17-methylmorphinan-6-one hydrochloride

on part of its chemical structure, referred to as its chemical group name.

In the United States, once a drug is marketed, a simpler **generic name** *(nonproprietary name)* is assigned to it by the U.S. Adopted Names (USAN) Council. Generic names are designed for easy pronunciation and memorization. However, they are still more complex than trade names. They are based on a fundamental pharmacological or chemical characteristic of the drug that will not be confused with other drug names. While there is only one generic name for each specific drug, there may be multiple brand names due to the different manufacturers who produce them. Generic names can be identified by class through their final syllables. For example, the "-cillin" drugs are antibiotics used to treat infections, while the *–statin* drugs are HMG-CoA reductase inhibitors used to treat high cholesterol. Table 1.4 lists classes of generic drugs and their uses.

A **prototype drug** is one to which all other drugs in a class are compared. It allows for predicting the actions and adverse effects of the drugs in their class. A prototype drug for a particular drug class is often the oldest and best understood, but this is not always the case. A newer drug with a more favorable safety profile is sometimes chosen as a prototype drug.

A **brand name** *(trade name)* is registered by the drug manufacturer with the U.S. Patent Office and approved by the FDA. This name is also known as the "proprietary name." The brand name can only be used by the manufacturer who registered it and is followed by the symbol "®," indicating its registration with the Patent Office. Brand names are usually short and easy to remember and refer to the entire formulation in which the drug is contained, not just the drug itself. The actual manufacturer suggests them for ease of use. Each company must register its brand name when different companies manufacture a generic drug. Examples of some common drug names, showing each type of name, are listed in Table 1.5.

YOU SHOULD REMEMBER

Many drug products contain multiple active ingredients. These are referred to as combination drugs. Remembering all generic names associated with a specific trade-name drug can be challenging. Generic names should be used to identify the active ingredients in a combination drug to prevent confusion with various trade names. The generic name is traditionally written in lowercase letters first, followed by the trade name in parentheses with the capitalized first letter. Examples include acetaminophen (Tylenol), ibuprofen (Motrin), alprazolam (Xanax), sildenafil (Viagra), and diazepam (Valium).

Check Your Knowledge

1. Which drug name can only be used by the manufacturer who registered it?
2. What is the description of a prototype drug?

DRUG SOURCES

Drugs may be derived from various sources, some of which are natural. The five primary sources of drugs include plant, animal, mineral, synthetic, and bio- or genetically engineered sources. Human tissues are being manipulated today to create new drugs (such as in stem cell research). Artificial agents are rapidly replacing agents derived from natural sources in today's pharmacology.

ANIMAL SOURCES

Animal body fluids and glands, including those of humans, can act as drugs. Examples of medicines derived from animal sources include insulin, pancreatin, pepsin, thyroid hormone, specific ACE inhibitors, forms of estrogen, and adrenaline.

PLANT SOURCES

Chemical and physical properties categorize plant sources. There is a long history of using various parts of plants for medicinal purposes. Alkaloids are alkaline, nitrogen-containing compounds that combine with acids to form salts. When hydrolyzed, *glycosides* produce sugar and nonsugar substances. Examples of drugs derived from plant sources include digitoxin, vincristine, atropine, morphine, quinine, reserpine, camphor, codeine, L-Dopa, pilocarpine, and caffeine.

TABLE 1.5

Examples of Trade and Generic Names for Common Medications

Trade Name	Generic Name	Trade Name	Generic Name
Advil®	Ibuprofen	Naprosyn®	Naproxen sodium
Aldomet®	Methyldopa	Nebcin®	Tobramycin
Alka-Seltzer®	Aspirin	Nizoral®	Ketoconazole
Allegra®	Fexofenadine HCl	Nolvadex®	Tamoxifen citrate
Amoxil®	Amoxicillin	Omnicef®	Cefdinir
Ativan®	Lorazepam	Pepcid®	Famotidine
Bactrim®	Sulfamethoxazole	Phenergan®	Promethazine
Benadryl®	Diphenhydramine	Prilosec®	Omeprazole
Biaxin®	Clarithromycin	Prinivil®	Lisinopril
Bufferin®	Aspirin	Procardia®	Nifedipine
BuSpar®	Buspirone	Prozac®	Fluoxetine
Cardizem®	Diltiazem	Proventil®	Albuterol
Catapres®	Clonidine	Retrovir®	Zidovudine
Ceclor®	Cefaclor	Rhinocort®	Budesonide, nasal
Cefzil®	Cefprozil	Robitussin®	Guaifenesin
Celexa®	Citalopram	Rocephin®	Ceftriaxone
Cipro®	Ciprofloxacin	Septra®	Sulfamethoxazole/trimethoprim
Claritin®	Loratadine	Synthroid®	Levothyroxine
Coumadin®	Warfarin	Tagamet HB®	Cimetidine
Demerol®	Meperidine	Tavist®	Clemastine fumarate
Depakene®	Valproic acid	Tegretol®	Carbamazepine
DiaBeta®	Glyburide	Terramycin®	Oxytetracycline HCl
Diabinese®	Chlorpropamide	Tobrex®	Tobramycin
Dilantin®	Phenytoin	Tums®	Calcium carbonate
Flagyl®	Metronidazole	Tylenol®	Acetaminophen
Fosamax®	Alendronate sodium	Valium®	Diazepam
Glucotrol®	Glipizide	Vancocin®	Vancomycin
Haldol®	Haloperidol	Vasotec®	Enalapril
Humulin®	Insulin (human)	Ventolin®	Albuterol
Inderal®	Propranolol	Vivarin®	Caffeine
Kantrex®	Kanamycin sulfate	Xanax®	Alprazolam
Keflex®	Cephalexin	Xylocaine®	Lidocaine HCl, local
Kefzol®	Cefazolin	Zantac®	Ranitidine
Klonopin®	Clonazepam	Zestril®	Lisinopril
Lanoxin®	Digoxin	Zithromax®	Azithromycin
Lasix®	Furosemide	Zocor®	Simvastatin
Lipitor®	Atorvastatin	Zoloft®	Sertraline HCl
Lopid®	Gemfibrozil	Zomig®	Zolmitriptan
Lopressor®	Metoprolol/tartrate	Zyban®	Bupropion HCl
Motrin®	Ibuprofen	Zyrtec®	Cetirizine HCl
Mycostatin®	Nystatin		

MINERAL SOURCES

Unlike plants and animals, mineral sources originate from the Earth and its soil, providing inorganic materials. These sources are utilized as they occur in nature. Sodium, iodine, potassium, iron, chloride, lithium carbonate, and gold are examples of minerals used to prepare medications. Gold, for instance, is used to treat severe rheumatoid arthritis that does not respond to other drugs.

SYNTHETIC SOURCES

Many drugs are prepared synthetically or semisynthetically through chemical reactions in a laboratory. Examples include specific antimicrobial agents, which are produced semisynthetically by chemically modifying substances from natural sources. Some human insulin products are semisynthetically modified to have the same chemical structure as human insulin despite being derived from animals. Other examples include barbiturates, aspirin, and sulfonamides. Many synthetic compounds are purer than those derived from natural sources. Synthetic penicillin has been stabilized for oral administration.

Most drugs used today are entirely synthesized and formed through chemical reactions in a laboratory. Their increased purity decreases the risk of hypersensitivity reactions. It also allows for better control of dosages, therapeutic effects, routes of administration, and adverse effects.

BIOGENETICALLY ENGINEERED SOURCES

Biotechnology is an emerging field in drug development that involves manipulating proteins to produce complex natural substances (such as hormones) or may include the genetic alteration of biological materials. It utilizes various studies, including molecular biology, genetic engineering, recombinant DNA technology, immunology, and pharmacology. Examples of genetically engineered agents include insulin, pituitary hormones, and erythropoietin.

The most significant potential for applying biotechnology lies in gene splicing, which involves genetically manipulating rapidly growing, nonpathogenic bacteria. This process enables the preparation of complex biological compounds that would otherwise be very difficult or costly to create using conventional methods. Gene splicing entails inoculating these bacterial organisms with plasmids – circular DNA molecules that carry genes the bacterium can perpetuate and duplicate. There are hundreds of different biotechnology products currently in development. Previously approved products of this type include human insulin, human growth hormone, tissue plasminogen activator, and the hepatitis B vaccine. Gene splicing utilizes *plasmids*, circular DNA molecules that carry genes that a bacterium can perpetuate. The bacterium then duplicates the genes it carries along with its chromosomes. Hundreds of biotechnology-produced drug products are in the final stages of testing.

Check Your Knowledge

1. List two examples of drugs derived from plant sources.
2. What is the meaning of gene splicing?

FURTHER READING

American Pharmacists Association. (2017). *Drug Information Handbook with International Trade Names Index — Adapted from the Drug Information Handbook*, 26th Edition. Lexi-Comp Inc.

American Society of Health-System Pharmacists, and Holdford, D.A. (2017). *Introduction to Acute & Ambulatory Care Pharmacy Practice*, 2nd Edition. ASHP Publications.

Ban, T.A., Healy, D., and Shorter, E. (2010). *The History of Psychopharmacology and the CINP, As Told in Autobiography: From Psychopharmacology to Neuropsychopharmacology in the 1980s and the Story of CINP*. Collegium Internationale Neuro-Psychopharmacologicum.

Davis, K.L., Charney, D., Coyle, J.T., and Nemeroff, C. (2002). *Neuropsychopharmacology: The Fifth Generation of Progress — An Official Publication of the American College of Neuropsychopharmacology*. Lippincott, Williams, and Wilkins.

Ebenezer, I. (2015). *Neuropsychopharmacology and Therapeutics*. Wiley-Blackwell.

Ghaemi, S.N. (2019). *Clinical Psychopharmacology: Principles and Practice*. Oxford University Press.

Iversen, L., Iversen, S., Bloom, F.E., and Roth, R.H. (2008). *Introduction to Neuropsychopharmacology*. Oxford University Press.

Jansen, S. (2016). *OTC Pocket Guide: OTC Pocket Guide — Quick Nonprescription Drug Reference*. CreateSpace Independent Publishing Platform.

Krinsky, D.L. (2017). *Handbook of Nonprescription Drugs: An Interactive Approach to Self-Care*. American Pharmacists Association.

Levin, B., Hanson, A., and Hurd, P.D. (2018). *Introduction to Public Health in Pharmacy*, 2nd Edition. Oxford University Press.

Li, J.J. (2015). *Top Drugs: Their History, Pharmacology, and Syntheses*. Oxford University Press.

Lovett, A.W., Peasah, S.K., Xiao, H., and Ryan, G.J. (2013). *Introduction to the Pharmacy Profession*. Jones & Bartlett Learning.

Marie, S. (2020). *The Art and Science of Psychopharmacology: Essential Tools for Treating Anxiety, Depression, Bipolar Disorder & Psychosis*. PESI Publishing & Media.

Pandit, N.K., and Soltis, R.P. (2011). *Introduction to the Pharmaceutical Sciences: An Integrated Approach*, 2nd Edition. Lippincott, Williams, and Wilkins.

Patrick, G. (2017). *An Introduction to Medicinal Chemistry*, 6th Edition. Oxford University Press.

Posey, L.M. (2009). *The Pharmacy Technician's Introduction to Pharmacy — APhA Pharmacy Technician Training Series*. American Pharmacists Association.

Robbins, T.W., and Sahakian, B.J. (2016). *Translational Neuropsychopharmacology* (Current Topics in Neurosciences Book 28). Springer.

Stahl, S.M. (2017). *Prescriber's Guide: Stahl's Essential Psychopharmacology*, 6th Edition. Cambridge University Press.

Tripathi, K.D. (2016). *Pharmacological Classification of Drugs with Doses and Preparations*, 5th Edition. Jaypee Brothers Medical Publishers Ltd.

Watkins, C.J. (2018). *Pharmacology Clear and Simple: A Guide to Drug Classifications and Dosage Calculations*, 3rd Edition. F.A. Davis Company.

2 Drug Development, Drug Approval, and Drug Regulation

LEARNING OBJECTIVES

After studying this chapter, readers should be able to

1. Review the investigational drugs for the new drug.
2. Explain the various types of phases used for conducting clinical studies.
3. Describe the three federal regulatory agencies in the United States.
4. Explain the primary purpose of drug regulations.
5. Describe the New Drug Application.
6. Explain the role of the Drug Enforcement Administration.
7. Summarize the role of the Food and Drug Administration.
8. Explain the role of the Centers for Disease Control and Prevention.
9. Describe the concept of drug recalls.
10. Explain the drug standards requirements for formulating drug substances.

OVERVIEW

Taking new drugs, treatments, or therapies to market is a demanding, expensive, and time-consuming journey from an initial idea to a product ready to launch. In some cases, the research and development process for new treatments and therapies can take over a decade and cost billions of dollars. A pharmaceutical company seeking FDA approval to sell a new prescription drug must complete a five-step process: discovery/concept, preclinical research, clinical research, FDA review, and FDA post-market safety monitoring. Drug regulations play an essential role in ensuring the safety and efficacy of approved drugs. They not only regulate the pricing of drugs but also their quality. The regulations are required for innovations and existing products to improve health status. Both products produced domestically and those imported from other countries need regulation. Every country has rules for innovation, manufacturing, drug testing, marketing, and post-marketing studies. The aim is to maintain the standards of the drug at every step to cater to the patient population of each country.

DRUG DEVELOPMENT

The development and testing of new drugs is a costly and lengthy process. It requires 6–12 years for a sufficient investigation to be completed, and costs commonly range close to $800 million for each new drug. **Investigational drugs** are those undergoing study to be marketed. A manufacturer must perform animal studies and clinically test a new drug on human subjects after filing a "Notice of Claimed Investigational Exemption for a New Drug" with the FDA. This complex form must include the following information:

- The chemical, biological, pharmacological, and toxicological properties of the new agent.
- Precise manufacturing and storage details.
- Each participating investigator's name and credentials.
- A signed investigator statement indicating awareness of the drug's nature, assurance that only authorized individuals will supervise the study and that all volunteers or clients undergoing testing are fully informed as to the nature of the study (and have signed a form with their written consent); also, consent forms must be read and signed by clients and witnesses.
- Protocols that identify how the drug will be administered to experimental subjects and include the specific observations or determinations required during the study.

Before testing on humans, preclinical testing of a new animal drug is required. Drugs are evaluated for pharmacokinetic properties, toxicities, and biological effects that may be useful. The preclinical period may last from 1 to 5 years. Once this is finished, the manufacturer may apply to the FDA for permission to begin clinical testing. Clinical trials may begin after the FDA approves an Investigational New Drug Application.

There are four typical phases used for performing clinical studies on human subjects before a drug is marketed, as follows:

- **Phase I (preclinical, lasting about 1 year):** The drug is evaluated in a group of 20–100 healthy human volunteers to determine toxicity and how it is metabolized and excreted; however, if the drug may cause severe adverse effects, the trial is performed using volunteer patients who have the disease under consideration. The primary objectives of Phase I drug testing are to assess drug metabolism, biological effects, and pharmacokinetics. Extensive testing of cultured cells and animals is used to gauge a drug's potential to harm humans. Results are inconclusive because the way a human responds to the drug is not always accurately reflected by laboratory testing methods.

DOI: 10.1201/9781003461913-3

- **Phase II (clinical, lasting approximately 2 years)**: A more detailed evaluation of the drug in healthy volunteer subjects and trials in small numbers of subjects with the disease state that the drug is intended to treat. The goals of Phase II drug testing are to evaluate therapeutic utility and determine the optimal dosage range. Up to several hundred patients are tested in this phase, and control groups are used that receive placebos – ideally, a **double-blind study** is used.
- **Phase III (review, lasting about 3 years)**: Broad clinical trials to evaluate the drug's effectiveness in treating the intended disease state. Nurses are most involved in this phase and may be involved in administering investigational medications to clients, following clinical protocols exactly. The goals of Phase III drug testing are to evaluate safety and effectiveness. Either hundreds or thousands of patients with the related disease are tested in this phase, and results are compared to other drugs that are already on the marketplace. After this phase is completed, the manufacturer applies to the FDA for conditional approval of an NDA. The FDA's *Critical Path Initiative* attempted to modernize *bioinformatics* (organizational properties of biological systems) to improve the safety, effectiveness, and manufacturability of new medical products.

  It affected the fields of **genomics**, **proteomics**, imaging, and **bioinformatics**. Proteomics is defined as the expression, localization, functions, and interactions of proteins produced by the genes of an organism. Updates to the drug approval process include the Prescription Drug User Fee Act, the FDA Modernization Act, and the FDA Amendments Act. The Prescription Drug User Fee Act required drug and biologic manufacturers to provide yearly product user fees. This income allowed the FDA to increase its workforce and restructure to be more efficient. As a result, the FDA approved twice as many new drugs as before 1992, and these approvals were nearly twice as fast. The FDA Modernization Act added more employees to the FDA's drug and biologic program and increased user fees. The FDA Amendments Act allowed more resources to be used for comprehensive new drug reviews.
- **Phase IV (post-marketing)**: Post-marketing surveillance of the drug's activity; prescribers are encouraged to report results of the clinical use of the drug that they have encountered; unexpected adverse effects are vital here.

Only individuals who have signed informed **consent forms** should receive investigational drugs. People who have volunteered to be tested can withdraw from the program at any time. Figure 2.1 shows drug development timelines with the four phases of drug approval.

YOU SHOULD REMEMBER

It takes approximately 11 years before a drug is submitted to the FDA for review. For every 5,000 chemicals that enter preclinical testing, only five will eventually be tested on humans. Only one of these chemicals will be approved.

Check your Knowledge

1. What are the four phases used for performing clinical studies of drugs?
2. Why is preclinical testing of new drugs required?
3. What are the consent forms?

DRUG APPROVAL

A new drug approval procedure is defined as a stepwise method provided by regulatory agencies of each country that sets guidelines for a drug manufacturer or a sponsor who wishes to seek marketing authorization in a specific country. The whole drug approval procedure has been divided into several stages depending on the regulatory process adopted in each country. Currently, different countries stick to their regulatory requirements as prescribed by the governing bodies for new drug approval, since it is tough to have a single regulatory entity to provide marketing authorization in all countries.

The primary purpose of regulations is to safeguard public health and ensure that the company's manufacturing and marketing of pharmaceutical products comply with regulatory norms. This calls for pharmaceutical products to be developed, formulated, produced, evaluated, and tracked, conforming to regulatory guidelines to ensure that they are effective and safe and that the patient's well-being is unaffected. However, this concern for safety and quality does not come cheap, so the **New Drug Application** (NDA) process in many developing countries remains less regulated. The FDA could approve products such as drugs, food, medical devices, vaccines, radiation-emitting products, tobacco products, and cosmetics.

DRUG REGULATION

The process of testing, developing, and marketing medicines must be regulated to protect the public's interests. Prescription drugs are among the most common healthcare interventions and have turned some once-fatal diseases into manageable conditions, but they have also been a growing source of controversy. Patients in the

The United States struggles with increasing costs and expresses concerns about why many conditions, such as **Alzheimer's disease**, remain without adequate therapeutic options. At the center of these debates lies the FDA, which monitors the prescription drug marketplace and enforces

CLINICAL TRIAL PHASES

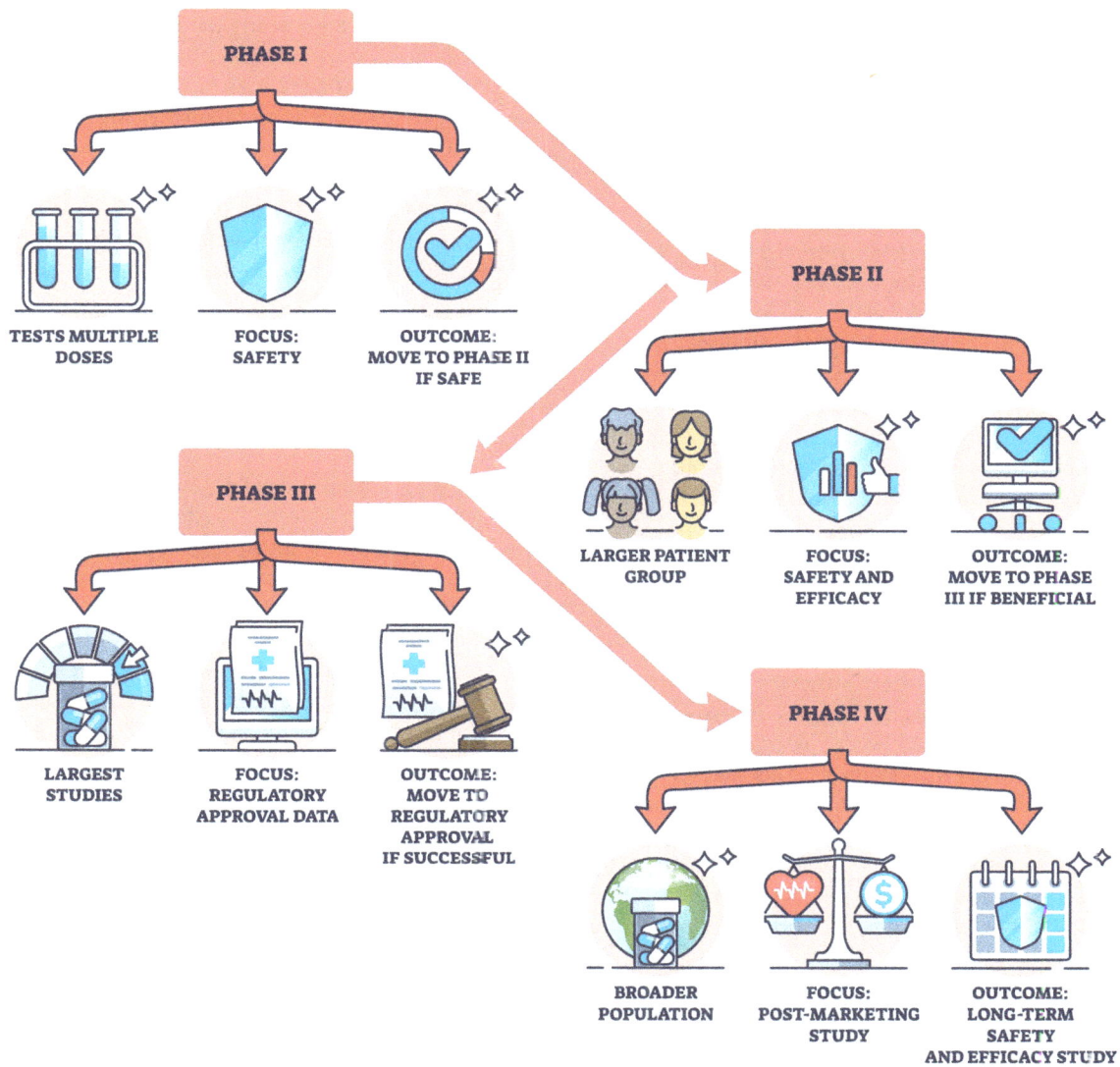

PHASE I

TESTS MULTIPLE DOSES

FOCUS: SAFETY

OUTCOME: MOVE TO PHASE II IF SAFE

PHASE II

LARGER PATIENT GROUP

FOCUS: SAFETY AND EFFICACY

OUTCOME: MOVE TO PHASE III IF BENEFICIAL

PHASE III

LARGEST STUDIES

FOCUS: REGULATORY APPROVAL DATA

OUTCOME: MOVE TO REGULATORY APPROVAL IF SUCCESSFUL

PHASE IV

BROADER POPULATION

FOCUS: POST-MARKETING STUDY

OUTCOME: LONG-TERM SAFETY AND EFFICACY STUDY

FIGURE 2.1   Drug development timelines with the four phases of drug approval.

basic rules and laws that affect how prescription drugs are discovered, developed, and sold.

The Controlled Substances Act (CSA) also provides a mechanism for substances to be controlled or decontrolled (removed from control). Proceedings to add, delete, or change the schedule of a drug or other substance may be initiated by the Drug Enforcement Administration (DEA), the Department of Health and Human Services (HHS), or by petition from any interested party, including:

- The manufacturer of a drug
- A medical society or association
- A pharmacy association
- A public interest group concerned with drug abuse
- A state or local government agency

Specific factors that schedule a drug or other substance should be considered, or whether a substance should be decontrolled or rescheduled. Schedules are usually created to enforce safety and quality standards across an industry. In other cases, they are designed to protect consumers in industries with low competition.

FEDERAL REGULATORY AGENCIES

The FDA, the Centers for Disease Control and Prevention (CDC), and the DEA are federal agencies that help secure Americans' health regarding diseases and drugs.

FOOD AND DRUG ADMINISTRATION

The FDA is a branch of the Department of Health and Human Services (HHS). The agency oversees all domestic and imported food, bottled water, and wine beverages with less than 7% alcohol. It is also responsible for cosmetics, medicines, medical devices, radiation-emitting products, and feed and drugs used for farm animals. Thus, it controls all drugs for legal use. The FDA initiates, implements, and enforces all laws related to the administration of drugs.

CENTERS FOR DISEASE CONTROL AND PREVENTION

The CDC is a federal agency that provides facilities and services for investigating, identifying, preventing, and controlling disease. It also oversees all foods and foodborne diseases. The CDC offers statistics and information to health professionals about treating common and rare diseases worldwide. Its primary function is to issue regulations for infection control. It was established in 1946 as the Communicable Disease Center and became the Centers for Disease Control in 1970; the words "and Prevention" were added in 1992, but Congress requested that "CDC" remain the agency's initials. This agency has also been deeply involved in the war against HIV infection and AIDS.

DRUG ENFORCEMENT ADMINISTRATION

The Drug Enforcement Administration (DEA) oversees controlled substances, which includes investigating and prosecuting individuals who grow or manufacture these substances for illegal distribution. The DEA;s mission is to enforce controlled substance laws and regulations, prosecuting individuals and organizations that develop, manufacture, or distribute illegal substances. The agency also targets individuals who use violence to coerce others into participating in their illicit activities and disseminates information about unlawful substances to educate and inform the public, enabling them to assist the DEA in its efforts. Furthermore, the DEA collaborates with other governments to enforce laws regulating global drug trafficking and related items.

YOU SHOULD REMEMBER

A regulatory agency is an organization designed to manage a specific area of human activity through a set of rules and regulations. It is usually created to enforce safety and quality standards across an industry or to protect consumers in industries with low competition.

Check Your Knowledge

1. What is the role of the FDA in regulating drugs?
2. What are the responsibilities of the CDC in drug development?
3. What is the mission of the DEA in regulating controlled substances?

DRUG RECALLS

Most **drug recalls** are voluntary; the manufacturer identifies an issue and recalls the affected drug. However, sometimes a drug is recalled after the FDA raises concerns.[4] Regardless of the reason for the recall, the FDA's role is to supervise a manufacturer's strategy and assess and ensure the appropriateness of its handling of the recall. Approximately 80% of Americans' medications contain some component manufactured abroad, primarily in China and India. While these manufacturing processes aid in keeping the costs of medications down, this has also created a supply chain that can be challenging to track, which may increase the number of drug recalls. The FDA's medical staff determines a product's health hazard potential and assigns a drug recall classification. Drug recalls are divided into three classes:

1. **Class I:** The use or exposure to the product will cause severe adverse reactions, injury, or death. This class is the most urgent and severe of the three types of FDA recalls. Class I recalls usually pertain to defective products that can cause serious health problems or death. An example would be an over-the-counter medication contaminated with a toxin. If a Class I recall occurs, the FDA will oversee the recall process and ensure the manufacturer takes sufficient steps to protect the public.
2. **Class II:** Using or exposing the product might cause severe injury or temporary illness. If the FDA expects a product's defect to only result in a short-term health issue, or if there is only a slight chance it could lead to a severe problem, the FDA will designate the recall as Class II. As is the case with Class I recalls, the FDA will take an oversight role to ensure its adequacy.
3. **Class III:** The use or exposure to the products is unlikely to cause injury or illness but violates FDA regulations. Class III recalls are the least serious of the three types of FDA recalls. They apply to minor product defects or errors unlikely to harm someone's health.

YOU SHOULD REMEMBER

FDA recalls apply to products subject to the FDA's jurisdiction. In addition to drugs, foods, and cosmetics, an FDA recall can apply to vaccines, human blood, blood products, human tissue used for transplantation, and medical devices.

DRUG STANDARDS

Drug standards are the requirements for formulating drug substances, **ingredients**, and dosage forms. Drugs stocked in the pharmacy must be compendial (listed), and a drug formulary or list of drugs stocked by the pharmacy must be maintained. These drug standards are contained in the United States Pharmacopeia (USP) and the **National Formulary (NF)**, published by the U.S. Pharmacopeial Convention, Inc. Pharmaceutical services must be under the general supervision of a licensed pharmacist. The pharmacist working within an institutional setting, such as a nursing home, must schedule regular visits to the facility to supervise the drug handling and administration procedures. At least monthly, they must review each patient's drug regimen and report any discrepancies or irregularities to the administrator and the medical director. This is a significant requirement in terms of patient safety and professional integrity.

FURTHER READING

Adams, D.G., Cooper, R.M., Hahn, M.J., and Kahan, J.S. (2015). *Food and Drug Law and Regulation*, 3rd Edition. Food and Drug Law Institute.

Ainsworth, S. (2020). *Neonatal Formulary: Drug Use in Pregnancy and the First Year of Life*, 8th Edition. OUP Oxford.

Endrenyi, L., Declerck, P., and Chow, S.C. (2017). *Biosimilar Drug Product Development* (Drugs and the Pharmaceutical Sciences Volume 216). CRC Press.

Friedhoff, L.T. (2009). *New Drugs: An Insider's Guide to the FDA's New Drug Approval Process for Scientists, Investors and Patients*. PSPG Publishing.

Guarino, R.A. (2016). *New Drug Approval Process* (Drugs and the Pharmaceutical Sciences Volume 190), 5th Edition. CRC Press.

Hernberg-Stahl, E., and Reljanovic, M. (2013). *Orphan Drugs: Understanding the Rare Disease Market and its Dynamics* (Series in Biomedicine Book 46). Woodhead Publishing.

Hutt, P., Merrill, R., and Grossman, L. (2013). *Food and Drug Law: Cases and Materials*, 4th Edition. Foundation Press.

Kanovsky, S.M., and Pines, W.L. (2020). *A Practical Guide to FDA's Food and Drug Law and Regulation*, 7th Edition. Food and Drug Law Institute.

King, T.I., and Baker, T.R. (2012). *FDA Drug Approval: Elements and Consideration* (Pharmacology Research, Safety Testing and Regulation). Nova Biomedical.

Meyers, A.S. (2016). *Orphan Drugs: A Global Crusade*. Meyers.

Ng, R. (2015). *Drug: From Discovery to Approval*, 3rd Edition. Wiley-Blackwell.

O'Donnell, J.J., Somberg, J., Idemyor, V., and O'Donnell, J.T. (2019). *Drug Discovery and Development*, 3rd Edition. CRC Press.

Potter, S.O.L. (2009). *Potter's Compend of Materia Medica Therapeutics and Prescription Writing*. University of Michigan Library.

Pryde, D.C., Palmer, M., Fox, D., Kent, A., and Boyd, R. (2014). *Orphan Drugs and Rare Diseases: RSC Drug Discovery* (Volume 38). Royal Society of Chemistry.

Teitelman, J., and Detweiler, K. (2020). *AAHA Guide to Safeguarding Controlled Substances*. AAHA Press.

The Law Library. (2018). *Exclusion of Orphan Drugs for Certain Covered Entities under 340B Program* (US Department of Health and Human Services (HHS) Regulation). The Law Library.

United States Department of Justice. (2019). *Pharmacist's Manual: An Informational Outline of the Controlled Substances Act*. Lulu.com

United States Government Accountability Office. (2013). *Health Care: Food and Drug Administration's Drug Approval Process*. BiblioGov

3 Pharmacokinetics

LEARNING OBJECTIVES

After studying this chapter, readers should be able to

1. Summarize the four primary processes of pharmacokinetics.
2. Discuss factors affecting drug absorption.
3. Review how drugs are distributed throughout the body.
4. Discuss how plasma proteins affect drug distribution.
5. Describe barriers to drug distribution.
6. Explain drug metabolism and its applications in pharmacology.
7. Describe time–response relationships.
8. Identify significant processes by which drugs are excreted.
9. Differentiate between loading and maintenance doses.
10. Describe the drug's half-life and bioavailability.

OVERVIEW

Pharmacokinetics studies drug movement throughout the body and what the body does to medication after administration. The body contains many barriers that medications must cross to exert their effects. These barriers include the stomach membranes, the portal vein, the liver, the systemic circulation, the target tissues, and the target cells. Numerous physiological processes occur as a medication attempts to reach target tissues and cells. Stomach acid, digestive enzymes, liver enzymes, and many other bodily substances break down drug molecules. A drug that is interpreted as foreign by the body may trigger an immune response. The kidneys, large intestine, and lungs excrete drugs. The four categories of pharmacokinetics are absorption, distribution, metabolism, and excretion.

CELL MEMBRANE AND DRUG TRANSPORT CHANNELS

A **plasma membrane** surrounds every cell (see Figure 3.1). It comprises a double phospholipid layer with proteins embedded inside it. Phospholipids are fats that contain phosphate. The plasma membrane has a round "head" section that is hydrophilic and a double "tail" that is hydrophobic. Its structure allows the easy passage of water molecules via **osmosis**.

There are three essential ways that drugs cross cell membranes. These include

CELL MEMBRANE

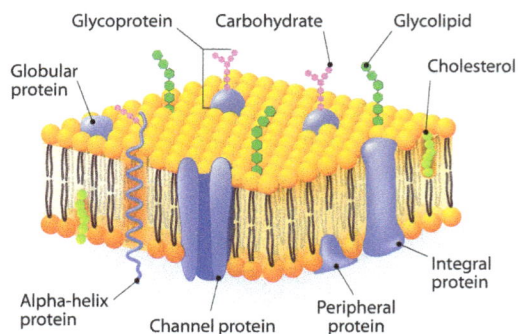

FIGURE 3.1 The plasma membrane (cell membrane).

- **Direct membrane penetration:** This is the most common method and is used by most drugs. They are too large to pass through channels or pores and lack transport systems. A drug must be lipophilic for direct membrane penetration. Since the membranes are composed chiefly of lipids, the drug must be able to dissolve into the lipids in the membrane. Molecules that are not lipid-soluble and cannot penetrate membranes include **polar molecules** and ions.
- **Passage through channels and pores**: Very few drugs cross membranes via this method. Channels are tiny and specific only for certain molecules. As a result, only compounds with a molecular weight of less than 200 can move through their particular channels. Small ions such as potassium and sodium can cross membranes using this method.
- Passage via a transport system: This may or may not require energy expenditure.

  Each transport system is selective for specific drugs, depending on its structure. Examples include transport systems that allow certain oral medications to move across the membranes separating the intestinal lumen from the blood. Without the kidneys' transport systems, renal excretion of many drugs would be prolonged.

PRIMARY PROCESSES OF PHARMACOKINETICS

The primary processes of *pharmacokinetics* are drug absorption, distribution, metabolism, and excretion (see Figure 3.2). These processes share a common component, which is the movement of drugs. Drugs may move

DOI: 10.1201/9781003461913-4

Pharmacokinetics 17

The Pharmacokinetics Parameters
ADME

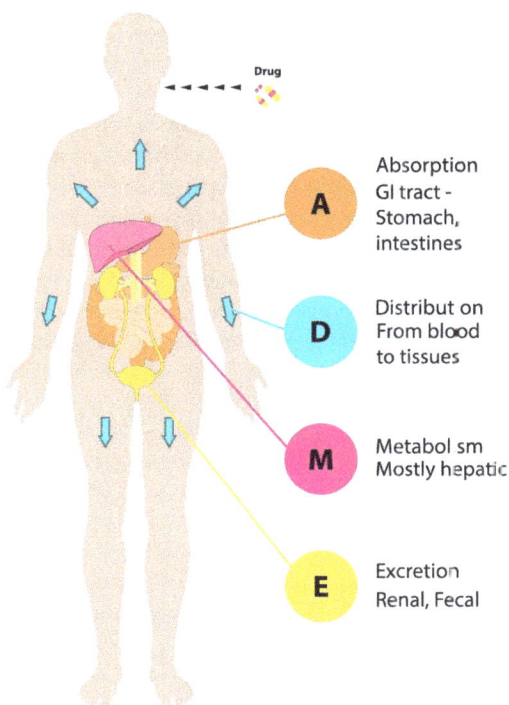

FIGURE 3.2 The four steps of pharmacokinetics.

to their target sites or move out of the body. Absorption is the process of drug movement from the administration site to the bloodstream. Distribution is how drugs are transported throughout the body. Metabolism is the process that changes the drug's activity, also known as **biotransformation**. Excretion is the process that removes medications from the body.

DRUG ABSORPTION

In pharmacokinetics, drug absorption is the movement of a drug into the bloodstream. This takes place in several phases and is related to the type of drug administration utilized. The rate of absorption determines how quickly a drug's effects will begin. The amount of absorption determines the intensity of the drug's effects. A drug must first dissolve before it can be absorbed. Therefore, the absorption rate is partially determined by the dissolution rate. A drug formulated to allow for rapid dissolution will have effects that occur more quickly than one formulated for slow dissolution.

A drug's absorption rate is greatly determined by the surface area available. Absorption is quicker over a larger surface area. Therefore, orally administered drugs are usually absorbed not from the stomach but from the small intestine. This is because the lining of the **microvilli** of the small intestine causes it to have a vast surface area, whereas the surface area of the stomach is much smaller.

Usually, highly lipid-soluble drugs are absorbed more quickly than drugs with lower lipid solubility. Lipid-soluble drugs easily cross membranes, separating them from the blood. Drugs of lower lipid solubility cannot do this quickly. Drug absorption is also influenced by pH partitioning. When the difference between the plasma pH and the pH at the administration site causes drug molecules to have a greater tendency for **ionization** in the plasma, absorption is enhanced. Some medications require strong acids to be absorbed. For example, an antibiotic like cefuroxime needs the pH of the stomach to be low.

There must be adequate blood flow to the administration site for a drug to be absorbed. For example, since large muscles have a high blood flow, intramuscular injections may be used since they maximize drug absorption. Conversely, the skin has slow absorption due to the poor blood supply in the **epidermis**, reducing the time for topical drugs to take effect. When needed, blood flow may be manipulated to slow absorption. A local anesthetic, if absorbed too quickly, may be toxic. Therefore, local anesthetics are often combined with a **vasoconstrictor** such as epinephrine to reduce blood flow to the treated region. This slows the absorption of the local anesthetic.

The **dissolution rate** is commonly modified by altering the particle size, reducing the dose required for the same therapeutic effect. Though this reduction in particle size increases the dissolution rate, solubility is unaffected. Dissolution may also be changed by choosing a **polymorph** of a drug compound.

The gastrointestinal (GI) epithelium is the primary barrier to absorption via the oral route, as the capillary walls offer significantly less resistance. To cross the GI epithelium, drugs must pass through the cell membrane. Intestinal absorption of certain drugs is reduced by **P-glycoprotein**, which pumps them out of the epithelial cells and back into the intestinal lumen. Significant variance exists in the rate and extent of drug absorption after oral administration. This is based on factors such as drug solubility and stability, pH of the stomach and intestine, gastric emptying time, food in the GI tract, coadministration of other drugs, and special coatings used in the drug formulation.

Drug absorption may be affected by interactions with certain foods. An example is when oral tetracyclines are administered with certain foods (such as dairy products). If food containing calcium, iron, or magnesium is consumed simultaneously with oral tetracycline, there will be a significant delay in the drug's absorption. Other oral medications experience slowed absorption when taken with a high-fat meal, which can slow stomach motility to a large degree.

YOU SHOULD REMEMBER

Most drugs are wholly absorbed when taken between meals. This is because the presence of food in the stomach may slow absorption.

Check Your Knowledge

1. What are the concepts of pharmacokinetics?
2. Why are orally administered drugs usually absorbed from the small intestine?
3. What is the function of P-glycoprotein in intestinal absorption?

PASSIVE TRANSPORT

Passive transport (diffusion) requires nearly no energy (see Figure 3.3). It depends upon the drug's lipid solubility and is the most common and essential method of drug movement through membranes in the body. Cell membranes have a fatty bilayer through which drugs must pass for diffusion. More lipid-soluble drugs diffuse more quickly.

ACTIVE TRANSPORT

Active transport requires energy to occur. It moves fluid particles through membranes from lower concentration to higher concentration regions. Active transport requires specific **carrier proteins** in the cell membranes to occur. Figure 3.4 illustrates active transport.

ABSORPTION OF MEDICATIONS THROUGH THE DIGESTIVE SYSTEM

Most orally administered drugs are absorbed through the digestive system. Oral administration is the most common route of drug administration and is convenient and economical. Oral medications are usually absorbed across the stomach or upper intestinal wall. They enter the blood vessels that are part of the *hepatic portal circulation*. This circulation carries blood directly to the liver. Drugs in this blood are immediately exposed to metabolism by liver enzymes via the *first-pass effect* (see Figure 3.5).

DRUG DISTRIBUTION

In pharmacokinetics, drug distribution describes the reversible transfer of a drug from one location to another within the body. It is based on vascular permeability, tissue blood flow, tissue perfusion rates, the ability of a drug to bind to plasma proteins and tissue, and **pH parturition**. Drugs may interact with blood components. They are physically or chemically changed before reaching their target tissues and cells. The simplest factor determining drug distribution

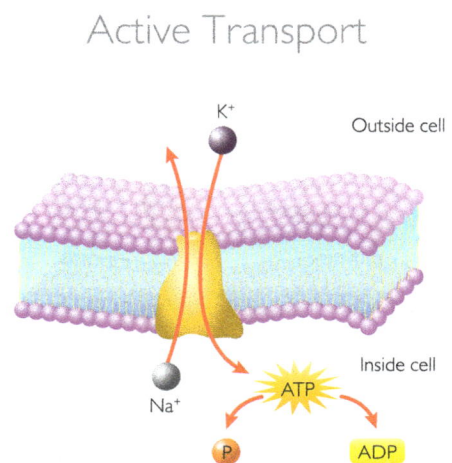

FIGURE 3.4   Active transport.

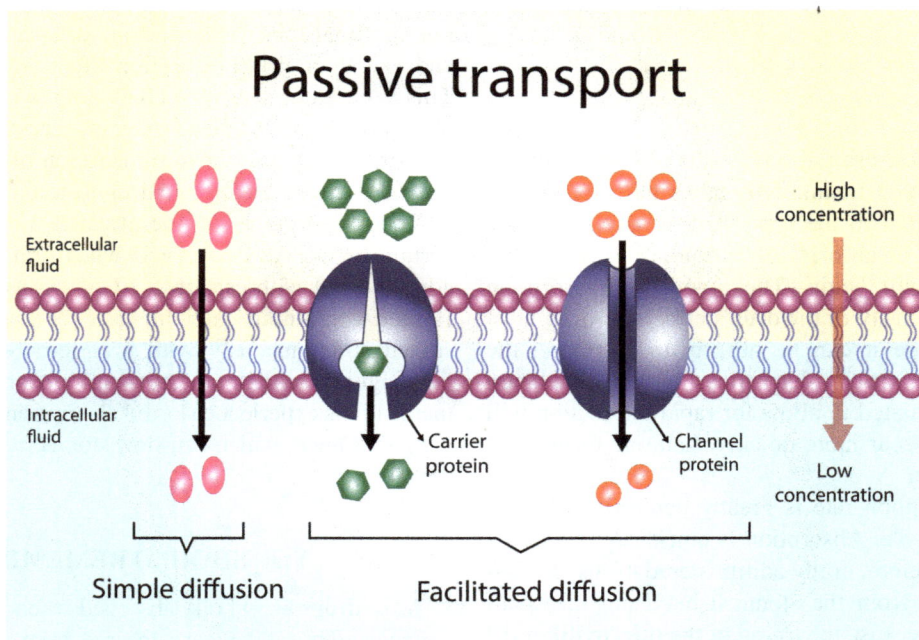

FIGURE 3.3   Passive transport (diffusion).

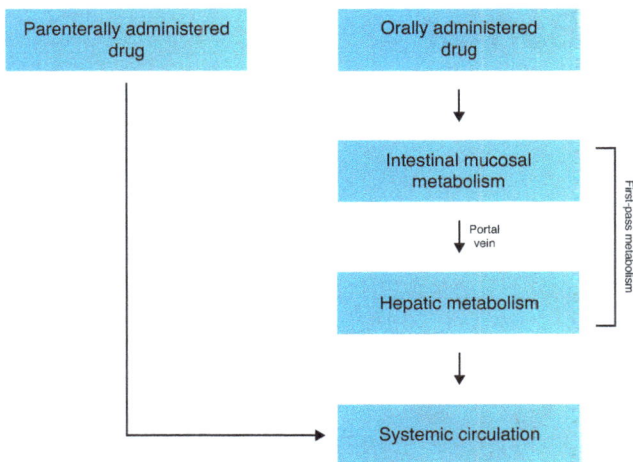

FIGURE 3.5   The first-pass effect.

is the amount of blood flow to body tissues. Organs with the most significant blood supply receive the highest exposure to absorbed drugs. These include the heart, kidneys, and liver.

BLOOD FLOW TO TISSUES

Blood flow is the movement of blood through a vessel, tissue, or organ and is usually expressed in blood volume per unit of time. The contraction of the ventricles of the heart initiates it. Ventricular contraction ejects blood into the major arteries, resulting in flow from regions of higher pressure to areas of lower pressure, as blood encounters smaller arteries and arterioles, then capillaries, the venules, and veins of the venous system.

Because of the unequal distribution of cardiac output to various body organs, blood flow rates to tissue capillaries are incredibly different. For example, blood flow is more significant to the brain, heart, kidneys, and liver than to other organs. Fat tissue has decreased perfusion, resulting in lower drug distribution. Blood flow must be considered locally as well. After a traumatic injury, the blood supply to a fractured bone may be much less because of damaged blood vessels. Perfusion can also be blocked by cellular debris related to inflammation. Antibiotics may be unable to reach the area of injury until blood flow is restored. Delivery of large concentrations of drugs to injured, necrotic, or abscessed areas with insufficient perfusion is always tricky.

DRUG SOLUBILITY

Solubility is the dissolution of a solute in a solvent to give a **homogeneous system**. It is one of the critical parameters for achieving the desired drug concentration in systemic circulation for a desired pharmacological response. The greater the lipid solubility of a drug, the faster it is redistributed to other tissues. When given by inhalation or intravenous routes, highly lipid-soluble drugs are first distributed to organs with higher blood flow. Lipid-soluble drugs are more widely distributed. Less vascular tissues take up the drug later as plasma concentration falls and the drug is withdrawn from these sites. If the site of action is in a highly perfused organ, the drug's redistribution results in drug action termination. When a particular drug is given repeatedly or continuously over long periods, the drug becomes longer-acting. Low perfusion and high-capacity sites are progressively filled with consistently high drug levels.

DRUG BINDING TO PLASMA PROTEINS

Drug–protein binding involves the reversible binding of a drug to a **plasma protein**, such as albumin. When drug–protein complexes are formed, they are too large to pass through capillary membranes. Therefore, they continue circulating in the bloodstream until the drug is released or displaced from the drug–protein complex. Drugs bound in complexes cannot reach their target cells or be removed by the kidneys. For example, warfarin is highly bound to plasma proteins and forms drug–protein complexes in the plasma. Just 1% is freely available to reach target cells.

Specific drugs have a greater affinity for these sites than others. A drug–drug or drug–food interaction can cause the medication to be displaced from plasma proteins. The medication is then free to reach high levels in the blood, and adverse effects may occur. For example, aspirin or cimetidine may displace warfarin from its protein-binding sites. Since the blood levels of free warfarin then significantly increase, the risk of bleeding becomes significant.

Due to the spaces between endothelial cells, most capillaries are relatively porous. As a result, many medications can travel through these spaces to exit the bloodstream and enter the interstitial fluid quickly. This is not true for all parts of the body, however. The brain and placenta are examples of structures that have particular anatomic barriers. These barriers prevent many medications from entering the bloodstream. The brain barrier is known as the **blood–brain barrier** (BBB).

BARRIERS TO DRUG DISTRIBUTION

The BBB is a selective semipermeable membrane between the blood and the interstitium of the brain, allowing cerebral blood vessels to regulate molecule and ion movement between the blood and the brain. Many drugs exit the blood circulation by moving through these spaces to enter the interstitial fluid quickly. Brain capillaries have endothelial cells sealed by tight junctions and a thick basement membrane (see Figure 3.6).

The BBB protects the brain and placenta from **pathogens** and toxic substances. However, highly lipid-soluble agents can cross this barrier easily. Examples of these agents include general anesthetics, antianxiety agents, sedatives, and anticonvulsants. Drugs that cannot cross this barrier easily include antitumor medications and antibiotics. As a result, brain tumors

Blood-brain Barrier

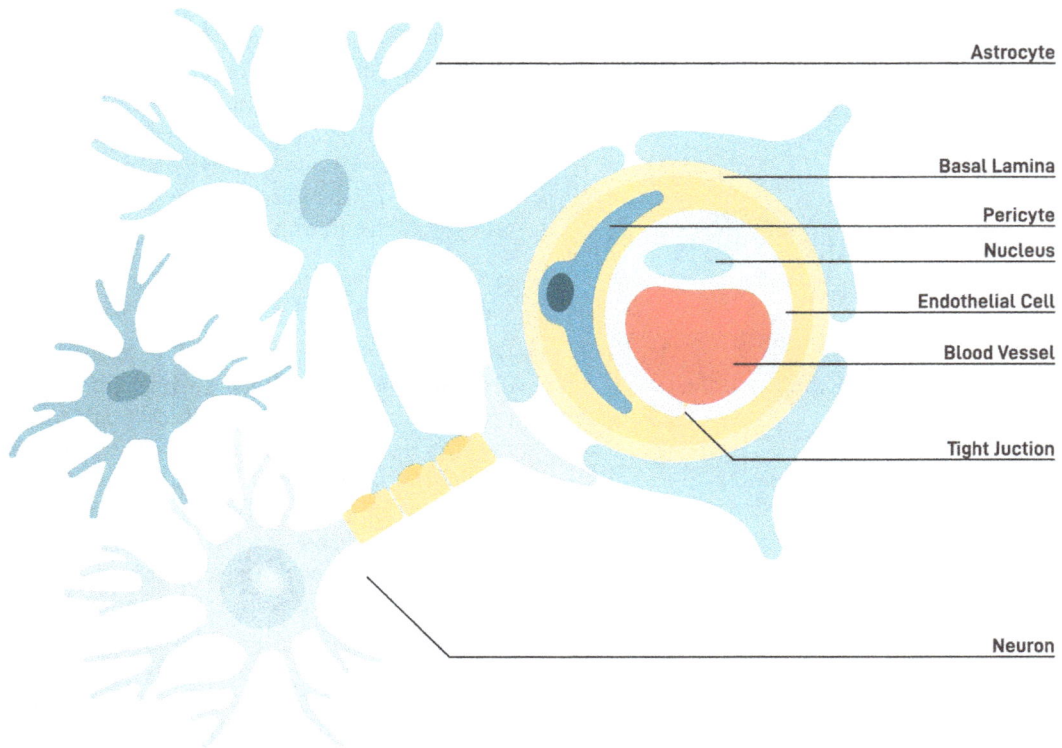

FIGURE 3.6   The blood–brain barrier.

and infections are difficult to treat in this manner. When an infection such as meningitis inflames the BBB, it becomes more permeable. As well as the tight junctions, the BBB is protected by *P-glycoprotein*. In the CNS, this substance pumps drugs back into the blood, limiting their access to the brain. Drugs cross the BBB of neonates more quickly because it is not fully developed.

The placental barrier, or the **fetal–placental barrier**, separates maternal and fetal blood. It prevents harmful substances from passing from the mother's bloodstream to the fetus. However, it does not provide protection against alcohol, caffeine, cocaine, or certain prescription medications. Therefore, pregnant women should consult medical professionals before taking any over-the-counter drugs and herbal supplements. Some compounds can cause congenital disabilities, which may include low birth weight, intellectual disability, and severe malformations. Illegal drugs, such as opioids, can make the baby drug-dependent. Additionally, the use of anesthetics or analgesics can depress respiration in neonates.

Check Your Knowledge

1. What type of binding occurs between drugs and proteins such as albumin?
2. Why is the BBB significant?
3. What is the function of the fetal–placental barrier?

DRUG METABOLISM

Drug metabolism refers to the biochemical modification of a drug, typically achieved through the actions of specialized enzymes. This process often transforms **lipophilic** compounds into forms that can be excreted more effectively. A water-soluble drug is less capable of entering tissues. Each chemical alteration of a drug leads to a functional change as well. The rate of drug metabolism is crucial in determining the duration and intensity of the drug's effects. Additionally, drug metabolism can lead to either intoxication or detoxification of the administered substance.

The primary site of drug metabolism is the **smooth endoplasmic reticulum** in liver cells. The liver is the first organ perfused by chemicals absorbed in the GI tract. Drugs enter hepatic circulation through the portal vein, where they undergo significant metabolism (known as the **first-pass effect**). This effect indicates that some medications may be converted to an inactive form during their initial passage through the liver before entering the systemic blood circulation. To circumvent the first-pass effect, drugs that are rapidly metabolized in the liver are often administered parenterally, allowing them to temporarily bypass the liver and achieve therapeutic levels in the systemic circulation. Sublingual administration also avoids the liver; nitroglycerin can be directly absorbed into systemic circulation, enabling it to reach its action sites before passing through the liver.

The hepatic microsomal enzyme system, also known as the P450 system, serves as the liver's primary mechanism for drug metabolism. **Cytochrome P450** is a crucial component of this system. It consists of 12 related enzyme families, three of which (CYP1, CYP2, and CYP3) are responsible for drug metabolism. The remaining nine metabolize endogenous compounds like steroids and fatty acids. Specific subforms of each cytochrome family metabolize only certain drugs and are identified by abbreviations such as CYP1A2, CYP2D6, and CYP3A4. Hepatic microsomal enzymes catalyze various reactions with drugs as **substrates**. **At** times, drug metabolism does not result in the breakdown of drugs into smaller molecules but rather in the formation of a molecule larger than the parent drug.

There is significant variation in metabolism among patients. This may be genetically determined. Diminished hepatic metabolic activity can result in altered drug action. Infants, for example, only develop a mature microsomal enzyme system once they reach 1 year of age. Therefore, those younger than 6 months have immature livers, and drugs must be administered with caution. In older adults, hepatic activity is typically reduced, meaning that lower doses of many medications are sufficient for treatment. Liver diseases also necessitate decreasing doses. Certain genetic disorders cause some patients to lack specific metabolic enzymes, which slows their metabolic rates. The use of alcohol and tobacco also impacts liver metabolism. Herbal supplements, such as St. John's Wort, an enzyme inducer, can enhance the response to certain medications. Even grapefruit juice can inhibit enzyme activity.

Some drugs influence the liver to enhance drug metabolism rates. For example, phenobarbital can double the liver's drug-metabolizing capacity after several days. It prompts the liver to produce drug-metabolizing enzymes (*induction*). This process leads to two therapeutic consequences. A drug can either increase its metabolism rate, necessitating a higher dosage to achieve therapeutic effects, or the faster metabolism of other concurrently used medications may also require increasing their dosages.

YOU SHOULD REMEMBER

When the stomach is empty, alcohol enters the bloodstream at a rate of 20%–25%, bypassing the first-pass effect. The rest is absorbed from the small intestine and metabolized by the liver.

Check Your Knowledge

1. Where is the principal site of drug metabolism?
2. What is the concept of the first-pass effect?
3. What is the function of the hepatic microsomal enzyme system?

DRUG EXCRETION

Drug excretion is the process of removal of drugs from the body. Drugs are eliminated from the body either unchanged or as metabolites. Water-soluble compounds are excreted more quickly than components with high lipid solubility, except for excretion via the lungs. Lipid-soluble drugs are readily eliminated once they are metabolized to more water-soluble compounds. Drug concentration in the blood and its duration of action are determined by the rate at which it is excreted. Drug action duration and intensity may increase when the liver or kidneys are compromised.

For the kidneys to eliminate drugs from the body, they must prevent them from being reabsorbed from the urine into the bloodstream. Each drug must be chemically changed into a compound that is less fat-soluble and, therefore, less capable of reabsorption. The process of converting fat-soluble drugs into water-soluble **metabolites** that can be excreted via the kidneys is carried out in the liver. The method of metabolism usually decreases a drug's pharmacological activity. Metabolites that remain in the body typically are pharmacologically inactive or less active than the parent drug. Many drugs increase the rate at which an enzyme system metabolizes them, increasing the speed of elimination from the body. Certain drugs can also stimulate their metabolism, explaining why increasing doses may need to be administered to produce the same effect that smaller doses previously produced.

In the kidneys, drugs are filtered at the **renal corpuscle** and may undergo reabsorption in the renal tubule. As in the rest of the body, non-ionized and lipid-soluble drugs cross renal tubular membranes more quickly, returning to the blood circulation. Water-soluble and ionized drugs mostly remain in the filtrate. Certain substances and drug–protein complexes are too large to be filtered in the **Bowman's capsule**. Therefore, they are sometimes secreted into the distal tube of the **nephron** (see Figure 3.7).

The three processes that occur for the urinary excretion of a drug are summarized as follows:

- **Glomerular filtration**: Renal excretion begins at the glomerulus of the kidney tubule. Fluid and small molecules (including drugs) are forced through the pores of the capillary walls. This process (glomerular filtration) moves medications from the blood into the urine inside the tubules. Blood cells and large molecules such as proteins do not undergo filtration because they are too large to pass through the capillary pores. Therefore, drugs bound to albumin remain behind in the blood.
- **Passive tubular reabsorption**: The vessels delivering blood to the glomerulus return near the renal tubule, distal to the glomerulus. Here, drug concentrations in the blood are lower than in the tubule. This concentration gradient drives the drugs from the tubule lumen back into the blood.

Nephron Structure

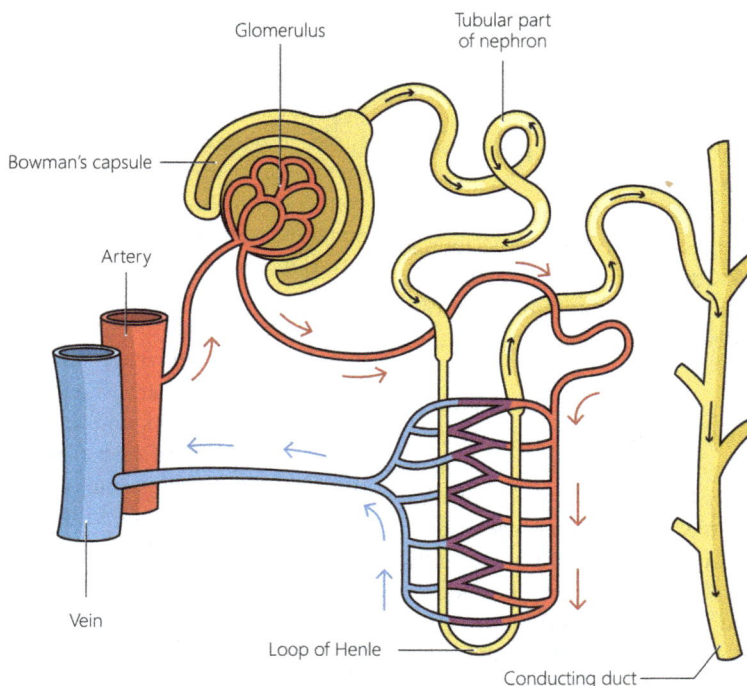

FIGURE 3.7 The structure of the nephron.

Lipid-soluble drugs undergo passive reabsorption from the tubule back into the blood. Lipid-soluble drugs can easily cross membranes that compose the tubular and vascular walls. In contrast, ions and polar compounds (not lipid-soluble) remain in the urine for excretion. The conversion of lipid-soluble drugs into more polar forms reduces the passive reabsorption of drugs, accelerating their excretion.

- **Active tubular secretion**: Drugs are pumped from the blood to the tubular urine by active transport systems in the kidney tubules. One pump class is for organic acids, and one is for organic bases. Tubule cells also contain P-glycoprotein, which can pump drugs into the urine. The capacity of these pumps is relatively high, and they play a significant role in the excretion of specific compounds.

Secretion mechanisms are less active in very young and very old patients. The pH of the filtrate in the renal tubule controls the renal excretion of drugs. Weak acids are excreted more significantly when the filtrate is slightly alkaline. They are ionized in an alkaline environment and remain in the filtrate to be excreted in the urine. Weakly essential drugs are excreted more when the filtrate is slightly acidic since they are ionized in that environment. It is possible to manipulate the pH of the filtrate to increase the speed of renal excretion. Patients with decreased renal function are more likely to develop toxicities.

Other possible sources of excretion include breath, saliva, sweat, tears, feces, breast milk, and bile. Many metabolites of drugs created in the liver are excreted into the intestinal tract via the bile. Drugs excreted in the sweat in the highest quantities include anti-epileptics and beta-lactam antibiotics. Many drugs can pass through breast milk. Lipid-soluble drugs easily enter the breast milk, but polar, ionized, or protein-bound drugs cannot. Nursing mothers should avoid many medicines (if possible) while breastfeeding. If a drug is required, the prescriber must assess that it will not enter the mother's breast milk sufficiently to harm the infant.

YOU SHOULD REMEMBER

Kidney-impaired older patients may retain drugs for extended periods, requiring dosage decreases to avoid toxicity. Even small changes in renal function can rapidly increase serum drug levels.

Clearance

Clearance explains a drug's elimination rate to its plasma concentration, expressed in units equaling the volume per unit of time. Renal clearance determines a drug's renal elimination mechanisms, which result from glomerular filtration, active tubular secretion, and reabsorption. The renal proximal tubule is the primary site of carrier-mediated transport from blood to urine. Therefore, clearance is the rate of elimination of a substance from the plasma concentration. Clearance is constant for any drug eliminated with **first-order kinetics**. Therefore, the elimination rate to plasma concentration ratio is the same, regardless of plasma concentration.

Drug clearances differ widely. They range from a small percentage of blood flow to a maximum of total blood flow to the organs of elimination. Clearance, therefore, depends on the effects of the drug and the condition of these organs. For each organ of elimination, the clearance of a particular drug is equivalent to the organ's ability to extract the drug, multiplied by the drug's delivery rate to the organ. Clearance is often *flow-limited*. This is especially true when the blood is completely cleared of a drug as it passes through an organ of elimination. For any drug that is nearly completely removed, this is a function of the amount of blood flow through the organ. Diseases and other drugs may significantly affect this function. However, it is essential to understand that clearance is not constant for drugs eliminated with zero-order kinetics. Zero-order kinetics is a state in which the rate of an enzyme reaction is independent of the concentration of the substrate. The following equation explains clearance.

$$\text{Clearance} = \frac{\text{Rate of elimination}}{\text{Plasma concentration}}$$

PULMONARY EXCRETION

Pulmonary excretion is a primary route for eliminating gases and some volatile compounds. Volatile drugs, such as gaseous anesthetics, alcohol, or drugs with high volatility, can be excreted via the lungs into expired air.

BILIARY EXCRETION

The biliary excretion of compounds can significantly impact the systemic exposure, pharmacological effects, and toxicity of certain drugs. Drugs excreted into bile often undergo some degree of reabsorption along the GI tract. Some examples of medications that are excreted are mycophenolic acid, warfarin, and digoxin. Penicillins are rapidly excreted, reabsorbed, and re-excreted in high concentrations, whereas streptomycin, neomycin, paromomycin, and chloramphenicol reach lower levels in the bile than in the plasma.

TIME–RESPONSE RELATIONSHIPS

The therapeutic response of a drug defines **time–response relationships**. Most drugs' therapeutic response depends on their concentration in the plasma. The concentration of a drug in its target tissue is a better predictor of drug action, but it is impossible to measure directly. The amount of a drug in a blood sample is not the same as

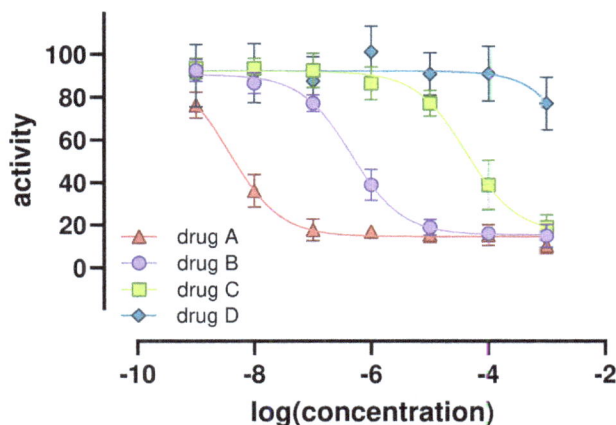

FIGURE 3.8   Dose–response relationships of drugs.

in the target tissue. Monitoring a drug's plasma level is vital when it has a low margin of safety. This **therapeutic drug monitoring** allows the prediction of drug action and toxicity. Drug doses can then be kept within predetermined therapeutic ranges. They can be adjusted upward or downward based on this monitoring and the patient's responses.

After a single dose of oral medication is given, measuring the drug's plasma level can illustrate various important pharmacokinetic principles (see Figure 3.8). It shows a sample drug's minimum effective concentration, therapeutic range, and toxic concentration.

DRUG PLASMA LEVELS

The **minimum effective concentration** of a drug is the amount required to produce a therapeutic effect. The therapeutic range is when a drug produces its desired therapeutic action. Once its levels peak, the drug plasma level slowly reduces, leaving the lower end of the therapeutic range because of excretion. A high drug dose can cause the plasma level to reach a toxic concentration. This is the level of a drug that results in serious adverse effects. Drugs differ in the width of their therapeutic range from those of other medications.

Since drug responses are related to drug concentrations at sites of action, and the site of action of most drugs is not in the blood, regulating plasma drug levels to control drug responses is a concept that must be understood. Usually, it is nearly impossible to measure drug concentrations at sites of action. However, it is generally not necessary to measure drug concentrations at actual sites of action to have a reason for adjusting drug dosage. For most drugs, there is a direct relationship between therapeutic and toxic responses and the amount of drug present in the plasma. It is possible to determine plasma drug concentrations that are highly predictive of therapeutic and toxic responses. Therefore, the dosing goal is "achieving specific drug plasma levels."

The two most crucial drug plasma levels are the minimum effective and toxic concentrations. The *minimum effective concentration* is the plasma drug level below which therapeutic effects will not occur. A drug must be present at or above the minimum effective concentration to be effective. Toxicity occurs when plasma drug levels are too high. The **toxic concentration** is the plasma level at which poisonous effects begin. Doses must be kept low enough so that the toxic concentration is not reached. Drug responses can only occur once plasma drug levels have reached the minimum effective concentration. Therefore, there is a latent period between drug administration and the onset of effects. The rate of absorption determines the extent of this delay. The duration of a drug's impact is largely determined by its metabolism and excretion. Therapeutic responses are maintained if drug levels remain above the minimum effective concentration.

YOU SHOULD REMEMBER

Metabolism and excretion are primarily responsible for causing plasma drug levels to fall. They are the main determinants of how long a drug's effects will continue.

DRUG HALF-LIFE

The length of time a drug concentration remains in the therapeutic range (window) is known as its *duration of action*. It is determined by how quickly the drug is metabolized and excreted. Duration of action is usually described as a *drug's half-life* or plasma **half-life**. This is the time needed after it is administered for a drug's plasma concentration to decrease by 50% of its steady state. This may also be described as the time the drug takes to lose one-half of its pharmacologic, physiologic, or radiologic activity.

A drug's half-life can range from a few minutes to several days. Some drugs, such as amiodarone, have extremely long half-lives of up to 55 days. The biological half-life of water in humans is between 7 and 14 days. Drugs with shorter half-lives are used for conditions and procedures with a brief duration. These include dental procedures and treating a simple headache. Short-acting drugs generally need to be administered every few hours. Medicines that have longer half-lives are used for conditions and procedures that have a long duration. These include heart failure and hypertension treatments, and preventing migraine headaches or seizures. Longer-acting drugs can be administered only once per day or even less frequently. However, it is essential to remember that the longer a drug stays active in the body, the higher the risk of long-term adverse effects. This is vital in patients with renal or hepatic impairment and diminished metabolism or excretion.

When a drug is discontinued, it usually takes about four half-lives to be considered "functionally" eliminated, with 94% elimination. For example, since the drug procaine has a half-life of just 8 minutes, it is considered functionally

eliminated in 32 minutes. The 6% remaining drug is too small to produce either beneficial or toxic effects. However, this determination of functional elimination does not apply to all drugs. For example, the half-life of morphine is approximately 3 hours. This means that the body's level of morphine will decrease by 50% every 3 hours, no matter how much morphine has been administered. The percentage never changes, but it is essential to understand that the amount of morphine lost is more significant when the total body stores are higher.

Overall, a drug's half-life determines its dosing interval. For a drug with a short half-life, the dosing interval must be correspondingly brief. If an extended dosing interval were used, the drug level would fall below the minimum effective concentration between doses, meaning the therapeutic effects would be lost. A drug with a longer half-life can have a larger dosing interval without losing its benefits.

BIOAVAILABILITY

A drug's **bioavailability** is the fraction of the administered dose that reaches systemic circulation. This may also be explained by its ability to reach systemic circulation from the administration site. Different formulations of the same drug can affect this.

Intravenous drugs differ in that they offer approximately 100% bioavailability. When using other routes to administer medicines, bioavailability is generally reduced by incomplete absorption, first-pass metabolism, and distribution into other tissues before it enters the systemic circulation. In the intestine, it is reduced by intestinal transporters, which is part of the concept of incomplete absorption. Drug responses can be variable due to enteric coatings, tablet disintegration time, and sustained-release formulations. Variability is more related to oral preparations than parenteral preparations. However, variability is usually minimal and lacks clinical significance.

Drugs with a **narrow therapeutic** range are of most significant concern. Even a relatively small change in drug levels can significantly change responses. Since slight declines could cause therapeutic failure, and small increases could cause toxicity, bioavailability differences may dramatically impact the patient's health.

LOADING DOSAGE

Most drugs are administered in multiple doses. The therapeutic range must be maintained when this occurs over an extended period. Ideally, the next drug dose should be administered before the drug plasma falls below the minimum effective concentration. Figure 3.9 shows a drug's accumulation in the bloodstream due to multiple dosing. Correct timing of doses results in a plateau in the drug plasma level being reached and maintained. The plateau can be reached more quickly by administering a **loading dose**, followed by regular maintenance doses. At the plateau drug plasma level, the amount of drug absorbed is equal to the

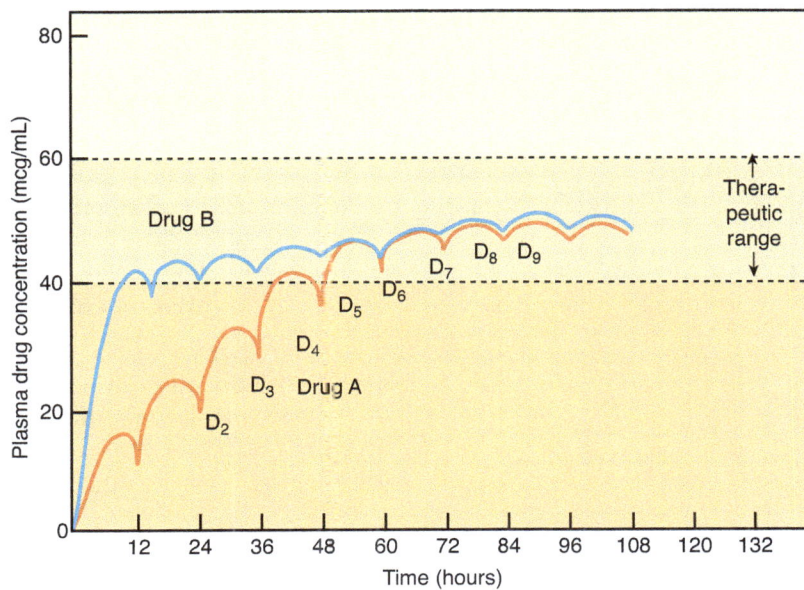

FIGURE 3.9 Multiple-dose drug administration.

amount excreted. This maintains the therapeutic range of the drug. It generally takes about four half-lives to reach this equilibrium. Continuous medication infusion allows the plateau to be reached quickly and maintained with slight fluctuations in drug plasma levels.

Plasma drug levels are not smooth or continuous. They have slight peaks and troughs. The highest level is called the *peak concentration*, while the lowest is the *trough concentration*. Peaks must not rise into the toxic range. It is also essential that troughs stay within the therapeutic range. Many drugs are available in extended-release or sustained-release formulations to minimize these occurrences. They slowly release the drug to maintain its therapeutic range. Fluctuations may also be reduced by changing the dosing schedule. For example, if 90 mg of a drug is ordered, it may be given as 45 mg bid or 30 mg tid to reduce the amount of variation between peaks and troughs.

A drug plateau may be reached more quickly by administering loading doses followed by regular maintenance doses. A loading dose is a large amount of drug administered 1–2 times to get a level in the bloodstream sufficient to produce a fast therapeutic response. A loading dose is significant for a drug with a prolonged half-life, especially when raising the drug plasma level quickly is critical. An example would be an antibiotic given to treat a severe infection.

If a therapeutic concentration must be achieved rapidly and the volume of distribution is large, the loading dose may also need to be significant. The equation used to calculate this is as follows:

$$\text{Loading dose} = \frac{\text{Volume of distribution} \times \text{Desired plasma concentration}}{\text{Bioavailability}}$$

When the loading dose is large, meaning that the volume of distribution is much larger than the blood volume, the dose should be given slowly. This prevents toxicity from excessively high plasma levels during the distribution phase. Administering a drug by continuous infusion helps to reduce fluctuations in drug levels. This allows plasma levels to be kept nearly constant. Another method is administering a depot preparation, which steadily and slowly releases the drug. A third method is to reduce the size of each dose and the dosing interval (keeping the total daily dose constant).

MAINTENANCE DOSAGE

Before plasma levels fall toward zero, intermittent **maintenance doses** are given. These keep the plasma drug concentration inside the therapeutic range. This method allows equilibrium to be reached nearly as rapidly as a continuous infusion. The maintenance rate of drug administration is equal to the rate of elimination at a *steady state*. The maintenance dose is a function of clearance. The equation used to calculate maintenance dosage is as follows.

$$\text{Maintenance dose} = \frac{\text{Clearance} \times \text{Desired plasma concentration}}{\text{Bioavailability}}$$

The dosing rate for the maintenance dosage is computed as the average dose per unit of time. The units must be constant throughout these calculations, such as mL/min or whatever units are used. Daily dose sizes may be calculated based on the dose per minute multiplied by 60 minutes per hour and then multiplied by 24 hours per day.

The number of doses to be given daily is usually determined by the drug's half-life and the difference between the minimum therapeutic and toxic concentrations. This is the drug's therapeutic window (see Chapter 4). If a concentration above the minimum therapeutic level must always be maintained, one of two methods is used. The first is to give larger doses at longer intervals. The second is to give smaller doses at shorter intervals. Smaller and more frequent doses should be given to prevent toxicity when the therapeutic window is small.

Check Your Knowledge

1. What are the differences between the loading dose and maintenance doses?
2. How do you calculate the loading dose?
3. How do you calculate the maintenance doses?

CLINICAL CASE STUDIES

Clinical Case Study 1

A 38-year-old man complained of strep throat. He is currently taking celecoxib for arthritis and nizatidine for heartburn. He starts oral cefuroxime for 10 days, but after 6 days, the patient calls his physician to say that the sore throat is just as bad as it has been. The physician evaluates the cefuroxime, which requires a lower stomach pH to be adequately absorbed. In this case, nizatidine has probably caused the stomach pH to increase, impairing the absorption of cefuroxime. There is evidence that acid-suppressing drugs can lower the bioavailability of cefuroxime by as much as 60%. Since the patient needed nizatidine for his heartburn, it would have been better to prescribe a macrolide antibiotic or a cephalosporin such as cephalexin, since these drugs do not require a lower stomach pH to be effective.

CRITICAL THINKING QUESTIONS

1. How does the pH of the stomach affect the absorption of oral medications?
2. Why does drug absorption occur in the small intestine rather than the stomach?
3. Why does drug absorption occur in larger muscles rather than the epidermis?

Clinical Case Study 2

A 34-year-old male with no known medical comorbidities consumed almost half a bottle of whiskey on his birthday. A few hours later, he ingested 140 g of acetaminophen. He was presented 48 hours later to the emergency department. His labs were significant for acetaminophen level, aspartate aminotransferase, and alanine aminotransferase; total bilirubin was also high, and arterial ammonia was consistent with acute liver failure. His mental status deteriorated rapidly to encephalopathy and required interventions to prevent further neurological complications and

multiorgan dysfunction. A CT head scan showed early signs of diffuse cerebral edema.

CRITICAL THINKING QUESTIONS

1. Where is the principal site of drug metabolism?
2. What is the hepatic microsomal enzyme system?
3. What are the variations in metabolism between patients?

Clinical Case Study 3

A 79-year-old woman was hospitalized with a 12-year history of type 2 diabetes, symptomatic diabetic nephropathy, peripheral vascular insufficiency, and chronic renal disease. Her medications included aspirin, insulin, gabapentin, atorvastatin, and vancomycin. She was required to have hemodialysis three times per week. The patient underwent dialysis while in the hospital, and her symptoms improved, though they recurred between dialysis sessions. It was determined that since gabapentin is excreted, the patient's reduced kidney capacity had elevated gabapentin, triggering her neurologic symptoms.

CRITICAL THINKING QUESTIONS

1. Where does most drug excretion occur, and what factors affect clearance?
2. What are the three processes for the urinary excretion of a drug?
3. How much clearance is reduced in humans by the age of 80?

FURTHER READING

Bertolini, J., Goss, N., and Curling, J. (2012). *Production of Plasma Proteins for Therapeutic Use*. Wiley.

Bonate, P.L. (2014). *Pharmacokinetic-Pharmacodynamic Modeling and Simulation*, 2nd Edition. Springer.

Chow, S.C., and Liu, J.P. (2008). *Design and Analysis of Bioavailability and Bioequivalence Studies* (Chapman & Hall/CRC Biostatistics Series Book 27), 3rd Edition. Chapman and Hall/CRC Press.

Coleman, M.D. (2020). *Human Drug Metabolism*, 3rd Edition. Wiley-Blackwell.

Dressman, J.B., and Reppas, C. (2010). *Oral Drug Absorption: Prediction and Assessment* (Drugs and the Pharmaceutical Sciences), 2nd Edition. Informa Healthcare.

Derendorf, H., and Schmidt, S. (2019). *Rowland and Tozer's Clinical Pharmacokinetics and Pharmacodynamics: Concepts and Applications*, 5th Edition. Lippincott, Williams, & Wilkins.

Fisher, J.W., Gearhart, J.M., and Lin, Z. (2020). *Physiologically Based Pharmacokinetic (PBPK) Modeling: Methods and Applications in Toxicology and Risk Assessment*. Academic Press.

Gabrielsson, J., and Hjorth, S. (2012). *Quantitative Pharmacology: An Introduction to Integrative Pharmacokinetic-Pharmacodynamic Analysis*. Swedish Pharmaceutical Press.

Jambhekar, S.S., and Breen, P.J. (2012). *Basic Pharmacokinetics*, 2nd Edition. Pharmaceutical Press.

Ma, S., and Chowdhury, S. (2020). *Identification and Quantification of Drugs, Metabolites, Drug Metabolizing Enzymes, and Transporters: Concepts, Methods, and Translational Sciences*, 2nd Edition. Elsevier Science.

Niazi, S.K. (2014). *Handbook of Bioequivalence Testing* (Drugs and the Pharmaceutical Sciences Volume 213), 2nd Edition. CRC Press.

Peters, S.A. (2012). *Physiologically-Based Pharmacokinetic (PBPK) Modeling and Simulations: Principles, Methods, and Applications in the Pharmaceutical Industry*. Wiley.

Rosenbaum, S.E. (2016). *Basic Pharmacokinetics and Pharmacodynamics: An Integrated Textbook and Computer Simulations*, 2nd Edition. Wiley.

Rowland, M., and Tozer, T.N. (2010). *Clinical Pharmacokinetics and Pharmacodynamics: Concepts and Applications*, 4th Edition. Lippincott, Williams, & Wilkins.

Rudek, M.A., Chau, C.H., Figg, W.D., and McLeod, H.L. (2016). *Handbook of Anticancer Pharmacokinetics and Pharmacodynamics: Cancer Drug Discovery and Development*, 2nd Edition. Humana Press.

Smith, D.A., and Fox, D. (2010). *Metabolism, Pharmacokinetics and Toxicity of Functional Groups: Impact of Chemical Building Blocks on ADMET* (Drug Discovery, Volume 1). Royal Society of Chemistry.

Southwood, R., Fleming, V.H., and Huckaby. (2018). *Concepts in Clinical Pharmacokinetics*, 7th Edition. American Society of Health Systems.

Tozer, T.N., and Rowland, M. (2015). *Essentials of Pharmacokinetics and Pharmacodynamics*, 2nd Edition. Lippincott, Williams, & Wilkins.

Wagner, J.G. (2019). *Pharmacokinetics for the Pharmaceutical Scientist*. CRC Press.

Wu, B., Lu, D., and Dong, D. (2020). *Circadian Pharmacokinetics*. Springer.

4 Pharmacodynamics

LEARNING OBJECTIVES

After studying this chapter, readers should be able to

1. Describe the mechanism of action of agonists and antagonist receptors.
2. Identify the differences between agonists and antagonists.
3. Review a drug's therapeutic index and its margin of safety.
4. Distinguish the significance of the dose–response relationship to clinical practice.
5. Compare and contrast potency and efficacy.
6. Describe the various factors that affect drug actions.
7. Explain the effect of drug actions on age, gender, and body weight.
8. Compare anaphylactic shock with an idiosyncratic reaction.
9. Describe synergism, potentiation, and teratogens.
10. Compare and contrast median lethal dose and median toxicity dose.

OVERVIEW

The term pharmacodynamics refers to how a drug changes in the body. It is the branch of pharmacology focusing on the mechanisms of drug action and how drug concentrations cause responses in the body. To understand how drugs can apply the principles of pharmacodynamics to the treatments required by the patients. It exhibits a unique affinity for a drug–receptor site, indicating the strength of its binding to the site. All drugs produce their effects by interacting with biological structures or targets at the molecular level to induce a change in how the target molecule functions regarding subsequent intermolecular interactions. The pharmacologic response depends on the drug binding to its target as well as the concentration of the drug at the receptor site. Each patient may respond to a given drug much differently from others. The fundamentals of pharmacodynamics include dose–response relationships, drug–receptor interactions, and therapeutic indexes.

MECHANISM OF DRUG ACTIONS AND RECEPTORS

In pharmacology, the mechanism of action is defined as the specific biochemical interaction a drug uses to produce its pharmacological effect. A mechanism of action usually involves enzymes or receptors to which drugs bind.

Pharmacodynamics studies the impact of drugs on the body or microorganisms inside the body. These biochemical and physiological effects include mechanisms of drug action and relationships between drug concentrations and effects.

Most drugs mimic or inhibit normal physiological, biochemical, and pathological processes, or inhibit vital processes of pathogenic microorganisms. They produce their actions by activating or inhibiting specific cellular receptors. There are seven main actions of drugs:

- Stimulation
- Depression
- Blocking or antagonizing
- Stabilization
- Exchanging or replacing
- Direct beneficial chemical reaction
- Direct harmful chemical reaction

The desired activity of a drug is primarily due to the successful targeting of one of the following:

- Disruption of cellular membranes
- Chemical reactions with downstream effects
- Interaction with enzyme, structural, or carrier proteins
- Interaction with ion channels
- Ligand binding to hormone, neuromodulator, or neurotransmitter receptors

RECEPTORS

A **receptor** is a cellular molecule to which a medication binds to produce its effects. It is the medication's specific target. Most receptors are proteins. Drugs bind to receptors and cause a change in physiology. The response of a drug is proportional to the concentration of receptors bound or occupied by the drug. This is known as *receptor theory*. The normal function of receptors is to bind endogenous molecules. These include hormones and neurotransmitters. A drug uses existing targets to cause its effects.

Certain drugs do not act through receptors. Instead, they act through simple physical or chemical interactions with other small molecules. These drugs include antacids, antiseptics, chelating agents, and saline laxatives. By direct chemical interaction with stomach acid, antacids neutralize gastric acidity. Antiseptics such as ethyl alcohol work by precipitating bacterial proteins. **Chelating agents** such as dimercaprol prevent toxicity from heavy metals by forming complexes with them. The laxative known as magnesium

DOI: 10.1201/9781003461913-5

sulfate causes water to be retained in the intestinal lumen through an osmotic effect. All of these drugs work without interacting with cellular receptors.

YOU SHOULD REMEMBER

Aspirin works by irreversibly inhibiting the enzyme cyclooxygenase. This action suppresses the production of prostaglandins and thromboxanes, reducing pain and inflammation.

YOU SHOULD REMEMBER

Drugs in the brain disrupt the signaling process mediated by neurotransmitters. Certain drugs, like marijuana and heroin, can activate neurons because their chemical structures resemble those of natural neurotransmitters in the body. This enables the drugs to bind to and stimulate the neurons.

Check Your Knowledge

1. What are the seven main drug actions?
2. What is the receptor theory?
3. What are three examples of drugs that do not require receptors?

Selectivity

Drug receptors are macromolecules involved in chemical signaling between cells and the inside of cells. They may be located on the cell surface membrane or inside the cytoplasm. *Activated* receptors regulate cellular biochemical processes. Molecules that bind to receptors are called **ligands**. A ligand may activate or inactivate a receptor. When activated, cell functions may be increased or decreased. Selectivity is the degree to which a drug acts on a particular site rather than others. It is possible mainly because drugs act through specific receptors. Selective drug action is likely due to many types of receptors regulating only a few processes. A drug's **affinity** is the probability of occupying a receptor at any moment. Medications with high affinity become firmly attached to their receptors.

Receptor Types

Two basic receptor types exist in the body. These are *alpha* and *beta* receptors. Each has several subtypes. Drugs that are selective for only one type have been developed over time, permitting more specific drug therapy and reducing adverse effects. The *alpha-1, alpha-2, beta-1,* and *beta-2* subtypes offer a high degree of specificity. Some drugs are independent of cellular receptors. They work by interacting with cellular membrane permeability, depressing membrane excitability, or altering cellular pump activity. These

nonspecific cellular responses are utilized by such agents as general anesthetics, osmotic diuretics, and ethyl alcohol.

EFFECTORS

Effectors are molecules that change cellular activity because of a drug–receptor interaction. Enzymes such as **adenylyl cyclase** are the best examples of effectors. Some receptors are also effectors. A single molecule may have a drug-binding site and an effector mechanism. An example is the **tyrosine kinase** effector, part of the insulin receptor molecule. Another example is a sodium–potassium channel, which is the effector of the nicotinic acetylcholine receptor.

AGONISTS

An **agonist** is a chemical that binds to a cell's receptor, triggering a response by that cell. Agonists may mimic the action of naturally occurring substances. A **physiological agonist** creates the same bodily responses but does not bind to the same receptors. However, it should be understood that an *inverse agonist* causes an action opposite to that of an agonist. It exerts the opposite pharmacological effect of receptor agonists. Examples of inverse agonists include diphenhydramine (antihistamines) and nalmefene. It acts at the H1 receptor, reversing the effects of histamine on capillaries to reduce allergic reaction symptoms. Receptors can be activated or inactivated by either **endogenous** or **exogenous** agonists and antagonists. Biological responses can, therefore, be stimulated or inhibited. The mechanism of action of some agonists is unknown.

An *endogenous agonist* is a naturally produced compound that binds to and activates a specific receptor. For example, dopamine is the endogenous agonist for dopamine receptors. A **superagonist** is a compound that can produce a greater maximal response than the endogenous agonist for a target receptor. It has an efficacy of more than 100%. **Full agonists** bind to and activate a receptor and display full efficacy for that receptor.

An example is morphine, which mimics the actions of endorphins at mu-opioid receptors in the central nervous system. **Partial agonists** only have partial efficacy at the receptor compared to full agonists. Co-agonists work together with other co-agonists to produce the desired effect. Irreversible agonists bind permanently to receptors so that the receptors remain permanently activated. The maximal effect produced by a partial agonist is lower than that of a full agonist. An example is pentazocine, which relieves much less pain than a full agonist such as meperidine.

Partial agonists can also act as antagonists. For example, if a patient is already on a full agonist for pain relief and is then given a hefty dose of pentazocine, it will occupy the opioid receptors and prevent meperidine from activating them. The patient will then experience less than adequate pain relief.

Examples of agonists include neurotransmitters, hormones, and all other endogenous regulators. They are classified as agonists because they activate the receptors to which they bind. They then mimic the actions of the body's regulatory molecules. An agonist is a drug with affinity and high intrinsic activity. Affinity allows them to bind to receptors; inherent activity enables them to activate receptor function after binding.

Many therapeutic agents function as agonists. Examples include dobutamine, which mimics norepinephrine's actions at heart receptors, and norethindrone, which activates receptors for progesterone. Agonists can either speed up or slow down various processes. For example, acetylcholine can activate specific heart receptors to cause **bradycardia**. Drugs that mimic this action could still be called agonists, even though they cause a decline in heart rate.

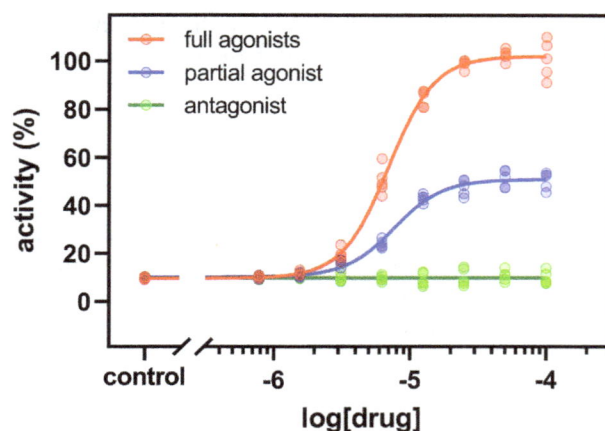

FIGURE 4.1   Antagonist, agonist, and partial agonist actions.

YOU SHOULD REMEMBER

Overdosage of diphenhydramine may lead to anticholinergic symptoms, seizures, and coma. A fatal outcome from a diphenhydramine overdose is not common. The patient initially experienced seizures, followed by cardiac conduction issues and hemodynamic compromise, which resulted in death despite life support measures.

ANTAGONISTS

Antagonism can actually "cancel" drug actions. For example, epinephrine can be antagonized by carvedilol. These opposing actions are based on the doses of the two drugs. Drugs with opposing actions are usually administered to reduce the adverse effects of one of the drugs. In many cases, a second drug does not reverse all the actions of the first drug but may selectively inhibit an undesirable action.

An antagonist does not provoke a biological response upon binding to a receptor. The drug *dimercaprol* is a typical example of a chemical antagonist. This drug chelates (combines to form a ring with) lead and other toxic metals. Another example is *pralidoxime*, which combines with the phosphorus in organophosphate cholinesterase inhibitors.

In other words, antagonists have an affinity but no efficacy for specific receptors. If there is no agonist present, administration of an antagonist will have no observable effect. If agonists activate receptors, the process will be stopped when an antagonist is administered, resulting in a discernible impact. Most drug antagonists achieve **potency** by competing with endogenous ligands or *substrates* at structurally defined receptor binding sites. Chronic use of antagonists is linked to neuronal death, with potent antagonists considered toxic. Figure 4.1 shows the actions of antagonists, agonists, and partial agonists.

Competitive antagonists reversibly bind to receptors at the same binding site as the endogenous ligand or agonist without activating the receptor. For example, the poison Curare is a competitive antagonist of *acetylcholine*. A *noncompetitive antagonist* binds to the receptor's active

site with irreversible or nearly irreversible action. Since competitive antagonists bind reversibly to receptors, the inhibition caused by competitive antagonists may be surmounted. When enough agonists are present, all receptors are occupied, and inhibition is overcome.

The binding of noncompetitive antagonists is irreversible. Therefore, their inhibition cannot be overcome. Thus, these agents are rarely used therapeutically. However, the effects of noncompetitive antagonists do not last forever. As receptors are naturally broken down and replaced, the effects of these antagonists wear off. They may take only a few days to lose their impact since receptor life cycles are often short.

Antagonists may also be used to treat drug overdoses. The administration of naloxone is used for an overdose of morphine, fentanyl, or heroin. It can quickly reverse the serious adverse effects of these drugs. An indirect drug interaction is exemplified by digoxin and furosemide. Digoxin has cardiotoxic impacts, such that furosemide can increase them, as it enhances the excretion of potassium ions. This interaction can be life-threatening.

YOU SHOULD REMEMBER

Drug overdose remains a significant public health issue in the United States, primarily driven by synthetic opioids like illicit fentanyl. Today, the FDA approved Narcan, a 4 mg naloxone hydrochloride nasal spray, for over-the-counter use – the first naloxone product approved for use without a prescription. Naloxone is a medication that quickly reverses the effects of opioid overdose and is the standard treatment for such cases. This action paves the way for life-saving medication to counter an opioid overdose.

THERAPEUTIC INDEX AND THERAPEUTIC WINDOW

The **therapeutic index** of a drug consists of its median effective dose (ED_{50}) value compared to its median lethal dose (LD_{50}) value. It is an estimate of a drug's safety.

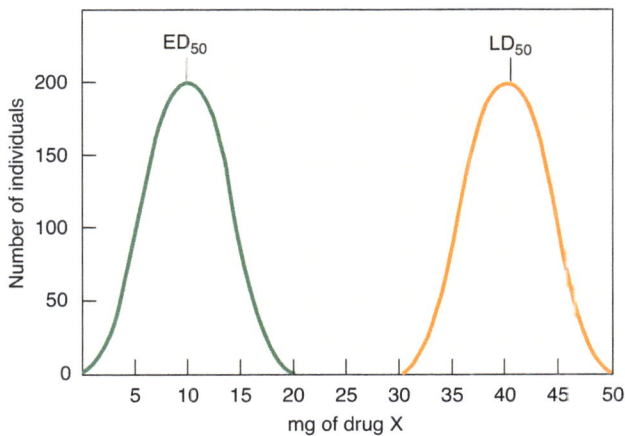

(a) Drug X : $TI = \dfrac{LD_{50}}{ED_{50}} = \dfrac{40}{10} = 4$

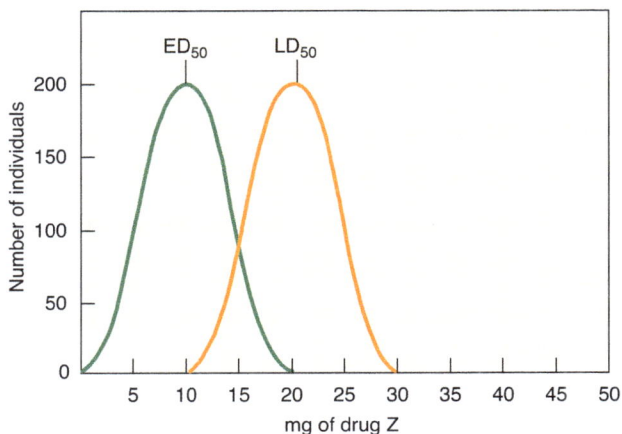

(b) Drug Z : $TI = \dfrac{LD_{50}}{ED_{50}} = \dfrac{20}{10} = 2$

FIGURE 4.2 A frequency distribution curve explaining the therapeutic index.

An extremely safe drug may have a substantial toxic dose and much smaller effective doses. A full range of toxic doses cannot be ethically studied in humans. Also, this estimate is a poor safety index even in animal testing because of the varying slopes of dose–response curves and other factors. The dose in the middle of this curve represents a drug's median effective dose (Figure 4.2).

To further understand the therapeutic index, look at the following formula:

$$\text{Therapeutic index} = \frac{\text{Median lethal dose } (LD_{50})}{\text{Median effective dose } (ED_{50})}$$

The curve for therapeutic effects overlaps with the curve for lethal effects. This shows that high doses that produce therapeutic effects in one person can cause the death of another person. For a drug to be completely safe, the highest dose needed to produce therapeutic effects must be well below the lowest dose required to produce death.

The **therapeutic window** is more clinically helpful in gauging a drug's safety. It describes the safe dosage range between a drug's minimum effective (therapeutic) and minimum toxic concentrations or doses. It determines an acceptable range of drug plasma levels when determining dosages. If a drug's average minimum therapeutic plasma concentration is 10 mg/L, and its toxic effects are seen at 30 mg/L, the therapeutic window is 10–30 mg/L. The therapeutic window, as well as the therapeutic index, depends on a specific toxic effect that is used in this determination. The minimum effective concentration usually determines the *desired trough levels* of the drug, given intermittently. The minimum poisonous concentration determines the *permissible peak levels* of the drug in the plasma.

The greater the difference between the median lethal dose (LD_{50}) and the median effective dose (ED_{50}), the greater the therapeutic index. A higher therapeutic index value indicates a safer drug for administration. Therapeutic indices can vary widely, ranging from 1:2 to over 1:100. Other estimates of drug safety are also available. The *median toxicity dose (TD_{50})* is the amount that causes toxicity in 50% of patients based on animal data or adverse effects observed in clinical trials. The *margin of safety (MOS)* is determined by dividing the lethal dose for 1% of animals by the dose that produces a therapeutic effect in 99% of the same population. Generally, a higher MOS value suggests a safer medication. However, this metric does not account for severe, nonlethal adverse effects resulting from lower drug doses.

Check Your Knowledge

1. How do you describe the therapeutic index and therapeutic window?
2. What are the meanings of lethal and median effective doses?
3. What is the median toxicity dose?

DOSE–RESPONSE RELATIONSHIPS

The dose–response relationship is the relationship between the size of an administered dose and the intensity of the response produced. This is an essential component of medication therapy. The dose–response relationship determines the minimum amount of a drug that can be used, the maximum response it can cause, and the dosage increase needed to achieve the desired response. The essential characteristics of the dose–response relationship are shown in Figure 4.3. Part A shows a linear coordinate plotting of dose–response data. Part B shows a semilogarithmic coordinate plotting of the same data.

Maximal Efficacy

Maximal efficacy is the most significant effect that a drug can produce. The height of the dose–response curve

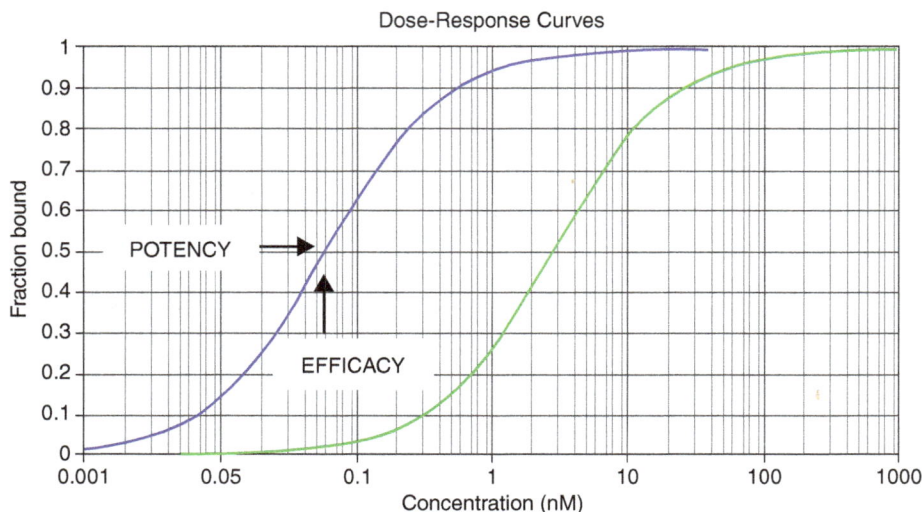

Dose-Response Curves

FIGURE 4.3 The dose–response curve.

indicates it. A drug with high maximal efficacy is not always preferred over a drug with lower efficacy. Each drug is tailored to the patient's individual needs. Drugs that produce extremely intense responses are more complex to tailor to individual patients.

An example is the diuretic furosemide, which has high maximal efficacy and can cause dehydration. A lower maximal efficacy of diuretics, such as hydrochlorothiazide, would be preferred if the patient only needed a modest volume of water to be mobilized. A drug's effectiveness is almost always more important than its potency.

POTENCY

Potency is the amount of drug needed to elicit a desired effect. However, potency is usually not a vital characteristic overall. It can be necessary if a drug is so weak that massive doses are required to achieve the desired effect. In this case, another drug with higher potency would be used in smaller amounts to achieve the desired effect. In pharmacology, potency refers to how much of a drug is needed for a desired effect compared to how much of another drug is required to produce the same effect.

Drug potency is, in comparison to other factors, not an important characteristic. It only means that a drug that is more potent than another can be administered in lower doses. It is only clinically crucial if a drug has such a low potency level that its doses must be massive. In such a case, another drug with higher potency would be used since it would not require such large doses. The potency of a drug does not imply anything about its maximal efficacy – these are entirely different qualities. One drug can be more effective than another while still being more potent. Two drugs with identical potencies may still have equal effectiveness. A drug's *potency* means how well it produces its effects at low doses.

VARIOUS FACTORS THAT AFFECT DRUG ACTIONS

Many different factors affect drug actions and the dose required for each patient. These include age, gender, weight, daily body rhythms, existing disease states, immunological factors, environmental factors, tolerance, cumulation, and interactions.

AGE

Generally, children or very old patients require less than the average adult dosage of a given drug. For children, several formulas are available to estimate the correct dosage of a drug compared to the standard adult dose. Since the kidneys and liver of infants are less effective than those of adults, drugs tend to accumulate in their bodies, which can lead to toxicity. Older patients, in particular, are at an increased risk of toxicity since they generally take more drugs for various conditions, increasing their likelihood of adverse outcomes.

GENDER

Men and women also have different responses to certain drugs. Since men have more vascular muscles, intramuscular injections have a faster action than they do for women. Since women have more fatty tissue, drugs that deposit fat may be released more slowly, with prolonged effects, compared to how they affect men. Also, women should always be asked about possible pregnancy when any drug is considered. Physiological factors can also influence drug effects. These include acid–base balance, **diurnal body rhythms** of the nervous and endocrine systems, electrolyte balance, and hydration.

Female patients usually require smaller doses than male patients. Exceptions to this rule include iron preparations and other blood products because of the blood females lose during menstruation. Women also have a lower concentration of fluids in their bodies, allowing an increased saturation of *receptor sites* and a higher concentration of a given drug in their brains. They generally have a higher degree of body fat, resulting in drugs being active for more extended periods. Examples of medicines affecting women for more extended periods include inhaled anesthetics. This is because they deposit fat and may cause drowsiness and sedation even weeks after surgery. Also, women should be asked about the possibility of becoming pregnant since pregnancy influences the use of certain drugs. Intramuscular injections act more quickly in men than in women because men have more vascular muscles than women.

Drugs with significant gender-related differences include digoxin, alcohol, opioid analgesics such as pentazocine and nalbuphine, and quinidine. For heart failure, digoxin may increase mortality in women while not causing an increase in mortality in men. Since women metabolize alcohol more slowly than men, a woman who consumes alcoholic beverages in the same amount as a man (even if the woman weighs the same amount) will show more signs of intoxication. Pentazocine and nalbuphine are much more effective in women than in men, meaning pain relief can be achieved with much lower doses. Quinidine has different cardiac effects in women than in men.

Weight

The average adult is considered to weigh 150 pounds, but there are significant variances in body weight among people of all races, heights, etc. Therefore, heavier patients require larger doses than fragile patients. The doses of many drugs are calculated on a weight basis. This is usually calculated as a specified number of grams or milligrams to be administered per pound or kilogram of body weight. The greater a patient's body mass, the more likely a drug is diluted throughout the system.

Prescribers often base dosage adjustments on body surface area to account for weight differences. Body surface area accounts for the patient's weight and the amount of fatty tissue they have. The percentage of body fat greatly influences drug absorption, changing the concentration of the drug at its sites of action. Therefore, dosage adjustments based on body surface area are more precise in controlling drug responses than those based only on body weight.

Diurnal Body Rhythms

Diurnal body rhythms are also commonly referred to as *circadian rhythms*. A circadian rhythm is any biological process with an endogenous oscillation of about 24 hours. They are adjusted to a patient's local environment by external cues, the most important of which is daylight. To be called circadian, a biological rhythm must repeat once daily, persist even without external cues, be adjustable to match the local time zone, and exhibit temperature compensation.

Circadian rhythms allow us to adapt to environmental changes and regulate internal metabolic processes. They are linked to the light (daytime)/dark (nighttime) cycle. Normal adults have a "built-in" day, which averages approximately 24 hours. While indoor lighting affects circadian rhythms, we are still "set" to a 24-hour clock, regulating periods of sleep and wakefulness.

Circadian rhythms are measured based on *melatonin* secretion by the pineal gland, core body temperature, and plasma levels of *cortisol*. Melatonin is absent from the system, or extremely low, during the day. It rises in dim light at about 9:00 pm every night. Other critical physiological changes that occur because of circadian rhythms include heart rate and the production of red blood cells.

In humans, the timing of medical treatment coordinated with the circadian rhythms may significantly affect the efficacy of the treatment and reduce toxicity or adverse reactions. Health problems associated with disturbances of these rhythms include delayed sleep phase syndrome, seasonal affective disorder, and others. Sleep–wake cycle reversal may be a sign of uremia, azotemia, or acute renal failure. When circadian rhythms are disrupted, there is usually an adverse effect on the body. Other physiological factors influencing drug actions include acid–base balance, hydration, and electrolyte balance.

YOU SHOULD REMEMBER

Jet lag is an adverse effect on the body caused by the disruption of circadian rhythms. Jet lag occurs when a person travels across several time zones and must attempt to adapt to their new surroundings quickly. Symptoms of jet lag include disorientation, fatigue, and insomnia.

Check Your Knowledge

1. What factors can affect drug actions?
2. Why do females require smaller doses than male patients?
3. Why do older patients require less than the average adult dosage?

Diseases

Disorders that affect pharmacodynamics include genetic mutations, malnutrition, thyrotoxicosis, myasthenia gravis, Parkinson's disease, and insulin-resistant diabetes mellitus. Hepatic and renal diseases are the primary disease states that affect pharmacodynamics. For example, changes in liver function affect the body's response to drugs due to pharmacodynamics and pharmacokinetics. Therapeutic effects may be reduced or enhanced. The patient's health status and existing disease states must always be considered when determining dosages for medicinal treatment.

Diseases such as liver cirrhosis may result in enhanced CNS effects of some drugs, enhanced renal adverse effects of others, and reduced sensitivity to drugs such as diuretics.

Since the liver is the primary site of drug detoxification and metabolism, drugs will accumulate to toxic levels when liver function is poor. If liver disease is present, drug dosages must be reduced to prevent their accumulation to harmful levels. This only applies to drugs that are eliminated mainly by the liver.

Likewise, since the kidneys heavily influence drug excretion, reduced kidney function may cause drugs to accumulate in the body. Drug dosages often have to be reduced until kidney function returns to normal.

IMMUNOLOGICAL FACTORS

Immunological factors concern the development of drug allergies. Once a patient is exposed to a drug's proteins, they may develop **antibodies** to the drug. Future exposures to the drug may cause the patient to experience a severe allergic reaction. Drug sensitivities can range from mild to severe. A skin rash may signify mild reactions. Severe reactions may cause **anaphylaxis**, shock, and even death.

ENVIRONMENTAL FACTORS

Environmental factors can also influence drug therapy. When a patient's response to a drug is not as expected, alterations may occur in the patient's environment. For instance, sedatives are more effective in a calm, relaxing setting without excessive stimulation. Sedatives are most effective when the environment reduces tension and stress. Additionally, other drugs can be impacted by temperature changes. Antihypertensives may be overly potent in warmer climates, as these climates naturally dissipate body heat through vasodilation. Since the vessels are already dilated, antihypertensives may become excessively effective.

TOLERANCE

Over time, the body can tolerate certain drugs, and its reaction to various concentrations is reduced. This may be linked to increased biotransformation of the drug. It can also be caused by increased resistance to drug effects and other pharmacokinetic factors (see **Chapter 3**). Tolerance causes a drug to become unable to cause the same reaction as it initially caused. Larger doses are required to achieve the therapeutic effect. Tolerance may be avoided by giving smaller doses or using a drug in combination with other medications. Resistance to drugs within the same class is known as **cross-tolerance**.

Pharmacodynamic tolerance is associated with the long-term administration of a drug. It is a decreased response to a drug because of the action of cellular mechanisms. Often, it is due to a downregulation of the number of receptors. Examples include heroin and morphine. The patient requires increased drug levels to produce effects previously produced at lower levels. The minimum effective concentration is abnormally high. Adaptive processes cause this type of tolerance in response to chronic receptor occupation.

Metabolic tolerance results from accelerated drug metabolism. It is caused by certain drugs (for example, barbiturates) that can induce the synthesis of hepatic drug-metabolizing enzymes. This causes the rates of drug metabolism to increase. The dosage must be increased to maintain therapeutic drug levels. Unlike pharmacodynamic tolerance, metabolic tolerance does not affect the minimum effective concentration.

Tachyphylaxis is reduced drug responsiveness caused by repeated dosing over a short time. It occurs quickly but is not a common mechanism of drug tolerance. Nitroglycerin is an example. When administered via a transdermal patch, its effects occur in less than 24 hours if the patch is left in place. The drug's effect is lost due to the depletion of a cofactor needed for nitroglycerin to act.

DRUG INTERACTIONS

Drugs can interact with other medications, herbal supplements, foods, and laboratory tests. Drug–drug interactions may occur with medications that have narrow safety margins. Interference with a drug's pharmacokinetics or pharmacodynamics can lead to serious outcomes (see **Chapter 6**).

Check Your Knowledge

1. What are the primary diseases that affect pharmacodynamics?
2. What causes metabolic tolerance?

ADVERSE DRUG REACTIONS AND ADVERSE EFFECTS

An *adverse drug reaction* is harm associated with a drug, at a typical dosage, during everyday use. Every drug has the potential to produce adverse events. Adverse drug reactions have occurred after only a single drug dose, due to prolonged drug administration, or because of the combination of two or more drugs. An *adverse drug event* refers to an injury caused by a drug at a regular or excessive dosage and any harm associated with its use (see **Chapter 6**).

ALLERGIC REACTIONS

Drug allergies make up 6%–10% of all adverse drug effects. Allergic reactions to drugs may require immediate medical attention. The signs of anaphylaxis are the same regardless of the drug that causes them. Allergic reactions will not occur on the first exposure to a substance. However, a first exposure allows the body to create antibodies and memory lymphocyte cells for the antigen. A cross-allergy is an allergic reaction to drugs with a similar structure and from the same pharmacologic class (see **Chapter 6**).

ANAPHYLACTIC SHOCK

Anaphylaxis is signified by symptoms such as an itchy rash, **pharyngeal edema,** and hypotension. When drugs are involved, the average onset of the condition is 5–30 minutes. The primary treatment for anaphylaxis is an injection of epinephrine.

Any medication may trigger anaphylaxis and anaphylactic shock. Other common causes include chemotherapy, vaccines, protamine, and herbal preparations. Certain medications can cause anaphylaxis by directly triggering **mast cell** degranulation. These include vancomycin, morphine, and X-ray contrast solutions. The frequency of a reaction to an agent partly depends on its frequency of use and intrinsic properties. For example, anaphylaxis to penicillins or cephalosporins only occurs after binding to blood proteins. Specific agents may bind more quickly than others. Anaphylaxis to penicillin occurs in one of every 2,000–10,000 administrations of this agent, with death occurring in less than one in every 50,000 administrations.

Biphasic anaphylaxis is the recurrence of symptoms within 1–72 hours, with no further exposure to the allergen. **Anaphylactoid reactions** do not involve an allergic reaction but are due to direct mast cell degranulation. Skin testing is available to confirm penicillin allergies. Other drugs require blood testing to determine potential allergies.

IDIOSYNCRATIC REACTIONS

An **idiosyncratic reaction** or response produces an unusual, unexpected response unrelated to a drug's pharmacologic action. This is not classified as an allergy since it is not immune-related. Idiosyncratic reactions vary from patient to patient and are both unpredictable and rare. Genetic differences, often related to enzymes, cause some. These can cause a drug to be transported by a metabolic pathway that leads to the accumulation of its metabolites. This accumulation can cause different, unexpected responses. With improved drug adverse effect reporting, idiosyncratic reactions are diminishing since more and more of these effects are now understood and even expected.

DRUG TOLERANCE

Drug tolerance is a standard in which a patient reacts to a specific drug and concentration progressively reduces. This requires an increase in drug concentration to achieve the desired effect. Drug tolerance may be related to both psychological and physiological factors. This occurrence is reversible, and its rate depends on the drug involved, dosage, and frequency. Differential development occurs due to the different effects of the same drug. **Dispositional tolerance** occurs when a decreased amount of a substance reaches the site it affects. It may be caused by an increase in the induction of the enzymes required for the degradation of the drug.

CUMULATIVE EFFECT

A drug can accumulate in the body over time, leading to toxic levels and adverse effects. The **cumulative effect** of a drug is a condition in which repeated administration of a drug may produce more pronounced effects than those produced by the initial dose. It is also called **cumulative action**. Cumulative effects are usually due to doses that are too frequent or because of the administration of a drug over a long period. If a drug is excreted slowly yet is administered frequently, it may accumulate in the body, producing toxicity. It is essential to closely monitor the frequency of administration of a drug and the length of time it is being administered.

The cumulative effect can be avoided by precisely following drug regimens. Unfortunately, many people do not comply with their drug regimens. Often, a patient consumes their entire daily intake of drugs at once so they do not forget to take each drug at different times during the day. Many drugs should be taken at specific times, and some should not be administered simultaneously. Other people realize they forgot to take a dose and then take two doses simultaneously later to make up for it.

ANTINEOPLASTIC DRUGS

Some drugs have been linked to causing cancer years or decades after their administration. Though the FDA does not approve most medications if they cause cancer in laboratory animals, sometimes approval is granted. This is because of the risk–benefit ratio. For example, a patient with a condition likely to cause premature death without drug treatment requires a drug with known cancer-causing effects that can still be curative. In this case, the benefits of taking the curative drug outweigh the risks of it causing cancer. A drug like this would be used only when all safer alternatives have been tried or no alternative exists. Only a few drugs increase the risk of acquiring cancer (see Table 4.1). These drugs have three primary categories: antineoplastics, hormones, hormone antagonists, and immunosuppressants.

Some **antineoplastic** drugs are known chemical carcinogens. **Mutations** in deoxyribonucleic acid (DNA) result from molecular damage caused by these drugs. Though some damage is naturally repaired, specific mutations persist and accumulate in cells during aging. The most significant cancer risk from antineoplastic therapy is **leukemia**. The benefit, however, of drug therapy combined with radiation therapy and surgery outweighs the small risk of developing cancer later in life. With immunosuppressants, **lymphoma** is the most significant type of cancer risk. Cancers from hormones or hormone antagonists often affect reproductive organs.

TERATOGENS

A **teratogen** is a substance that disrupts normal fetal development and leads to congenital disabilities. Drugs, alcohol,

TABLE 4.1

Potential Cancer-Causing Drugs

Drugs	Type of Cancer Caused
Antineoplastics	Leukemia, urinary
Chlorambucil	cancer
Cyclophosphamide	
Dacarbazine	
Doxorubicin	
Etoposide	
Nitrosoureas	
Teniposide	
Hormones and hormone antagonists	Breast, hepatic, and uterine cancers
Anabolic steroids	
Estrogen replacement therapy and oral contraceptives	
Tamoxifen	
Immunosuppressants	Lymphoma, skin
Azathioprine	cancer
Cyclosporine	

chemicals, and toxic substances serve as examples of teratogens. For instance, in the early 1960s, a drug known as thalidomide was used to treat morning sickness. Fetal exposure during this early stage of development resulted in cases of phocomelia, a congenital malformation where the hands and feet are attached to shortened arms and legs.

POTENTIATION

Potentiation is the enhancement of one drug by another so that the combined effect is greater than the sum of the effects of either drug. In potentiation, drugs act synergistically, yet this phenomenon is not identical to **synergism**. An example of drug potentiation is when the antihistamine *Phenergan* is given with the analgesic narcotic *Demerol* to intensify its effects, allowing less of the narcotic to be administered. Additional examples of potentiation include

- **Diazepam with tofisopam**: increased anticonvulsant effects of the diazepam
- **Benzodiazepines with mu-agonist opioids**: greatly enhanced analgesia.

ADDITIVE EFFECT

The **additive effect** involves two drugs from a similar therapeutic class. They may interact to produce a combined **summation response**. Therapeutically, the additive effect may be used for such reasons as the treatment of hypertension, as when diuretics and beta-blockers are used together. The diuretic will lower systolic blood pressure (BP), while the beta-blocker may produce even more BP-lowering effects. The first reason for two drugs to be used this way is to reduce the potential for adverse effects by keeping

their dosages low. The second reason is that a single drug may need to be more vital to achieve the desired outcome. Also, diuretics and beta-blockers act at significantly different receptors. Diuretics act in the renal tubules, while beta-blockers act in cardiac muscle.

PLACEBO EFFECT AND PSYCHOLOGICAL FACTORS

A **placebo** is a preparation that has no pharmacologic activity. A patient's response to a placebo is based on their belief that it is a medication with beneficial properties. Placebos are most commonly used as control preparations during clinical trials. They help determine the actual effect of a drug being tested compared to a group of patients who receive only a placebo. Still, they may believe it is the exact medication. The results of both groups are then compared.

The placebo effect occurs when psychological factors influence a patient's response to a drug. It does not concern the drug's biochemical or physiologic properties. The psychological factors affecting a patient's response to a drug cannot be accurately assessed. However, it is believed that a patient's response to a drug is at least partially related to the placebo effect. Also, a better patient attitude influences compliance with a drug regimen. The presence of a placebo response does not mean that the original pathology of the patient was "conceived" by the patient. Newer evidence suggests that placebo responses may be much smaller than previously thought.

Placebo responses can also be harmful instead of beneficial. If the patient is convinced that their medication is potentially harmful or ineffective, the placebo effect may detract from the positive impact of the medication. Also, a patient receiving a placebo who believes it is a dangerous experimental drug may exhibit "adverse effects" even though the placebo contains no medication.

SYNERGISM

Synergism occurs when drugs interact to enhance or magnify one or more adverse effects. The negative impact of synergism is a form of contraindication. An example is when alcohol is taken with a barbiturate. Since these substances have similar actions, their effects may be greatly exaggerated, resulting in possible coma and death. Additional examples of synergism include

- **Verapamil with propranolol**: causing bradycardia and circulation defects
- **ACE inhibitors with aldosterone antagonists**: additive potassium-sparing effects
- **Nonreversible MAO inhibitors with SSRIs:** fever, chills, diarrhea, ataxia, confusion, and potentially, death

However, synergism can have pharmacologic benefits. When treating certain infections, for example, two antibiotics may be combined to increase antimicrobial actions against various pathogens.

Check Your Knowledge

1. What can be the result of the cumulative effect of a drug?
2. What types of cancers may occur by using antineoplastic medications during pregnancy?
3. What are the differences between a drug allergy and a drug intolerance?

CLINICAL CASE STUDIES

Clinical Case Study 1

A 56-year-old woman was brought to the emergency department. Because the patient might have taken too much diphenhydramine, she was complaining of blurry vision, confusion, urinary retention, and hallucinations. She developed hypotension and cardiac arrest from an overdose of the drug. Her arterial blood gases revealed acidosis. The EKG showed QRS widening, QT prolongation, and a flattening of the T-wave. She developed seizures and coma.

CRITICAL THINKING QUESTIONS

1. What are the signs and symptoms of diphenhydramine overdose?
2. What is diphenhydramine classified as, and how does it work?

Clinical Case Study 2

A 16-year-old was taken to the emergency department after developing a severe headache, vomiting, and a high fever. The physical examination revealed a stiff neck and rashes on the body. After a spinal tap and blood tests, he was diagnosed with bacterial meningitis. The physician ordered two different antibiotics to be more effective for his treatment.

CRITICAL THINKING QUESTIONS

1. What does the term "the combination of two drugs" mean?
2. What is the purpose of a combination of two or more drugs?

FURTHER READING

Derendorf, H., and Schmidt, S. (2019). *Rowland and Tozer's Clinical Pharmacokinetics and Pharmacodynamics: Concepts and Applications*, 5th Edition. Lippincott, Williams, and Wilkins.

Ensom, M.H.H., Kiang, T.K.L., and Wilby, K.J. (2016). *Pharmacokinetic and Pharmacodynamic Drug Interactions Associated with Antiretroviral Drugs*. Adis.

Gabrielsson, J., and Hjorth, S. (2017). *Quantitative Pharmacology: An Introduction to Integrative Pharmacokinetic-Pharmacodynamic Analysis*, 2nd Edition. Swedish Pharmaceutical Press.

Gabrielsson, J., and Weiner, D. (2002). *Pharmacokinetic and Pharmacodynamic Data Analysis: Concepts & Applications*, 3rd Edition. Swedish Pharmaceutical Press.

Jann, M.W., Penzak, S.R., and Cohen, L.J. (2016). *Applied Clinical Pharmacokinetics and Pharmacodynamics of Psychopharmacological Agents*. Adis.

Kwon, Y. (2013). *Handbook of Essential Pharmacokinetics, Pharmacodynamics, and Drug Metabolism for Industrial Scientists*. Springer.

Mager, D.E., and Kimko, H.H.C. (2016). *Systems Pharmacology and Pharmacodynamics* (AAPS Advances in the Pharmaceutical Sciences Series 23). Springer.

Meibohm, B. (2006). *Pharmacokinetics and Pharmacodynamics of Biotech Drugs: Principles and Case Studies in Drug Development*. Wiley-Blackwell.

Mrsny, R.J., and Daughterty, A. (2019). *Proteins and Peptides: Pharmacokinetic, Pharmacodynamic, and Metabolic Outcomes* (Drugs and the Pharmaceutical Sciences). CRC Press.

Nightingale, C.H., Ambrose, P.G., Drusano, G.L., and Murakawa, T. (2019). *Antimicrobial Pharmacodynamics in Theory and Clinical Practice (Infectious Disease and Therapy)*, 2nd Edition. CRC Press.

Owen, J.S., and Fiedler-Kelly, J. (2014). *Introduction to Population Pharmacokinetic/Pharmacodynamic Analysis with Nonlinear Mixed Effects Models*. Wiley.

Rosenbaum, S.E. (2016). *Basic Pharmacokinetics and Pharmacodynamics: An Integrated Textbook and Computer Simulations*, 2nd Edition. Wiley.

Rowland, M., and Tozer, T.N. (2010). *Clinical Pharmacokinetics and Pharmacodynamics: Concepts and Applications*, 4th Edition. Lippincott, Williams, and Wilkins.

Rudek, M.A., Chau, C.H., Figg, W.D., and McLeod, H.W. (2014). *Handbook of Anticancer Pharmacokinetics and Pharmacodynamics* (Cancer Drug Discovery and Development), 2nd Edition. Humana Press.

Schuttler, J., and Schwilden, H. (2008). *Modern Anesthetics* (Handbook of Experimental Pharmacology 182). Springer.

Shaw, L.M., Schentag, J.J., Evans, W.E., and Burton, M.E. (2005). *Applied Pharmacokinetics & Pharmacodynamics: Principles of Therapeutic Drug Monitoring*, 4th Edition. Lippincott, Williams, and Wilkins.

Tozer, T.N. (2012). *Introduction to Pharmacokinetics and Pharmacodynamics: The Quantitative Basis of Drug Therapy*. Lippincott, Williams, and Wilkins.

Tozer, T.N., and Rowland, M. (2015). *Essentials of Pharmacokinetics and Pharmacodynamics*, 2nd Edition. Lippincott, Williams, and Wilkins.

Udy, A.A., Roberts, J.A., and Lipman, J. (2018). *Antibiotic Pharmacokinetic/Pharmacodynamic Considerations in the Critically Ill*. Adis.

Vinks, A.A., Derendorf, H., and Mouton, J.W. (2014) *Fundamentals of Antimicrobial Pharmacokinetics and Pharmacodynamics*. Springer.

5 Pharmacogenomics and Drug Toxicity

LEARNING OBJECTIVES

After studying this chapter, readers should be able to

1. Explain pharmacogenetics and genetic variants.
2. Compare ototoxicity and neurotoxicity.
3. Describe the characteristics and symptoms of hepatotoxicity.
4. List some drugs that may cause nephrotoxicity.
5. Review certain medicines that can cause cardiotoxicity.
6. List some medications that may cause rhabdomyolysis.
7. Explain mechanical, chemical, or physiologic antidotes.
8. Describe poisons and their specific antidotes.

OVERVIEW

Pharmacogenomics is a rapidly growing field of medicine that investigates how a person's genetic makeup may affect how their body processes certain medications. It involves genetic testing that looks for changes in specific genes. Pharmacogenomics is part of the field of precision medicine. This personalized treatment is based on a person's genes, environment, and lifestyle. Therefore, it can help healthcare providers prescribe a medication that leads to fewer side effects or that may work better for patients. Specific drug therapies are individualized if a genetic variation that affects drug disposition is known. This may improve drug efficacy and reduce adverse effects. Drug toxicity refers to the level of damage that a compound can cause to an organ. It involves the accumulation of an excessive amount of any medication in the blood. The medication's effects are more intense at toxic levels, and adverse effects may be severe or lethal. Drug toxicity can occur when a dose is too high or when drug clearance via the liver or kidneys is insufficient to remove the drug from the blood circulation.

PHARMACOGENETICS

The advent of new technologies and discoveries has influenced pharmacology more than ever before. **Pharmacogenetics** refers to genetic differences in metabolic pathways affecting individual drug responses, including therapeutic and adverse effects. In the future, drugs may be customized for patients with specific genetic similarities. It is hoped that advances in pharmacogenetics will eliminate idiosyncratic responses. Analyzing a patient's DNA may help determine which drug should be prescribed.

Pharmacogenetics is occasionally used interchangeably with pharmacogenomics, which refers to somatic mutations in tumor DNA that modify drug responses. Subtle genetic variations in drug-metabolizing enzymes have already been identified. These genetic differences in enzymes significantly influence drug-induced toxicity. In the future, pharmacogenetics may allow for customizations to each patient's molecules. This could radically transform the field of pharmacology in the years to come.

Genetic variants can significantly modify drug responses by influencing drug metabolism. As a result, a drug's metabolism may either accelerate or slow down. Consequently, the patient may experience reduced benefits from the drug or increased toxicity. If a drug has a low therapeutic index, even minor increases in drug levels can result in toxicity. Similarly, relatively small decreases in drug levels can lead to therapeutic failure. Drugs known to exhibit varying effects based on genetics include isoniazid (INH), warfarin, codeine, mercaptopurine, and fluorouracil.

Genetic variations can change the structure of drug receptors and other target molecules. This can have various effects on drug responses. Variants such as these have been documented in normal cells, viruses, and cancer cells.

YOU SHOULD REMEMBER

Pharmacogenetics starts with an unexpected drug response and evaluates its genetic cause. At the same time, pharmacogenomics begins by looking for genetic variations within a population that may explain specific observed responses to a therapeutic drug.

DRUG TOXICITY OF SPECIFIC ORGANS

Drug toxicity involves the accumulation of an excessive amount of any medication in the bloodstream. The medication's effects are more intense at toxic levels, and adverse effects may be severe or lethal. Drug toxicity and drug overdose are often confused. One difference is that drug toxicity generally occurs over time, while drug overdose happens when too much of a substance is consumed at once. Drug toxicity is typically accidental, while drug overdose can be either accidental or intentional. Drug toxicity can occur as a result of the over-ingestion of medication, causing too much of the drug to be in a person's system at once. This can happen if the dose taken exceeds the prescribed amount or if the prescribed dosage is too high.

DOI: 10.1201/9781003461913-6

With certain medications, drug toxicity can also occur as an adverse drug reaction. In this case, the usual therapeutic dose of the drug can cause unintentional, harmful, and unwanted side effects. Sometimes, the **threshold** between effective and toxic doses is very narrow. A therapeutic dose for one patient might be detrimental to another patient. Drugs with a longer half-life can also accumulate in a patient's blood circulation and increase over time, resulting in drug toxicity.

Drug toxicity symptoms can differ depending on the medications patients are taking. For example, lithium toxicity may cause mild symptoms when it is consumed once, causing nausea, vomiting, diarrhea, weakness, and dizziness. In more severe cases, the symptoms of acute lithium toxicity include tremors, ataxia, nystagmus, seizures, muscle cramps, and coma.

Drug toxicity can affect many body organs. However, the liver and kidneys are common organs affected by chemical toxicity. The kidneys are responsible for filtration; so, unsurprisingly, harmful agents in the blood may accumulate there. The liver is the primary site of drug biotransformation and detoxification. Clinical presentations of toxic liver injury range from indolent, often asymptomatic progression of impairment of hepatic function to rapid development of hepatic failure. Neurotoxicants can also cause significant toxicity in the CNS.

NEUROTOXICITY

Neurotoxicity refers to brain or peripheral nervous system damage caused by exposure to natural or artificial toxic substances. These toxins can alter the nervous system's activity by disrupting or killing nerves. **Neurons** are essential for transmitting and processing information in the brain and other areas of the nervous system. Neurotoxicity is a significant cause of **neurodegenerative** disorders. It is a relatively common adverse effect of specific drugs. The differences between therapeutic doses and doses that produce adverse effects may be insignificant. Medications with the potential for neurotoxicity include antiseizure, antianxiety, antipsychotic, and antidepressant agents. Some examples of neurotoxic agents are amitriptylinoxide, dibenzepin, opipramol, and oxaprotiline. Amphetamines are the most commonly used illicit drugs after cannabis. Other common drugs of abuse include cocaine, opiates, and alcohol. Chemotherapy drugs, radiation, mercury, lead, pesticides, and certain foods must also be considered neurotoxic. Aminoglycosides, tetracyclines, clindamycin, erythromycin, polymyxins, ethambutol, INH, and chloramphenicol may cause severe neurotoxicity.

The signs and symptoms include **hallucinations**, depression, sedation, **mania**, behavioral changes, seizures, imbalance, hearing loss, resting tremors, and visual alterations. Other symptoms include suicidal ideation, **delirium,** cognitive decline, sexual dysfunction, and seizures.

YOU SHOULD REMEMBER

The effects of neurotoxicity depend on various factors, including the characteristics of the neurotoxin, the dose a person has been exposed to, the ability to metabolize and excrete the toxin, the capacity of affected mechanisms and structures to recover, and the vulnerability of the cellular target.

OTOTOXICITY

Ototoxicity refers to inner ear damage that develops as a side effect of certain medications. This condition can lead to issues with hearing and balance, functions that the inner ear regulates. More than 200 known ototoxic medications are available on the market. These drugs are used to treat severe infections and illnesses, including cancer and heart disease. The most commonly known ototoxic drugs include certain aminoglycosides such as gentamicin, streptomycin, neomycin, kanamycin, tobramycin, and vancomycin, as well as chemotherapy agents like cisplatin and carboplatin. Other ototoxic substances include macrolide antibiotics (like azithromycin), antituberculosis agents (such as viomycin), salicylates, loop diuretics (including furosemide and ethacrynic acid), antimalarials, quinine, and acetaminophen. Environmental toxins can also contribute to ototoxicity, potentially resulting in permanent hearing issues. These toxins include mercury, lead, manganese, carbon disulfide, and carbon monoxide.

Ototoxicity symptoms can start suddenly or develop slowly over time. Usually, the first signs of ototoxicity are **tinnitus**, hearing loss, **hyperacusis**, aural fullness, and **vertigo.** Symptoms can develop rapidly or gradually and can be either reversible or irreversible. Even minimal to mild hearing loss can hamper speech, language, cognitive, and social development in children, which may lead to poor educational performance and psychosocial functioning.

YOU SHOULD REMEMBER

Ototoxic antibiotics should be avoided during pregnancy since they can harm the fetal labyrinth of the inner ear. Older adults and individuals with preexisting hearing loss should avoid ototoxic medications if other effective alternatives are available. The lowest effective dosage of ototoxic drugs should be utilized, and levels should be closely monitored, primarily for aminoglycosides.

HEPATOTOXICITY

Hepatotoxicity is a condition characterized by damage to the liver caused by exposure to harmful substances, such as certain medications, alcohol, nutritional supplements, or chemicals. It is one of the most common adverse drug effects since the liver detoxifies chemicals that enter the body. Hepatotoxic drugs may cause minor and reversible symptoms or even be fatal. In some cases, toxic hepatitis develops within hours or days of exposure to a toxin. In other cases, regular use may take months before signs and symptoms appear.

More than 1,000 medications and herbal compounds are known to cause hepatotoxicity, including INH, amoxicillin–clavulanate, sulfamethoxazole–trimethoprim, statins, ciprofloxacin, anabolic steroids, amiodarone, phenytoin, valproate, and methotrexate. The risk factors associated with the development of drug-induced hepatotoxicity include older age, female sex, and increased body mass index.

The symptoms of toxic hepatitis often go away when exposure to the toxin stops. But toxic hepatitis can permanently damage the liver, leading to irreversible **cirrhosis** and, in some cases, liver failure, which can be life-threatening. Symptoms include right upper quadrant pain, **jaundice,** fever, itching, **anorexia**, bloating, weight loss, fatigue, nausea, or vomiting. In patients with preexisting liver disease, extreme care must be taken when administering potentially hepatotoxic drugs.

NEPHROTOXICITY

The kidney is the main organ that the human body requires to achieve and perform different vital functions, including **homeostasis**, detoxification, regulation of extracellular fluids, and excretion of toxic metabolites. Various mechanisms can cause nephrotoxicity, including renal tubular toxicity, **glomerulonephritis**, and **crystal nephropathy**. In detecting early renal damage, blood urea and serum creatinine can evaluate nephrotoxicity and renal dysfunction. The most common drugs that generate crystals are sulfonamides, methotrexate, ampicillin, acyclovir, ciprofloxacin, and triamterene.

Nephrotoxicity occurs when a drug, chemical, or toxin damages the kidneys, resulting in possible chronic kidney disease. Aminoglycosides, NSAIDs, diuretics, proton pump inhibitors (Nexium, Prilosec), and angiotensin-converting enzyme inhibitors can cause kidney damage. Certain medications administered in hospital settings, such as aminoglycosides and vancomycin, can also damage the kidneys. Numerous drugs used to manage and treat multiple diseases, including hypertension, diabetes, and other pathologies, are also nephrotoxic.

Acute kidney injury occurs when the kidneys cannot filter waste products from the blood, causing waste to build up in the blood circulation. It can range from mild to severe. If severe, ongoing, and untreated, it can be fatal. However, it can also be reversed. The signs and symptoms of acute kidney injury may include nausea, loss of appetite, fatigue, confusion, **edema**, itching, irregular heartbeat, chest pain, and seizures.

CARDIOTOXICITY

Cardiotoxicity refers to heart dysfunction, including electrical or muscle damage, leading to heart toxicity. This can result in heart failure and arrhythmia. Cardiotoxicity may stem from certain drugs or anticancer therapies and can develop years after cancer treatment, particularly in adults who underwent treatment during childhood. Some types of cancer treatments carry a higher risk of cardiotoxicity, with symptoms potentially emerging years posttreatment.

Chronically administered medications, such as neurologic or psychiatric agents, can induce cardiotoxicity. The highest incidence of cardiotoxicity occurs when trastuzumab is combined with anthracyclines. **Some** medicines including alkylating antineoplastic agents, cytarabine, fluorouracil, dasatinib, and cardiac glycosides like digoxin and digitalis may particularly affect heart function. Digoxin is used to treat heart failure. It has a narrow therapeutic window, with a therapeutic range of 0.5–2 ng/mL and a toxic concentration of 2.5 ng/mL or more.

MUSCLE TOXICITY

Various drugs can damage the skeletal muscles due to their direct toxic effects on the **myocytes,** exacerbated by the poisonous effects of exercise. This damage results in the lysis of myocytes. **Myoglobin** causes renal damage through direct toxicity and tubular obstructions. Many drugs produce rhabdomyolysis, including statins, alcohol, heroin, ketamine, and cocaine. **Rhabdomyolysis** is a rare condition where muscle fibers break down, posing life-threatening risks after an injury or excessive exercise without adequate rest. Some medications such as antidepressants, antipsychotics, and antiviral drugs can also induce rhabdomyolysis. Additionally, statins may lead to this condition; substances like alcohol, cocaine, heroin, and LSD are toxic to the body and can damage muscles.

Check Your Knowledge

1. What are the signs and symptoms of neurotoxicity?
2. What are the common drugs that cause hepatotoxicity?
3. What are the signs and symptoms of acute kidney injury from drug toxicity?
4. What drugs may cause cardiotoxicity?
5. What drugs can result in rhabdomyolysis?

ANTIDOTES

Antidotes are drugs or substances that neutralize the effects of specific poisons. There are mechanical, chemical, or physiologic antidotes. A mechanical antidote is *activated charcoal,* which absorbs poisons in the GI tract and prevents them from being absorbed. Chemical antidotes neutralize toxins. Physiological antidotes directly oppose the actions

TABLE 5.1
Poisons and Their Specific Antidotes

Poison	Name of Antidote	Classification of Antidote
Acetaminophen	Acetylcysteine	Mucolytic agent
Benzodiazepines	Flumazenil	Benzodiazepine antagonist
Carbon monoxide	Oxygen	Nonmetallic element
Cyanide	Amyl nitrate	Nitrate inhalant
Iron	Deferoxamine	Heavy metal antagonist
Methanol	Ethanol	Organic chemical
Opiates	Naloxone	Opioid antagonist
Organophosphates	Atropine or pralidoxime	(Atropine): anticholinergic or anti-parasympathetic (parasympatholytic) drug; (Pralidoxime): oxime drug

- Dimercaprol for arsenic, gold, or inorganic mercury poisoning
- Flumazenil for benzodiazepine overdose
- Pralidoxime for poisoning by anticholinesterase nerve agents

of a poison. Specific antidotes are available for just a few poisons (see Table 5.1). Specific systemic antidotes reduce concentrations of toxic substances by combining with them or increasing their excretion rate. Other systemic antidotes compete with poisons for their receptor sites. Antidotes have different mechanisms of action. Poisons and their specific antidotes are described and listed in Table 5.1.

CLINICAL CASE STUDIES

Clinical Case Study 1

An 84-year-old woman with a history of bipolar depression on lithium for 25 years, well-controlled hypertension, congestive heart failure, and chronic kidney disease was brought to the emergency department with progressively worsening resting tremors, ataxia, confusion, and cognitive decline for 6 weeks. She has been taking spironolactone 25 mg for 1 week recently. During the past several days, she began worsening resting tremors and visual hallucinations, both of which had initially manifested the previous year. In addition, she was experiencing confusion and disorientation, raising concerns about possible dementia. In the following 2 weeks, the tremors continued to worsen, and she began having auditory and visual hallucinations.

CRITICAL THINKING QUESTIONS

1. What medications have the potential for neurotoxicity?
2. What antibiotic may cause neurotoxicity?
3. What are the signs and symptoms of neurotoxicity?

Clinical Case Study 2

A 73-year-old white man with a history of hypertension, hypothyroidism, and borderline type 2 diabetes was admitted to the local hospital. His medications included levothyroxine, amlodipine, and chlorthalidone. He denied the use of nonsteroidal anti-inflammatory drugs (NSAIDs) and herbs. Still, he did report taking numerous over-the-counter dietary supplements for energy, aging, joint pains, and sexual performance. He mentioned that he started taking all the supplements 2 years ago on the recommendation of his chiropractor. His nephrologist ruled out glomerulonephritis. However, the renal biopsy showed severe acute tubular necrosis.

CRITICAL THINKING QUESTIONS

1. What mechanisms can cause nephrotoxicity?
2. What drugs may result in nephrotoxicity?
3. What are the signs and symptoms of acute kidney injury?

FURTHER READING

Altman, R.B., Flockhart, D., and Goldstein, D.B. (2012). *Principles of Pharmacogenetics and Pharmacogenomics*. Cambridge University Press.

American Society of Health-System Pharmacists. (2017). *Concepts in Pharmacogenomics*, 2nd Edition. American Society of Health-System Pharmacists.

Benfenati, E. (2016). *In Silico Methods for Predicting Drug Toxicity* (Methods in Molecular Biology 1425). Humana Press.

Cobert, B., Gregory, W.W., and Thomas, J.L. (2019). *Cobert's Manual of Drug Safety and Pharmacovigilance*, 3rd Edition. WSPC.

Coleman, M.D. (2020). *Human Drug Metabolism*, 3rd Edition. Wiley-Blackwell.

Cox Gad, S. (2016). *Drug Safety Evaluation* (Pharmaceutical Development Series), 3rd Edition. Wiley.

Dasgupta, A., and Langman, L.J. (2019). *Pharmacogenomics of Alcohol and Drugs of Abuse*. CRC Press.

Schuttler, J., and Schwilden, H. (2008). *Modern Anesthetics* (Handbook of Experimental Pharmacology 182). Springer.

6 Drug Interactions and Adverse Drug Effects

LEARNING OBJECTIVES

After studying this chapter, readers should be able to

1. Explain synergistic and antagonistic drugs.
2. Compare and contrast adverse effects and side effects.
3. Describe the four types of allergic reactions.
4. Distinguish between type II and type III hypersensitivity.
5. Describe idiosyncratic drug reactions and polypharmacy.
6. Explain three broad categories of drug–drug interactions.
7. Describe food–drug interactions and give three examples.
8. Identify the teratogenic effects on the fetus.
9. Explain the causes and risk factors for drug–disease interactions
10. Summarize teratogens and their impact on pregnancy.

OVERVIEW

Drug–drug interactions can be defined as two different medications altering each other's effects on the body, referring to the capability of changing one or more pharmacological responses as a direct consequence of being exposed to one or more drugs taken together or operating concurrently in an organ system. A systematic understanding of drug interactions, particularly absorption, elimination, transport, and drug metabolism, may help prevent adverse effects. Predicting pharmacodynamic interactions often requires a deeper understanding of the underlying mechanisms. Electronic prescribing systems are beneficial. Adverse drug reactions (ADRs) are unexpected and unplanned responses to drugs, even when average doses are used. The FDA mandates reporting severe or unexpected events within 2 weeks of their occurrence. Unfortunately, insufficient attention has been paid to the incidence of ADRs in children of all ages.

DRUG INTERACTIONS

Drug interactions occur when two or more drugs influence each other's activity. This action may be synergistic, antagonistic, or result in a new effect that neither drug produces. Drugs, foods, and herbs may interact with each other. Drug interactions often arise from accidental misuse or a lack of knowledge about their active ingredients. The interaction between two drugs may also increase the risk of adverse effects.

Drug interactions may arise from various processes. These processes include alterations in pharmacokinetics, which consequently affect the pharmacodynamic properties of the drugs (see Chapter 4). Factors or conditions predisposing a patient to drug interactions include aging, **polypharmacy**, genetic factors, liver or kidney diseases, steep dose–response curves, and rates of hepatic metabolism.

A drug may prevent another drug from binding to plasma proteins during drug distribution. It can also displace another drug from its binding sites. Then, the amount of unbound or free drugs increases. The drug serum concentration can then be raised to toxic levels. Examples of drugs that can displace other drugs from protein-binding sites include nonsteroidal anti-inflammatory drugs, phenylbutazone, and sulfonamides. Another reaction can occur if a drug alters plasma pH. When it increases, acidic drugs become ionized, and the reverse is true for alkaline (basic) drugs.

During drug metabolism, hepatic enzyme activity can be increased or decreased. Increased drug metabolism promotes the inactivation and excretion of drugs administered concurrently. This results in diminished drug effectiveness. Care must be taken if the inducing drug is discontinued since hepatic enzymes will return to baseline levels within days to weeks. Then, a new dose of the other drug must be adjusted to avoid toxicity. Also, prodrugs may be activated by metabolism. When the metabolism of prodrugs is increased, the therapeutic response is likely to increase rather than decrease.

Hepatic metabolism inhibitors will cause drug interactions that oppose the drug interactions of inducers. A drug that inhibits the hepatic enzyme CYP3A4 will decrease the metabolism of other drugs that are substrates for this enzyme.

A drug interaction may occur during drug excretion if a substance alters the **glomerular filtration rate**. The amount of fluid filtered by the kidneys per minute is directly related to **cardiac output**. Drugs that increase cardiac output influence the glomerular filtration rate, accelerating the excretion of other drugs. Another interaction may occur if a drug modifies the secretion or reabsorption of another drug in the renal tubule. Renal elimination can also be influenced by drugs that change the pH of the renal tubule filtrate, causing drugs to become more or less ionized. Therapeutically, these drug interactions can be manipulated to increase urine alkalinity or acidity. Other drugs may interact due to biliary excretion; however, this is far less common than drug interactions involving the kidneys.

DOI: 10.1201/9781003461913-7

ADVERSE DRUG EFFECTS

Adverse effects can range from mild to severe and may even be life-threatening. They are distinct from **side effects** and may be genetically influenced. Factors such as medication errors, poor judgment, or suicide attempts can lead to a drug **overdose**. Any drug can potentially be overdosed based on its *margin of safety* or *therapeutic window, which* refers to the difference between effective and toxic doses. These margins are relevant for all drugs, whether prescribed or over-the-counter (OTC). An overdose frequently occurs with drugs that have large margins of safety, while adverse effects may arise from therapeutic doses or when the margin of safety is small or nonexistent.

ADVERSE DRUG REACTIONS

Adverse drug reactions (ADRs) are unexpected and unplanned responses to drugs, even when average doses are administered. All adverse events must be reported once the drug is introduced to the market. The FDA mandates that severe or unexpected events be reported within 15 days of their occurrence. ADRs can occur at any age and can range from minor to severe and even lethal. Additionally, ADRs lead to more than 1.3 million emergency department visits annually in the United States, with approximately 350,000 patients requiring hospitalization for further treatment. Allergies and intolerances may arise in other patients; these are also classified as reportable ADRs. The most common ADRs are fatigue, drowsiness, skin reactions, nausea, vomiting, diarrhea, constipation, and alopecia.

The signs and symptoms of an ADR manifest soon after the first dose or following chronic use. In older adults, subtle ADRs can lead to functional deterioration, changes in mental status, depression, confusion, failure to thrive, and loss of appetite. Symptoms of allergic ADRs include itching, rash, upper or lower airway edema with difficulty breathing, and hypotension. Idiosyncratic ADRs can present nearly any symptom or sign and are usually unpredictable.

ADRs are increasingly common and are a significant cause of morbidity and mortality. They are classified into five categories: A, B, C, D, and E. Type A reactions are predictable from the known pharmacology of a drug and are associated with high morbidity and low mortality rates. Type B reactions are idiosyncratic, bizarre, or novel responses that cannot be predicted from the known pharmacology of a drug and are associated with low morbidity and high mortality. Type C reactions persist for a relatively long time; an example is osteonecrosis of the jaw with bisphosphonates. Type D reactions, also known as delayed reactions, become apparent after using a medicine, and their timing can make them more challenging to detect. Type E reactions are end-of-use events and are associated with the withdrawal of a drug; an example is insomnia, anxiety, and perceptual disturbances following the withdrawal of benzodiazepines. Age, gender, disease states, pregnancy, ethnicity, and polypharmacy influence susceptibility to ADRs.

Modifying the dose or eliminating or reducing triggering factors may suffice for managing dose-related ADRs from medications. In cases of allergic and idiosyncratic ADRs, the drug should typically be discontinued and not administered again. Switching to a different drug class is often necessary for allergic ADRs and may occasionally be required for dose-related ADRs. For instance, opioid-induced constipation can be alleviated with the use of an opioid receptor antagonist such as lubiprostone (Amitiza).

The prevention of ADRs requires an understanding of the medication and its potential reactions. Computer-based analysis should be utilized to investigate possible drug interactions, and this investigation should be repeated whenever medications are changed or added. Medications and initial doses must be carefully selected for older adults, as various genes have been identified as being associated with ADRs.

ALLERGIC DRUG REACTIONS

Allergic drug reactions are undesired and unintended responses to drugs that occur at usual therapeutic doses. They involve unique biochemical mechanisms and immunologic amplification. Fewer than 15% of all ADRs are allergic. Allergic drug reactions are classified into four types of immune responses. The most common immunologic reaction is an IgE-mediated type I reaction. Allergies may occur when a drug acts as an **allergen**. After a patient is sensitized, subsequent exposure to the drug can produce various kinds of allergic reactions. Some drugs may cause minor skin rashes and hives. Symptoms can occur immediately or hours after receiving the drug. Serum sickness is a delayed reaction that occurs a week or more after exposure to a drug. A patient's medical history and skin tests can help predict drug allergies.

Drug allergies can also lead to autoimmune reactions. Hypersensitivity reactions require the host's pre-sensitized state and may result in sudden death. For instance, a penicillin injection can trigger a severe allergic reaction. However, when a patient takes penicillin orally, an allergic response may cause nausea and vomiting. The four types of allergic reactions include the following:

Type I hypersensitivity encompasses **atopic diseases** and heightened IgE-mediated immune responses, including asthma, rhinitis, conjunctivitis, and dermatitis. Allergic diseases represent immune reactions to foreign, drug, and food allergens. Type I is characterized by immediate hypersensitivity or anaphylaxis resulting from IgE production following exposure to an antigen. For instance, when fragments of penicillin function as **haptens**, they trigger the immune system. Repeated exposure to the antigen leads to degranulation of **mast cells** and release of histamine, leukotrienes, or other inflammatory mediators.

Anaphylactic shock is a life-threatening emergency that arises from **anaphylaxis**. This condition can lead to hypotension, airway narrowing, **dyspnea**, and wheezing. Additional signs and symptoms may include hives, flushed skin, tachycardia, swollen tongue or lips, nausea, vomiting, **dysphasia**, and abdominal pain. Treatment options include epinephrine, antihistamines, corticosteroid injections, and oxygen therapy.

Type II **hypersensitivity** refers to an antibody-mediated immune reaction in which antibodies (IgG or IgM) are directed against cellular or extracellular matrix antigens, leading to cellular destruction, functional loss, or tissue damage. This damage can occur through multiple mechanisms. Type II responses arise when a drug binds to red blood cells, which are recognized by IgG or other antibodies. Cell lysis is initiated via **complement fixation** and cytolysis by **cytotoxic T cells** or macrophages. The most common causes include penicillin, thiazides, cephalosporins, and methyldopa.

Type III **hypersensitivity** describes an inappropriate or overreactive immune response to an antigen that produces undesirable effects. This type of allergy involves a complex–mediated response that necessitates the formation of IgG or IgM against antigens. Examples include systemic lupus erythematosus, post-streptococcal **glomerulonephritis**, rheumatoid arthritis, farmers' lung, and **serum sickness**. Symptoms can vary depending on the tissues affected, including joint pain and swelling, rashes, fever, and kidney damage.

Type IV **hypersensitivity** refers to delayed-type hypersensitivity responses caused by the activation of cytotoxic T cells and T helper cells, which provoke an inflammatory reaction against exogenous or endogenous antigens. Other cells, such as monocytes, eosinophils, and neutrophils, may also be involved in certain situations. After exposure to an antigen, an initial local immune and inflammatory response attracts leukocytes. The antigen engulfed by macrophages and monocytes is presented to T cells, which then become sensitized and activated. These cells release cytokines and chemokines, potentially causing tissue damage and various illnesses. Examples of type IV hypersensitivity reactions include contact dermatitis and drug hypersensitivity.

When the immune system attacks its cells, *autoimmunity* develops. Drugs that can trigger autoimmunity include the antihypertensives *hydralazine* and *methyldopa*, isoniazid, and the antiarrhythmic *procainamide*. Vancomycin flushing syndrome, previously known as **Red Man Syndrome**, is an anaphylactoid reaction caused by the rapid infusion of the glycopeptide antibiotic vancomycin. This rare syndrome may also be caused by rifampicin, amphotericin B, ciprofloxacin, and teicoplanin. The syndrome involves the degranulation of mast cells, which occurs independently of preformed complement or IgE. Typically, wheals and **urticaria** develop on the neck, arms, and upper trunk. The syndrome usually resolves after the drug's infusion rate is reduced or discontinued.

YOU SHOULD REMEMBER

A drug allergy can be deadly if patients experience severe symptoms, such as swelling of the tongue or throat, difficulty breathing, tachycardia, lightheadedness, or loss of consciousness.

Check Your Knowledge

1. What is the description of allergens?
2. What are the causes of allergy?
3. What are the four types of allergic reactions?

IDIOSYNCRATIC DRUG REACTIONS

An *idiosyncratic drug reaction* is a unique, strange, or unpredictable reaction that is rare. Enzyme deficiencies due to genetic or hormonal variations may cause it. The most common form of idiosyncratic reaction is a skin rash. For example, carisoprodol may induce transient dizziness, **quadriplegia**, or temporary vision loss. Patients must disclose a history of allergic reactions or susceptibility to specific medications, vaccines, serum, or blood transfusions before treatment. Those predisposed to allergic reactions should wear an alert bracelet or necklace. The term idiosyncratic drug reaction is used to designate an ADR that does not occur in many patients treated with that drug, and these events do not involve the drug's therapeutic effect.

DRUG–DRUG INTERACTION

Drug–drug interaction (DDIs) is the clinical response to the administration of a combination of drugs that differs from the expected effects of each drug when taken alone. It may result in ADRs, reduced therapeutic benefit, or harm to the patient. More than 100,000 potential types of drug interactions have been documented, although most do not lead to adverse effects. The incidence can range from 5% to 10% in patients. However, not all drug interactions are harmful. Some drug interactions are utilized therapeutically. For example, local anesthetics often combine epinephrine with lidocaine to prolong lidocaine's effects. Reversing agents, such as naloxone (Narcan), are administered after surgery to counteract the effects of narcotics. Some drugs are prescribed concurrently for cancer therapy to maximize therapeutic effects at multiple sites of cancer cell growth. Beneficial drug interactions can enhance treatment.

DDI may involve prescription or OTC drugs. Types of drug–drug interactions include duplication, antagonism, and alterations in how the body processes one or both drugs. When two drugs with the same effect are taken, their adverse effects may be intensified. Two drugs with opposing actions can interact, reducing the effectiveness of either one or both. Vitamins, minerals, herbs, or amino acids may interact with certain medications. Alcohol influences bodily processes and interacts with many medications. For instance, consuming alcohol with the antibiotic metronidazole can result in flushing, headache, palpitations, nausea, and vomiting.

DDI may decrease a drug's effectiveness, leading to unexpected side effects or enhancing the impact of a specific medication. Some drug interactions can even be harmful to patients. Drug interactions can be classified into three main categories: drug–drug interactions, drug–food and beverage interactions, and drug–condition interactions.

DDI is more common due to growing populations, longer lifespans, and **polypharmacy**. These interactions can be pharmacodynamic or pharmacokinetic. A pharmacokinetic drug–drug interaction occurs when one drug affects another drug's absorption, distribution, metabolism, or excretion. Such interactions can alter the concentrations of either active drug in the body. If the same P450 enzyme metabolizes two medicines, there may be competitive or irreversible inhibition of the enzyme, influencing drug concentrations. Additionally, drug metabolism may be induced, potentially leading to toxicities or faster clearance and diminished therapeutic effects.

Examples of DDI include clozapine, erythromycin, and specific antifungal interactions with oral contraceptives. Additionally, drugs can interact with foods, beverages, and herbal supplements. One example of a supplement that interacts with nonsteroidal anti-inflammatory drugs is *ginkgo biloba*. Since both substances inhibit platelet aggregation, using them together may increase the risk of bleeding.

FOOD–DRUG INTERACTION

Food–drug interactions can occur with both prescription and OTC medications. This includes antacids, vitamins, iron supplements, herbs, dietary supplements, and beverages. Some nutrients can influence the metabolism of certain drugs by binding to the medication's components. This can reduce their absorption or accelerate their elimination. For instance, the acidity of fruit juice may diminish the effectiveness of antibiotics such as penicillin. Dairy products can impact the infection-fighting properties of tetracycline. Additionally, certain green leafy vegetables can negate the effects of warfarin, an anticoagulant medication.

An antidepressant, such as a monoamine oxidase inhibitor, is dangerous when mixed with foods or drinks that contain **tyramine**. These include red wine, beer, chocolate, processed meat, avocados, and cheeses. Grapefruit juice can interfere with some blood pressure and organ transplant medicines by increasing their metabolic breakdown. Some medicines are only absorbed with a full meal or even a meal high in fat content. For example, ginkgo biloba may increase the impact of anticoagulants, while calcium supplements may reduce the absorption of thyroid supplements.

DRUG–DISEASE INTERACTION

DDIs may occur when a drug affects a preexisting condition or disease. Some disorders can interact with medications, increasing the risk of adverse effects. For example, aspirin can increase bleeding in patients with **peptic ulcer** disease, and people with hypertension may be at greater risk for tachycardia from oral decongestants found in OTC cough, cold, and allergy products. Sometimes, helpful drugs for one disease can be harmful for another disorder. For instance, some beta-blockers taken for heart disease or

hypertension may worsen asthma or make it difficult for diabetes patients to recognize when their blood sugar is too low. Aging can also elevate drug interaction risks. For example, sedatives increase the risk of falls in older adults, and lower doses of narcotics are often more effective for pain relief. Additionally, the anticoagulant warfarin can cause more bleeding in elderly patients and usually requires a lower dose as well.

Genetic factors are becoming increasingly recognized as contributors to ADRs and metabolic DDIs. The CYP enzymes 2D6, 2C19, 2C9, 1A2, and 3A4 exhibit genetic **polymorphism**, which refers to genetic differences that result in functional or nonfunctional enzymes within a population. Individuals with genetically determined low enzyme activity are considered poor metabolizers, while others produce more enzymes than most of the population and are termed "ultra-metabolizers." For CYP1A2 and 3A4, most individuals produce intermediate amounts of the enzyme; these make up the so-called normal population. Few individuals exhibit very low or very high activity. Genetic variations of CYP2D6, 2C19, and 1A2 are substantial and may also account for differences in drug response among individual patients.

Other factors that can increase the risk for DDIs include patient-specific conditions such as kidney impairment or liver failure, which may reduce the body's ability to eliminate drugs. Comorbid diseases or conditions, such as malnutrition, severe heart failure, and dehydration, could also increase DDI risk. Drug-specific factors such as the dose of a drug or drugs with a high risk for toxicity (e.g., digoxin, cyclosporine, and warfarin) can increase DDI risk.

TERATOGENIC EFFECTS

Teratogens are substances that can harm the **fetus** during pregnancy. Studies have shown that teratogens cause congenital disorders and increase the chance of miscarriage, **stillbirth**, or other pregnancy complications. Teratogenic agents may interfere with normal prenatal development, causing the formation of one or more developmental abnormalities in the fetus. Examples of teratogens include medications (thalidomide, some antibiotics, and antidepressants), alcohol, cigarette smoke, illegal drugs (heroin and cocaine), certain vaccines, some viral infections (cytomegalovirus and the rubella virus), X-ray radiation, and environmental factors. A *teratogenic effect* is a drug-induced congenital disability, such as congenital absence of any part of a limb, abnormal limb size, abnormal smallness of the eyes, and blindness (see Figure 6.1). There are many causes of congenital malformations. They include genetic, environmental, and combinations of both. Genetic factors include chromosomal abnormalities and single-gene defects. Ecological factors include chemicals, ionizing radiation, maternal metabolic disorders, intrauterine infections, and drugs.

Amelia of the upper limb (Q71.0)   Amelia of the lower limb (Q72.0)

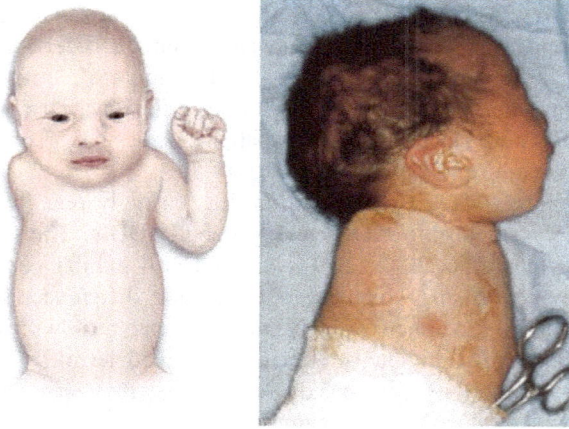

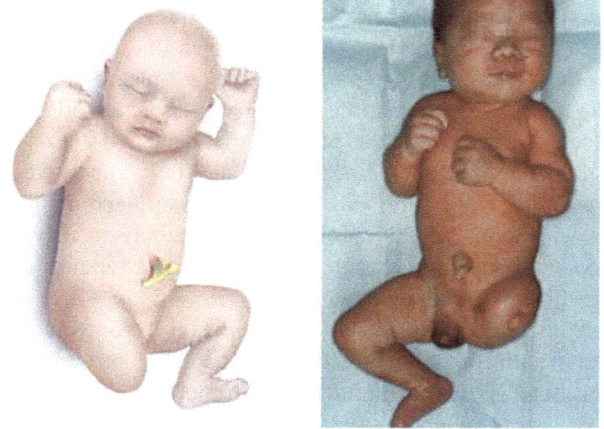

FIGURE 6.1   Teratogenic effects.

POLYPHARMACY EFFECT OF DRUG INTERACTIONS

In the US, 48% of the population takes at least one prescription medicine, and more than 10% take five or more prescription medications. Given the thousands of prescription and OTC drugs and dietary supplements available, the number of possible combinations and potential DDIs is surprising. Fortunately, many potential DDIs do not have a noticeable clinical effect. For the interaction to be clinically significant, it must cause a change in the expected response to the drug. DDIs may cause different reactions in different patients and are affected by many factors. Logically, the risk of a DDI is higher with an increased number of drugs a patient is taking. Older adults who have more chronic diseases are at particular risk for DDIs. Patients older than 74 years of age are six times more likely to be taking potentially interacting drugs.

Check Your Knowledge

1. What are some examples of allergic drug reactions?
2. Where can drug–drug interactions occur in the body?
3. How many types of allergic drug reactions are known?
4. What are some examples of drug–disease interactions?
5. What are the polypharmacy effects of drug interactions?

CLINICAL CASE STUDIES

Clinical Case Study 1

A 72-year-old man has hypercholesterolemia and has been taking a statin for years. His blood tests revealed he needs a higher dose to manage his cholesterol. Soon after switching to the higher dose, he began to experience symptoms related to liver disease, which he did not have before. The physician reduced his statin to a lower dose, and the symptoms subsided.

CRITICAL THINKING QUESTIONS

1. How can drug–drug interactions occur?
2. How can drugs that are helpful for one disease be harmful to another disease?
3. How can food–drug interactions occur?

Clinical Case Study 2

A 12-year-old girl with strep throat was brought to a pediatrician's office to get a penicillin IM injection. The medical assistant questioned the child's mother about her child having any allergies to penicillin; the mother said that, as far as she knows, there have been no allergies to any medications. The inexperienced medical assistant administered penicillin G procaine without any allergy test. Fifteen minutes later, the child went into anaphylactic shock. The pediatrician immediately began to treat the patient.

CRITICAL THINKING QUESTIONS

1. What is anaphylactic shock?
2. What are the signs and symptoms of anaphylactic shock?
3. What is the treatment for anaphylactic shock?

FURTHER READING

Aronson, J.K. (2015). *Meyler's Side Effects of Drugs—The International Encyclopedia of Adverse Drug Reactions and Interactions*, 16th Edition. Elsevier.

Barenholtz Levy, D. (2023). *Maybe It's Your Medications—How to Avoid Unnecessary Drug Therapy and Adverse Drug Reactions*. Skyhorse.

Bowen, I.H., Corrigon, D., Cubbin, J.I., de Smet, P.A.G.M., Hansel, R., Sonnenborn, U., Westendorf, J., Winterhoff, H., and Woerdenbag, H.J. (2012). *Adverse Effects of Herbal Drugs*, 2nd Edition. Springer-Verlag.

Brinker, F. (2010). *Herbal Contraindications and Drug Interactions—Plus Herbal Adjuncts with Medicines*, 4th Edition. Eclectic Medical Publications.

Grover, J.K. (2018). *Adverse Drug Reactions*. CBS Publishers and Distributors Pvt Ltd.

Kiang, T.K.L., Wilby, K.J., and Ensome, M.H.H. (2016). *Pharmacokinetic and Pharmacodynamic Drug Interactions Associated with Antiretroviral Drugs*. Adis.

Lee, A., and Cuthbert, M. (2023). *Adverse Drug Reactions*, 3rd Edition. Pharmaceutical Press.

Ozkaya, E., and Didem Yazganoglu, K. (2014). *Adverse Cutaneous Drug Reactions to Cardiovascular Drugs*. Springer.

Pai, M.P., Kiser, J.J., Gubbins, P.O., and Rodvold, K.A. (2018). *Drug Interactions in Infectious Diseases—Antimicrobial Drug Interactions* (Infectious Disease Series), 4th Edition. Humana Press.

Piscitelli, S.C., Rodvold, K.A., and Pai, M.P. (2011). *Drug Interactions in Infectious Diseases* (Infectious Disease Series), 3rd Edition. Humana Press.

Preskorn, S.H. (2018). *Drug-Drug Interactions with an Emphasis on Psychiatric Medications*. Professional Communications, Inc.

Preston, C.L. (2019). *Stockley's Drug Interactions—A Source Book of Interactions, Their Mechanisms, Clinical Importance and Management*, 12th Edition. Pharmaceutical Press.

Talbot, J., and Aronson, J.K. (2011). *Stephens' Detection and Evaluation of Adverse Drug Reactions—Principles and Practice*, 6th Edition. Wiley-Blackwell.

7 Dietary Supplements and Herbal Remedies

LEARNING OBJECTIVES

After studying this chapter, the reader should be able to

1. Review dietary supplements and herbal remedies.
2. Explain the contraindications of black cohosh.
3. Discuss the indications of chamomile and cinnamon.
4. Identify the indications of saw palmetto and goldenseal.
5. Describe the adverse effects of dong quai and echinacea.
6. Explain the primary uses of ginkgo and ginseng.
7. Describe the benefits of using garlic and St. John's wort.
8. Identify the adverse effects of Ma huang.
9. Describe the numerous safety concerns with ephedra.
10. Discuss the indications of valerian and yohimbe.

OVERVIEW

Plant-based products used to treat diseases or maintain health are called herbal, botanical, or phytomedicines. An herbal supplement is a product made from plant sources and used only for internal use. Herbal supplements come in all forms. They may be dried, chopped, powdered, or in capsule or liquid form. The practice of using herbal supplements dates back thousands of years. Today, the use of herbal supplements is prevalent in the U.S. Approximately 76% of people take dietary supplements. It is estimated that more than 170 million Americans take nutritional supplements. One out of four patients using prescription medication also consumes some dietary supplements. Around the world, about 80% of individuals routinely use nutritional supplements and herbal remedies to treat or prevent illnesses. Some believe that dietary supplements and herbal products are harmless and helpful. However, since the FDA does not regulate them, there can be contaminants and variances in quality of which consumers need to be made aware. Sometimes, the preparations may cause significant toxicity and adverse effects. Therefore, drug interactions and drug toxicity must always be considered.

DIETARY SUPPLEMENTS

Dietary supplements encompass vitamins, minerals, herbs, amino acids, and enzymes. For a more detailed discussion of vitamins and minerals, refer to Chapter 28. People should consult their healthcare provider before taking dietary supplements, as some may have adverse effects or interact with other dietary supplements and medications. Dietary supplements are also referred to as nutritional supplements. It is crucial to determine when patients are using these products while obtaining a drug history. When patients transition from one healthcare setting to another, such as from home to the hospital or from home to the outpatient surgery setting, identifying dietary supplement use is particularly important to minimize potentially dangerous drug interactions and toxicity.

Some dietary supplement preparations may be manufactured using various plant parts (such as roots, leaves, seeds, and flowers) that can have different effects, potencies, and toxicities. Some supplement–drug interactions may relate to interference with liver metabolism pathways. Hepatic enzymes such as CYP450 can be effective. Drugs and nutritional supplements are associated with altered drug absorption and adverse effects, which can be devastating for patients.

HERBAL REMEDIES

Herbal remedies are plants used as medicines. People use them to help prevent or cure diseases, relieve symptoms, boost energy, promote relaxation, or aid in weight loss. However, unlike drugs, herbal remedies cannot be regulated or tested. The most commonly used herbal remedies are available, and they explain benefits, contraindications, safety, and drug interaction concerns.

BLACK COHOSH

Black cohosh is a small plant with star-shaped white flowers. It contains chemicals known as phytoestrogens, which have properties similar to estrogens. Black cohosh is found in the eastern United States and is commonly used to treat menopausal symptoms in women, such as hot flashes, night sweats, vaginal dryness, irritation, and palpitations. It is also indicated for the treatment of **dysmenorrhea**, arthritis pain, and infertility.

Pregnant and breastfeeding women should avoid black cohosh because it may stimulate contractions and cause premature labor. A rare case report of liver failure occurred after taking black cohosh for several weeks. Adverse effects include nausea, vomiting, headache, **palpitations**, hypertension, and rash. Black cohosh may contain small amounts of salicylic acid; therefore, those with aspirin allergies should not take this herb.

DOI: 10.1201/9781003461913-8

FIGURE 7.1   Chamomile plant.

CHAMOMILE

Chamomile is one of the oldest medicinal herbs known to humanity. It is derived from a daisy-like plant in the *Asteraceae family* (see Figure 7.1). It serves as both a flavoring agent and an herbal remedy. Its primary applications include treating nausea, diarrhea, anxiety, and mouth ulcers resulting from chemotherapy or other treatments. Additionally, it has been utilized for insomnia, gingivitis, and skin irritations. The dry powder possesses antioxidant and anti-inflammatory properties. The main constituents of chamomile flowers are polyphenol compounds.

Chamomile is likely safe when used in amounts commonly found in teas. It may be safe for short-term oral use and medicinal purposes. The long-term safety of using chamomile on the skin for medicinal purposes is not well understood. Adverse effects are uncommon and may include allergic reactions, dizziness, and nausea. Rare cases of anaphylaxis could occur.

Interactions between chamomile and nonsteroidal anti-inflammatory drugs, warfarin, and cyclosporine may occur. Chamomile may also interact with vitamin B12 and sleep-enhancing **herbal supplements**. Chamomile is contraindicated in patients with a history of, or current, breast, uterine, or ovarian cancers. Women with endometriosis or uterine fibroids should also avoid it.

CINNAMON

Cinnamon is a spice obtained from specific types of trees. It can be extracted from the cinnamon tree's bark, roots, leaves, flowers, and fruits (see Figure 7.2). There are many types of cinnamon. Ceylon cinnamon, grown primarily in Sri Lanka, is known as "true" cinnamon. Cassia cinnamon *(Cinnamomum aromaticum) is* grown in southeastern Asia. Cinnamon has a long history of being used in traditional medicine in various parts of the world, including China, India, and Iran. It is used in cooking and baking or added to many foods.

Cinnamon lowers blood cholesterol, increases insulin production, and reduces blood glucose levels. The distinctive qualities of cinnamon arise from its essential oil and its primary component, *cinnamaldehyde*. Cinnamon is cultivated in Vietnam, China, Sri Lanka, Indonesia, and other countries in Southeast Asia. There are many indications for using cinnamon, which include memory problems, bad breath, acne, toothache, coagulation issues, vomiting, diarrhea, flatulence, the common cold, influenza, sore throat, menstrual cramps, amenorrhea, hemorrhoids, heart disease, and increased **libido**.

Adverse effects of excessive cinnamon include hypoglycemia, liver damage, increased mouth sores, and dyspnea. Cinnamon may interact with medications for diabetes mellitus, heart disease, and liver disease, leading to intensified effects of these medications.

YOU SHOULD REMEMBER

Little is known about whether using cassia cinnamon during pregnancy or breastfeeding is safe. Ceylon cinnamon may be unsafe for use during pregnancy if consumed in amounts greater than those commonly found in foods. Little is known about whether using Ceylon cinnamon during breastfeeding in such amounts is safe.

Check Your Knowledge

1. What are the adverse effects of Black cohosh?
2. What are the primary indications for chamomile?
3. What are the indications and adverse effects of cinnamon?

FIGURE 7.2 Cinnamon.

Dong Quai

Dong Quai (*Angelica sinensis*) root has been used for over a thousand years as a spice, tonic, and medicine in China, Japan, and Korea. It remains a common component in traditional Chinese medicine (TCM), typically combined with other herbs. It is frequently prescribed to treat dysmenorrhea, **amenorrhea**, migraine headaches, hypertension, and premature ejaculation. Dong Quai can also aid in opening blood vessels. It is sometimes referred to as the "female ginseng." Dried herbs (raw root) may be boiled or soaked in wine before consumption. Dosage and administration can include tablets, capsules, and powders. It is administered as an injection in hospitals or health centers in China and Japan.

The yellow-brown root of the plant grows at high altitudes in the cold, damp mountains of China, Korea, and Japan (see Figure 7.3). Adverse effects of dong quai include increased skin sensitivity to sunlight, burping, flatulence, and hypertension. Taking high doses of dong quai for more than 6 months may increase the risk of developing cancer. It should be avoided during pregnancy because it can affect the uterus and increase the risk of congenital disabilities. It may also be linked to bleeding disorders, endometriosis, uterine fibroids, and cancers of the breast, uterus, and ovaries. Dong quai has a significant interaction with warfarin and can increase the risks of bruising and bleeding. Other interactions include anticoagulant/antiplatelet drugs and estrogens.

FIGURE 7.3 Dong Quai roots.

Dong quai should be avoided by individuals with chronic diarrhea, abdominal bloating, or those at risk for hormone-related cancers, such as breast, ovarian, and uterine cancers. In very high doses, dong quai may increase sensitivity to sunlight and lead to skin inflammation or rashes. Individuals taking dong quai are advised to avoid sun exposure or use sunscreen.

ECHINACEA

Echinacea is one of the most popular herbs in America today. It is a Native American medicinal plant named for the prickly scales on its large conical seed head, resembling the spines of an angry hedgehog. This herb is purple cone-flower (see Figure 7.4). Native Americans have employed it for centuries to treat various ailments. Today, it is commonly used for the common cold or flu. However, it is also used to treat pain, inflammation, migraines, scarlet fever, syphilis, malaria, blood poisoning, and diphtheria. Other applications include preventing upper respiratory infections and treating cancer, genital herpes, and otitis media in children.

Adverse effects of echinacea include anaphylaxis, and individuals with asthma appear to be at a higher risk for this event. Other side effects include rash, urticaria, dyspepsia, nausea, and vomiting. Echinacea seems to modulate immune function, so it should not be administered to patients undergoing treatment for cancer and HIV or those taking other immunosuppressive drugs. Individuals with leukemia, diabetes, multiple sclerosis, tuberculosis, HIV or AIDS, any autoimmune diseases, or potentially liver disorders should avoid taking echinacea.

GARLIC

Garlic is the edible bulb of a plant in the lily family (see Figure 7.5). Different garlic preparations include raw garlic, garlic oil, garlic powder tablets, and aged garlic extract. People worldwide, including Egyptians, Greeks, Romans, Chinese, and Japanese, have traditionally used it for health purposes. Currently, garlic is most commonly promoted as a dietary supplement for cardiovascular conditions, including

FIGURE 7.4   Echinacea plant.

FIGURE 7.5   Garlic bulb.

hypertension and hypercholesterolemia. Its use has grown globally. Other indications for garlic include treating **alopecia**, breast disease, and upper respiratory infections. A garlic-rich diet seems to confer a lower risk of developing colon, prostate, stomach, and breast cancer. Garlic's antibacterial properties and antioxidants can improve skin health by killing acne-causing bacteria.

Adverse effects of garlic include nausea, vomiting, flatulence, **tachycardia**, flushing, breath and body odor, and insomnia. Large quantities of garlic are contraindicated during pregnancy because they may cause fetal death or stimulate labor. Additionally, large amounts should be avoided during breastfeeding as they may cause colic in infants or pose dangers to children through unknown mechanisms. Garlic is contraindicated for individuals with known hypersensitivity or gastritis. Furthermore, patients should not use garlic before or after surgery due to the risk of bleeding.

YOU SHOULD REMEMBER

Crushing, chopping, or mincing garlic effectively releases the compounds that benefit your health. Eating raw garlic can also offer additional advantages over cooked garlic, including relaxing smooth muscles in blood vessels, dilating them, and potentially lowering blood pressure.

Check Your Knowledge

1. What are the contraindications of dong quai?
2. What are the primary indications of echinacea?
3. What are the indications of garlic?

GINGER

Ginger (*Zingiber officinale*) is one of the most commonly consumed foods worldwide. It belongs to the plant family that includes **cardamom** and turmeric. Indians and Chinese have used ginger as a tonic root for over 5,000 years to treat various illnesses, and this plant is now cultivated across the humid tropics, with India being the largest producer. Ginger is available in numerous forms, including fresh, dried, pickled, preserved, crystallized, candied, and powdered or ground. Its flavor is peppery and slightly sweet, with a strong and spicy aroma (see Figure 7.6).

Ginger is used as an antiemetic to treat nausea associated with pregnancy, motion sickness, and chemotherapy. It also promotes blood flow to the gallbladder and has anti-inflammatory and antiplatelet activity. The root is available in several forms: tablets, capsules, fresh or dried root, topical cream, tea, and liquid extract. Several components of the root exhibit pharmacologic activity, including gingerol and shogaol.

The adverse effects of ginger include heartburn, dyspepsia, and diarrhea at recommended doses, as well as some rashes from topical cream. Due to its antiplatelet activity, ginger should be used cautiously in patients taking other antiplatelet medications such as NSAIDs, aspirin, and warfarin. It is contraindicated in patients with heart disease, pregnant or breastfeeding women, and individuals with diabetes.

GINKGO

Ginkgo is one of the oldest living tree species in the world and has a long history in TCM. Members of the royal court were given ginkgo nuts for senility. Other historical uses of ginkgo include treating bronchitis, asthma, and urinary tract disorders. It is also known as *Ginkgo biloba,* one of the most common herbs sold in the United States. Today, the extract from ginkgo leaves is promoted as a dietary supplement for many conditions, including anxiety, allergies, dementia, eye problems, peripheral artery disease, and tinnitus. The seeds are toxic, can cause seizures, and are not used medicinally.

Ginkgo may be used to treat morning sickness, Alzheimer's dementia, and **premenstrual syndrome,** and to improve memory. The adverse effects of ginkgo include bleeding, particularly for patients on anticoagulant medications, who may have an increased risk of bleeding. Other adverse effects include headache, nausea, vomiting, and diarrhea. The Ginkgo biloba tree is shown in Figure 7.7.

GINSENG

Ginseng (*Panax ginseng*) is the root of plants in the genus Panax. The part of the plant most frequently used for health purposes is the root. Asian ginseng is native to China and Korea and has been utilized for health-related purposes in TCM for thousands of years (see Figure 7.8). When used short-term as part of a specific multi-ingredient topical skin application, Asian ginseng is likely safe. Its safety after prolonged repetitive topical use has not been determined. Asian ginseng may be unsafe when taken orally during pregnancy, as one of its chemicals has been found to cause congenital disabilities in animals. Limited information exists on whether it is safe to use Asian ginseng while breastfeeding.

Ginseng's pharmacological effects have been utilized to treat cancer, diabetes, and cardiovascular diseases. It enhances immune and brain function, reduces stress, and boosts antioxidant activities. The side effects of

FIGURE 7.6   Ginger root.

FIGURE 7.7   Ginkgo biloba tree.

FIGURE 7.8   Ginseng root.

ginseng include rash, nausea, vomiting, and abdominal pain. There are reports of hypoglycemia, hypertension, and breast tenderness with very high doses of ginseng. Other side effects include insomnia, headaches, tachycardia, and vaginal bleeding. Ginseng may also induce mania and psychosis.

GOLDENSEAL

Goldenseal is a perennial herb in the buttercup family. The dried root is commonly used in the US (see Figure 7.9). It ranks among the five top-selling herbs in the United States, yet little scientific evidence is available regarding its efficacy. Goldenseal is native to the Ohio River Valley, and the plant produces a fruit similar to a raspberry. Nonetheless, the root is utilized in herbal medicine to treat menstrual issues, muscle spasms, hypertension, poor appetite, sciatic pain, and infections, particularly in the eyes and skin. Berberine salts are a significant component of goldenseal and contribute to some of its benefits. Adverse effects may include nausea, vomiting, headache, leukopenia, hypotension, and bradycardia. Goldenseal is considered unsafe at high doses or with long-term use, as the active constituent berberine can lead to significant toxicity.

Goldenseal is contraindicated during pregnancy, breastfeeding, and for infants because it may cause brain damage. It may also elevate the risk of bleeding. Goldenseal should be discontinued at least 2 weeks before surgery due to these bleeding risks.

Check Your Knowledge

1. What are the indications of ginger?
2. What are the adverse effects of ginkgo?
3. What are the indications of goldenseal?

FIGURE 7.9   Goldenseal root.

KAVA KAVA

Kava Kava is an herbal remedy made from the roots of *Piper methysticum*, a plant found on the islands of the Pacific Ocean. It has been used as a ceremonial drink in the Pacific Islands for hundreds of years, and some people report effects similar to alcohol. The roots are chewed or ground into a pulp and added to cold water. The resulting thick brew has been likened to France's social equivalent of wine. Kava is most frequently used to treat anxiety, reduce stress, alleviate insomnia, and manage **Parkinson's disease.**

Taking kava may lead to rashes, fatigue, nausea, jaundice, and fever. High doses and prolonged use of kava could result in dry mouth, dizziness, partial hearing loss, restlessness, alopecia, contact dermatitis, tremors, and stomach upset. The FDA has issued a warning against its use. Most liver damage is associated with heavy or chronic use and can cause cirrhosis and liver failure. Individuals who use kava tea long-term in South Pacific cultures may develop yellow skin and discoloration of hair and nails.

MA HUANG

Ma huang (*Ephedra sinica*) is another herb used for thousands of years in Chinese medicine (see Figure 7.10). In recent years, many dietary supplements in the form of

FIGURE 7.10   Ma huang herb.

capsules, tablets, drinks, and powders marketed in the West have contained Ma huang or ephedrine. The FDA has banned the use of this plant. It has been extensively used for many centuries by Chinese physicians for various respiratory illnesses, especially asthma, due to its bronchodilation effects. Ma huang was most commonly used as a source of the alkaloid ephedrine and was included in various dietary supplements and other compounds. The use of Ma huang or its alkaloid, ephedrine, in nutritional supplements has faced intense criticism because of severe adverse effects.

In Germany, ephedra is approved for treating mild bronchospasm. Ma huang was banned by the FDA in 2004. Numerous safety concerns surround ephedra's side effects due to the herb's sympathomimetic properties. Ephedra has reported more severe adverse events than any other herbal preparation, including hypertension, tachycardia, myocardial infarction, stroke, and dysrhythmia. Other side effects consist of insomnia, dizziness, urinary retention, anxiety, nausea, headache, and nervousness. Because Ma huang contains ephedrine, it should be avoided in patients with hypertension, as well as those with liver or kidney diseases. It is also contraindicated during pregnancy, lactation, and in small children.

MILK THISTLE

Milk thistle is a plant named for the white veins on its large, prickly leaves. It has been used for over 2,000 years to treat liver diseases. One of the active ingredients in milk thistle, silymarin, is extracted from the plant's seeds. Silymarin is believed to possess antioxidant properties. Milk thistle is available in oral capsules, tablets, and liquid extracts. People primarily use the supplement to address liver conditions.

Milk thistle is native to Kashmir but is also found in North America (see Figure 7.11). It is used to treat alcoholic cirrhosis and hepatitis. The active ingredient in milk thistle, found in the seeds, is thought to have antioxidant properties. Milk thistle is generally considered a safe compound with few adverse effects. These effects may include dyspepsia, nausea, abdominal pain, diarrhea, and rash. It should not be used during pregnancy or lactation.

SAW PALMETTO

Saw palmetto is a palm-like plant with berries (see Figure 7.12). These berries were a staple food and medicine for the Native Americans of the southeastern United States. In the early 1900s, men used the berries to treat urinary tract conditions and increase sperm production, thus boosting libido. Today, the primary use of saw palmetto is to treat benign prostatic hyperplasia (BPH). Researchers are still determining exactly how saw palmetto works. However, it contains plant-based chemicals that may be effective for BPH. This herb may influence testosterone levels in the body and reduce the amount of an enzyme that promotes the growth of prostate cells. Saw palmetto also has an anti-inflammatory effect on the prostate.

It can be purchased in dried berries, powdered tablets, capsules, liquid tinctures, and **lipotropic extracts**. The product label should indicate that the contents are standardized and contain 85% to 95% fatty acids and sterols. Read labels carefully and buy only from reputable companies.

More than 2 million men in the United States have used saw palmetto to treat BPH, and it is commonly recognized as a first-line treatment in Germany, Austria, and Italy. Standard doses for BPH are 160 mg taken twice daily. Saw palmetto may produce clinical effects similar to those of finasteride, which is prescribed for BPH. It can enhance urinary flow and reduce nocturia.

Saw palmetto is generally safe when used as directed. However, adverse effects of saw palmetto may include an increased risk of bleeding for those with bleeding disorders or who use anticoagulants. Because saw palmetto affects androgen activity, it should be avoided during pregnancy due to its potential impact on the fetus. Additionally, saw palmetto can inhibit the action of androgen medications such as testosterone and methyltestosterone.

The adverse effects include nausea, diarrhea, headache, and dizziness. There have been two reports of liver damage and one report of pancreatic damage in people who took saw palmetto. However, there is insufficient information to determine whether saw palmetto was the cause of these effects.

FIGURE 7.11 Milk thistle seeds.

FIGURE 7.12   Saw palmetto berries.

ST. JOHN'S WORT

St. John's wort has a history of medicinal use dating back to ancient Greece, where it was employed for various illnesses, including nervous disorders. It also has antibacterial, antioxidant, and antiviral properties. Its anti-inflammatory properties have been applied to the skin to help treat wounds and burns. St. John's wort is one of the most commonly purchased herbal products in the United States.

St. John's wort is a native plant with yellow, star-shaped flowers (see Figure 7.13). It is commonly found in Europe, the United States, Africa, and Asia. The leaves can be dried, made into tea, or prepared as a dried extract in pills or capsules. It is a popular treatment for anxiety, depression, **obsessive–compulsive disorder**, attention-deficit hyperactivity disorder, **premenstrual syndrome**, eczema, and pain. St. John's wort is likely safe when used in doses of up to 900 mg daily for less than 12 weeks. Adverse effects may include insomnia, restlessness, dizziness, nausea, diarrhea, allergic reactions, **xerostomia**, erectile dysfunction, and orgasm dysfunction. It is unsafe to use during pregnancy.

YOU SHOULD REMEMBER

St. John's wort is available in various forms, including capsules, tablets, tinctures, teas, and oil-based skin lotions. Chopped or powdered versions of the dried herb are also available. Most products are standardized to contain 0.3% hypericin.

Check Your Knowledge

1. What are the adverse effects of Ma huang?
2. What are the indications of milk thistle?
3. What are the primary indications of saw palmetto?
4. What are the indications of St. John's wort?

VALERIAN

Valerian has been used to treat anxiety, insomnia, and restlessness since the 2nd century A.D. It became popular in Europe during the 17th century. It has also been suggested for treating stomach cramps. This herb is found in Europe and parts of Asia and grows in North America. Its dark green leaves are pointed at the tips and hairy on the underside. Small, sweet-smelling, white, light purple, or pink flowers bloom in June. The root is light grayish-brown and has little odor when fresh (see Figure 7.14). Valerian can be used to treat insomnia, anxiety, depression, and menopausal symptoms. The root and stem of the plant are used in preparations that include teas, tablets, tinctures, powders, and capsules. Valerian may also act as a smooth muscle relaxant and a vasodilator. Daily use of valerian may be more beneficial for improving sleep quality than "as needed." A newer use of valerian is as a memory enhancement product for individuals with Alzheimer's.

Adverse effects include depression, drowsiness, dry mouth, headache, sedation, tachycardia, and **dyspepsia**. Women who are pregnant or nursing should not take valerian without medical advice, as the potential risks to the fetus or infant have not been evaluated.

FIGURE 7.13   St. John's wort.

FIGURE 7.14   Valerian root.

YOHIMBE

Yohimbe is an evergreen tree native to central and western Africa. Its bark contains a compound called yohimbine.

The bark has been used traditionally as an aphrodisiac, to enhance sexual performance, and to treat erectile dysfunction. However, the bark can cause hypertension, tachycardia, and anxiety. It can also interact with certain medications for depression. Taking it in high doses or for an extended period can be hazardous. Yohimbe is used for erectile dysfunction, weight loss, athletic performance, diabetic neuropathy, angina, and hypertension. Yohimbe may cause agitation, anxiety, insomnia, skin flushing, rash, dizziness, rapid heartbeat, increased urination, headache, nausea, and vomiting. More severe side effects can include chest pain, heart attack, heart rhythm problems such as atrial fibrillation, kidney failure, and seizures.

CLINICAL CASE STUDIES

Clinical Case Study 1

A 72-year-old man was diagnosed with atrial fibrillation and hypercholesterolemia. He began anticoagulant therapy with warfarin. The patient has had a history of hypertension for the past 20 years and uses raw garlic in his diet every day. Current daily medications include enalapril, warfarin, atenolol, and pravastatin.

CRITICAL THINKING QUESTIONS

1. What are the advantages of garlic in the diet?
2. What are the adverse effects of garlic?
3. What herbal remedies are contraindicated and might interact with anticoagulants?

Clinical Case Study 2

A 67-year-old woman was transferred to the emergency department. She has been increasingly confused and sweaty over the past 2 days. Her medical history reveals significant depression, hypertension, and coronary artery disease. She also complains of headaches, shivering, and palpitations. The patient admits to using St. John's wort. Her current daily medications include atenolol, atorvastatin, citalopram, and ramipril.

CRITICAL THINKING QUESTIONS

1. What is the characteristic of the St. John's wort plant?
2. What are the indications of St. John's wort?
3. What are the adverse effects of St. John's wort?

FURTHER READING

Balch, P.A. (2010). *Prescription for Nutritional Health — A Practical A-to-Z Reference to Drug-Free Remedies Using Vitamins, Minerals, Herbs & Food Supplements*, 5th Edition. Avery.

Brinker, F. (2010). *Herbal Contraindications and Drug Interactions plus Herbal Adjuncts with Medicines*, 4th Edition. Eclectic Medical Publications.

Chandler Goldstein, M., and Goldstein, M.A. (2020). *Dietary Supplements: Fact versus Fiction*. Greenwood.

Enna, S.J., and Norton, S. (2012). *Herbal Supplements and the Brain: Understanding Their Health Benefits and Hazards*. F.T. Press — Science.

Gaby, A.R., and the Healthnotes Medical Team. (2006). *A-Z Guide to Drug-Herb-Vitamin Interactions*, 2nd Edition. Harmony.

Harrod Buhner, S. (2012). *Herbal Antibiotics: Natural Alternatives for Treating Drug-Resistant Bacteria*, 2nd Edition. Storey Publishing.

Haycock, B.B., and Sunderman, A.A. (2016). *Dietary Supplements* (Nutrition and Dietetics Practice Collection). Momentum Press — Health.

Jafari, M. (2021). *The Truth about Dietary Supplements: An Evidence-Based Guide to a Safe Medicine Cabinet*. Archangel Ink.

Jesson, L.E., and Tovino, S.A. (2010). *Complementary and Alternative Medicine and the Law*. Carolina Academic Press.

Kailin, D.C. (2006). *Quality in Complementary & Alternative Medicine*. CMS Press.

Micozzi, M.S. (2018). *Fundamentals of Complementary, Alternative, and Integrative Medicine*, 6th Edition. Saunders.

Physicians' Desk Reference, Inc. (2008). *PDR for Nonprescription Drugs, Dietary Supplements, and Herbs*, 30th Edition. Physicians' Desk Reference Inc.

Sarubin Fragakis, A., and Thompson, C.A. (2007). *The Health Professional's Guide to Popular Dietary Supplements*, 3rd Edition. American Dietetic Association.

Talbott, S.M., and Hughes, K. (2006). *The Health Professional's Guide to Dietary Supplements*. Lippincott, Williams, & Wilkins.

Termini, R.B. (2019). *Food and Drug Law: Federal Regulation of Drugs, Biologics, Medical Devices, Foods, Dietary Supplements, Personal Care, Veterinary and Tobacco Products*, 10th Edition. Forti Publications.

Williams, K.M. (2021). *Dr. Sebi's Book of Remedies*. Sebi Books.

8 Medication Errors

LEARNING OUTCOMES

After studying the chapter, readers should be able to

1. Review the three steps of the medication process in which errors can occur.
2. Describe high-alert medicines that can cause harm to patients.
3. Summarize the factors that can contribute to medication errors.
4. Discuss the MedWatch program related to medication errors.
5. Discuss the importance of reporting medication errors.
6. Describe polypharmacy in relation to medication errors.
7. Explain the concept of reducing medication errors.
8. Review unexpected events or medication administration errors.

OVERVIEW

Medication errors frequently occur in various settings. These medical mistakes are primarily observed in children and older adults. Key factors contributing to medication errors include prescribing, dispensing, administration, and patient compliance. It is essential to ensure that all medications intended for patient use are thoroughly understood to prevent mistakes. Medication reconciliation, a safety strategy that compares patients' medication lists with those maintained by their healthcare providers, can help reduce drug–drug interactions. The seven "rights" of drug administration must be strictly followed. Accurate documentation is critical, and medical staff are legally and ethically obligated to report medication errors to enhance patient outcomes.

THE THREE PHASES OF MEDICATION ERRORS

A **medication error** occurs when an incorrect or inappropriate drug is administered and can happen in various ways. Medication errors may result from manufacturing mistakes. The morbidity and mortality costs associated with medication errors are estimated to reach 77 billion dollars annually in the United States. Approximately 9,000 patients die from medication errors each year in the United States. The three phases in which errors often occur include

- **Prescribing the medication**: Errors mainly occur when choosing a drug, its dose, and the dosing schedule or dosage.

- **Dispensing medication may** lead to errors caused by inadequate or inexperienced pharmacy staff and counseling. All dispensed medications should be double-checked against the **medication administration record** (MAR). This helps prevent mistakes before drugs are dispensed.
- **Administering the medication:** Errors are often attributed to the person giving it. They may be responsible for poor communication and misreading prescriptions or drug labels. Other contributing factors include similar labeling and packaging of products and drugs with look-alike names.

Medication errors involve the dose or dosing regimen, particularly concerning analgesics or antibiotics. These errors may arise from the growing number of prescriptions dispensed each year. The current shortage of pharmacists and pharmacy technicians also contributes to medication errors. The rising use of over-the-counter drugs and various herbal products can also lead to medication errors.

High Alert Medications

High-alert medications are agents that pose a heightened risk of causing significant patient harm when used in error. While mistakes may not be more common with these drugs, the consequences of such errors can be far more devastating for patients. The five medications most commonly associated with severe adverse events are insulin, IV narcotics, IV heparin, IV potassium concentrates, and IV **hypertonic** sodium chloride solutions. Most of these drugs are available as **floor stock** in nursing areas, requiring careful double-checking of their use.

Confusing abbreviations, acronyms, and symbols are often used. These include "U" for "unit," "IU" for "international unit," trailing zeros such as X.0 mg, and the lack of a leading zero, such as X mg, the < sign for "less than," the > sign for "greater than," and abbreviated drug names. Writing out words, phrases, and drug names is always the best way to prevent errors.

FACTORS CONTRIBUTING TO MEDICATION ERROR

Medication errors can occur at any time or place. They are the most common cause of morbidity and preventable hospital deaths. Studies indicate that medical errors are now the third-leading cause of death in the United States, following strokes and diabetes. Medication errors can lead to extended hospital stays, increased costs, and patients being separated from their families and daily activities. High error rates can harm the reputations of healthcare staff and their facilities, potentially resulting in financial penalties and lawsuits.

However, medical errors can occur in almost any healthcare setting, including hospitals, clinics, surgery centers, medical offices, nursing homes, pharmacies, and patients' homes. All healthcare facilities should aim to reduce medication errors to the lowest possible level. Each mistake should be thoroughly documented and investigated to prevent future errors. Investigations should promote the accurate reporting of what occurred without the fear of penalties. The following factors may contribute to medication errors:

- **Miscommunication**: poor handwriting, similar drug names, and dosing unit confusion
- **Use of incorrect abbreviations**: many abbreviations can be mistaken for different units and different doses
- **Lack of appropriate labeling**: by the manufacturer or the pharmacist
- **Missing information**: lack of patient information (such as allergies, diseases, drug history, laboratory information, and drug information or warnings)
- **Environmental factors**: noise, lighting, stress, and fatigue that affect healthcare providers

Formulation, packaging, labeling, and distribution errors can occur during drug manufacturing. Mistakes may also result from undiscovered toxicities. Medication errors are preventable and represent the single largest source of medical errors. These errors may happen due to the seven "rights" of drug administration:

- Wrong patient
- Incorrect route
- Incorrect drug
- Incorrect dose
- Incorrect time
- Incorrect technique
- False information in the patient's chart

YOU SHOULD REMEMBER

Many factors can lead to medication errors, including misinterpretation of handwriting, miscalculations, misunderstandings of verbal or phone orders, and incorrect administration.

YOU SHOULD REMEMBER

Common errors include dosing errors and prescribing medications to which the patient has had an allergic response.

Check Your Knowledge

1. What is the description of medication errors?
2. What is the concept of high-alert medications?
3. What factors may cause medication errors?

DOCUMENTATION

Documenting medication errors is both a legal and ethical obligation and a valuable opportunity to enhance patient safety and improve the quality of care. This practice provides accurate and timely information to patients and their families, facilitating communication with physicians and other team members regarding the error and its causes. Furthermore, it enables the identification and analysis of factors contributing to the error, including human, system, or environmental elements. Adhering to the organization's policies and procedures is essential. Organizations may utilize various forms, systems, or channels, including incident reports, electronic health records, or quality improvement databases.

The information in the documentation must be relevant and accurate. This encompasses the date, time, and location of the error; the patient's name, identification number, and diagnosis; the medication name, dosage, route, frequency, and indication; the type and severity of the error; any actual or potential harm to the patient; actions taken to address the mistake; causes and contributing factors of the error; and recommendations for improvement. When reporting medication errors, use clear, objective language that avoids blaming or criticizing anyone. Documenting medication errors is not meant to assign fault or punish anyone, but to learn from the mistake and prevent it from recurring. Steer clear of words or phrases that imply judgment and focus instead on descriptive language presenting facts. Additionally, avoid vague or ambiguous terms, opting for precise language instead.

Accurate documentation in medical records and error reports is essential for legal reasons. This practice enhances medication administration processes and ensures that patient safety is upheld. If there is an attempt to delay corrective action or conceal a mistake, the legal consequences may be more severe. The same applies if interventions are not recorded in the patient's chart.

Medical facilities implement quality improvement programs to manage medication errors. These programs can notify practitioners of medication error trends and aid in developing specific solutions. Many facilities utilize **root cause analysis** to prevent future mistakes by examining what occurred, why it happened, and what steps can be taken to avoid a recurrence. The initial cause analysis helps assess whether the recurrence risks have been mitigated.

REPORTING MEDICATION ERRORS

Healthcare professionals are ethically and legally responsible for reporting all medication errors. Patients may require lifesaving interventions, ongoing supervision, and treatment. The FDA coordinates the reporting of medication errors through its **MedWatch program** or the *FDA Safety Information and Adverse Event Reporting Program.* This program provides essential information about medical products, including safety concerns regarding prescription and over-the-counter drugs, biologics, medical devices, radiation-emitting devices, and specialized nutritional products. Medical professionals should utilize the MedWatch database to report medication errors and situations that could lead to mistakes. Reporting can be done anonymously, directly to the FDA, via the internet, or by telephone.

Check Your Knowledge

1. What is the purpose of documenting medication errors?
2. What is the MedWatch program?

UNEXPECTED EVENTS

Unexpected events refer to unwanted, uncomfortable, or dangerous effects that drugs may produce. These events can lead to severe physical or psychological injury or even death. The incidence and severity of these events vary based on patient characteristics such as age, sex, genetic background, coexisting disorders, and geographic factors, as well as drug-specific factors like the type of drug, routes of administration, treatment duration, and dosage.

Incidence is higher among older adults and those taking multiple medications. According to the National Electronic Injury Surveillance System, the therapeutic use of diabetes medications and anticoagulants can lead to unexpected events in older adults. A comprehensive medication history can help prescribers understand patients' previous experiences with drug treatments, particularly in identifying any prior unexpected events that may warrant caution against re-exposure to the drug.

REDUCING MEDICATION ERRORS

Most healthcare facilities store medications in locked, automated, and computerized cabinets that require a specific code for access. **Automation** helps maintain an accurate inventory of medication supplies. In larger facilities, risk management departments work to minimize medication errors and evaluate associated risks. Risk management specialists track information, investigate incidents, identify problems, and provide recommendations to improve patient care accuracy.

Medication reconciliation, a safety strategy that compares patients' medication lists with those maintained by healthcare providers treating them, can reduce the risk of drug–drug interactions. Collaboration among staff members and risk management specialists aids in modifying policies and procedures. Errors can be minimized by properly storing medications, keeping reference materials up to date, and avoiding the following:

- Use of expired medications
- Transfer of medications between containers
- Overstocking
- Use of dangerous abbreviations

Medication errors may relate to professional practice, healthcare products, procedures, and systems. Guidelines for risk reduction include

- Using an adequate number of healthcare staff members who are trained to prepare, dispense, and administer medications to patients
- Using standardized measurement systems for inpatients and outpatients
- Using error-tracking systems that consistently target and monitor common patient errors
- Using electronic medical records and e-prescribing so that automatic surveillance measures can find potential medication errors, especially if the information is lacking or conflicting with other information
- Compiling medication profiles for all patients with updated allergy histories after each encounter
- Defining a strategy for drug ordering, dispensing, and administration that includes reviews of original drug orders

CLINICAL CASE STUDIES

Clinical Case Study 1

For postoperative pain, a 6-month-old infant was ordered morphine "0.5 mg IV," as needed. A secretary recorded the order in the MAR as "5 mg." An inexperienced nurse followed the directions on the MAR without question and gave the infant a 5 mg IV morphine dose instead of 0.5 mg. About an hour after the second dose, the infant developed respiratory and cardiac arrest. If the physician had written "0.5 mg," the secretary would probably not have misunderstood the intended amount, and the nurse would have administered the correct dose.

CRITICAL THINKING QUESTIONS

1. What are the three phases of medication errors?
2. What are the five most common medications associated with severe adverse events?
3. What are the seven "rights" of drug administration?

FURTHER READING

Cohen, M.R. (2006). *Medication Errors*, 2nd Edition. American Pharmacists Association.

Curren, A.M., and Witt, M.H. (2014). *Curren's Math for Meds: Dosages and Solutions*, 11th Edition. Cengage Learning.

Guerra, T. (2017). *How to Pronounce Drug Names: A Visual Approach to Preventing Medication Errors*. Lulu.com.

Haroutounian, S. (2019). *Preventing Medication Errors at Home*. Oxford University Press.

Institute of Medicine, Board on Health Care Services, Committee on Identifying and Preventing Medication Errors, Cronenwett, L.R., Bootman, J.L., Wolcott, J., and Aspden, P. (2007). *Preventing Medication Errors* (Quality Chasm Series). National Academies Press.

Kelly, W.N. (2006). *Prescribed Medications and the Public Health: Laying the Foundation for Risk Reduction*. CRC Press.

Kukhet, G. (2017). *Your License Is on the Line: Medication Errors – Dispensing Without Error Dietary Supplements, OTC Products, Being an Active Member, and Medical Abbreviations*. CreateSpace Independent Publishing Platform.

Larson, C.M., and Saine, D. (2013). *The Medication Safety Officer's Handbook*. American Society of Health-System Pharmacists.

McAslan, M.S. (2012). *Read the Prescription Label: and Other Tips to Prevent Deadly and Costly Medication Errors*. Balboa Press.

Persaud, R., and Tabanao, R. (2019). *Medication Administration: A Guide for Unregulated Healthcare Providers*. Tellwell Talent.

Pronsky, Z.M., Elbe, D., Ayoob, K., Crowe, J.P., Epstein, S., and Roberts, W.H. (2015). *Food Medication Interactions*, 18th Edition. FMI.

Solanki, N., and Shah, C. (2014). *Study of Medication Errors in Western Population of India: Root Cause Analysis of Medication Errors at Multispecialty Hospital*. Lap Lambert Academic Publishing.

Spath, P.L. (2011). *Error Reduction in Health Care: A Systems Approach to Improving Patient Safety*, 2nd Edition. Jossey-Bass.

United States Congress House Committee. (2015). *Issues Relating to Medication Errors: Hearing Before the Subcommittee on Health of the Committee on Ways and Means, House of Representatives*. Palala Press.

Workman, M.L., and LaCharity, L.A. (2015). *Understanding Pharmacology: Essentials for Medication Safety*, 2nd Edition. Saunders.

Part II

Drugs Affecting the Nervous System

9 Drugs Affecting the Central Nervous System

LEARNING OUTCOMES

After studying this chapter, readers should be able to

1. Describe the classifications of seizures.
2. Compare the characteristics of partial and absence seizures.
3. Explain the adverse effects of antiseizure drugs.
4. Review classifications of migraines.
5. Explain the primary signs and symptoms of Parkinson's disease (PD).
6. Describe the contraindications of dopamine agonists for PD.
7. Explain the indications of clonazepam.
8. Explain the contraindications of entacapone. Discuss the etiology and risk factors of Alzheimer's disease (AD).
9. Describe the mechanism of action for anticholinergic agents.
10. Explain the indications and adverse effects of acetylcholinesterase inhibitors.

OVERVIEW

The nervous system is a complex part of the human body. However, significant progress continues within the scientific discipline of neuroscience. Many medications are available for various disorders of this system. Pharmacotherapy for nervous system disorders is extensive. This chapter discusses antiseizure treatments, antimigraine medications, and the treatment of PD and AD. A seizure is an abnormal, uncontrolled electrical discharge in the brain's cortical gray matter, transiently interrupting normal function. Seizures are absent, partial, febrile, tonic–clonic, and myoclonic. A migraine is a complicated condition characterized by a severe headache, most often unilateral on one side of the head. The pain is much worse than a typical headache and is usually accompanied by various neurological symptoms. PD is the second most common degenerative disorder affecting the nervous system. It develops insidiously and slowly progresses over several months to years. AD involves the destruction of memory, cognitive skills, and the ability to perform even the simplest tasks. Senile plaques, beta-amyloid deposits, and tangles of neurofibrils are found in the cerebral cortex. Treatment for each of these disorders is specific.

CLASSIFICATIONS OF SEIZURES

Seizures have various classifications, including absence seizures, partial seizures, generalized seizures, tonic–clonic seizures, and myoclonic seizures. This condition is also known as epilepsy. A seizure is described as a chronic neurological disorder characterized by recurrent seizures resulting from the excessive excitability of neurons in the brain. Seizures can involve a brief loss of consciousness or cause violent **convulsions**. Sometimes, patients may experience memory, mood, and learning difficulties. It is essential to understand that a *seizure* encompasses all forms of epileptic events, while a *convulsion refers explicitly* to the abnormal motor activity that occurs during a seizure attack.

ABSENCE SEIZURES

Absence seizures begin on both sides of the brain simultaneously, causing lapses in awareness. The older term for absence seizures was "petit mal seizures". Absence seizures start and end quickly, lasting only a few seconds. They can be so brief that they are often mistaken for "daydreaming" and may not be detected for months. Most patients with absence seizures continue to move somewhat, with subtle clonic eyelids, facial muscles, or fingers. Small, synchronized movements of both arms may occur at a rate of about three per second. Absence seizures are a common form in childhood, typically beginning between the ages of four and puberty. They may occur several hundred times daily, making the child appear inattentive.

PARTIAL SEIZURES

A **partial seizure** can be classified as either simple or complex. A *simple partial seizure* affects only one area of the brain, does not result in loss of consciousness, and typically lasts 1–2 minutes. A *complex partial seizure* is more common in adults with epilepsy. During this type of seizure, the person cannot speak, control movements, or recall the seizure after it ends.

GENERALIZED SEIZURES

Generalized seizures are abnormal electrical activities that begin simultaneously in both brain hemispheres. These common and dramatic forms were formerly known

DOI: 10.1201/9781003461913-11

as grand mal seizures. The signs and symptoms include unconsciousness, groaning, crying out, shaking, sudden collapse, spasms, stiffening, loss of bowel or bladder control, stoppage of breathing, and **cyanosis**. Generalized seizures are diagnosed through the patient's electroencephalogram, magnetic resonance imaging, medical history, and blood tests.

TONIC–CLONIC SEIZURES

A **tonic–clonic seizure** begins with little warning in both hemispheres but may also occur on one side before spreading to the entire brain. The patient may experience **apathy**, depression, and irritability. They may also exhibit jerking movements and muscle stiffening. The individual may cry out due to whole-body muscle spasms. The lateral part of the tongue is often bitten. Apnea and cyanosis may occur as a result of spasms in the respiratory muscles.

The patient cannot remember this event and falls asleep for several hours, frequently awakening with a headache. Severe muscle contractions during this event may cause fractures of the spinal vertebral bodies and skin **abrasions**. Tonic–clonic seizures can last from 1 to 3 minutes. If the seizure exceeds 5 minutes, it should be considered a medical emergency due to the risk of status epilepticus.

MYOCLONIC SEIZURES

A **myoclonic seizure** is a brief, shock-like jerk of the muscles. It occurs in various epilepsy syndromes with unique characteristics. During this type of seizure, the patient is usually awake and able to think clearly. The seizures typically last about 1–2 seconds. Myoclonic seizures cause abnormal movements on both sides of the body simultaneously.

ANTISEIZURE DRUGS

Choosing the proper seizure medication for a patient is challenging. No single medication is effective and may lead to various adverse effects. Antiseizure medications (ASMs) must consider which side effects should be avoided in specific cases. Narrow-spectrum ASMs are effective primarily for certain types of seizures, such as absence or myoclonic seizures. Broad-spectrum ASMs, on the other hand, are effective for a broader range of seizures, including focal plus absence myoclonic seizures. A summary of ASMs is provided in Table 9.1.

CLINICAL INDICATIONS

It is essential to accurately diagnose the specific types of seizures occurring. The selection of medications is usually based on established efficacy in the diagnosed seizures, patient responsiveness, anticipated toxicities, and possible drug interactions. Treatments may involve combinations of

TABLE 9.1
Antiseizure Drugs

Narrow-Spectrum AEDs	Broad-Spectrum AEDs
Phenytoin (Dilantin)	Valproic acid (Depakote)
Phenobarbital (Luminal)	Lamotrigine (Lamictal)
Carbamazepine (Tegretol)	Topiramate (Topamax)
Oxcarbazepine (Trileptal)	Zonisamide (Zonegran)
Gabapentin (Neurontin)	Levetiracetam (Keppra)
Pregabalin (Lyrica)	Clonazepam (Klonopin)
Lacosamide (Vimpat)	Rufinamide (Banzel)
Vigabatrin (Sabril)	Fosphenytoin (Cerebyx)
Tiagabine (Gabitril)	

agents, which can be effective when an initial drug or drugs are insufficient. The drugs of choice for partial seizures and generalized tonic–clonic include phenytoin, lamotrigine, carbamazepine, and valproic acid.

Valproic acid and ethosuximide are preferable to clonazepam because they typically cause only mild sedation. However, ethosuximide should be considered carefully due to potential gastrointestinal (GI) adverse effects. Valproic acid is the preferred choice if the patient experiences concurrent generalized tonic–clonic or myoclonic seizures. The effectiveness of lamotrigine, levetiracetam, and zonisamide for absence seizures is evident.

MECHANISM OF ACTION

The mechanism of action of antiseizure drugs involves suppressing action potentials. Other mechanisms include sodium channel blockade, calcium channel blockade, GABA-related targets, and glutamate synapses. Sometimes, several mechanisms work in concert. The mechanisms of action of antiepileptic drugs that affect GABA are illustrated in Figure 9.1.

ADVERSE EFFECTS

The adverse effects of antiseizure drugs include nausea, vomiting, diarrhea, loss of appetite, weight loss, **glossitis**, gingival hyperplasia, skin rash, headache, dizziness, nervousness, **insomnia**, tremor, double or blurred vision, **nystagmus**, cognitive impairment, and difficulty speaking.

CONTRAINDICATIONS

The contraindications of antiseizure drugs include **osteomalacia**, diabetes mellitus, hypoalbuminemia, pancytopenia, **thrombocytopenia**, alcoholism, myasthenia gravis, **glaucoma,** hypercholesterolemia, **porphyria**, myocardial infarction, dysrhythmias, coronary artery disease, heart failure, liver injury, kidney failure, systemic lupus erythematosus, mood changes, suicidal thoughts, and depression.

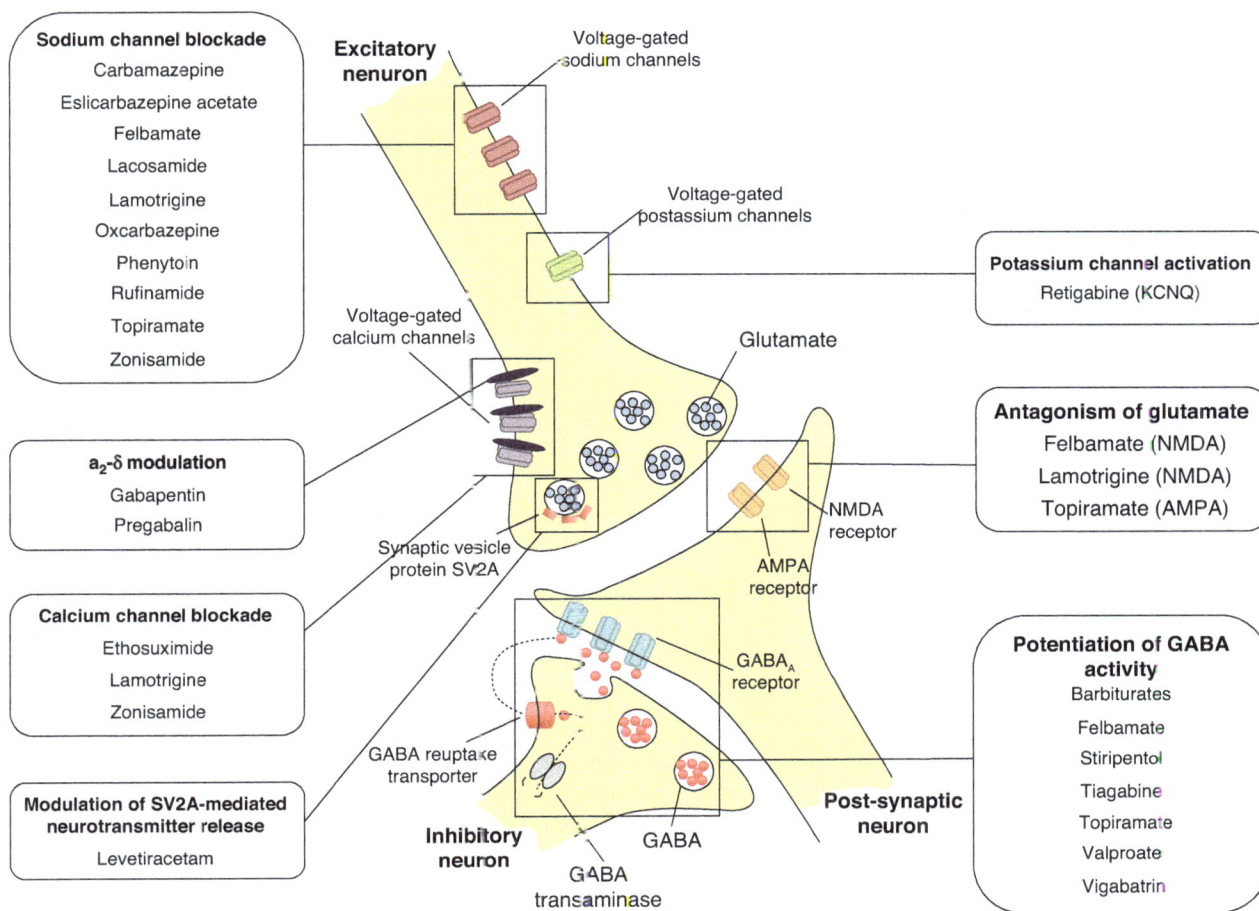

FIGURE 9.1 Action of antiepileptic drugs that affect GABA.

Check Your Knowledge

1. What are the drugs of choice for partial seizures?
2. What are the adverse effects of ASMs?
3. What are the contraindications for antiseizure medicines?

CLASSIFICATIONS OF MIGRAINES

Migraines are categorized into several types. The most common include migraines with an **aura**, migraines without an aura, silent migraines, and chronic migraines. This chapter will focus on migraines with and without an aura.

MIGRAINE WITH AURA

Migraine with aura is less common than migraine without aura. Symptoms begin 30 minutes before the headache starts. The visual symptoms include flashing lights, wavy lines, or temporary vision loss; there may also be an inability to speak, numbness, and tingling. People aged 50 or older may experience the aura without other symptoms. A migraine with aura usually lasts for about an hour. Relaxation techniques, regular sleep and eating routines, and adequate hydration are helpful without the need for medications.

MIGRAINE WITHOUT AURA

Migraine without aura is the most common type of migraine, affecting 75% of patients. It causes a severe headache on one side of the head. Bright lights, loud sounds, and physical activity can worsen this condition. Anxiety, depression, obesity, and snoring are among the risk factors.

PHARMACOTHERAPY FOR MIGRAINE

For migraines, different goals include stopping an acute migraine while it progresses and preventing or reducing the frequency of migraines. Several triptans are available, each with distinct pharmacokinetics and routes of administration. The choice of triptan should be individualized based on the patient's migraine characteristics, the route of administration, pharmacokinetics, and cost. Treatments for migraines without aura include over-the-counter painkillers or triptans. Table 9.2 summarizes the triptans and ergot alkaloids used to treat acute migraines. Pharmacotherapy can be more effective when it begins before the pain becomes severe.

For mild migraines, NSAIDs should be tried first, and acetaminophen should be combined with an NSAID and caffeine. Triptans and ergot alkaloids may be used when a mild migraine persists and is not relieved by NSAIDs. Triptans or ergot alkaloids are the preferred agents for moderate migraines. If these are contraindicated or ineffective, then dopamine agonists can be utilized. For severe migraines, serotonin agonists are usually administered. Narcotic analgesics can alleviate migraine pain that has not responded to any other medications.

TABLE 9.2

Triptans and Ergot Alkaloids Used to Treat Acute Migraines

Generic Name	Trade Name
Triptans	
Almotriptan	Axert
Eletriptan	Relpax
Frovatriptan	Frova
Naratriptan	Amerge
Rizatriptan	Maxalt
Sumatriptan	Imitrex, Imitrex Statdose, Tosymra
Zolmitriptan	Zomig, Lasmiditan
Ergot alkaloids	
Dihydroergotamine	Migranal
Ergotamine	Ergomar

PARKINSON'S DISEASE

PD is the second most common **degenerative disease** affecting the nervous system. It develops insidiously and progresses slowly over several months to several years. Approximately 1 million people in the United States have PD, and PD is more prevalent in older males than females. There is often a family history indicating a genetic link. Other potential causes include head trauma, brain infections or tumors, and exposure to neurotoxins. However, the most frequent reason is the use of antipsychotic and anesthetic drugs.

The primary signs and symptoms of PD include muscle rigidity, **bradykinesia**, tremor, and postural instability. Muscle rigidity leads individuals to have difficulty moving their limbs. The rigidity of facial muscles results in a lack of expression or alterations in expression. When the pharyngeal muscles are involved, chewing and swallowing can become challenging. Arm swinging is diminished during walking, with the arms held close to the sides of the body instead. *Bradykinesia* refers to an uncontrolled slowness of voluntary movement and highly noticeable speech, resulting in difficulties with speaking, chewing, and swallowing. Individuals may struggle to initiate activities and control fine muscle movements. They shuffle their feet while walking and no longer take normal strides, which makes walking much more challenging.

Tremors of the head and hands can lead to continuous shaking or motions even at rest. The tremors may become so severe that holding objects becomes challenging. As PD progresses, pill-rolling motions occur, with the thumbs and forefingers rubbed together in circular movements. The significant features of PD are illustrated in Figure 9.2.

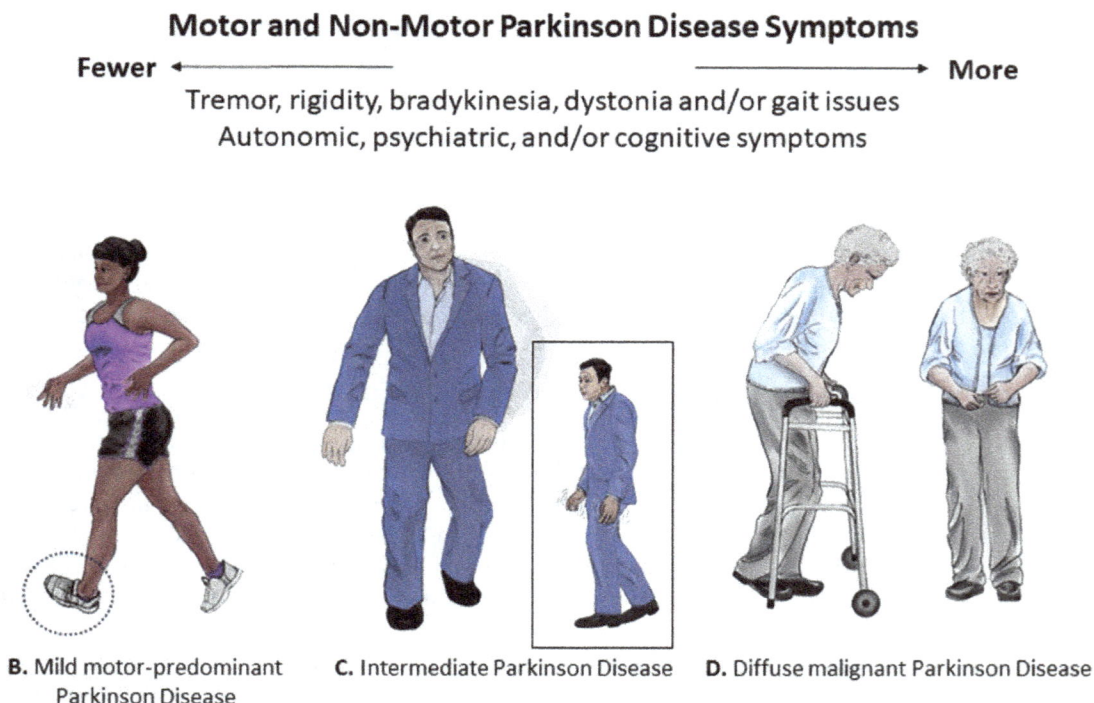

Motor and Non-Motor Parkinson Disease Symptoms

Fewer ◄——————————   ——————————► More

Tremor, rigidity, bradykinesia, dystonia and/or gait issues
Autonomic, psychiatric, and/or cognitive symptoms

B. Mild motor-predominant Parkinson Disease

C. Intermediate Parkinson Disease

D. Diffuse malignant Parkinson Disease

FIGURE 9.2   Significant features of Parkinson's disease.

YOU SHOULD REMEMBER

For patients with dystonia in PD, skeletal muscle relaxants and botulinum toxin injections may be necessary. Neuropathic pain is less common and is treated with gabapentin, pregabalin, duloxetine, venlafaxine, spinal cord stimulation, and nerve blocks.

PHARMACOTHERAPY FOR PD

PD is one of the chronic degenerative conditions affecting the nervous system. Currently, there is no cure for PD, but some drugs offer benefits in controlling motor symptoms. While these drugs can lead to significant improvements in motor function, they may also produce problematic adverse effects, particularly as the disease progresses.

DOPAMINERGIC DRUGS

Dopaminergic agents activate **dopamine receptors** and are more commonly used than anticholinergic agents. Mild symptoms are typically treated with selegiline or rasagiline, while more severe symptoms are managed with either carbidopa–levodopa or a dopamine agonist. The dopaminergic agents used to treat PD are summarized in Table 9.3.

Clinical Indications

Carbidopa–levodopa is indicated for the treatment of PD, post-encephalitis Parkinsonism, and symptomatic Parkinsonism following carbon monoxide or manganese intoxication. *Carbidopa–levodopa–entacapone* is intended explicitly for treating PD to alleviate nausea and vomiting. *Apomorphine* is indicated for PD and alcoholism as an emetic and is also utilized for erectile dysfunction. *Bromocriptine* is indicated for PD, pituitary tumors, hyperprolactinemia, neuroleptic malignant syndrome, and

TABLE 9.3

Dopaminergic Agents Used to Treat Parkinson's Disease

Generic Name	Trade Name
Carbidopa–levodopa	Sinemet
Carbidopa–levodopa	Parcopa
Carbidopa–levodopa	Sinemet CR
Carbidopa–levodopa	Rytary ER
Carbidopa–levodopa/entacapone	Stalevo
Carbidopa	Lodosyn
Apomorphine	Apokyn
Bromocriptine	Parlodel
Pramipexole	Mirapex
Pramipexole	Mirapex ER
Ropinirole	Requip
Ropinirole	Requip XL
Rotigotine	Neupro

adjunctively for type 2 diabetes. *Pramipexole and ropinirole* are indicated for PD and restless legs syndrome. *Rotigotine* is intended for use in early and late-stage PD and restless legs syndrome, reducing the on–off periods associated with PD.

Mechanism of Action

The dopamine agonists work by imitating dopamine's actions when levels are low. They activate postsynaptic dopamine receptors in the corpus striatum and may allow for reduced maintenance doses of levodopa, even in combination products like carbidopa–levodopa. The exact mechanisms of action of *pramipexole* and *ropinirole* for PD remain unknown.

Adverse Effects

The adverse effects of *carbidopa–levodopa* and *carbidopa* alone include dyskinesias, nausea, vomiting, constipation, headaches, and asthenia. The adverse effects of the *dopamine agonists* encompass abnormal involuntary movements, vertigo, peripheral edema, GI ulcers, pulmonary fibrosis, psychosis, hallucinations, hypotension, myocardial infarction, congestive heart failure, cardiac fibrosis, tachycardia, and pericardial effusion. Additional adverse effects of the *dopamine agonists* include insomnia, impulse control disorder, confusion, the on–off phenomenon, dizziness, drowsiness, asthenia, fainting, visual disturbances, ataxia, and shortness of breath. With *rotigotine*, unexpected sleep attacks may occur, which can result in injury to the patient.

Contraindications

The contraindications of dopamine agonists include known hypersensitivity, **narrow-angle glaucoma**, suspicious undiagnosed skin lesions, or a history of **melanoma**. *Apomorphine* is contraindicated with adrenergic receptor antagonists, alcohol, and dopamine antagonists. Contraindications of *bromocriptine* include uncontrolled hypertension, use of other medications that lower blood pressure, and a history of psychosis or cardiovascular disease. *Pramipexole* is contraindicated with recurrent daytime sleep episodes, orthostatic hypotension, rhabdomyolysis, drowsiness, hallucinations, sleep apnea, dyskinesia, breastfeeding, uncontrolled impulsive behaviors, and kidney disease. *Ropinirole* is contraindicated in patients with psychosis, **bradycardia**, orthostatic hypotension, drowsiness, hallucinations, dyskinesia, unruly impulsive behaviors, and kidney failure.

Pregnancy Category: Levodopa, C

Check Your Knowledge

1. What are dopaminergic agents?
2. What are the clinical indications for carbidopa–levodopa?
3. What are the contraindications for dopamine agonists?

TABLE 9.4
Anticholinergics Used for Parkinson's Disease

Generic Name	Trade Name
Benztropine	Cogentin
Trihexyphenidyl	Artane

TABLE 9.5
MAO-B Inhibitors Used for Parkinson's Disease

Generic Name	Trade Name
Rasagiline	Azilect
Safinamide	Xadago
Selegiline	Eldepryl
Selegiline	Zelapar

ANTICHOLINERGIC DRUGS

Anticholinergic agents block receptors for acetylcholine. These agents are rarely effective for bradykinesia and other disabilities and can be used either as monotherapy or in combination with other antiparkinsonian drugs. Anticholinergics used for PD include benztropine and trihexyphenidyl (see Table 9.4).

Clinical Indications

Benztropine and trihexyphenidyl are indicated for PD and dyskinesias caused by specific agents. They are also used to treat drug-induced **extrapyramidal** symptoms, manage acute dystonic reactions, and prevent dystonic reactions. Trihexyphenidyl effectively alleviates spasms, stiffness, tremors, and issues with muscle control.

Mechanism of Action

Anticholinergic agents block muscarinic receptors in the striatum. *Benztropine exhibit*s antimuscarinic, antihistaminic, and local anesthetic effects, competing with acetylcholine in the central nervous system. Additionally, anticholinergic agents can block dopamine reuptake and storage, prolonging dopamine's effects. *Trihexyphenidyl* has an atropine-like action, producing antispasmodic effects on parasympathetic innervation. Its precise mechanism of action in PD remains unknown.

Adverse Effects

Adverse effects of anticholinergics include dry mouth, blurred vision, nervousness, nausea, constipation, and urinary retention. Patients with preexisting cognitive deficits and older adults are at a higher risk of central anticholinergic adverse effects.

Contraindications

Benztropine is contraindicated in patients with myasthenia gravis, glaucoma, coronary artery disease, dysrhythmia, heart failure, chronic obstructive pulmonary disease, **achalasia**, hiatal hernia, stomach ulcers, ulcerative colitis, **toxic megacolon**, intestinal paralysis, liver disease, kidney failure, enlarged prostate, and dysuria. *Trihexyphenidyl* is contraindicated in patients with known hypersensitivity, glaucoma, tardive dyskinesia, cardiac disease, hypertension, prostatic hypertrophy, and liver or kidney impairment.

Check Your Knowledge

1. What are some examples of anticholinergic drugs?
2. What are the clinical indications for benztropine and trihexyphenidyl?
3. What are the contraindications for benztropine?

MONOAMINE OXIDASE TYPE B INHIBITORS

Monoamine oxidase type B (MAO-B) is an enzyme in the body that breaks down several chemicals in the brain, including dopamine. Therefore, using a medication that blocks the effect of an MAO-B inhibitor makes more dopamine available for the brain to use, slightly improving many of the motor symptoms of PD. The use of MAO-B to treat PD is summarized in Table 9.5.

Clinical Indications

Rasagiline can be used as monotherapy to treat the signs and symptoms of early PD or as an adjunct therapy for more advanced cases. Safinamide is indicated for idiopathic PD. Selegiline is utilized as an adjunct in the treatment of PD. It is also approved for off-label use as a palliative treatment for dementia associated with AD.

Mechanism of Action

The MAO-B inhibitors have a complex mechanism of action. Rasagiline prevents the breakdown of dopamine by irreversibly binding to MAO-B. Safinamide has similar effects but is reversible, unlike rasagiline and selegiline. Additionally, it blocks sodium and calcium channels.

Adverse Effects

The adverse effects of rasagiline include hallucinations, dizziness, fainting, sore throat, tachycardia, stomach pain, muscle pain, chest pain, coughing, dysphagia, loss of appetite, nausea, dysuria, sweating, liver abnormalities, and unusual bleeding or bruising. The common adverse effects of safinamide include fatigue, dizziness, insomnia, orthostatic hypotension, nausea, and headache. The adverse effects of selegiline include facial dyskinesias of the face, lips, tongue, arms, or legs.

Contraindications

Rasagiline is contraindicated in individuals with bipolar disorder, psychosis, dyskinesia, hypertension, orthostatic hypotension, malignant melanoma, and liver diseases. Safinamide is contraindicated for those with hypersensitivity and liver injury. Selegiline should be avoided in patients with melanoma and liver abnormalities.

YOU SHOULD REMEMBER

Some MAO-B inhibitors interact with foods high in tyramine, potentially raising blood pressure to dangerous levels. These foods include cheeses, beer, fava beans, dried or cured meats, and soybeans.

Check Your Knowledge

1. What are some examples of MAO-B inhibitors used for PD?
2. What are the adverse effects of the mechanism of rasagiline?
3. What are the contraindications for rasagiline?

CATECHOL-O-METHYLTRANSFERASE INHIBITORS

The *catechol-O-methyltransferase (COMT) inhibitors* extend the duration of levodopa's effects by inhibiting its metabolism, thereby preventing a brief impact. The primary COMT inhibitors are listed in Table 9.6.

Clinical Indications

Entacapone is used in combination with levodopa for motor symptoms. Opicapone is indicated as adjunctive therapy for PD along with levodopa and carbidopa. Tolcapone has the same indications, but it should only be used after all other Parkinson's medications have been tried.

Mechanism of Action

Entacapone's mechanism of action involves its ability to inhibit COMT and modify the plasma pharmacokinetics of levodopa. *Opicapone effectively and selectively reversibly blocks COMT*, limited to outside the CNS. It dissociates slowly from COMT, providing a duration of action longer than 24 hours, despite having a short blood plasma half-life.

TABLE 9.6

Catechol-O-Methyltransferase (COMT) Inhibitors

Generic Name	Trade Name
Entacapone	Comtan
Opicapone	Ongentys
Tolcapone	Tasmar

The precise mechanism of action of *tolcapone* remains unknown, although it may relate to its capacity to inhibit COMT and alter the plasma pharmacokinetics of levodopa.

Adverse Effects

The adverse effects of entacapone include discolored urine, diarrhea, and increased dyskinesia. Opicapone may cause hypotension, dizziness, dyskinesia, constipation, and weight loss. Negative impacts of tolcapone include the same ones as for entacapone, in addition to risks of liver damage.

Contraindications

Entacapone should be avoided in patients with known hypersensitivity to it. Opicapone is contraindicated in patients with **pheochromocytoma**, **paraganglioma,** or nontraumatic rhabdomyolysis. Tolcapone must be avoided in patients with liver disease or known hypersensitivity to this agent.

AMANTADINE

Amantadine was initially used to treat influenza A; however, it is no longer recommended. It is an antidyskinetic agent that can treat early symptoms and help reduce tremors. Amantadine (Symmetrel) is available in 100 mg tablets or capsules and 50 mg/5 mL syrup.

Clinical Indications

In 1973, the FDA approved amantadine for treating PD. The immediate-release form of amantadine is indicated as monotherapy for tremors and stiffness. It is also used in combination therapy with levodopa for managing dyskinesias.

Mechanism of Action

The mechanism of action of amantadine in PD is uncertain. It may act as an anticholinergic by binding to the sigma-1 receptor. Therefore, it increases dopamine release and blocks dopamine reuptake.

Adverse Effects

Adverse effects of amantadine include dry mouth, nausea, constipation, dysrhythmia, confusion, dizziness, leg discoloration, orthostatic hypotension, insomnia, paranoia, hallucinations, and urinary retention.

Contraindications

Amantadine should be avoided in chronic kidney disease and used cautiously in individuals with prostate hyperplasia or glaucoma. Additionally, live attenuated vaccines are contraindicated during amantadine treatment.

ALZHEIMER'S DISEASE

AD destroys memory, thinking skills, and the ability to perform even the simplest tasks. Senile plaques, beta-amyloid deposits, and tangles of neurofibrils in the cerebral cortex characterize it. These tangles can only be identified through biopsy or autopsy, as they are not visible in computed tomography or magnetic resonance imaging scans. Approximately 80% of older adults have AD, and around 6.2 million Americans are affected by the condition.

ETIOLOGY AND RISK FACTORS

The cause of AD is unknown, but several factors may contribute to dementia. However, AD can be influenced by genetic and environmental factors. The formation of plaques around the brain's neurons and twisted strands of protein (tangles) accumulates inside the brain's nerve cells. These changes to the neurons in the brain lead to injury of the brain tissue and necrosis of the neurons, resulting in dementia. AD is a slowly progressive disease, with changes occurring years before signs and symptoms appear.

DIAGNOSIS

To diagnose AD, obtain blood work (vitamin B12 and thyroid levels) and assess changes in alcohol, water, and medication intake. There may be increased concentrations of tau protein in the brain and cerebrospinal fluid, along with reduced levels of choline acetyltransferase and **somatostatin**.

PHARMACOTHERAPY FOR ALZHEIMER'S DISEASE

Unfortunately, there is no cure for AD. Current pharmacotherapeutic techniques focus on delaying progression by managing behavioral symptoms, preserving mental capacities, and slowing the emergence of illness symptoms. Two classes of medications that help delay the progression of dementia are cholinesterase inhibitors and N-methyl-D-aspartate (NMDA) antagonists. There are also newer drugs called monoclonal antibodies (Aduhelm and Leqembi) that remove specific forms of beta-amyloid.

ACETYLCHOLINESTERASE INHIBITORS

Acetylcholinesterase inhibitors reduce acetylcholine levels and can increase cognitive symptoms. These agents include donepezil (Aricept), rivastigmine (Exelon), galantamine (Razadyne), and tacrine (Cognex). Today, tacrine is seldom used due to its excessive toxicity. Overall, acetylcholinesterase inhibitors have only a modest effect on AD, do not cure the disease, and may improve cognition.

Clinical Indications

Donepezil, galantamine, and *tacrine* are indicated for the treatment of mild to moderate dementia associated with AD.

Rivastigmine is also used for PD. Tacrine is rarely used in the United States.

Mechanism of Action

Acetylcholinesterase inhibitors prevent acetylcholinesterase from breaking down acetylcholine into choline within the CNS, autonomic ganglia, and neuromuscular junctions.

Adverse Effects

The adverse effects of acetylcholinesterase inhibitors include loss of appetite, nausea, vomiting, diarrhea, agitation, headache, **asthenia**, dizziness, muscle cramps, insomnia, weight loss, **arthralgia**, and dyspepsia.

Contraindications

Acetylcholinesterase inhibitors should be avoided in patients with hypersensitivity, epilepsy, atrioventricular heart block, sinus bradycardia, asthma, chronic obstructive pulmonary disease, GI bleeding, urinary bladder obstruction, and lightheadedness.

NMDA RECEPTOR MODULATORS

The NMDA receptor modulator, such as memantine, slows the progression of AD. Memantine is the only FDA-approved NMDA antagonist. Drug therapy for dementia associated with AD is summarized in Table 9.7.

Clinical Indications

Memantine is used to treat moderate to severe dementia associated with AD. It can slow the neurotoxicity involved in Alzheimer's and other neurodegenerative diseases.

Mechanism of Action

Memantine inhibits the excitatory neurotransmitter glutamate in the brain, which plays a role in normal learning and memory. When glutamate binds to NMDA receptors, calcium enters neurons to facilitate communication. In AD, excessive amounts of glutamate are present, causing calcium to flow into nerve cells and leading to cell death due to toxicity.

TABLE 9.7

Drug Therapy for Dementia of Alzheimer's Disease

Generic Name	Trade Name
Cholinesterase inhibitors	
Donepezil	Aricept
Galantamine	Razadyne, Razadyne ER
Rivastigmine	Exelon
NMDA antagonist	
Memantine	Namenda, Namenda XR

Adverse Effects

Adverse effects of memantine include nausea, vomiting, diarrhea, weight loss, headache, dizziness, insomnia, confusion, excitement, and bradycardia. Less common adverse effects are fatigue, coughing, and dyspnea. Memantine appears to have no significant toxicities.

Contraindications

Memantine should be avoided in patients with known hypersensitivity to the drug. It is also contraindicated in individuals with a history of myocardial infarction, hypertension, heart failure, epilepsy, or kidney or liver disease.

Check Your Knowledge

1. What are some examples of drug therapy for dementia?
2. What is the mechanism of action for memantine?
3. What are the contraindications for memantine?

CLINICAL CASE STUDIES

Clinical Case Study 1

A 21-year-old woman had a history of absence seizures that started when she was ten. At that time, she was diagnosed with generalized epilepsy. Valproic acid was initiated, and she had no seizures for 6 years. When she was 15, she started taking oral contraceptives and had daily absence seizures and a few generalized tonic–clonic seizures. Lamotrigine was tried, but the seizures worsened. Eventually, seizure control was provided by a combination of valproic acid and levetiracetam.

CRITICAL THINKING QUESTIONS

1. What are the differences between an absence and a generalized seizure?
2. What are the drugs of choice for generalized tonic–clonic seizures?
3. What are the adverse effects of ASMs?

Clinical Case Study 2

A 32-year-old woman experienced her first migraine, lasting 3 days. She had nausea, vomiting, and severe headaches. Her mother has a history of migraines. Taking oral contraceptives worsened the headaches significantly. The patient always experienced an aura before each migraine. Regularly consuming alcohol triggered an attack. She was diagnosed with "migraine with aura" and started on appropriate treatment.

CRITICAL THINKING QUESTIONS

1. What are the signifying factors of migraines?
2. How is "migraine with aura" described?
3. What are the pharmacotherapies used for severe migraines?

Clinical Case Study 3

A 79-year-old woman was diagnosed with PD 9 years ago. Recently, her balance and strength have decreased dramatically. The patient was transferred to a rehabilitation facility, where she stayed for several weeks, followed by 4 weeks of physical therapy. The patient has severe depression because of her PD. She regularly uses a cane and protective devices in her bathroom to avoid falling. The patient has fallen three times in the last month while at home.

CRITICAL THINKING QUESTIONS

1. What are the signs and symptoms of PD?
2. What are the adverse effects of carbidopa–levodopa and carbidopa?
3. What are the contraindications of dopamine agonists?

FURTHER READING

Ahlskog, J.E. (2009). *Parkinson's Disease Treatment Guide for Physicians*. Oxford University Press.
Ahlskog, J.E. (2015). *The New Parkinson's Disease Treatment Book: Partnering to Get the Most from Your Medications*, 2nd Edition. Oxford University Press.
Bazil, C.W. (2004). *Living Well with Epilepsy and Other Seizure Disorders: An Expert Explains What You Need to Know* (Living Well). HarperResource.
Betts, T., and Greenhill, L. (2005). *Managing Epilepsy with Women in Mind*. CRC Press.
Bhatia, K., Chaudhuri, K.R., and Stamelou, M. (2017). *International Review of Neurobiology — Parkinson's Disease*, Volume 132. Academic Press.
Brain, S.D., and Geppetti, P. (2019). *Calcitonin Gene-Related Peptide (CGRP) Mechanisms* (Handbook of Experimental Pharmacology, 255). Springer.
Devinsky, O. (2007). *Epilepsy: A Patient and Family Guide*, 3rd Edition. Demos Health.
Griffiths, R. (2019). *Parkinson's Disease: An In-Depth Metabolic Guide*. Griffiths.
Grossinger, R., Podoll, K., and Dahlem, M. (2006). *Migraine Auras: When the Visual World Fails*. North Atlantic Books.
Khan, M.O.F., and Philip, A.E. (2020). *Medicinal Chemistry of Drugs Affecting the Nervous System (for Pharmacy Students)*, Book 2. Bentham Books.
Lee, R.S. (2019). *Headaches: Amazing Natural remedies to Alleviate Migraines, Cluster, Sinus, Tension, and Rebound Headaches*. Maria Fernanda Moguel Cruz.

Lindop, F., and Skelly, R. (2022). *Parkinson's Disease: A Multidisciplinary Guide to Management*. Elsevier.

Maassen van den Brink, A., and MacGregor, E.A. (2019). *Gender and Migraine* (Headache Series). Springer/European Headache Federation.

Obeso, J.A., Horowski, R., and Marsden, C.D. (2013). Continuous dopaminergic stimulation in Parkinson's disease. *Journal of Neural Transmission — Supplementum* 27.

Okun, M.S. (2015). *10 Breakthrough Therapies for Parkinson's Disease*. Books4Patients.

Patsalos, P.N., and St. Louis, E.K. (2018). *The Epilepsy Prescriber's Guide to Antiepileptic Drugs*, 3rd Edition. Cambridge University Press.

Tiberi, M. (2015). *Dopamine Receptor Technologies* (Neuromethods, 96). Humana Press.

Vine, J.M. (2017). *A Parkinson's Primer: An Indispensable Guide to Parkinson's Disease for Patients and Their Families*. Paul Dry Books.

Weiner, W.J., Shulman, L.M., and Lang, A.E. (2013). *Parkinson's Disease: A Complete Guide for Patients and Families* (Press Health Book), 3rd Edition. Johns Hopkins University Press.

10 Drugs Affecting the Autonomic Nervous System

LEARNING OUTCOMES

After studying the chapter, readers should be able to

1. Describe the classifications of catecholamines.
2. Summarize the clinical signs of alpha-2 adrenergic agonists.
3. Explain the clinical indications of beta-1 adrenergic agonists.
4. Review the classifications of adrenergic antagonists.
5. Describe the mechanisms of action of the alpha-adrenergic antagonists.
6. Discuss contraindications for the use of alpha-receptor antagonists.
7. Explain the various types of cholinergic agonists.
8. Explain the pharmacotherapy of myasthenia gravis.
9. Describe the contraindications of atropine.
10. Identify the expression of a ganglionic blocker.

OVERVIEW

The autonomic nervous system (ANS) is a component of the peripheral nervous system, playing a crucial role in maintaining homeostasis. Adrenergic agonists stimulate adrenergic receptors, simulating certain aspects of sympathomimetics. They are categorized into catecholamines and noncatecholamines. Adrenergic compounds inhibit the action of epinephrine, norepinephrine (NE), and other catecholamines, which control autonomic outflow and some functions of the central nervous system (CNS) at the adrenergic receptors or inhibit their release. Adrenergic antagonists can act directly or indirectly and are also referred to as sympatholytics. There are five adrenergic receptors divided into two groups: the first group consists of the beta (β) adrenergic receptors, which include β_1, β_2, and β_3 receptors. The second group comprises the alpha (α) adrenoreceptors with α_1 and α_2 receptors. Most symptoms produced by adrenergic antagonists involve parasympathetic activation. Cholinergic agonists are a group of medicines that mimic the actions of acetylcholine (ACh), acting directly on the ACh receptor. Some drugs are specific to muscarinic receptors, while others are typical for nicotinic receptors.

AUTONOMIC NERVOUS SYSTEM

The ANS is a component of the peripheral nervous system. It plays a crucial role in maintaining **homeostasis** and significantly influences the stability of the human body's internal environment. The ANS consists of motor neurons that innervate smooth muscle, cardiac muscle, and glands. Signals continually move from the visceral organs into the CNS, which uses these signals to support the body's needs. Blood is directed to areas that require more; the heart rate is increased or decreased, blood pressure (BP) is adjusted, body temperature is modified, and stomach secretions are regulated. The actions of the ANS are primarily unnoticed. The term "autonomic" means that this subdivision of the peripheral nervous system is somewhat independent. Also known as the involuntary nervous system or general visceral motor system, it has two branches: the sympathetic and parasympathetic divisions, which have opposing effects.

The ANS is divided into cholinergic and adrenergic divisions. Axons of preganglionic and postganglionic parasympathetic neurons release ACh, while axons of preganglionic sympathetic neurons also release ACh. In contrast, only the axons of postganglionic sympathetic neurons release NE. The effects of localized cholinergic discharge are discrete and short-lasting due to high cholinesterase concentrations at the cholinergic nerve endings.

The upper part of the brain coordinates and regulates somatic and autonomic motor activities. Skeletal muscles are active and work hard, requiring additional oxygen and glucose. Autonomic control mechanisms increase the heart rate and dilate the airways to meet these needs and maintain homeostasis.

ADRENERGIC AGONISTS

The **adrenergic agonists** are categorized as **catecholamines** and noncatecholamines. The catecholamines include NE, epinephrine, and dopamine. NE is the primary neurotransmitter of the sympathetic nervous system, while epinephrine serves as the primary hormone of the adrenal medulla. Dopamine acts as the biochemical precursor to NE in the CNS. Noncatecholamines exhibit somewhat similar actions to catecholamines but are more selective for receptor sites. The medications in the adrenergic agonist class mimic the sympathetic nervous system and are also called **sympathomimetics**.

CLASSIFICATIONS OF ADRENERGIC AGONISTS

Adrenergic agonists are classified as *catecholamines* or *noncatecholamines*. The *noncatecholamines* include ephedrine, phenylephrine, and terbutaline. There are two principal types of adrenoreceptors: alpha (αα) and beta (ββ). These receptors have been further subdivided into the main subtypes of α_1, α_2, β_1, β_2, and β_3. The activation of alpha

DOI: 10.1201/9781003461913-12

TABLE 10.1

Classifications of the Generic and Trade Names of Adrenergic Agonists

Generic Name	Trade Name
Dopamine	Intropin
Epinephrine	EpiPen, Adrenalin
Norepinephrine	Levophed
Metaraminol	Aramine
Midodrine	ProAmatine
Oxymetazoline	Afrin
Phenylephrine	Nasop
Apraclonidine	Iopidine Eye
Clonidine	Catapres
Dexmedetomidine	Precedex
Lofexidine	Lucemyra
Methyldopa	generic only
Tizanidine	Zanaflex
Dobutamine	Dobutrex
Isoproterenol	Isuprel
Albuterol	Ventolin
Arformoterol	Brovana
Levalbuterol	Xopenex
Metaproterenol	Alupent
Olodaterol	Striverdi Respimat
Salmeterol	Serevent Diskus
Amphetamine	Evekeo
Ephedrine	generic only
Methylphenidate	Ritalin
Pseudoephedrine	Sudafed

---

and beta receptors through the administration of adrenergic agonists produces effects consistent with sympathetic (fight-or-flight) stimulation. *Nonselective adrenergic agonists* activate both alpha and beta receptors. The generic and trade names of adrenergic agonists are listed in Table 10.1.

ALPHA-1 ADRENERGIC AGONISTS

Alpha-adrenergic receptors play an essential role in regulating BP. Alpha-1 receptors are the classic postsynaptic alpha receptors on vascular smooth muscle. They determine both arteriolar resistance and venous capacitance, thereby affecting BP. Additionally, they are found on the muscles of the iris, the smooth muscle of the gastrointestinal (GI) tract, and in the male and female reproductive tracts and sphincters of the urinary bladder. Stimulation of the alpha-1 receptors induces vasoconstriction of blood vessels, **mydriasis**, decreased GI motility, contraction of the external sphincter of the bladder, reduced bile secretion, and stimulation of sweat glands. Examples of alpha-1 adrenergic agonists include *midodrine, metaraminol,* and *phenylephrine.*

Clinical Indications

Alpha-1 adrenergic agonists are utilized in the management of various disorders, including vasodilatory shock,

orthostatic hypotension, hypoperfusion, **septic shock**, cardiopulmonary arrest, heart failure, nasal congestion, and subconjunctival hemorrhage (red eye). *Midodrine* is particularly indicated for symptomatic orthostatic hypotension. *Metaraminol* is frequently used to treat and prevent hypotension resulting from bleeding, shock due to brain damage, and spinal anesthesia. *Phenylephrine* injections are indicated for hypotension caused by shock or anesthesia. An ophthalmic formulation is designed to dilate the pupils and induce vasoconstriction, an intranasal formulation addresses congestion, and a topical formulation is used for hemorrhoids. The drug also has off-label use in the treatment of priapism.

Mechanism of Action

The mechanism of action of *midodrine* involves its conversion into the active metabolite *desglymidodrine*, which exerts its effects by activating the alpha-adrenergic receptors in the arteriolar and venous vasculature. This process increases vascular tone and elevates BP. *Metaraminol* acts through peripheral vasoconstriction, leading to an increase in both systolic and diastolic BP. Additionally, the drug indirectly promotes the release of NE from its storage sites. *Phenylephrine* mediates vasoconstriction and mydriasis depending on its route and administration site. Systemic exposure raises systolic and diastolic BP along with peripheral vascular resistance. The resultant increase in BP stimulates the **vagus nerve**, triggering reflex bradycardia.

Adverse Effects

The adverse effects of alpha-1 adrenergic agonists include hypertension, urinary retention, blurred vision, constipation, and sweating. *Midodrine* may cause chills, numbness, tingling, scalp itching, and skin rashes. *Metaraminol* can lead to headaches, nausea, vomiting, anxiety, dizziness, or nervousness. *Phenylephrine* can cause insomnia, nausea, lightheadedness, and tremors.

Contraindications

Alpha-1 adrenergic agonists should be avoided in cases of severe coronary or cardiovascular disease, **glaucoma**, hypovolemia, urinary retention, **pheochromocytoma**, sensitivity to adrenergic substances, **Graves' disease**, and tachycardia in pregnant or lactating women or children.

ALPHA-2 ADRENERGIC AGONISTS

Alpha-2 receptors are located on presynaptic neurons and are found both in the brain and the periphery. In the brain stem, they modulate sympathetic outflow. Their function in the periphery is not yet fully understood, but they may contribute to controlling sympathetic tone and local and regional blood flow. The centrally acting agonists of this group stimulate receptors in the CNS to decrease sympathetic nervous system activity and lower BP and heart rate. Examples of these agonists include *clonidine, dexmedetomidine, lofexidine,* and *tizanidine.*

Clinical Indications

Alpha-2 adrenergic agonists are indicated for treating hypertension, either alone or in combination with diuretic medications. They are also utilized epidurally as adjunct therapy for severe pain. *Clonidine* is additionally employed to manage menopausal flushing, attention-deficit/hyperactivity disorder, opioid, alcohol, and nicotine withdrawal, as well as specific pain conditions. *Dexmedetomidine* is frequently used in intensive care settings to achieve light to moderate sedation and demonstrates analgesic properties.

Since it is opioid-sparing, it does not cause significant respiratory depression, making it suitable for use in mechanically ventilated adults. This drug is also indicated for **colonoscopy** as an adjunct to other sedatives to reduce their dosages in the subarachnoid or epidural space to produce anesthesia for leg surgeries and to treat the adverse cardiovascular effects of acute amphetamine or cocaine intoxication and overdose. *Lofexidine* is effective in treating opioid withdrawal over 14 days and for women experiencing postmenopausal hot flashes. *Tizanidine* is indicated for muscle spasticity resulting from spinal cord injury or multiple sclerosis.

Mechanism of Action

Alpha-2 adrenergic agonists stimulate alpha-2 adrenergic receptors in the CNS, inhibiting sympathetic vasomotor centers. They reduce plasma concentrations of NE, decrease systolic BP and heart rate, and inhibit **renin** release from the kidneys. *Clonidine* crosses the blood–brain barrier, stimulating alpha-2 receptors in the **brainstem** to decrease peripheral vascular resistance and lower BP. It is specific to presynaptic alpha-2 receptors in the vasomotor center. *Dexmedetomidine* acts similarly to clonidine. *Lofexidine* binds to the same receptors, reducing NE release and decreasing **sympathetic tone**. *Tizanidine* has an unclear mechanism of action.

Adverse Effects

Adverse effects of alpha-2 adrenergic agonists include hypotension, peripheral edema, rash, pruritus, dry mouth or eyes, drowsiness, dizziness, constipation, **impotence**, nausea, vomiting, hepatitis, hallucinations, and depression. *Clonidine* can cause fatigue and irritability. *Dexmedetomidine* may induce hypertension when administered via rapid IV or bolus. *Lofexidine* might cause insomnia and bradycardia. *Tizanidine* can lead to hepatitis, hyperlipidemia, hepatic failure, **exfoliative dermatitis**, bradycardia, angioedema, and anaphylactoid reactions. Moderate adverse effects include blurred vision, depression, hallucinations, cystitis, hypotension, constipation, excitability, anemia, **leukopenia**, **thrombocytopenia**, and jaundice.

Contraindications

Alpha-2 adrenergic agonists should be avoided in children, as well as in pregnant and lactating patients, those with hypertension associated with **preeclampsia**, and individuals with hepatitis, cirrhosis, **polyarteritis nodosa**, pheochromocytoma, **scleroderma**, or **blood dyscrasia**. *Clonidine* is contraindicated in cases of depression, dry eye, myocardial infarction, and stroke. *Dexmedetomidine* has no absolute contraindications but should be used cautiously in higher doses if bradycardia is present. *Lofexidine* is not recommended for patients taking beta-blockers or angiotensin-converting enzyme inhibitors due to decreased BP or heart rate risks. *Tizanidine* should be avoided in patients with known hypersensitivity. It must be used with caution in individuals with hypotension, renal impairment or failure, hepatic disease, when driving or operating machinery, psychosis, during pregnancy, breastfeeding, and in older patients. All of these medications should never be discontinued abruptly.

Check Your Knowledge

1. What are the clinical indications of alpha-1 adrenergic agonists?
2. What are the adverse effects of alpha-1 adrenergic agonists?
3. What are the clinical indications of alpha-2 adrenergic agonists?
4. What are the contraindications of alpha-2 adrenergic agonists?

BETA-1 ADRENERGIC AGONISTS

Beta-1 receptors are found in the myocardium, smooth muscle of the GI tract, sphincters, and renal arterioles. Two primary drugs in this category are *dobutamine*, a selective beta-1 adrenergic agonist, and *isoproterenol*, a nonselective beta-1 adrenergic agonist. The beta-1 subtype is primarily located in the heart, kidneys, and **adipocytes**. Stimulation of the beta-1 subtype leads to increased heart rate and cardiac contractility. Beta-3 adrenoreceptors are situated in the heart, uterus, bladder, and adipocytes, where they regulate the inhibition of cardiac contractility, bladder relaxation, uterine function, GI smooth muscle function, and lipolysis.

Clinical Indications

Beta-1 adrenergic agonists are used to treat circulatory shock, hypotension, and cardiac arrest. *Dobutamine* is indicated when parenteral therapy is necessary for **inotropic** support in the short-term management of **cardiac decompensation** due to depressed contractility, which may stem from organic heart disease or cardiac surgeries. *Isoproterenol* is utilized to address heart block and episodes of Adams–Stokes syndrome not associated with ventricular tachycardia or fibrillation. It is also employed in emergencies for cardiac arrest until electric shock can be administered, for bronchospasm during anesthesia, and as an adjunct treatment for hypovolemic, septic, or cardiogenic shock, low cardiac output, and congestive heart failure. Additionally, the drug can enhance the condition of intestinal stem cells through beta-2 adrenoreceptors following chemotherapy.

Mechanism of Action

Beta-1 adrenergic agonists increase the heart's contraction rate and strength, promote lipolysis in adipose tissue, reduce digestion and GI motility, and enhance glomerular filtration. *Dobutamine* is a synthetic catecholamine that targets beta-1 receptors to boost myocardial contractility and stroke volume, thereby increasing cardiac output. It also affects beta-2 and alpha-1 receptors. *Isoproterenol* is a beta-1 adrenergic agonist with additional beta-2 activity.

Adverse Effects

The common adverse effects of beta-1 adrenergic agonists include hypertension, tachycardia, and constipation. *Dobutamine* may also cause **ectopic heartbeat** and phlebitis. *Isoproterenol* can lead to nervousness, headaches, dizziness, nausea, palpitations, angina, pulmonary edema, ventricular arrhythmias, tachyarrhythmias, difficulty breathing, sweating, mild tremors, weakness, flushing, and pallor. Additionally, isoproterenol has been associated with insulin resistance, which can result in diabetic **ketoacidosis**.

Contraindications

Dobutamine is contraindicated in patients with idiopathic hypertrophic subaortic stenosis and in those who have previously shown signs of hypersensitivity to the drug. *Isoproterenol* is contraindicated in patients with tachyarrhythmias, heart block due to an overdose of digoxin, ventricular arrhythmias requiring inotropic therapy, or in patients with angina.

BETA-2 ADRENERGIC AGONISTS

Beta-2 adrenoreceptors are found in the smooth muscle of the bronchioles, vascular smooth muscle, heart, uterus, bladder, adipocytes, eyes, and the blood vessels supplying the brain.

Clinical Indications

Beta-2 adrenergic agonists are utilized in patients with chronic obstructive pulmonary disease (COPD), premature labor, circulatory shock, and peripheral vascular disease. *Albuterol* is specifically indicated for the relief and prevention of bronchospasm, as well as for acute prophylaxis against triggers that induce bronchospasm. *Salmeterol* is recommended for moderate-to-severe persistent asthma following prior treatment with a short-acting beta-agonist in combination with inhaled corticosteroids. This drug is also utilized for COPD, with or without corticosteroids, and can serve as monotherapy for exercise-induced bronchospasm. *Indacaterol* is indicated for treating COPD.

Mechanism of Action

Stimulation of beta-2 receptors results in bronchodilation, vasodilation of blood vessels, and relaxation of the uterus during pregnancy. *Albuterol* activates the receptors, leading to the activation of adenyl cyclase and an increase in intracellular concentrations of cAMP. *Salmeterol* stimulates the beta-2 receptors in the bronchial musculature, causing them to relax. This medication is 10,000 times more lipid-soluble than albuterol. Additionally, salmeterol dissolves in the lipid bilayer of the cell membrane. *Indacaterol* activates its receptors to relax the bronchial airways.

Adverse Effects

The adverse effects of beta-2 adrenergic agonists include **muscle stiffness**, tremors, and **hyperthermia**. These agents can also cause hyperglycemia. *Albuterol may* lead to headaches, dry mouth, palpitations, tachycardia, flushing, anxiety, and insomnia. *Salmeterol* might cause migraine headaches, dizziness, and hypokalemia. *Indacaterol* can result in headaches, dizziness, nausea, itching, rash, palpitations, tachycardia, upper respiratory tract infections, and joint pain.

Contraindications

Beta-2 adrenergic agonists are contraindicated in individuals who are sensitive to this agent, as well as those with diabetes mellitus, cardiac arrhythmias, coronary artery disease, hypertension, digoxin intoxication, **hyperthyroidism**, preeclampsia, **eclampsia**, intrauterine infection, hypovolemia, and angle-closure glaucoma. *Albuterol* should be avoided in patients with arrhythmias. *Salmeterol* is contraindicated as a *monotherapy* for asthma and as a primary treatment for **status asthmaticus** or COPD. *Indacaterol* is contraindicated in patients with known hypersensitivity.

Check Your Knowledge

1. What are the clinical indications of beta-1 adrenergic agonists?
2. What are the adverse effects of beta-2 adrenergic agonists?
3. What are the contraindications of beta-2 adrenergic agonists?

BETA-3 ADRENERGIC AGONISTS

The beta-3 adrenergic receptor is the least studied isotype of the beta-adrenergic subfamily. Beta-3 adrenergic agonists bind selectively to the beta-3 adrenergic receptors. New aspects of β_3 adrenergic receptors have recently been described, involving them in urine concentrating mechanisms, fat mass reduction, and inflammatory processes, thus opening the way for new potential therapeutic applications. *Mirabegron (Myrbetriq)* and *vibegron (Gemtesa)* were approved for the treatment of overactive bladder.

Clinical Indications

Mirabegron is indicated for the treatment of overactive bladder. *Vibegron* is indicated for the treatment of overactive bladder with symptoms of urge urinary incontinence, urgency, and **polyuria** in adults.

Mechanism of Action

Mirabegron affects the **detrusor muscles** of the urinary bladder, decreasing their contraction so that it can store more urine at a given time. The drug also influences non-voiding contractions by reducing their frequency. *Vibegron* causes a conformational change in its target receptors, inducing activation of adenylate cyclase via G proteins, which promotes cAMP formation. *Solabegron* works similarly to mirabegron and vibegron.

Adverse Effects

Mirabegron commonly causes bladder pain, bloody or cloudy urine, problems with urination, dizziness, urgency, blurred vision, headaches, back pain, nervousness, and heart irregularities. *Vibegron* may cause dry mouth, headaches, bronchitis, constipation, diarrhea, and nausea. This agent can also cause upper respiratory and urinary tract infections.

Contraindications

Mirabegron and *vibegron* should be avoided with known hypersensitivity, severe uncontrolled hypertension, and pregnancy. They are also contraindicated in patients with severe liver disease, chronic kidney disease, and urinary bladder blockage. The contraindications of *solabegron* are unknown.

ALPHA-ADRENERGIC ANTAGONISTS

Alpha-adrenergic antagonists are a type of BP medicine that lowers BP by keeping NE from constricting the smooth muscles in the walls of arteries and veins.

CLINICAL INDICATIONS

Alpha-receptor antagonists are utilized to manage hypertension and to treat patients with peripheral vascular disease, pheochromocytoma, and urinary retention. *Doxazosin* is prescribed for hypertension, **benign prostatic hyperplasia**, and perioperative management of pheochromocytoma. *Terazosin* is commonly used to alleviate symptoms of an enlarged prostate. *Phentolamine* is primarily employed to address hypertensive emergencies caused by pheochromocytoma, as well as for cocaine-induced cardiovascular complications in patients who do not respond fully to benzodiazepines, nitroglycerin, and calcium channel blockers. *Phenoxybenzamine* can be used to manage hypertension resulting from pheochromocytoma and benign prostatic hyperplasia.

MECHANISM OF ACTION

Alpha-receptor antagonists selectively inhibit the actions of alpha adrenoreceptors, leading to vasodilation in arterioles and a reduction in BP. Dilating veins indirectly decreases BP by lowering venous return to the heart, reducing cardiac output, and systemic BP.

ADVERSE EFFECTS

The adverse effects of alpha-receptor antagonists include postural hypotension, dizziness, nasal congestion, **delayed orgasm**, and fatigue. Patients should avoid driving for at least 12 hours after taking the first dose. Female patients should not breastfeed while taking these drugs or use OTC medications for coughs, colds, or allergies. *Doxazosin* may cause **dyspnea**, nausea, abdominal pain, and inflammation. Severe adverse effects include arrhythmias and priapism. *Prazosin* and *terazosin* can cause headaches and palpitations, while *phentolamine* and *phenoxybenzamine* lead to orthostatic hypotension.

CONTRAINDICATIONS

Alpha-receptor antagonists are contraindicated in patients known to be hypersensitive to these drugs. *Doxazosin* should be avoided in cases of orthostatic hypotension, liver disease, and cataract surgery. *Prazosin* is contraindicated for those with hypotension. *Terazosin* is not recommended for patients with priapism. *Phentolamine* is contraindicated in patients with myocardial infarction, coronary insufficiency, and angina. *Phenoxybenzamine* is also contraindicated in those with coronary artery disease, hypotension, pneumonia, and decreased kidney function.

BETA-ADRENERGIC ANTAGONISTS

Beta-blocking antagonists inhibit the effects of epinephrine (adrenaline). They induce bradycardia with less force and lead to hypotension. Beta-blockers also assist in dilating veins and arteries to enhance blood flow.

CLINICAL INDICATIONS

Beta-blocking antagonists are indicated to treat hypertension, MI, tachyarrhythmias, and angina pectoris. Some examples of beta-blocking antagonists include *metoprolol, propranolol, carvedilol, nebivolol, bisoprolol,* and *atenolol*.

MECHANISM OF ACTION

Beta-blocking antagonists primarily affect cardiac muscle by competitively blocking beta-adrenergic receptors in the heart. *Metoprolol* blocks beta-1 adrenergic receptors in heart muscle cells, which decreases the action potential. This reduction lowers sodium ion uptake, prolongs repolarization, and slows the release of potassium ions. *Propranolol* exerts its effects by competitively blocking beta-adrenergic stimulation in the heart, typically induced by epinephrine and NE.

Carvedilol is a nonselective adrenergic blocker with alpha-1 adrenergic receptor antagonist properties. It has a lesser impact on heart rate than purely selective beta-blockers. *Nebivolol* is a highly selective beta-1 adrenergic receptor antagonist with weak beta-2 adrenergic receptor antagonist activity. Its effects lead to decreased resting heart rate, exercise heart rate, and myocardial contractility.

Bisoprolol is effective through antagonism of beta-1 adrenoreceptors, resulting in lower cardiac output. *Atenolol* is a cardioselective beta-blocker that binds to the beta-1 adrenergic receptors. *Sotalol* inhibits beta-1 adrenoreceptors in the myocardium and rapid potassium channels to slow repolarization, lengthen the QT interval, and slow and shorten the conduction of action potentials via the atria.

ADVERSE EFFECTS

The adverse effects of beta-receptor antagonists include insomnia, dizziness, fatigue, and diarrhea.

CONTRAINDICATIONS

Beta-blockers should be avoided in patients with known hypersensitivity, heart block, heart failure, and cardiogenic shock. These agents are contraindicated in individuals with a history of asthma or COPD unless no alternative is available. Breastfeeding should be avoided while taking beta-blockers.

CHOLINERGIC AGONISTS

Cholinergic agonists are also known as *parasympathomimetics* or *cholinomimetics*. They stimulate the parasympathetic nervous system by mimicking the actions of ACh. ACh is found at the ganglia and the terminal nerve endings of the parasympathetic system. Direct-acting cholinergic agonists include pilocarpine, carbachol, cevimeline, and bethanechol. Indirect-acting cholinergic agonists encompass neostigmine, physostigmine, donepezil, pyridostigmine, galantamine, and rivastigmine.

CLASSIFICATIONS OF CHOLINERGIC AGONISTS

The two types of cholinergic receptors include **muscarinic receptors** (which innervate smooth muscle and slow the heart rate) and **nicotinic receptors** (which affect the skeletal muscles). Many cholinergic drugs are nonselective and can impact both the muscarinic and nicotinic receptors. The classifications of the generic and trade names of cholinergic agonists are summarized in Table 10.2.

Check Your Knowledge

1. What are the clinical indications of beta-blocking antagonists?
2. What are the adverse effects of beta-blocking antagonists?
3. What are the other names of cholinergic
4. What are the trade names of carbachol and neostigmine?

TABLE 10.2

Classifications of Generic and Trade Names of Cholinergic Agonists

Generic Name	Trade Name
Bethanechol	Urecholine
Carbachol	Miostat
Cevimeline	Evoxac
Pilocarpine	Salagen, Isopto-Carpine
Donepezil	Aricept
Galantamine	Razadyne
Neostigmine	Bloxiverz
Physostigmine	Antilirium
Pyridostigmine	Mestinon
Rivastigmine	Exelon
Aclidinium	Tudorza Pressair
Atropine	AtroPen
Benztropine	Cogentin
Cyclopentolate	Cyclogyl
Darifenacin	Enablex
Dicyclomine	Bentyl
Fesoterodine	Toviaz
Hyoscyamine	Anaspaz
Ipratropium	Atrovent
Cantil	Gastrointestinal ulcers
Oxybutynin	Ditropan
Scopolamine	Transderm-Scop

MUSCARINIC AGONISTS

The muscarinic agonists directly activate cholinergic receptors at the neuroeffector junctions of the parasympathetic nervous system. ACh has little medicinal value since acetylcholinesterase (AChE) quickly destroys it in the synapses and has many adverse effects. Muscarinic agonists are used as medications that are not easily destroyed by AChE. They have a longer duration of action compared to ACh. These agonists exhibit poor absorption across the GI tract and generally do not cross the blood–brain barrier. Additionally, they only slightly affect ACh nicotinic receptors in the **ganglia**. Few muscarinic agonists are widely used in pharmacotherapy due to their potentially serious adverse effects.

Nearly all widespread effects of these agonists result from parasympathetic activation. These drugs can increase the tone of smooth muscles and contractions in the GI tract. When administered before meals, the enhanced peristalsis they induce allows for quicker stomach emptying. These agonists are never prescribed for individuals with suspected GI tract obstructions, as the heightened peristalsis could potentially injure or rupture the mucosa. Additionally, these agents stimulate smooth muscles in the urinary tract to enhance ureteral peristalsis and accelerate bladder emptying. Consequently, they are highly effective for treating urinary retention. However, they are contraindicated in cases of obstructive uropathy, since increased contractions of the smooth muscle in the ureter and bladder could exacerbate pain and bleeding.

These agonists stimulate most exocrine glands, increasing digestive, lacrimal, salivary, and sweat secretions. Therefore, they are used for **xerostomia**, such as in **Sjögren's syndrome**. In the eyes, muscarinic agonists cause the iris muscles to contract, resulting in **miosis**; however, this is not a therapeutic goal and is thus considered an adverse effect. Even so, these agonists contract the ciliary muscle, which blurs vision but allows fluid to drain from the eye's anterior chamber, reducing intraocular pressure. Therefore, topical agonists effectively treat glaucoma, though safer alternatives are available. Muscarinic agonists also contract bronchial smooth muscle, so they are not used in patients with a history of asthma.

When muscarinic receptors are activated in the cardiovascular system, the heart rate slows and BP decreases. **Baroreceptors** in the carotid arteries and aortic arch signal the vasomotor center of the medulla to increase heart rate, leading to **reflex tachycardia**. This can also occur with any other drug that causes a decrease in BP. To prevent bradycardia, the oral or parenteral administration of cholinergic agonists must be closely monitored. BP is monitored to avoid hypotension, especially in patients with preexisting cardiovascular disease. If a patient has hyperthyroidism, atrial fibrillation is a possible outcome. Cholinergic agonists are not used if severe heart disease is present. The central muscarinic agonists used in medicine include *bethanechol, carbachol, cevimeline,* and *pilocarpine.*

CLINICAL INDICATIONS

Muscarinic agonists are parasympathomimetic drugs indicated for glaucoma, urinary retention, ileus, and Alzheimer's disease (AD). *Bethanechol* is utilized for nonobstructive urinary retention with bladder atony. *Carbachol* is **administered ocularly to induce miosis and lower intraocular pressure in the treatment of glaucoma.** *Cevimeline* is prescribed for xerostomia due to **Sjögren's syndrome**. When taken orally, it enhances saliva flow to alleviate dry mouth. *Pilocarpine* **ophthalmic** is indicated for glaucoma.

MECHANISM OF ACTION

Muscarinic agonists mimic the action of ACh on muscarinic receptors, leading to cardiac slowing and contraction of smooth muscles in the intestinal tract, bronchioles, detrusor muscle, urethra, and iris muscle. Consequently, they increase the secretion of salivary, gastric acid, and airway mucosal glands.

ADVERSE EFFECTS

Adverse effects of muscarinic agonists include sweating, flushing, blurred vision, miosis, **sialorrhea**, nausea, vomiting, and abdominal cramping. Severe adverse effects can consist of bradycardia, orthostatic hypotension, **syncope**, tachycardia, and asthma.

CONTRAINDICATIONS

Muscarinic agonists are contraindicated in cases of bowel obstruction, active ulcers, or inflammatory bowel disease. They are also contraindicated in urinary obstruction, recent bladder surgery, asthma, or COPD. *Additionally, m*uscarinic agonists are contraindicated in the presence of recent MI, bradycardia, hyperthyroidism, Parkinson's disease, epilepsy, and peritonitis.

YOU SHOULD REMEMBER

Muscarinic agonists imitate the effects of ACh on muscarinic receptors and are used to treat conditions such as ileus, urinary retention, AD, and glaucoma.

Check Your Knowledge

1. What are the functions of the muscarinic agonists?
2. What are the adverse effects of muscarinic agonists?
3. What are the contraindications of the muscarinic agonists?

AChE INHIBITORS

AChE inhibitors are indirect-acting medications that inhibit AChE, the enzyme responsible for breaking down ACh. Consequently, ACh can remain and accumulate in the synaptic cleft longer. As a result, irreversible AChE inhibitors, such as nerve gas and certain insecticides, have limited medical uses. Examples of AChE inhibitors include donepezil, neostigmine, pyridostigmine, and rivastigmine.

Clinical Indications

Reversible AChE inhibitors have been used in the diagnosis and treatment of various diseases, including myasthenia gravis, AD, postoperative ileus, bladder distention, glaucoma, and as an **antidote** for anticholinergic overdose.

Mechanism of Action

AChE inhibitors bind to and reversibly inactivate cholinesterases, inhibiting ACh hydrolysis, thereby increasing ACh concentrations at cholinergic synapses. They also act as potent agonists of the alpha-1 receptor. Some agents bind to AChE and have short durations of action; this action is *reversible by AChE inhibitors.*

Adverse Effects

The adverse effects of AChE inhibitors include involuntary muscle contractions, nausea, vomiting, increased salivation, and miosis. The serious adverse effects of AChE inhibitors encompass cardiac abnormalities such as bradycardia, atrioventricular block, torsades de pointes, hypotension, dyspnea, seizures, cholinergic crisis, bronchospasm, and respiratory muscle paralysis. They may also lead to diarrhea, loss of appetite, insomnia, and unusual tiredness or weakness.

Contraindications

AChE inhibitors must be avoided in cases of known hypersensitivity and used cautiously during anesthesia, in individuals with coronary artery disease, peptic ulcer disease, and GI bleeding. They are also contraindicated for patients with seizure disorders, asthma, COPD, urinary tract obstruction, and diabetes mellitus.

YOU SHOULD REMEMBER

AD is linked to the loss of cholinergic neurons in the brain and a reduction in ACh levels. The primary therapeutic focus in AD treatment strategies is the inhibition of brain AChE.

NICOTINIC AGONISTS

Nicotinic agonists are drugs that mimic the action of ACh at nicotinic **ACh receptors**. They can produce widespread, nonselective effects on the ANS. Therefore, the only widely used drug that activates ganglionic receptors is nicotine. The respiratory mucosa, GI tract, and skin absorb these drugs. Nicotine stimulates parasympathetic and sympathetic responses since it acts at the ganglia. As nicotine is absorbed, the CNS becomes more alert, and heart rate and BP increase. They may decrease soon due to the baroreceptor reflex.

Nicotine is used as *nicotine replacement therapy* to help individuals cease using tobacco products. It is available in forms such as chewing gum (Nicorette), transdermal patches (Habitrol), and nasal spray (Nicotrol). These methods alleviate unpleasant withdrawal symptoms, including headaches, lack of concentration, insomnia, depression, and food cravings. Nicotine is gradually reduced over about 8–12 weeks. Regular use of nicotine replacement therapy can nearly double a patient's chances of quitting tobacco products, but considerable motivation and ongoing follow-up are necessary.

CHOLINERGIC ANTAGONISTS

Cholinergic antagonists block the effects of ACh. There are two types of cholinergic antagonists: muscarinic antagonists and nicotinic antagonists. The muscarinic antagonists have more clinical indications than the nicotinic antagonists. They are also summarized in Table 10.3.

MUSCARINIC ANTAGONISTS

Muscarinic antagonists have been used for centuries to treat various conditions. A primary source was the *Atropa belladonna* plant, employed initially to dilate the eyes or

---

TABLE 10.3

Classifications of the Generic and Trade Names of Muscarinic Antagonists

Generic Name	Trade Name
Aclidinium	Tudorza Pressair
Atropine	AtroPen
Benztropine	Cogentin
Cyclopentolate	Cyclogyl
Darifenacin	Enablex
Dicyclomine	Bentyl
Fesoterodine	Toviaz
Hyoscyamine	Anaspaz
Ipratropium	Atrovent
Cantil	Gastrointestinal ulcers
Oxybutynin	Ditropan
Scopolamine	Transderm-Scop

---

redden the cheeks as an early form of female "beauty" techniques. The plant is shown in Figure 10.1. However, it is highly toxic and has been associated with suicides. With the development of newer, safer medications, the use of muscarinic antagonists has declined. They are now primarily prescribed when a safer and preferred drug is contraindicated or poorly tolerated by the patient.

Clinical Indications

Muscarinic antagonists are used to treat COPD, exacerbations of asthma, peptic ulcers, irritable bowel syndrome, pupil dilation during ophthalmic examinations, bradycardia, the reduction of salivary and respiratory secretions during anesthesia, and to reverse poisoning from muscarinic mushrooms or organophosphates.

Mechanisms of Action

The mechanisms of action of muscarinic antagonists are exemplified by the prototype drug, atropine, which blocks muscarinic receptors. Atropine increases the heart rate to treat bradycardia. Muscarinic antagonists can block high doses of nicotinic receptors in skeletal muscle and ganglia.

Adverse Effects

The adverse effects of muscarinic antagonists include dry eyes, xerostomia, urinary retention, hyperthermia, photophobia, and increased intraocular pressure. Other adverse effects are blurred vision, mydriasis, headaches, coughing, constipation, insomnia, pharyngitis, sinusitis, and allergic conjunctivitis. Severe adverse effects include worsening of glaucoma, dysrhythmias, and paralytic ileus. *Atropine* can lead to **ventricular fibrillation**, delirium, and coma.

Contraindications

Muscarinic antagonists should be avoided in patients with acute angle-closure glaucoma. Atropine must be used carefully in cardiovascular conditions as it increases the heart rate. It should be handled with caution in cases of

FIGURE 10.1 Atropa belladonna plant.

hyperthyroidism, ulcerative colitis, and paralytic ileus. Muscarinic antagonists must be administered cautiously to patients with Down syndrome.

GANGLIONIC BLOCKERS

The nicotinic antagonists, also known as ganglionic blockers, act at the autonomic ganglia. They reduce hypertension and are used in emergencies. Ganglionic blockers interrupt the transmission of nerve impulses at nicotinic receptors in the autonomic ganglia. Since both the sympathetic and parasympathetic nervous systems have ganglia, these blockers are nonselective. The parasympathetic system is more significantly affected because most organs' dominant baseline autonomic tone is parasympathetic. The only therapeutic effect is vasodilation, and these blockers can cause severe hypotension, necessitating the availability of emergency kits containing agents such as epinephrine to reverse extreme hypotension. *Mecamylamine* is a long-acting nicotinic receptor antagonist initially used for severe hypertension. Safer antihypertensive drugs have replaced this drug in clinical use.

Mecamylamine has recently been approved for nicotine dependence as it reduces the brain's psychologically rewarding effects of nicotine. It is also used for **Tourette's syndrome** when that condition is unresponsive to other medications. Common adverse effects include fatigue, weakness, headaches, blurred vision, mydriasis, sedation, decreased libido, impotence, and urinary retention. Serious adverse effects of mecamylamine include orthostatic hypotension, precipitation of angina, adynamic ileus, and choreiform movements. It is contraindicated in known hypersensitivity, coronary insufficiency, glaucoma, pyloric stenosis, and uremia.

NEUROMUSCULAR BLOCKERS

Neuromuscular blockers are essential for many hospitalized patients, but their use is controversial. They are frequently used in surgical situations and rapid sequence **intubation**. Succinylcholine does not enter the CNS or cause loss of consciousness or anesthesia. The patient can still experience pain and is fully aware of their surroundings. It is administered as part of balanced anesthesia. Neuromuscular blockers induce muscle paralysis during surgical procedures that require an extended duration. The generic and trade names of neuromuscular blockers are summarized in Table 10.4.

Clinical Indications

The primary indication for the prototype drug, succinylcholine, as an adjunct to surgical anesthesia, is to achieve complete skeletal muscle relaxation of the abdominal muscles.

TABLE 10.4

Generic and Trade Names for Neuromuscular Blockers

Generic Name	Trade Name
Succinylcholine	Anectine
Mivacurium	(Generic only)
Atracurium	Tracrium
Cisatracurium	Nimbex
Vecuronium	Norcuron
Pancuronium	Pavulon

Neuromuscular blockers are also used to facilitate better insertion of endotracheal tubes and to assist with mechanical ventilation. They are ideal for short procedures such as **electroconvulsive shock** therapy or intubation.

Mechanism of Action

Nondepolarizing neuromuscular blockers are competitive ACh antagonists that directly bind to nicotinic receptors on the postsynaptic membrane. They block the binding of ACh, preventing the motor endplate from depolarizing, which leads to muscle paralysis.

Adverse Effects

The adverse effects of neuromuscular blockers include increased salivation, weakness, rashes, and flushing. Severe adverse effects include anaphylaxis, respiratory depression, prolonged apnea, bradycardia, hypotension, cardiac arrest, **malignant hyperthermia**, and muscle paralysis. In some patients, succinylcholine has induced hyperkalemia.

Contraindications

Succinylcholine is contraindicated in patients with a personal or family history of malignant hyperthermia, during the acute phases of multiple injuries, significant burns, or denervation of skeletal muscle, as well as in those with skeletal muscle myopathies or hypersensitivity to succinylcholine. It should be used cautiously in patients with electrolyte imbalances, kidney disease, fractures, muscle spasms, or heart failure. Careful monitoring for pulmonary or metabolic disorders is essential to avoid respiratory depression and acidosis. It is also contraindicated in patients with narrow-angle glaucoma or subarachnoid hemorrhage. *Mivacurium, rocuronium, vecuronium*, and *pancuronium* are contraindicated in cases of known hypersensitivity.

KEY TERMS

Adipocytes; Adrenergic agonists; Baroreceptors; Blood dyscrasia; Cardiac decompensation; Catecholamines; Colonoscopy; Detrusor muscles; Dyspnea; Eclampsia; Ectopic heartbeat; Exfoliative dermatitis; Ganglia; Glaucoma; Graves' disease; Homeostasis; Hyperthermia; Hyperthyroidism; Impotence; Inotropic; Intubation; Ketoacidosis; Leukopenia; Malignant hyperthermia; Muscarinic receptors; Mydriasis; Nicotinic receptors; Orthostatic hypotension; Peritonitis; Pheochromocytoma; Polyarteritis nodosa; Preeclampsia; Priapism; Prostatic hyperplasia; Reflex tachycardia; Renin; Scleroderma; Septic shock; Sialorrhea; Sjögren's syndrome; Status asthmaticus; Sympathomimetics; Sympathetic tone; Syncope; Thrombocytopenia; Tourette's syndrome; Vagus nerve; Ventricular fibrillation; Xerostomia

CLINICAL CASE STUDIES

Clinical Case Study 1

A 71-year-old man with hypertension and type 2 diabetes is evaluated for his general health. His sister had recently died of a heart attack. He has been smoking for 45 years. His BP was 167/84 mm Hg during the visit. The patient is currently taking clonidine (alpha-2 adrenergic agonist) for his hypertension and describes often having a headache, shortness of breath, facial flushing, and clammy skin.

CRITICAL THINKING QUESTIONS

1. What are the indications of alpha-2 adrenergic agonists?
2. What are the adverse effects of alpha-2 adrenergic agonists?
3. What are the contraindications of alpha-2 adrenergic agonists?

Clinical Case Study 2

A 69-year-old woman was hospitalized after her physician diagnosed her with COPD. The patient has had hypertension for several years and is currently taking amlodipine. She has been a smoker for the past 38 years. Examination revealed her to have a respiratory rate of 32 breaths per minute, without cyanosis. The patient was given nebulized ipratropium, salbutamol, and normal saline.

CRITICAL THINKING QUESTIONS

1. What are the indications of muscarinic antagonists?
2. What are the adverse effects of muscarinic antagonists?
3. What are the contraindications of muscarinic antagonists?

FURTHER READING

Beller, J., and Briggs, J. (2020). *Dementia Overview — 2021 Update — Dementia Types, Symptoms, & Risk Factors. Dementia & Alzheimer's Handbook for Seniors, Caregivers, & Medical Professionals*. Jerry Beller Health Research.

Bonisch, H., Finberg, J.P.M., Fleming, W.W., Graefe, K.H., Langer, S.Z., et al. (2012). *Handbook of Experimental Pharmacology*. Springer.

Chen Chang, T., Ramulu, P., and Hodapp, E. (2016). *Clinical Decisions in Glaucoma*, 2nd Edition. Ta Chen Chang.

Collins, W.J. (2013). *The Effects of Certain Parasympathomimetic Substances on the Emotions of Normal and Psychotic Individuals*. Literary Licensing, LLC.

Cozart, D. (2017). *Alzheimer's Treatment: Acetylcholinesterase Inhibitors*. Cozart.

Dauncey, E.A., and Larsson, S. (2018). *Plants That Kill: A Natural History of the World's Most Poisonous Plants*. Princeton University Press.

Delius, W., Gerlach, E., Grobecker, H., and Kubler, W. (2011). *Catecholamines and the Heart: Recent Advances in Experimental and Clinical Research*. Springer-Verlag.

Dos Santos, G.A.A. (2021). *Pharmacological Treatment of Alzheimer's Disease: Scientific and Clinical Aspects*. Springer.

Eiden, L.E. (2013). *Catecholamine Research in the 21st Century: Abstracts and Graphical Abstracts, 10th International Catecholamine Symposium, 2012*. Academic Press

Elisa, S., Nagelhout, J.J., and Heiner, J.S. (2020). *Current Anesthesia Practice: Evaluation & Certification Review*. Elsevier.

Fung, A. (2010). *Ophthalmic Clinical Examination – The Sydney Eye Hospital Registrars' Manual*. Fung.

Hai, C.M. (2016). *Vascular Smooth Muscle: Structure and Function in Health and Disease*. World Scientific (WSPC).

Juris, E., Carden, E., and Toussaint, C. (2014). *Positive Options for Complex Regional Pain Syndrome (CRPS): Self-Help and Treatment* (Positive Options for Health). Hunter House.

Medifocus.com Inc. (2018). *Medifocus Guidebook on Reflex Sympathetic Dystrophy*. CreateSpace Independent Publishing Platform.

Mena, C., and Jayasuriya, S. (2019). *Peripheral Vascular Disease: A Clinical Approach*. Lippincott, Williams, and Wilkins.

Mitchell, I., and Govias, G. (2021). *Asthma Education: Principles and Practice for the Asthma Educator*, 2nd Edition. Springer.

Schisler, J.C., Lang, C.H., and Willis, M.S. (2016). *Endocrinology of the Heart in Health and Disease: Integrated, Cellular, and Molecular Endocrinology of the Heart*. Academic Press.

Tabery, H.M. (2013). *Keratoconjunctivitis Sicca and Filamentary Keratopathy: In Vivo Morphology in The Human Cornea and Conjunctiva*. Springer.

Tirosh, O. (2014). *Liver Metabolism and Fatty Liver Disease* (Oxidative Stress and Disease Volume 35). CRC Press.

Wahli, W., and Guillou, H. (2021). *Metabolism and Metabolomics of Liver in Health and Disease*. Mdpi AG.

11 Psychotherapeutic Agents

LEARNING OBJECTIVES

After studying this chapter, readers should be able to

1. Explain the causes of major depressive disorder (MDD).
2. Review the classifications of antidepressants.
3. Explain the most common anxiety disorders.
4. Summarize the primary medications for anxiety and sleep disorders.
5. Describe the contraindications to benzodiazepines.
6. Describe the classifications of bipolar disorder.
7. Explain the pharmacotherapy used to treat bipolar disorder.
8. Describe the risk factors for schizophrenia.
9. Identify the four symptoms involved in schizophrenia.
10. Explain the primary adverse effects of medications used for schizophrenia.

OVERVIEW

MDD is one of the most common mental illnesses, affecting more than 21 million American adults each year. There is a persistent sense of hopelessness and despair. Major depression manifests through symptoms that interfere with the ability to sleep, eat, work, study, and enjoy previously pleasurable activities. Depression can result in various emotional and physical problems with long-term consequences. Individuals with anxiety disorders often experience intense, excessive, and persistent worry and fear about everyday situations. Bipolar disorder is a severe mental illness characterized by unusual shifts in mood, ranging from extreme highs to lows. Patients with bipolar disorder also encounter changes in their energy, thinking, behavior, and sleep patterns. Schizophrenia impacts how people think, feel, and behave. Individuals with schizophrenia often appear to have lost touch with reality, causing significant distress for both the patient and their family members. Positive symptoms of schizophrenia include **delusions** and hallucinations, while negative symptoms encompass a blunted affect and lack of motivation.

DEPRESSION

MDD is a mental illness characterized by a low mood lasting at least 2 weeks in most situations. It is often accompanied by low self-esteem, a loss of interest or pleasure in activities once enjoyed, low energy, and unexplained pain. Depressive disorders occur at all ages but are more commonly developed during mid-teenage years, twenties, or thirties. The precise cause remains unknown; however, it may stem from heredity, reduced levels of certain neurotransmitters, psychosocial factors, or changes in neuroendocrine function. Patients suffering from anxiety are more likely to develop depression.

Additionally, depression can occur in individuals with other mental disorders. Females are at a higher risk than males for depressive disorders. Changes in neurotransmitter levels are likely involved in depression, with examples including abnormal regulation of catecholaminergic, cholinergic, glutamatergic, and serotonergic neurotransmission.

PHARMACOTHERAPY

Medications and psychotherapy are effective for most individuals with depression. Numerous types of antidepressants and classifications are available to treat depressive disorders (see Table 11.1).

SELECTIVE SEROTONIN REUPTAKE INHIBITORS

Selective serotonin reuptake inhibitors (SSRIs) are prescribed as the first medications for depressive disorders. They are considered first-line drugs because of their high safety, efficacy, and tolerability.

Clinical Indications

The primary indication for SSRIs is MDD. However, they are frequently prescribed for generalized anxiety disorder, **bulimia nervosa**, and **posttraumatic stress disorder** (PTSD).

Mechanism of Action

SSRIs inhibit the presynaptic uptake of serotonin and norepinephrine after they have been released from the synaptic cleft. This prevention of reuptake prolongs the duration of these monoamines in the central nervous system (CNS) synaptic cleft.

Adverse Effects

The common adverse effects of SSRIs include insomnia, dry mouth, nausea, constipation, dizziness, loss of appetite, headache, and sexual dysfunction, particularly with paroxetine. These side effects are usually mild and resolve after the first few weeks of treatment.

Contraindications

SSRIs have contraindications that include known hypersensitivity, recent acute myocardial infarction, heart failure, abnormal bleeding, suicidal thoughts, unusual behavior changes, bradycardia, hypokalemia, hypomagnesemia, seizure disorders, and mania.

DOI: 10.1201/9781003461913-13

YOU SHOULD REMEMBER

Prozac can stimulate patients with depression; therefore, it should be given in the morning. Meanwhile, Paxil may induce sedation and should be administered at night, although it might also cause insomnia.

TABLE 11.1

Classification of Antidepressants

Generic Name	Trade Name
Selective Serotonin Reuptake Inhibitors	
Citalopram	Celexa
Escitalopram	Lexapro
Fluoxetine	Prozac
Fluvoxamine	Luvox
Paroxetine	Paxil
Sertraline	Zoloft
Vilazodone	Viibryd
Vortioxetine	Trintellix
Serotonin Modulators	
Buspirone	BuSpar
Mirtazapine	Remeron
Nefazodone	Serzone
Serotonin–Norepinephrine Reuptake Inhibitors	
Desvenlafaxine	Pristiq
Duloxetine	Cymbalta
Levomilnacipran	Fetzima
Sibutramine	Effexor
Venlafaxine	
Norepinephrine–Dopamine Reuptake Inhibitors	
Bupropion	Wellbutrin
Dexmethylphenidate†	Focalin
Methylphenidate†	Ritalin
Heterocyclic Antidepressants	
Amitriptyline	Elavil
Amoxapine	Asendin
Clomipramine	Anafranil
Desipramine	Norpramin
Doxepin	(Generic only)
Imipramine	Tofranil
Maprotiline	Ludiomil
Nortriptyline	Pamelor
Protriptyline	Vivactil
Trimipramine	Surmontil
Monoamine Oxidase Inhibitors	
Isocarboxazid	Marplan
Phenelzine	Nardil
Selegiline	Eldepryl
Tranylcypromine	Parnate

Check Your Knowledge

1. What are the causes of MDD?
2. What are the clinical indications of SSRIs?
3. What are the adverse effects of SSRIs?

SEROTONIN MODULATORS

Serotonin modulators, such as 5-hydroxytryptamine₂ blockers, have multimodal actions specific to the serotonin neurotransmitter system. They modulate one or more serotonin receptors and inhibit the reuptake of serotonin.

Clinical Indications

Serotonin modulators are used to treat major depression, generalized anxiety, bulimia, obsessive–compulsive disorder, and bipolar depression.

Mechanism of Action

Serotonin modulators inhibit presynaptic alpha-2-adrenergic receptors, leading to increased release of serotonin and norepinephrine. They activate the sympathetic nervous system and function as potent antagonists of the H1 histamine receptors. *Trazodone* has a multifunctional mechanism of action. At lower doses, it induces hypnotic effects, while at higher doses, it facilitates the blockade of the serotonin transporter, resulting in antidepressant effects. Its controlled-release formulation enhances the tolerability of higher doses.

Adverse Effects

The adverse effects of serotonin modulators may include nausea, vomiting, diarrhea, severe allergic reactions, **euphoria** or sadness, increased talkativeness, dizziness, seizures, fever, chills, sore throat, and coughing. They may also cause dyspnea, irritability, agitation, aggressiveness, hostility, heightened depression, suicidal thoughts, slurred speech, severe weakness, and loss of coordination.

Contraindications

The serotonin modulators are contraindicated in individuals with known hypersensitivity, dehydration, hypercholesterolemia, hyperglycemia, hyponatremia, bipolar disorder, **neuroleptic malignant syndrome**, angina, stroke, hypotension, liver disease, seizures, extreme restlessness, kidney impairment, **neutropenia**, and muscle pain or tenderness associated with increased creatine kinase. They are also contraindicated for patients with suicidal thoughts and a risk of angle-closure glaucoma.

SEROTONIN–NOREPINEPHRINE REUPTAKE INHIBITORS

Serotonin–norepinephrine reuptake inhibitors (SNRIs) are antidepressant medications used for various conditions beyond depression. Along with the SSRIs, SNRIs are classified as second-generation antidepressants, which have supplanted mainly first-generation antidepressants in recent years. The SNRIs comprise *desvenlafaxine, duloxetine, levomilnacipran, venlafaxine,* and *vortioxetine.*

Clinical Indications

SSRIs are used to treat MDD in adults. They are also indicated for the treatment of fibromyalgia, generalized anxiety disorder, diabetic neuropathy, and chronic muscle or joint pain.

Mechanism of Action

SNRIs block serotonin and norepinephrine reuptake and are considered vital for mood regulation. The human serotonin transporter and norepinephrine transporter are membrane proteins that retrieve neurotransmitters from the synaptic cleft back into the presynaptic nerve terminal.

Adverse Effects

SNRIs may cause nausea, vomiting, constipation, decreased appetite, dizziness, fatigue, **hyperhidrosis**, sexual dysfunction, and **xerostomia**. Other adverse effects include abnormal dreams, blurred vision, **mydriasis**, tremors, **vertigo**, and yawning.

Contraindications

SNRIs are contraindicated in patients with known hypersensitivity to these drugs. They should be avoided in conjunction with MAO inhibitors or within 2 weeks of discontinuing them. SNRIs are also contraindicated in individuals with seizure disorders, glaucoma, bipolar disorder, suicidal behavior, tachycardia, hypertension, urinary tract blockage, alcoholism, liver disease, kidney impairment, pregnancy, and breastfeeding.

Check Your Knowledge

1. What are the clinical indications of SSRIs?
2. What are the adverse effects of SNRIs?
3. What are the contraindications of SNRIs?

NOREPINEPHRINE–DOPAMINE REUPTAKE INHIBITORS

Norepinephrine–dopamine reuptake inhibitors (NDRIs) are a class of medications that effectively treat depression. They serve as the second-line treatment for clinical depression but are more commonly used to address attention-deficit/hyperactivity disorder and narcolepsy. The most widely used NDRI is *bupropion.*

Clinical Indications

NDRIs are indicated for depressive disorders, bipolar disorder, smoking cessation, and obesity. They are also used to treat attention-deficit hyperactivity disorder (ADHD), narcolepsy, Parkinson's disease, and alcohol cessation.

Mechanism of Action

NDRIs act as reuptake inhibitors by blocking the actions of the norepinephrine and dopamine transporters. This leads to increased extracellular neurotransmitter concentrations and adrenergic and dopaminergic neurotransmission.

Adverse Effects

NDRIs may lead to dry mouth, nausea, constipation, loss of appetite, abdominal pain, weight loss, insomnia, anxiety, excessive sweating, and tremors. In rare cases, they may cause serious adverse effects, such as seizures and Stevens–Johnson syndrome. Additional adverse effects of NDRIs can include **akathisia**, irritability, dyskinesia, lethargy, dizziness, palpitations, blurred vision, diplopia, mydriasis, skin rash, urticaria, and fever.

Contraindications

The contraindications of NDRIs include alcohol withdrawal, anorexia nervosa, bulimia nervosa, and benzodiazepine withdrawal. Bupropion should not be used alongside MAOIs, and there should be at least 2 weeks between discontinuing either type of medication before starting the other. NDRIs must be used cautiously if there is liver damage, severe kidney disease, or severe hypertension. They are also contraindicated in patients with known hypersensitivity, agitation, or glaucoma. NDRIs should be avoided in the first trimester of pregnancy due to their association with increased congenital heart defects.

MONOAMINE OXIDASE INHIBITORS

Monoamine oxidase inhibitors (MAOIs) were the first type of antidepressant developed. Currently, they are typically used only when other antidepressant medications have failed, partly due to the increased risk of hypertensive reactions with tyramine-rich foods (e.g., cured meats, alcohol, cheese, and smoked fish) and the heightened risk of **serotonin syndrome** when taken with other medications that elevate serotonin levels.

Clinical Indications

MAOIs represent a distinct class of antidepressants that target various forms of depression, panic disorder, social phobia, and depression with atypical features. Although MAOIs were the first antidepressants introduced, they are not the preferred choice for treating mental health disorders due to numerous dietary restrictions, side effects, and safety concerns. MAOIs are considered only when all other medications have failed. Another indication for their use is Parkinson's disease.

Mechanism of Action

MAOIs increase serotonin, norepinephrine, and dopamine levels by inhibiting an enzyme known as monoamine oxidase.

Adverse Effects

The most common adverse effects of MAOIs include dry mouth, nausea, diarrhea, constipation, drowsiness, insomnia, dizziness, headaches, and tremors. Additionally, MAOIs may cause chills, cold sweats, confusion, faintness, lightheadedness, tachycardia, hypomania, **paresthesia**, weight loss, sexual dysfunction, hypertension, rash, and urinary retention.

Contraindications

MAOIs are contraindicated in patients with known hypersensitivity, hypertension, stroke, congestive heart failure, a history of migraines, liver disease, **pheochromocytoma**, and severe kidney impairment. *Tranylcypromine* is contraindicated in porphyria and the consumption of tyramine-containing foods.

Check Your Knowledge

1. What are the clinical indications of MAOIs?
2. What are the adverse effects of MAOIs?
3. What are the contraindications for MAOIs?

SLEEP DISORDERS

Insomnia refers to difficulty falling or staying asleep. Sleep disorders rank among the most common conditions in medicine and psychiatry, carrying the potential to impair one's quality of life significantly. They may also arise from a variety of medical and psychiatric disorders. Sleep plays a crucial role in preventing Alzheimer's disease and other dementias. Older adults who sleep less than 6 hours at night are more likely to develop dementia. **Sleep apnea** and **restless legs syndrome** are considered other sleep disorders. The psychiatric manifestations of sleep disorders often include anxiety, depression, and cognitive difficulties.

ANXIETY DISORDERS

Anxiety disorders are the most prevalent mental health conditions, often leading to various signs and symptoms, including physical, somatic, cognitive, and behavioral manifestations. The experience of anxiety activates the sympathetic nervous system. Physical symptoms of the fight-or-flight response may include tachycardia, shortness of breath, hypertension, dry mouth, and excessive sweating. The most common anxiety disorders include *generalized anxiety disorder, panic disorder, social anxiety disorder*, obsessive–compulsive disorder, and PTSD.

Generalized anxiety disorders are characterized by excessive and irrational worry about various events. They can interfere with daily functioning, leading to anxiety about money, work, health, life, and death, as well as concerns about family and friends. **Panic disorder** involves recurrent panic attacks, typically accompanied by fears of future episodes or behavioral changes to avoid situations that might provoke an attack. Social anxiety disorder (**social phobia**) is marked by fear and anxiety in social situations, resulting in extreme distress and a reduced ability to function in daily life. **Obsessive–compulsive disorder** compels affected individuals to have specific recurrent thoughts (*obsessions*) and a need to engage in certain repetitive behaviors, known as *compulsions*, often in an attempt to manage these obsessive thoughts.

PHARMACOTHERAPY OF ANXIETY AND SLEEP DISORDERS

Benzodiazepines are the most prominent anti-anxiety medications; however, they are also used to treat sleep disorders. Table 11.2 lists the generic and trade names of benzodiazepines, as well as nonbenzodiazepine anxiolytics.

BENZODIAZEPINES

Benzodiazepines are primary medications for anxiety and sleep disorders. They belong to two classifications: sedative-hypnotics and anxiolytics. Sedative-hypnotics are used to induce sleep, while anxiolytics relieve anxiety. Additionally, benzodiazepines are classified as psychoactive drugs and are commonly referred to as minor **tranquilizers**. The classification of benzodiazepines is detailed in Table 11.3.

TABLE 11.2

Generic and Trade Names of Benzodiazepines and Nonbenzodiazepines

Generic Name	Trade Name
Alprazolam	Xanax
Chlordiazepoxide	Librium
Clonazepam	Klonopin
Clorazepate	Tranxene
Diazepam	Valium
Lorazepam	Ativan
Oxazepam	Serax
Estazolam	Prosom
Flurazepam	Dalmane
Quazepam	Doral
Temazepam	Restoril
Triazolam	Halcion
Buspirone	BuSpar
Eszopiclone	Lunesta
Meprobamate	Miltown
Ramelteon	Rozerem
Suvorexant	Belsomra
Tasimelteon	Hetlioz
Zaleplon	Sonata
Zolpidem	Ambien
Zopiclone	Zimovane

TABLE 11.3
Benzodiazepines

Generic Name	Trade Name
Short-Acting	
Midazolam	Versed
Oxazepam	Serax
Triazolam	Halcion
Intermediate-Acting	
Alprazolam	Xanax
Lorazepam	Ativan
Temazepam	Restoril
Long-Acting	
Clonazepam	Klonopin
Diazepam	Valium
Flurazepam	Dalmane

LIMBIC SYSTEM

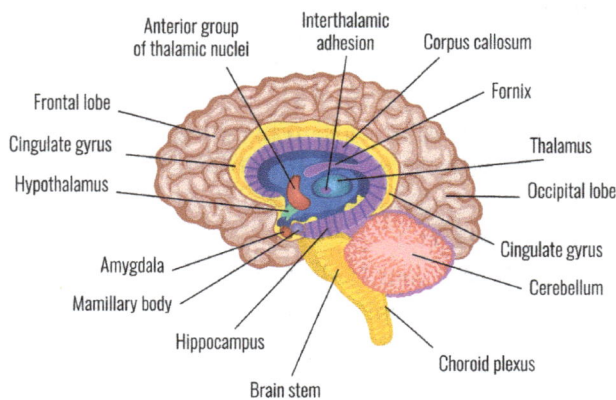

FIGURE 11.1   The structure of the thalamus, hypothalamus, and limbic system.

Clinical Indications

Benzodiazepines are commonly used to treat anxiety, depression, sedation, muscle relaxation, insomnia, and acute seizures. They are often combined with analgesics, anesthetics, and neuromuscular blockers as part of balanced anesthesia and moderate sedation. Additionally, these medications help to address and prevent alcohol withdrawal symptoms. Lower doses are given to older adults because they are more sensitive to CNS depressive effects.

Mechanism of Action

The mechanism of action of benzodiazepines relates to their ability to decrease CNS activity. The thalamus, hypothalamus, and limbic system may be affected (see Figure 11.1). Benzodiazepines **potentiate** the action of gamma-aminobutyric acid (GABA).

Receptors. They bind to receptors on chloride ion channels, enhancing the effects of GABA and facilitating the opening of these channels.

Adverse Effects

Benzodiazepines often have minimal adverse effects while demonstrating significant effectiveness and safety. However, the most common side effects associated with benzodiazepines include blurred vision, sleepiness, drowsiness, slurred speech, confusion, light-headedness, memory loss, nausea, constipation, and dry mouth.

Contraindications

Benzodiazepines are contraindicated for patients with known drug allergies, COPD, narrow-angle glaucoma, psychosis, and during pregnancy. Lactating women should also avoid them. *Triazolam* is often used for the short-term treatment of insomnia and jet lag. Prolonged use beyond 6 months may lead to tolerance and dependence.

Check Your Knowledge

1. What are the two classes of benzodiazepines?
2. What are the clinical indications of benzodiazepines?
3. What is the mechanism by which benzodiazepines act?

BARBITURATES

Barbiturates are sedative-hypnotic medications that cause sleep and relaxation. They are less common today because of the risk of misuse and specific adverse effects. However, they can be a backup treatment if other agents are ineffective. The classifications of barbiturates are listed in Table 11.4.

TABLE 11.4
Barbiturates

Generic Name	Trade Name
Ultrashort-Acting	
Methohexital	Brevital
Thiopental	Pentothal
Short-Acting	
Pentobarbital	Nembutal
Secobarbital	Seconal
Intermediate-Acting	
Butabarbital	Butisol
Long-Acting	
Phenobarbital	(Generic only)
Mephobarbital	Mebaral

Clinical Indications

Barbiturates can treat insomnia, seizures, and migraines. *Phenobarbital* is utilized for various types of seizures except for absence seizures and is the first-line choice for neonatal seizures. It is also indicated for benzodiazepine or alcohol detoxification, insomnia, and anxiety. *Secobarbital* is used to treat epilepsy, insomnia, and anxiolysis. *Thiopental* is applied in anesthesia, for medically induced coma, status epilepticus, **tranquilizers**, and lethal injection. *Pentobarbital* is employed to reduce intracranial pressure in **Reye's syndrome** and traumatic brain injury and for inducing coma in stroke. Barbiturates are no longer as widely utilized due to their potential for addiction and low therapeutic index.

Mechanism of Action

Barbiturates act on the **brainstem**, prolonging the inhibitory effects of both GABA and glycine. They increase the duration for which GABA-mediated chloride ion channels remain open. Additionally, barbiturates have the potential to block glutamic acid, an excitatory neurotransmitter.

Adverse Effects

Barbiturates may cause hypotension, drowsiness, headache, nausea, and skin rash. Severe adverse effects include fainting, apnea, confusion, hallucinations, and coma. *Phenobarbital* may also lead to **ataxia** and **nystagmus**.

Contraindications

Barbiturates should be avoided in individuals with known drug allergies, **dyspnea**, and severe liver or kidney diseases. They should be used cautiously in older patients due to their sedative properties and the increased risk of falling. *Phenobarbital* should be avoided in patients with acute **intermittent porphyria**.

ATYPICAL DRUGS

Atypical drugs are the latest agents that can treat various conditions. They include **melatonin** agonists and **orexin** antagonists (see Table 11.5).

TABLE 11.5
Atypical Sedative-Hypnotics

Generic Name	Trade Name
Melatonin Agonists	
Agomelatine	Valdoxan
Ramelteon	Rozerem
Tasimelteon	Hetlioz
Orexin Antagonists	
Lemborexant	Dayvigo
Suvorexant	Belsomra

Clinical Indications

Atypical drugs are indicated for treating psychosis, mania, depression, aggression, anxiety, sleep disorders, and **autism**. They are the preferred choice for acute psychoses.

Mechanism of Action

Melatonin acts by binding to the melatonin 1 and 2 receptors. Upon binding to these receptors, melatonin initiates the physiological processes that promote sleep. Suvorexant is a potent orexin receptor antagonist that inhibits these receptors, thereby inducing sleep.

Adverse Effects

The common adverse effects of atypical drugs include irritability, headache, drowsiness, depression, nausea, and abdominal pain. Older adults may face an increased risk of falling. Severe adverse effects of atypical drugs include hypersensitivity, sleepwalking, sleep-driving, sleep-eating, and engaging in sexual intercourse while asleep.

Contraindications

Melatonin agonists are contraindicated in patients with renal diseases, hepatic impairments, alcoholism, sleep apnea, suicidal thoughts, depression, and those who drive or eat while asleep. Suvorexant should be avoided in patients with COPD, sleep apnea, **cataplexy**, and depression. There are no proven contraindications for *tasimelteon*. Orexin antagonists should be avoided in patients with narcolepsy and neurological disorders. Atypical drugs are contraindicated in pregnancy (category C).

YOU SHOULD REMEMBER

Currently, the FDA has approved nine atypical antipsychotic drugs: aripiprazole, asenapine, clozapine, iloperidone, olanzapine, paliperidone, quetiapine, risperidone, and ziprasidone.

Check Your Knowledge

1. What are the indications of atypical drugs?
2. What are the adverse effects of atypical drugs?
3. What are the contraindications of atypical drugs?

BIPOLAR DISORDER

Bipolar disorder is characterized by extreme mood swings, which can range from intense highs (mania) to profound lows (depression). The causes of bipolar disorders are thought to include genetics, stress, and brain structure or function. Symptoms may persist for several weeks to 6 months, with depressive episodes typically lasting longer than manic episodes. During a manic phase, the individual may experience high energy levels and engage in highly pleasurable but risky activities. Severe bipolar episodes may feature psychotic symptoms such as insomnia, hallucinations, delusions, agitation, talkativeness, racing thoughts, and euphoria. Bipolar I *disorder* involves one or more episodes of mania, and most patients diagnosed with this type will experience both manic and depressive episodes. Bipolar II disorder is marked by depressive episodes that alternate with **hypomanic episodes**, although a complete manic episode never occurs.

PHARMACOTHERAPY

Pharmacotherapy for bipolar disorder includes mood stabilizers like lithium and specific antiseizure medications (see **Chapter 9**), a second-generation antipsychotic, or a combination of both. No single medication is universally effective for bipolar disorder. A list of common drugs used to treat bipolar disorders can be found in Table 11.6.

LITHIUM

Lithium is a powerful medication with antimanic properties that helps reduce feelings of mania, excitement, elevated mood, and distraction. It also aids in treating bipolar episodes. Most patients respond well to lithium, which offers better protection against manic rather than depressive episodes.

Clinical Indications

Lithium controls the manic episodes of bipolar disorder and helps prevent or lessen their intensity. It can be used in adults and children over age 7. Mood stabilizers indicate manic or agitated episodes. Antidepressants and psychotherapy are employed to address depressive episodes. However, psychotherapy can also assist in managing all aspects of the disorder.

TABLE 11.6
Common Drugs Used for Bipolar Disorder

Generic Name	Trade Name
Lithium	Lithobid, Eskalith
Aripiprazole	Abilify
Asenapine	Saphris
Brexpiprazole	Rexulti
Cariprazine	Vraylar
Clozapine	Clozaril
Iloperidone	Fanapt
Lurasidone	Latuda
Olanzapine	Zyprexa
Olanzapine–fluoxetine	Symbyax
Paliperidone	Invega
Quetiapine	Seroquel
Risperidone	Risperdal, Rykindo
Ziprasidone	Geodon

Mechanism of Action

Lithium is broadly distributed in the CNS and interacts with various neurotransmitters and receptors, reducing the release of norepinephrine while enhancing serotonin synthesis.

Adverse Effects

Lithium may cause nausea, vomiting, diarrhea, dry mouth, dizziness, headache, **alopecia**, **polydipsia**, and polyuria. It can also lead to cognitive impairment and hypothyroidism. The medication may exacerbate **acne**, **psoriasis**, kidney damage, and tremors.

Contraindications

Lithium should be avoided in cases of significant renal diseases, dehydration, sodium depletion, Addison's disease, heart failure, and untreated hypothyroidism. For patients with less severe renal impairment, careful consideration of the risks and benefits of lithium treatment is necessary, as there is a high risk of toxicity associated with these conditions.

Check Your Knowledge

1. What are the clinical indications of lithium?
2. What are the adverse effects of lithium?
3. What are the contraindications for lithium?

SCHIZOPHRENIA

Schizophrenia is a severe mental health condition that affects how patients think, feel, and behave. It can lead to delusions, hallucinations, disorganized thinking, and behavior. Hallucinations occur when individuals see things or hear voices that others do not perceive. Delusions involve strong convictions about things that are false. Patients with schizophrenia often lose touch with reality, which can make daily life very challenging. The exact causes of schizophrenia remain unknown; however, researchers believe that a combination of genetics, brain chemistry, and environmental factors may contribute. Schizophrenia is a chronic condition that progresses through several stages, exhibiting varied durations and phasic patterns. Schizophrenic patients often struggle with cognitive tasks due to disturbances in cortical information processing, which can manifest as abnormalities in **electroencephalogram** signals.

PHARMACOTHERAPY

Pharmacotherapy is essential for treating acute symptoms, inducing remission, and preventing relapse. Schizophrenia requires lifelong treatment, even when symptoms have subsided. Treatment with medications and psychosocial therapy can help manage the condition. In some cases, temporary or permanent hospitalization may be necessary. Medications are the cornerstone of schizophrenia treatment, with three classes utilized: first-generation drugs, second-generation drugs, and newer medications.

FIRST-GENERATION DRUGS

First-generation antipsychotics are classified into two categories: *phenothiazines* and nonphenothiazines.

- **Phenothiazines:**
  - chlorpromazine (Thorazine)
  - fluphenazine (Prolixin)
  - mesoridazine (Serentil)
  - perphenazine (Trilafon)
  - prochlorperazine (Compazine)
  - thioridazine (Mellaril)
  - trifluoperazine (Stelazine)
- **Nonphenothiazines:**
  - droperidol (Inapsine)
  - haloperidol (Haldol)
  - loxapine (Loxitane)
  - molindone (Moban)
  - pimozide (Orap)
  - thiothixene (Navane)
- **New Agents:**
  - asenapine (Saphris)
  - iloperidone (Fanapt)
  - lurasidone (Latuda)

Clinical Indications

First-generation antipsychotics are indicated for **schizoaffective disorders** and psychotic disorders. These drugs are also used for teenagers with early-onset schizophrenia.

Mechanism of Action

The mechanism of action of first-generation antipsychotics involves the postsynaptic blockade of dopamine receptors, which reduces dopaminergic neurotransmission in dopamine pathways.

Adverse Effects

First-generation drugs can cause photosensitivity, orthostatic hypotension, seizures, sedation, drowsiness, and constipation. Severe adverse effects include **agranulocytosis**, **pancytopenia**, anaphylaxis, tardive akathisia, dyskinesia, hypothermia, and **adynamic ileus**. Nonphenothiazines commonly lead to tremors, drowsiness, sedation, orthostatic hypotension, and weight gain. Severe adverse effects of nonphenothiazines include laryngospasm, hepatotoxicity, kidney failure, respiratory depression, and sudden death.

Contraindications

First-generation antipsychotics are contraindicated for individuals with a history of severe allergies, prostatic hypertrophy, narrow-angle glaucoma, heart failure, and seizure disorders.

YOU SHOULD REMEMBER

Schizophrenia consists of three stages: prodromal, active, and residual. The prodromal stage includes nonspecific symptoms such as lack of motivation, social isolation, and difficulty concentrating. Prodromal symptoms are not always obvious.

Check Your Knowledge

1. What are the new drugs to treat schizophrenia?
2. What are the adverse effects of first-generation antipsychotic drugs?
3. What are the contraindications of first-generation antipsychotic drugs?

SECOND-GENERATION DRUGS

Second-generation antipsychotic medications, commonly referred to as "atypical antipsychotic drugs," have replaced traditional agents as the first-line therapy in the treatment of schizophrenia. The second-generation antipsychotics are listed in Table 11.7.

Clinical Indications

Second-generation antipsychotics, except for clozapine, are indicated for the treatment and maintenance of schizophrenia and schizoaffective disorders. They can also be used to treat acute agitation. Furthermore, antipsychotics can be used in children with severe autism. Risperidone and olanzapine help control aggression in children.

Mechanism of Action

The mechanism of action primarily involves the postsynaptic blockade of dopamine receptors in the brain. It also inhibits serotonin antagonist receptors, which appears to reduce negative symptoms better than typical antipsychotics.

TABLE 11.7

Second-Generation Antipsychotics

Generic Name	Trade Name
Aripiprazole	Abilify
Asenapine	Saphris
Brexpiprazole	Rexulti
Cariprazine	Vraylar
Clozapine	Clozaril, Versacloz
Iloperidone	Fanapt
Lurasidone	Latuda
Olanzapine	Zyprexa
Paliperidone	Invega
Quetiapine	Seroquel
Risperidone	Risperdal
Ziprasidone	Geodon

Adverse Effects

Second-generation antipsychotics may lead to hypotension, blurred vision, tardive dyskinesia, tremors, drowsiness, sedation, restlessness, muscle spasms, diabetes mellitus, dry mouth, constipation, and weight gain.

Contraindications

Second-generation antipsychotics are contraindicated in patients with known hypersensitivity, diabetes mellitus, suicidal thoughts, seizures, heart failure, Alzheimer's disease, **metabolic syndrome X**, and severe hepatic impairment. *Clozapine* may cause rash, swelling, and **angioedema**. It also carries black box warnings due to risks including neutropenia, agranulocytosis, seizures, myocarditis, cardiomyopathy, orthostatic hypotension, bradycardia, and syncope.

CLINICAL CASE STUDIES

Clinical Case Study 1

A 26-year-old woman has experienced depression and anxiety for 7 years. Her medication includes citalopram, an SSRI, which helps improve her mood and reduce stress. The patient reports having no panic attacks, and these medications have significantly aided her focus on daily living activities.

CRITICAL THINKING QUESTIONS

1. What are the most common anxiety disorders?
2. What is the primary indication for SSRI medications?
3. What are the adverse effects of SSRIs?

Clinical Case Study 2

A 19-year-old man with schizophrenia has been taking quetiapine since his diagnosis a year ago. The patient visits his physician and admits that he has not been following his dosage regimen exactly due to significant weight gain and vomiting. His medication has been switched to risperidone because this drug is a better fit for him. His medication use is carefully monitored over time, which appears to be successful during the first few months.

CRITICAL THINKING QUESTIONS

1. What are the subdivisions of the first-generation antipsychotics?
2. What are the adverse effects of first-generation drugs?
3. What are the contraindications of second-generation drugs?

FURTHER READING

De Girolamo, G., McGorry, P.D., and Sartorius, N. (2019). *Age of Onset of Mental Disorders: Etiopathogenetic and Treatment Implications.* Springer.

Fang, Y. (2019). *Depressive Disorders: Mechanisms, Measurement and Management* (Advances in Experimental Medicine and Biology, 1180). Springer.

Goodwin, F.K., and Redfield Jamison, K. (2007). *Manic-Depressive Illness: Bipolar Disorders and Recurrent Depression*, 2nd Edition. Oxford University Press.

Granholm, E.L., McQuaid, J.R., and Holden, J.L. (2016). *Cognitive-Behavioral Social Skills Training for Schizophrenia: A Practical Treatment Guide.* The Guilford Press.

Hales, D.J., Hyman Rapaport, M., and Moeller, K. (2012). *Focus: Major Depressive Disorder.* American Psychiatric Publishing.

Heldt, J.P. (2017). *Memorable Psychopharmacology.* CreateSpace Independent Publishing Platform.

Herzberg, D. (2010). *Happy Pills in America: From Miltown to Prozac.* JHUP.

Jjemba, P.K. (2018). *Pharma-Ecology: The Occurrence and Fate of Pharmaceuticals and Personal Care Products in the Environment*, 2nd Edition. Wiley.

Jordan, A. (2018). *Antidepressants: History, Science, and Issues* (The Story of a Drug). Greenwood.

Kennedy, S.H., Lam, R.W., Nutt, D.J., and Thase, M.E. (2007). *Treating Depression Effectively: Applying Clinical Guidelines, 2007.* CRC Press.

Kim, Y.K. (2021). *Major Depressive Disorder: Rethinking and Understanding Recent Discoveries* (Advances in Experimental Medicine and Biology, 1305). Springer.

Leonhard, K., Beckmann, H., and Cahn, C.H. (2012). *Classification of Endogenous Psychoses and their Differentiated Etiology*, 2nd Edition. Springer-Verlag.

Lieberman, J.A. (2020). *Textbook of Schizophrenia*, 2nd Edition. American Psychiatric Association Publishing.

McGlothlin, W.H. (2012). *Amphetamines, Barbiturates, and Hallucinogens: An Analysis of Use, Distribution*, and Control (Volume 1). University of California Libraries.

Morrison, P., Taylor, D.M., and McGuire, P. (2020). *The Maudsley Guidelines on Advanced Prescribing in Psychosis.* Wiley-Blackwell.

Pratt, C.W., Gill, K.J., Barrett, N.M., and Roberts, M.M. (2016). *Psychiatric Rehabilitation*, 3rd Edition. Academic Press.

Preston, J.D., O'Neal, J.H., Talaga, M.C., and Moore, B.A. (2021). *Handbook of Clinical Psychopharmacology for Therapists*, 9th Edition. New Harbinger Publications.

Rose, J. (2020). *Schizophrenia: A Guide to the Symptoms, Management, and Treatment of Schizophrenia.* Ingram Publishing.

Small, C. (2015). *Valium and Other Antianxiety Drugs* (Dangerous Drugs). Cavendish Square Publishing.

12 Local and General Anesthetics

LEARNING OUTCOMES

After studying this chapter, readers should be able to

1. Describe the indications for local anesthesia.
2. Identify various types of regional anesthesia administration.
3. Review infiltration anesthesia and epidural anesthesia.
4. Explain the indications for spinal anesthesia.
5. Summarize the classification of general anesthesia drugs.
6. Describe the various stages of anesthesia.
7. Explain the contraindications of general anesthesia.
8. Describe the indications for nitrous oxide.
9. Explain the advantages of combining IV and inhaled anesthetics.
10. Review the symptoms of anesthesia adjuncts.

OVERVIEW

Anesthesia refers to the loss of sensation or awareness. Anesthetics cause either a local or general loss of sensation and pain. Anesthesia involves using medications to prevent pain during surgery and other procedures. These medications are known as anesthetics. They may be administered through injection, inhalation, topical lotion, spray, eye drops, or skin patches. Anesthesia can be applied in minor procedures, such as filling a tooth. It may also be utilized during childbirth or a colonoscopy.

Additionally, it is employed in both minor and major surgeries. Local anesthesia provides reversible nerve blockade, resulting in the loss of pain sensation. Local anesthetics are categorized as esters or amides. Topical application and direct infiltration anesthetize the immediate area. Regional blocks anesthetize larger areas via a nerve or field block.

LOCAL ANESTHETICS

Local anesthetics prevent nerves in a specific body part from sending signals to the brain, eliminating the pain sensation. However, patients may still experience some pressure or movement. It takes a few minutes for the feeling to diminish in the area where a local anesthetic is administered. Medications available for local anesthesia are classified as ester or amide (see Table 12.1). **Amide** local anesthetics are generally more effective than **ester** local anesthetics but may cause more toxic reactions than ester variants. Amides are highly stable in solution, while esters are unstable. Amino-esters are hydrolyzed in plasma by the enzyme pseudocholinesterase, whereas

---

TABLE 12.1

Classifications of Ester and Amide Drugs

Ester Group	Amide Group
Benzocaine	Bupivacaine
Chloroprocaine	Dibucaine
Cocaine	Etidocaine
Procaine	Lidocaine
Proparacaine	Mepivacaine
Tetracaine	Prilocaine
	Ropivacaine

---

amide compounds are enzymatically degraded in the liver and excreted in the urine.

Specific preparations are available for topical administration, **infiltration**, and various regional techniques, such as field block, regional block, epidural block, and spinal block (see Table 12.2). Some patients may be more sensitive to one group of local anesthetics and may tolerate others better. Table 12.3 lists the generic and trade names for local anesthetics.

YOU SHOULD REMEMBER

Procaine (like cocaine) has the advantage of constricting blood vessels, which reduces bleeding. However, hypersensitivity may develop in patients, leading to anaphylactic symptoms such as urticaria, pruritus, erythema, laryngeal edema, tachycardia, nausea, vomiting, dizziness, syncope, perspiration, and hyperthermia. Treatment includes cessation of procaine and a tailored regimen of intramuscular epinephrine, supplemental oxygen, intravenous corticosteroids, resuscitative fluids, beta-agonists, and supportive care.

CLINICAL INDICATIONS

Local anesthetics are used to remove moles, warts, hemorrhoids, fissures, and for ontological examinations, cataract removal, **cystoscopy**, and catheterization. They are also used for dental procedures such as fillings and wisdom tooth removal. Dentists commonly use two types of injections: infiltration and nerve block (see Figure 12.1). The effects of the injections can last for several hours. The most common anesthetic agent used by dentists is 2% *lidocaine*. The available dosages of lidocaine injectable solution include 0.4%, 0.5%, 0.8%, 1%, 1.5%, 2%, 4%, and 5%.

DOI: 10.1201/9781003461913-14

TABLE 12.2

Specific Methods for Delivering Local and Regional Anesthesia

Methods or Technique	Description
Topical (surface) anesthesia	A drug is applied to or sprayed directly to the surface of skin or mucous membranes and does not penetrate into deeper layers. This provides relief for minor skin irritation or before a needlestick.
Infiltration anesthesia	A drug is injected into deeper skin layers and moves to surrounding areas, providing pain relief during minor skin surgeries or dental procedures, or before a deep needlestick.
Nerve block	A drug is injected around a peripheral nerve; all regions that are innervated by the nerve lose sensation, as in dental procedures or regional anesthesia
Spinal anesthesia	A drug is injected into the cerebrospinal fluid, most often via the lumbar region; this blocks sensation to larger areas of the lower abdomen or pelvis.
Epidural anesthesia	A drug is injected into the epidural space that surrounds the spinal cord, blocking sensation to many nerve roots that supply limited portions of the chest, limbs, abdomen, or pelvis; this is often used for urologic and obstetric procedures.

TABLE 12.3

Generic and Trade Names of Local Anesthetics

Generic Name	Trade Names
Articaine	Septocaine
Benzocaine	Dendracin
Bupivacaine	Marcaine
Chloroprocaine	Nesacaine
Cocaine	Goprelto
Dibucaine	Nupercainal
Etidocaine	Duranest
Lidocaine	Xylocaine
Mepivacaine	Carbocaine
Prilocaine	Citanest
Procaine	Novocaine
Proparacaine	Alcaine
Ropivacaine	Naropin
Tetracaine	Niphanoid

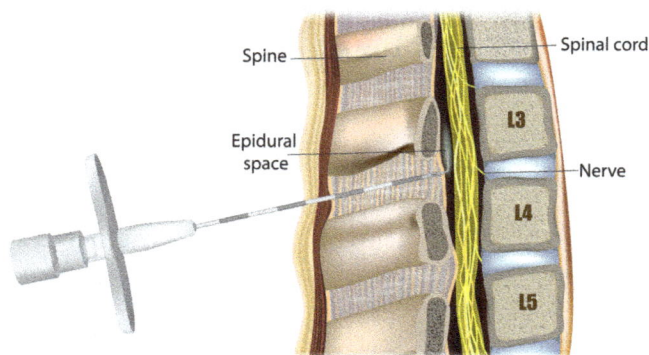

FIGURE 12.1   Nerve block injection.

MECHANISM OF ACTION

The local anesthetics inhibit voltage-dependent sodium channels, reducing the **influx** of sodium ions and preventing membrane depolarization while blocking the conduction of **action potentials**. Since drug molecules must cross lipid membranes to reach the cytoplasm, more lipid-soluble forms achieve effective intracellular concentrations faster than ionized forms. However, when the ionized drugs are inside axons, they can block the channels more effectively. Therefore, ionized drug forms are essential. Esters such as *procaine* primarily act by blocking sodium influx.

ADVERSE EFFECTS

The adverse effects of local anesthetics are generally insignificant. However, an overdose of these agents may result in dizziness, lightheadedness, confusion, anxiety, excitement, nervousness, restlessness, convulsions, visual abnormalities, headaches, fever, numbness, sweating, tinnitus, dyspnea, shivering, dysrhythmia, and weakness.

CONTRAINDICATIONS

Local anesthetics are contraindicated in individuals with a history of hypersensitivity to local anesthetics, sunscreens, sulfa drugs, or hair dyes.

Check Your Knowledge

1. What are the classifications of local anesthetics?
2. What are some specific methods for delivering local anesthetics?
3. What are the indications for local anesthetics?

SPINAL ANESTHESIA

Spinal anesthesia is a type of regional anesthesia characterized by a temporary loss of sensation in the abdomen and lower part of the body. The patient remains awake during the procedure, during which an anesthetic is injected into the **subarachnoid** space between the L3 and L4 vertebrae. Various sites along the spinal column can be used for a nerve block. In **epidural block anesthesia**, the local anesthetic is introduced into the epidural space, located just posterior to the spinal cord or **dura mater** (see Figure 12.2).

STAGES OF ANESTHESIA

General anesthesia is the medication that patients receive before surgery. It induces a deep, sleep-like state in patients. There are four stages of general anesthesia, which include the following:

- **Stage I (induction):** The earliest stage is when the patient takes the medication until they fall asleep. The general sensation is diminished, but the patient may stay awake; this stage continues until consciousness is entirely lost.
- **Stage II (excitement):** The patient may become delirious, heart and respiratory rates may become irregular, while blood pressure could increase. Administering intravenous agents may help to calm the patient. The patient should progress into stage III quickly to avoid feelings of panic. Close monitoring will be required for irregular heart rate and increased blood pressure.

- **Stage III (surgical anesthesia):** The skeletal muscles relax, **delirium** stabilizes, and cardiopulmonary effects occur. The patient is motionless, and eye movements are limited. The surgery begins, and the patient remains in this stage until the procedure is completed.
- **Stage IV (overdose):** This stage is known as **medullary paralysis** and should be avoided; if breathing and cardiac function cease, death may occur.

Check Your Knowledge

1. At what stage of local anesthetics does the patient remain awake?
2. At what stage do local anesthetics cause the patient to become delirious?
3. What stage of local anesthetics can be dangerous and potentially lead to death?

CLASSIFICATIONS OF MEDICATIONS IN GENERAL ANESTHESIA

The five classifications of medications used in general anesthesia are intravenous (IV) anesthetics, inhalational anesthetics, IV sedatives, synthetic opioids, and neuromuscular blocking drugs. Each type is based on the sedation and analgesia required for the intended procedure. General anesthesia is employed for major surgical procedures. **Regional anesthesia** affects an entire limb or a relatively large portion of the body, with fewer adverse effects than general anesthesia. Examples of regional anesthesia include epidural and spinal anesthesia. Monitored anesthesia care

Epidural Anesthesia

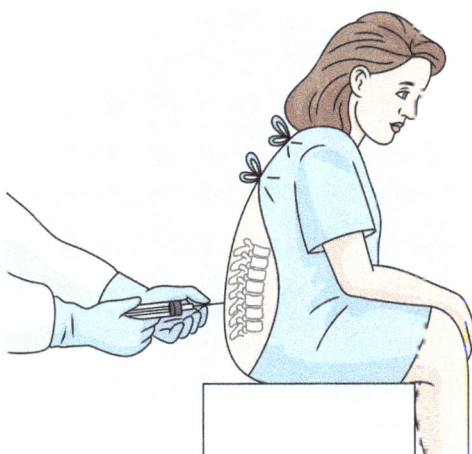

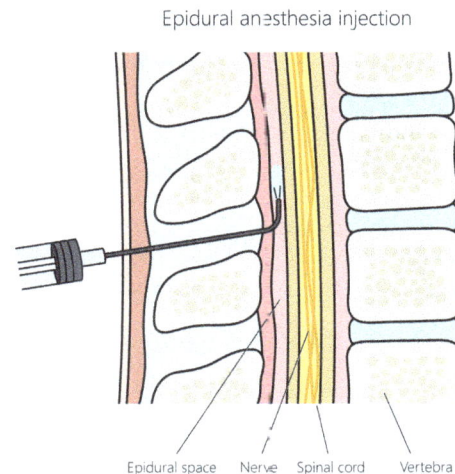

Epidural anesthesia injection

Epidural space   Nerve   Spinal cord   Vertebra

FIGURE 12.2   Epidural block anesthesia.

utilizes analgesics, sedatives, and low doses of other drugs, allowing the patient to remain responsive and breathe without assistance as a procedure is performed. Three subtypes of monitored anesthesia care include the following:

- **Anxiolysis (minimal sedation):** The patient can respond to verbal commands and maintains normal airway, cardiovascular, and ventilation functions.
- **Conscious sedation (moderate sedation):** The patient can respond to verbal commands, light, and touch. Airway, cardiovascular, and ventilation functions remain normal.
- **Deep sedation and analgesia:** Repeated or painful stimulation can arouse the patient, but airway and ventilation interventions may be needed. Cardiovascular functions are primarily adequate.

Sedative and hypnotic drugs are utilized for pre-operative and dental procedures. They possess sedative, hypnotic, anxiolytic, anticonvulsant, and muscle relaxant properties. Benzodiazepines are frequently used for patients on mechanical ventilation. Propofol and ketamine are classified as miscellaneous drugs used as intravenous anesthetics. Both have short durations of action and rapidly induce anesthesia. Intravenous anesthetics (opioids, benzodiazepines, and miscellaneous drugs) are detailed in Table 12.4.

CLINICAL INDICATIONS

Alfentanil is indicated as a component of balanced anesthesia. It assists with intubation, promotes anesthesia induction, and serves as an infusion to maintain anesthesia. It is indicated for managing severe postoperative pain when administered as an epidural injection. *Fentanyl* is given

---

TABLE 12.4

Intravenous Anesthetics

Generic Name	Trade Name
Opioids	
Alfentanil	Alfenta
Fentanyl	Sublimaze
Remifentanil	Ultiva
Sufentanil	Sufenta
Benzodiazepines	
Diazepam	Valium
Midazolam	Versed
Lorazepam (in the hospital)	
Miscellaneous	
Etomidate	Amidate
Ketamine	Ketalar
Methohexital	Brevital
Propofol	Diprivan

---

with oxygen for high-risk patients, including those undergoing open-heart surgery or complex neurologic or orthopedic procedures. Fentanyl is prescribed for treating pain in patients who are receiving continuous opioid therapy and have developed tolerance. *Remifentanil* is provided during the induction and maintenance of general anesthesia to offer pain relief after surgery and can be used alone or in combination with *midazolam. Sufentanil* is administered as part of balanced anesthesia or as a primary anesthetic, and 100% oxygen is given concurrently. It is sometimes indicated, along with *bupivacaine,* as an epidural anesthetic during labor and delivery. Sufentanil is 5–10 times more potent than fentanyl.

Benzodiazepines are primarily indicated for treating anxiety, but high doses can lead to sedation and unconsciousness. They are utilized as part of balanced anesthesia, commonly administered alongside inhalation anesthetics. Sometimes, they are given orally as premedication before surgery to help relax the patient. The benzodiazepine most frequently used for surgery is midazolam, although diazepam and lorazepam are also standard options. Midazolam is approved for the induction of amnesia, installation, maintenance of general anesthesia, and sedation.

Etomidate is indicated for the induction of general anesthesia for short medical and surgical procedures. It is also used as a supplement to anesthesia and nitrous oxide. *Ketamine* is indicated for surgeries and diagnostic procedures that do not require relaxation of the skeletal muscles. This agent is also administered before other general anesthetics are given. *Methohexital* is a barbiturate that is often indicated as a supplemental form of anesthesia with nitrous oxide. *Propofol* is now the most commonly used IV anesthetic due to its high effectiveness and relative safety. It is used for both the induction and maintenance of general anesthesia.

MECHANISM OF ACTION

Opioid agonists act on the mu and kappa receptors in the same way as morphine. However, fentanyl has a much faster onset of action. *Midazolam* affects the hypothalamus, thalamus, and limbic regions, producing CNS depression and skeletal muscle relaxation. The exact mechanism of action of propofol is unknown, but it is believed to activate gamma-aminobutyric acid receptors, resulting in a general inhibition of the CNS.

ADVERSE EFFECTS

Etomidate may cause injection site pain, myoclonus, involuntary movements, tonic movements, ocular movements, and postoperative nausea and vomiting. *Ketamine* may cause nystagmus, increased oxygen demand and pulmonary arterial pressure, disorientation, hallucinations, nightmares, **vivid dreams**, abnormal electroencephalogram results, increased salivation, and with rapid injection or high doses, respiratory depression.

Methohexital may cause hypotension, laryngospasm, bronchospasm, apnea, respiratory depression, visual hallucinations, vivid dreams, and arrhythmias. *Propofol* can also cause hypotension and respiratory depression. Additionally, it may lead to anaphylaxis, rhabdomyolysis, metabolic acidosis, kidney failure, and heart failure.

CONTRAINDICATIONS

There are no absolute contraindications to general anesthesia other than patient refusal or the lack of availability of trained personnel. However, some relative contraindications do exist. Intravenous anesthetics are contraindicated in cases of known drug allergies, narrow-angle glaucoma, acute porphyria, and if the patient has a known susceptibility to malignant hyperthermia.

YOU SHOULD REMEMBER

Alfentanil, etomidate, ketamine, and propofol are commonly used during rapid sequence tracheal intubation. Succinylcholine and rocuronium are widely employed neuromuscular blocking agents. Specific induction agents and paralytic drugs may be more beneficial in specific clinical situations than others.

Check Your Knowledge

1. What are the five classes of medications utilized in general anesthesia?
2. What are the indications for alfentanil and fentanyl?
3. What are the side effects of etomidate and ketamine?

CLINICAL CASE STUDIES

Clinical Case Study 1

A 21-year-old man received procaine local anesthetic to extract a tooth from the upper left portion of his mouth. He developed hypotension, laryngeal edema, tachycardia, nausea, vomiting, and dizziness. The dentist suspected an allergic reaction and administered an EpiPen along with oxygen. The office also called 911 to transfer him to the local hospital for further evaluation.

CRITICAL THINKING QUESTIONS

1. What are the differences between esters and amides as local anesthetics?
2. What are the signs and symptoms of allergic reactions to procaine?

3. What are the trade names of bupivacaine, chloroprocaine, etidocaine, lidocaine, and procaine?

Clinical Case Study 2

A 41-year-old obese man with type II diabetes mellitus was scheduled for shoulder surgery. He had undergone another surgical procedure 2 years prior, without any complications, and he had no history of drug allergies. The patient was premedicated with oral hydroxyzine for this surgery and was then given IV propofol and sufentanil. Tracheal intubation was facilitated with IV succinylcholine.

CRITICAL THINKING QUESTIONS

1. What are the indications of remifentanil and midazolam?
2. What drugs can cause muscle relaxation for tracheal intubation?
3. What are the adverse effects of methohexital and propofol?

FURTHER READING

Andropoulos, D.B., and Gregory, G.A. (2020). *Gregory's Pediatric Anesthesia,* 6th Edition. Wiley–Blackwell.

Chan, V., Finucane, B.T., Grau, T., Walji, A., Tsui, B.C.H., Chan, C.T.S., Bhargava, R., Dillane, D., Ganapathy, S., Lou, L., and Noga, M. (2008). *Atlas of Ultrasound- and Nerve Stimulation-Guided Regional Anesthesia.* Springer.

Chestnut, D.H., Wong, C.A., Tsen, L.C., Ngan Kee, W.D., Beilin, Y., Mhyre, J.M., and Bateman, B.T. (2019). *Chestnut's Obstetric Anesthesia: Principles and Practice*, 6th Edition. Elsevier.

Chiumello, D. (2016). *Topical Issues in Anesthesia and Intensive Care.* Springer.

Davis, P.J., and Cladis, F.P. (2016). *Smith's Anesthesia for Infants and Children*, 9th Edition. Elsevier.

Dosch, M. (2012). *Atlas of Neural Therapy: With Local Anesthetics*, 3rd Edition. Thieme.

Ehrenwerth, J., Eisenkraft, J.B., and Berry, J.M. (2020). *Anesthesia Equipment: Principles and Applications*, 3rd Edition. Saunders.

Elisha, S., Nagelhout, J.J., and Heiner, J.S. (2020). *Current Anesthesia Practice — Evaluation & Certification Review.* Elsevier.

Fellahi, J.L., and Leone, M. (2018). *Anesthesia in High-Risk Patients.* Springer.

Fernando, R., Sultan, P., and Phillips, S. (2021). *Quick Hits in Obstetric Anesthesia.* Springer.

Grant, S.A., and Auyong, D.B. (2016). *Ultrasound Guided Regional Anesthesia*, 2nd Edition. Oxford University Press.

Gravlee, G.P., Shaw, A.D., and Bartels, K. (2018). *Hensley's Practical Approach to Cardiothoracic Anesthesia*, 6th Edition. Lippincott, Williams, and Wilkins.

Hanke, C.W., Sommer, B., and Sattler, G. (2012). *Tumescent Local Anesthesia.* Springer.

Huntoon, M., Benzon, H., Nauroze, S., and Deer, T.R. (2011). *Spinal Injections and Peripheral Nerve Blocks*, Volume 4. Elsevier.

Malamed, S.F. (2019). *Handbook of Local Anesthesia*, 7th Edition. Elsevier.

Parmley, R.T., and Adriani, J. (2013). *Saddle Block Anesthesia: American Lecture Series*, Number 258. Literary Licensing, LLC.

Sairyo, K. (2021). *Transforaminal Full-Endoscopic Lumbar Surgery under the Local Anesthesia — State of the Art.* Springer.

Silvestri, E., Martino, F., and Puntillo, F. (2018). *Ultrasound-Guided Peripheral Nerve Blocks.* Springer.

13 Substance Abuse

LEARNING OBJECTIVES

After studying this chapter, readers should be able to

1. Explain substance abuse and misuse
2. Review substances that cause brain stimulation.
3. Identify the medication that causes hallucinations.
4. Review the most psychoactive agent found in cannabinoids.
5. Identify substances that depress brain activity.
6. Explain the signs and symptoms of cocaine use.
7. Discuss the indications for medical marijuana.
8. Describe from the effects of nicotine withdrawal.
9. Explain the physical effects of alcoholism.
10. Summarize the signs and symptoms of alcohol withdrawal.

OVERVIEW

Substance abuse, also known as drug abuse, refers to the use of a drug in amounts or through methods that are harmful to the individual. It represents a form of substance-related disorder. Drug abuse is a term frequently used in public health, medical, and criminal justice contexts. The stigma surrounding it is magnified by negative perceptions that impact individuals on antidepressants or atypical antipsychotics. Addiction is a brain disease that influences behavior. Approximately $532 billion is spent annually on addiction-related costs in the United States alone. Provisional data from the CDC's National Center for Health Statistics indicate that there were an estimated 107,543 drug overdose deaths in the United States during 2023. Substance abuse affects individuals of all ages. For instance, babies born to mothers who are addicted to drugs or alcohol are often born prematurely and are underweight. Affected adolescents tend to perform poorly in school, have more unplanned pregnancies, and experience a higher incidence of disease and violence. Illicit fentanyl, primarily manufactured in foreign clandestine labs and smuggled into the United States through Mexico, is being distributed across the country and sold on the illegal drug market. Fentanyl is being mixed with other illicit drugs to enhance potency, sold as powders and nasal sprays, and increasingly pressed into pills designed to resemble legitimate prescription opioids.

SUBSTANCE ABUSE AND MISUSE

Substance abuse and misuse affect every aspect of an individual's life and become problematic when an individual experiences significant distress. People with substance dependence exhibit tolerance, engage in fewer social activities, express a desire to quit, and experience withdrawal symptoms. Addiction is defined as the progressive, chronic abuse of a substance despite serious health and social consequences. Various factors, such as availability, dosage, cost, and method of administration, influence addiction. Sedatives and hypnotics promote relaxation in small doses, but higher doses can induce sleep and confusion. Excessive amounts can lead to coma and even death. Any combination of these agents can be lethal due to their **additive effects**.

Substance use disorder (SUD) or **substance misuse** is a treatable mental disorder that affects a person's brain and behavior. It leads to their inability to control the use of substances such as legal or illegal drugs, alcohol, or medications. Symptoms can range from moderate to severe, with addiction being the most serious form of SUD. Prescription opioids are commonly misused and may include taking these drugs in larger doses than prescribed, taking someone else's prescription drugs, or simply taking drugs to induce euphoria or pleasant effects, rather than for therapeutic reasons.

EFFECTS OF SUBSTANCE ABUSE

SUD can lead to both short- and long-term adverse health effects, which may be physical or mental and can range from moderate to severe. Substance abuse can contribute to the development of anxiety and depression. Using stimulants can increase anxiety, panic disorder, and insomnia, while alcohol, opioids, and sedatives may also induce depression. Hallucinogens and stimulants can trigger psychotic behaviors. SUD may arise from family problems, unemployment, poor decisions, financial loss, cognitive dysfunction, imprisonment, and sexual relationships that otherwise might not have occurred.

Intravenous drug misuse can be linked to HIV, viral hepatitis (B or C), and *Staphylococcal endocarditis*. Heart attacks and strokes may be associated with cocaine use. Alcohol and smoking correlate with more severe drug abuse. Certain drug overdoses can lead to respiratory depression, confusion, seizures, coma, and death. Drug misuse may contribute to violence, suicide, and motor vehicle accidents.

Check Your Knowledge

1. What are the various factors that may influence addiction?
2. What is the description of SUDs?
3. What are the complications of intravenous drug misuse?

CENTRAL NERVOUS SYSTEM STIMULANTS

CNS stimulants increase brain activity and are available by prescription to treat narcolepsy, ADHD, and obesity. Stimulant drugs are often used illegally to "get high." They enhance the effects of dopamine, norepinephrine, and serotonin in the central and peripheral nervous systems. Overdoses of stimulants may cause seizures and cardiac arrest. Examples of stimulants include cocaine, amphetamines, methylphenidate, and caffeine. Long-term abuse can lead to brain and heart damage.

COCAINE

Cocaine is a highly addictive, illegal stimulant drug derived from coca leaves, a plant native to South America. It is classified as a Schedule II substance due to its high potential for misuse, yet it can be legally prescribed for specific medical purposes. Most users inhale the drug by "snorting" it, taking it orally, or injecting it. The rock form of cocaine is often referred to as *crack* or *crack cocaine* and is smoked.

Cocaine is addictive because it directly impacts the brain's reward pathway, a system that reinforces pleasurable experiences and motivates users to repeat them. Cocaine blocks this transporter from removing dopamine, leading to the accumulation of dopamine in the synapse and producing an amplified signal to the receiving receptors. It also establishes a strong association between cocaine use and feelings of **euphoria**.

Cocaine induces intense euphoria, analgesia, decreased appetite, heightened sensory perception, and illusions of physical strength. Larger doses amplify these effects and can lead to tachycardia, mydriasis, hyperthermia, and sweating. When the euphoric feelings diminish, symptoms such as insomnia, agitation, depression, and extreme distrust of others may arise. Other symptoms include **rhinorrhea**, reddening of the surrounding area, and deterioration of the nasal cartilage. Overdose may cause seizures, dysrhythmias, stroke, and respiratory arrest.

YOU SHOULD REMEMBER

Chronic cocaine use can lead to myocarditis, heart attacks, strokes, and even sudden death. Snorting cocaine may cause perforation of the nasal septum. Smoking crack cocaine can result in a chronic cough, lung damage, and respiratory infections. Prolonged cocaine use can lead to neurological damage, memory loss, cognitive impairment, and seizures.

AMPHETAMINES

Amphetamines are stimulants that affect the brain and are used to treat conditions such as depression, ADHD, narcolepsy, obesity, and binge eating disorder. At high doses, amphetamines induce alertness, self-confidence, euphoria, and a sense of empowerment. Prolonged use typically leads to anxiety and restlessness. Newer medications are much safer and have replaced amphetamines, except for certain specific conditions. Illegally manufactured amphetamines remain popular in substance abuse.

Dextroamphetamine (Dexedrine) is used for short-term weight loss after other methods have failed and for narcolepsy. The effects of methamphetamine are similar to those of cocaine but last much longer. Amphetamines are generally taken orally or injected; however, the addition of "ice," the crystallized methamphetamine hydrochloride, allows it to be smoked. Unlike other hallucinogens, it is a Schedule II drug (marketed as Desoxyn). Methamphetamine is often obtained from illegal "meth" laboratories.

METHYLPHENIDATE

Methylphenidate (Ritalin) is related to amphetamines and stimulates the CNS. It is used for ADHD and narcolepsy. When taken by adults, Ritalin typically causes the opposite effects, especially in those without a diagnosis of ADHD. Methylphenidate is classified as a Schedule II drug, producing effects similar to those of amphetamines and cocaine. Illegally, the drug can be administered by inhalation or injection. The adverse effects of methylphenidate include headache, vomiting, dizziness, hypertension, and loss of appetite.

CAFFEINE

Caffeine is a psychoactive substance that stimulates the brain and is the most widely consumed psychoactive substance globally. It promotes wakefulness, enhances physical performance, and improves cognition. This stimulant is naturally found in plants' seeds, fruits, and leaves. Caffeine increases the effectiveness of over-the-counter pain relievers and is often included in their formulations. Consuming two or three cups of coffee daily is beneficial for heart health, while excessive intake can be harmful. The substance also causes **diuresis** and increases mental alertness, restlessness, irritability, nervousness, and insomnia. Its physical effects include bronchodilation, elevated blood pressure, increased stomach acid production, and heightened blood glucose levels. Repeated use can lead to physical dependence and tolerance, while withdrawal may result in fatigue, headaches, depression, and impaired physical and mental performance.

Check Your Knowledge

1. What are the symptoms of cocaine misuse?
2. What are the indications of amphetamines in medicine?
3. What are the physical effects of caffeine?

HALLUCINOGENS

Hallucinogens are psychedelic drugs that can potentially change how people see, hear, taste, smell, or feel. They also influence mood and thought. Although hallucinogens are diverse, they produce similar effects, such as altering consciousness. They are classified as Schedule I drugs with no approved medical use. Hallucinogens include lysergic acid diethylamide (LSD), mescaline, ketamine, and phencyclidine.

LYSERGIC ACID DIETHYLAMIDE

LSD is a synthetic drug that has been abused for its hallucinogenic properties since the 1960s. If consumed in a sufficiently large dose, LSD produces delusions and visual hallucinations that distort the user's sense of time and identity. It is a potent hallucinogen with a high potential for abuse and currently has no accepted medical use in treatment in the United States. LSD may cause visual hallucinations, deep personal insight, hyperthermia, tachycardia, hypertension, pupil dilation, sweating, loss of appetite, insomnia, dry mouth, and tremors. While under the influence, the user may experience impaired depth and time perception, accompanied by a distorted perception of the shape, size, movement, color, sound, touch, and body image. The ability to make sound judgments and perceive common dangers is impaired, making the user susceptible to personal injury. Tolerance may develop, but little or no dependence has been observed.

MESCALINE

Mescaline is a naturally occurring psychedelic drug with hallucinogenic effects similar to LSD. It is found in the peyote cactus, native to Mexico and Central America, and is used by Native Americans as part of traditional religious ceremonies. As a party drug, it produces hallucinations and alters consciousness for 4–6 hours. It may also cause anxiety, tachycardia, dizziness, diarrhea, vomiting, and headaches, but it is not physically addictive. Mescaline can be a powder, tablet, capsule, or liquid, and peyote buttons can be chewed to achieve the desired effects.

KETAMINE

Ketamine is a dissociative anesthetic with some hallucinogenic effects. It distorts the perception of sight and sound, causing the user to feel disconnected and out of control. Ketamine is called a "dissociative anesthetic hallucinogen" because it allows patients to feel detached from their pain and surroundings. This substance is highly effective for brief medical procedures that do not require skeletal muscle relaxation and can be used as a pre-anesthetic agent to initiate general anesthesia when combined with other general anesthetic agents. Ketamine can be misused as a powder that is snorted, mixed into drinks, or smoked. It is classified as a dissociative anesthetic, not a psychedelic drug like LSD, DMT, and psilocybin mushrooms.

PHENCYCLIDINE

Phencyclidine (PCP), also known as "Angel Dust," produces a prolonged, **trancelike state** that can cause severe brain damage. It is an addictive drug, and its use often leads to psychological dependence, cravings, and compulsive behavior. PCP causes unpleasant psychological effects, and users usually become violent or suicidal. This agent poses particular risks for young people. Even moderate use of the drug can negatively impact the hormones associated with normal growth and development. PCP use can also hinder the learning process in teenagers. It has a distinctive bitter taste and is available in tablets, capsules, and colored powders, which can be taken orally or snorted. A liquid form is also available and is highly flammable. Unlike other hallucinogens, this Schedule II drug is illegal.

High doses of PCP can cause seizures, coma, and even death. At high doses, PCP's effects may resemble the symptoms associated with schizophrenia, including delusions and paranoia. Long-term use of PCP can lead to memory loss, difficulty with speech or thought, depression, and weight loss. These problems can persist for up to a year after an individual has stopped using PCP.

Check Your Knowledge

1. What is the description of hallucinogens?
2. What drugs can cause hallucinations?
3. What are the adverse effects of phencyclidine?

CENTRAL NERVOUS SYSTEM DEPRESSANTS

CNS depressants induce relaxation or sedation and include barbiturates, benzodiazepines, nonbarbiturate hypnotics, opioids, and gamma-hydroxybutyrate (GHB). The CNS depressants authorized for medical use are controlled substances due to their potential for abuse.

BARBITURATES AND NONBARBITURATE SEDATIVE-HYPNOTICS

There are two major types of sedative drugs: barbiturates and nonbarbiturate sedative-hypnotics. They are used to treat sleep disorders and certain types of epilepsy. Individuals with addictions often switch between these drugs and amphetamines to manage their sleep patterns and wakefulness. Some sedatives can last an entire day, and higher doses may produce effects similar to alcohol intoxication. Overdoses pose significant dangers, potentially leading to respiratory arrest, coma, and death. Sedative-hypnotic drugs are discussed in more detail in **Chapter 9**.

BENZODIAZEPINES

Benzodiazepines are CNS depressants that are widely prescribed and have replaced barbiturates mainly for various conditions due to their higher safety margin. They are typically used for anxiety but also for muscle relaxation and seizure prevention. Their potential for abuse is significant. Benzodiazepines may cause individuals to appear distant from others, lightheaded, disoriented, or sleepy. Withdrawal symptoms are not as severe as those from barbiturates or alcohol. However, benzodiazepines can remain in circulation for several weeks because, as a class, they have an average half-life of about 12–40 hours, making abuse extremely dangerous. Benzodiazepines are also discussed in more detail in **Chapter 9**.

OPIOIDS

Opioids and opiates are used to treat severe pain, diarrhea, and persistent cough and are often referred to as *narcotics*. Opiates include morphine, codeine, and opium. Synthetic versions include meperidine, methadone, oxycodone, fentanyl, and illegal drugs such as heroin. Injecting heroin induces euphoria and sedation. Opioid overdose can lead to death due to opioids' effects on the brain, which regulates breathing. An opioid overdose may cause pinpoint pupils, unconsciousness, and dyspnea. Opioid addiction can develop swiftly, and withdrawal is often unpleasant. Methadone and buprenorphine are frequently used to treat patients addicted to opioids. Opioids are discussed in more detail in **Chapter 28**.

GAMMA HYDROXYBUTYRATE

GHB is an illicitly marketed substance that has recently gained popularity among bodybuilders and partygoers as a recreational drug. GHB is a potent CNS depressant that can be lethal when combined with alcohol or other depressants. This agent may cause euphoria, increased sex drive, or tranquility. Adverse effects include loss of consciousness, **amnesia**, hallucinations, nausea, sweating, and coma. GHB is available in liquid or powder form and has anabolic effects that assist in muscle building. Currently, there is no accepted medical use for GHB, and the U.S. Food and Drug Administration has prohibited its manufacture and sale.

People who use GHB regularly can develop a tolerance to the effects of the drug, meaning they may need to take more to achieve the desired effect. Regular use can also lead to physical dependence. Those who are physically dependent on GHB will experience withdrawal symptoms if they abruptly stop using it. Withdrawal symptoms may include anxiety, tremors, insomnia, hallucinations, and hypertension. Individuals who are physically dependent on GHB should seek medical help to manage withdrawal. GHB withdrawal can be life-threatening.

YOU SHOULD REMEMBER

Fentanyl is a potent synthetic opioid used for surgery or to manage severe pain. Most fentanyl-related harms and overdoses are associated with illegally manufactured fentanyl. Fentanyl analogs have significantly contributed to the alarming rise in drug overdose deaths in the U.S.

YOU SHOULD REMEMBER

Nitazenes are synthetic opioids that can be up to 43 times stronger than fentanyl. These drugs are being mixed into common street drugs, unknown to users. Unfortunately, nitazenes do not always respond to the use of Narcan, so overdoses may be untreatable.

CANNABINOIDS

Cannabinoids are derived from the hemp plant (*Cannabis sativa*) and are typically smoked to experience their effects. The cannabis plant comprises approximately 540 chemical substances.

Examples of **cannabinoids** include marijuana, hashish, and hash oil. The most psychoactive agent in these substances is delta-9-tetrahydrocannabinol (THC). Additionally, there are synthetic cannabinoids and **designer drugs** that have been chemically altered to evade federal regulatory laws. Synthetic cannabis is sold as *Spice* and K2. It binds to the same receptors in the brain that natural cannabinoids do, with potentially more dangerous consequences, including psychosis, paranoia, and hallucinations.

MARIJUANA

Marijuana is a natural substance and the most commonly used illicit drug in the United States. It is regularly referred to as "weed," "pot," "grass," "dope," or "reefer." It contains substantial amounts of THC, which is primarily responsible for its effects on a person's mental state. Some cannabis plants contain very little THC, and under U.S. law, these plants are considered "industrial hemp" rather than marijuana. Marijuana decreases motor activity and coordination, causing paranoia, euphoria, confusion, increased thirst, and hunger. Daily use increases the risk of lung cancer and respiratory conditions, often resulting in a lack of motivation. Marijuana produces little physical dependence or tolerance, with nearly nonexistent withdrawal symptoms. The metabolites of THC can remain in the body for many months (or years, depending on usage), so tests can easily show that a person has been using marijuana or has quit before the tests. Additionally, THC is detectable in the urine for several days after using marijuana.

Medical marijuana may assist in treating certain rare forms of disease. Conditions addressed with medical marijuana include glaucoma, weight loss associated with HIV/AIDS, nausea and vomiting related to cancer, **amyotrophic lateral sclerosis**, epilepsy, **Crohn's disease**, **multiple sclerosis**, muscle spasms, seizures, severe and chronic pain, and severe nausea. Prolonged marijuana use can lead to various effects on the body and brain, such as altered perception, mood changes, impaired cognition, tachycardia, orthostatic hypotension, and bloodshot eyes.

CIGARETTE SMOKING

Smoking cigarettes is a leading cause of preventable disease and death in the United States. It harms almost all organs in the body and can lead to many diseases, including heart attack, hypertension, stroke, **emphysema**, cancer, type 2 diabetes, cataracts, and immune system deficiency. Nicotine addiction is the most common of all, likely because it is highly reinforcing. The physical adverse effects of smoking are often severe later in life, including cancer, **emphysema**, and heart attacks. About 14% of Americans smoke, and because it is still legal, many do not perceive smoking as an addiction.

Smoking enhances alertness, focus, and relaxation but can lead to lightheadedness. Nicotine directly stimulates the brain and boosts the **basal metabolic rate**. Both physical and psychological dependence can develop relatively quickly. Withdrawal can result in anxiety, cravings, poor attention, insomnia, gastrointestinal disturbances, irritability, and headaches.

In the United States, over 480,000 people die each year from cigarette smoking and secondhand smoke exposure. Almost 50% of Americans who continue to smoke will die from a smoking-related disease. Smoking is the primary cause of lung cancer deaths, and other cancers are also significantly more common due to smoking. Passive smoking can affect a fetus, leading to low birth weight, intellectual disabilities, neonatal death, and **stillbirth**. Children of smokers are more likely to experience asthma, pneumonia, **bronchitis**, growth retardation, and **sudden infant death syndrome**.

ALCOHOL ABUSE

Alcohol use disorder (AUD) is a medical condition characterized by an impaired ability to stop or control alcohol use despite adverse social, occupational, or health consequences. It includes conditions that some people refer to as alcohol abuse, alcohol dependence, alcohol addiction, and the colloquial term alcoholism. Considered a brain disorder, AUD can be mild, moderate, or severe. Lasting changes in the brain caused by alcohol misuse perpetuate AUD and increase individuals' vulnerability to relapse. According to the 2023 National Survey on Drug Use and Health, 28.1 million adults aged 18 and older had AUD in the United States. Among youth, an estimated 757,000 adolescents ages 12–17 had AUD during this time frame.

Ethyl alcohol is a commonly abused drug that is legal and readily available. The abuse of alcohol has enormous economic, health, and social consequences. In small quantities, red wine has been proven to reduce the risk of heart attack and stroke; however, long-term alcohol abuse can have devastating effects. Alcohol can depress the brain and readily cross the blood–brain barrier. The results depend on the amount consumed, which may include relaxation, sedation, impaired memory, loss of coordination, and decreased judgment. Alcohol mimics the effects of antidepressants, antianxiety agents, and sedatives.

Symptoms of alcohol overdose include mental confusion, difficulty remaining conscious, vomiting, seizures, dyspnea, bradycardia, clammy skin, and hypothermia. Alcohol overdose can lead to permanent brain damage, coma, or death. Alcohol should never be combined with other CNS depressants because the effects are cumulative, potentially leading to profound sedation and coma. Withdrawal from alcohol can be severe and possibly life-threatening, causing insomnia, anxiety, or seizures. Long-term alcoholics may experience **delirium tremens**. This condition results in **paranoia**, confusion, disorientation, agitation, hallucinations, and fear. Chronic alcohol abuse primarily damages the liver and causes cirrhosis. **Cirrhosis** of the liver can lead to liver failure and death (see Figure 13.1). Impairment makes the liver more sensitive to all substances it metabolizes.

YOU SHOULD REMEMBER

A new report from the Surgeon General's office in the United States describes the scientific evidence that alcohol consumption increases the risk of at least seven types of cancer, including those of the mouth, throat, larynx, esophagus, colorectal region, liver, and breast. The report also includes recommendations to reduce alcohol-related cancers.

Check Your Knowledge

1. What are the adverse effects of marijuana?
2. What are the medical indications of marijuana?
3. What are the symptoms of alcohol overdose?

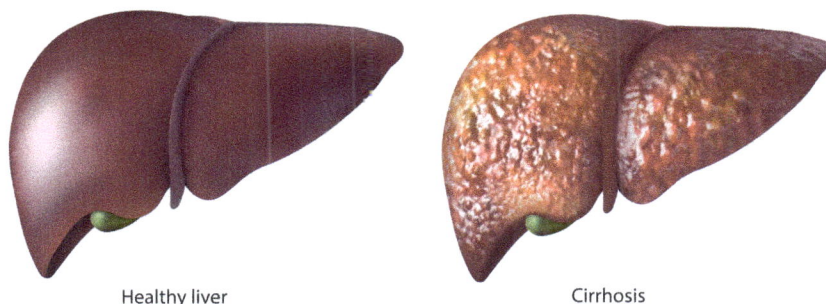

Healthy liver

Cirrhosis

FIGURE 13.1 Cirrhosis of the liver.

PREVENTION AND TREATMENT FOR SUBSTANCE ABUSE

Prevention is the key to substance abuse. Drug detoxification is the first step when treating addiction patients. Withdrawal may cause physical and psychological symptoms and can require hospitalization. Detoxification involves overcoming physical withdrawal when a person with an addiction stops taking a drug, and medical management may be necessary to prevent severe health problems. Medications can help modify brain chemistry to treat specific SUDs. They can also relieve cravings and withdrawal symptoms. To treat opioid addiction, methadone, buprenorphine, and naltrexone are used. For alcohol treatment, acamprosate and disulfiram are available. For tobacco, a nicotine patch, nasal spray, or gum can be used.

CLINICAL CASE STUDIES

Clinical Case Study 1

A 16-year-old boy exhibits signs of irritability and depression. He comes home later than expected. After several weeks, his mother received a phone call from the school, where the boy had been found having seizures on the classroom floor. The principal stated that he was transported to the emergency department. After examining, testing, and confirming that the boy had no history of epilepsy, the physician informed the mother that there was cocaine in his system.

CRITICAL THINKING QUESTIONS

1. What symptoms are associated with cocaine addiction?
2. What is the mechanism of action of cocaine in the body?
3. What are the adverse effects of cocaine on the body?

Clinical Case Study 2

A 45-year-old man was diagnosed with AIDS more than 10 years ago. Following the advice of his support group, he started using marijuana when his medical treatment for AIDS began. To avoid any risk related to inhaling contaminated marijuana, he purchased a cannabis tincture.

CRITICAL THINKING QUESTIONS

1. What are the effects of marijuana on the body?
2. What are the indications for the medical use of marijuana?
3. What are the adverse effects of marijuana?

FURTHER READING

Brooks, F., and McHenry, B. (2015). *A Contemporary Approach to Substance Use Disorders and Addiction Counseling*, 2nd Edition. American Counseling Association.

Connors, G.J., DiClemente, C.C., Marden Velasquez, M., and Donovan, D.M. (2015). *Substance Abuse Treatment and the Stages of Change — Selecting and Planning Interventions*, 2nd Edition. Guilford Press.

Coughlin, G., Kimbrough, S.S., and Kimbrough, L.L. (2008). *Patient Records and Addiction Treatment*, 4th Edition. Southworth Press.

Eskapa, R., Christian, C., and Sinclair, D. (2012). *The Cure for Alcoholism: The Medically Proven Way to Eliminate Alcohol Addiction*. BenBella Books.

Fisher, G., and Harrison, T. (2017). *Substance Abuse: Information for School Counselors, Social Workers, Therapists, and Counselors*, 6th Edition. Pearson.

Koob, G.F., Arends, M.A., and Le Moal, M. (2014). *Drugs, Addiction, and the Brain*. Academic Press.

Kouimtsidis, C., Reynolds, M., Drummond, C., Davis, P., and Tarrier, N. (2007). *Cognitive-Behavioral Therapy in the Treatment of Addiction — A Treatment Planner for Clinicians*. Wiley.

MacMillan, T., and Sisselman-Borgia, A. (2018). *New Directions in Treatment, Education, and Outreach for Mental Health and Addiction* (Advances in Mental Health and Addiction). Springer.

Marden Velasquez, M., Crough, C., Stokes Stephens, N., and DiClemente, C.C. (2015). *Group Treatment for Substance Abuse: A Stages-of-Change Therapy Manual*, 2nd Edition. Guilford Press.

Perkinson, R.R. (2016). *The Alcoholism and Drug Abuse Client Workbook*. Sage Publications, Inc.

Pichot, T., Smock, S.A., and Nelson, T.S. (2011). *Solution-Focused Substance Abuse Treatment*. Routledge.

Powell, D.J., and Brodsky, A. (2004). *Clinical Supervision in Alcohol and Drug Abuse Counseling: Principles, Models, Methods*. Jossey-Bass.

Reiter, M.D. (2019). *Substance Abuse and the Family: Assessment and Treatment*, 2nd Edition. Routledge.

Ruiz, P., and Strain, E. (2011). *Lowinson and Ruiz's Substance Abuse: A Comprehensive Textbook*, 5th Edition. Wolters Kluwer Health/Lippincott, Williams, & Wilkins.

Smith, R.L. (2015). *Treatment Strategies for Substance and Process Addictions*. American Counseling Association.

Yalisove, D. (2010). *Developing Clinical Skills for Substance Abuse Counseling*. American Counseling Association.

14 Drugs Affecting Coronary and Heart Disorders

DOI: 10.1201/9781003461913-16

LEARNING OUTCOMES

After studying the chapter, readers should be able to

1. Review the physiology of coronary arteries and the heart.
2. Explain angina pectoris and its etiology.
3. Describe the significant anti-ischemia drugs.
4. Summarize the classification of drugs for angina pectoris.
5. Discuss the contraindications of nitroglycerin.
6. Explain the indications of beta-blockers.
7. Describe the mechanism of action of calcium channel blockers (CCBs).
8. Discuss subacute and chronic myocardial infarction (MI).
9. Review the signs and symptoms of MI.
10. Summarize the goals of pharmacotherapy for MI.

OVERVIEW

Coronary heart disease is a type of heart disease in which the heart's arteries cannot deliver enough oxygen-rich blood to the heart. It is also sometimes referred to as coronary artery disease (CAD) or ischemic heart disease. Cardiovascular disorders affect the heart and blood vessels and are often related to atherosclerosis and arteriosclerosis. As plaque builds up in the arterial walls, it narrows the arteries, making it more difficult for blood to flow normally. The heart muscle requires a constant supply of oxygen-rich blood. CAD that narrows one or more of these arteries can block blood flow, resulting in angina or MI. Discomfort may not always be felt in the chest; it can also occur in the jaw or arm when a blood clot forms, potentially blocking blood flow entirely, which can lead to MI or stroke. Other chronic cardiovascular disorders include hypertension and heart failure. Many medications are available to treat CAD, including statins, niacin, fibrates, and bile acid sequestrants. Other treatments include nitroglycerin, aspirin, beta-blockers, CCBs, angiotensin-converting enzyme inhibitors, and angiotensin II receptor blockers. Drugs used for MI include morphine, thrombolytics or fibrinolytics, and beta-blockers.

PHYSIOLOGY OF THE HEART AND CORONARY ARTERIES

The heart wall consists of three layers: the outer **epicardium**, the middle **myocardium**, and the **endocardium**. The myocardium contains cardiac muscle fibers and is responsible for the contraction and relaxation that enable the heart to pump. The two atria are thinner than the ventricles because the force required for atrial contractions is much less than that needed for ventricular contractions. The right ventricle's walls pump blood to the lungs. In contrast, the left ventricle, which has much thicker walls, must generate sufficient force to pump oxygenated blood throughout the rest of the body (see Figure 14.1).

When the right ventricle contracts, deoxygenated blood flows through the **pulmonary valve** into the lungs via the pulmonary trunk. The **pulmonary trunk** divides into the left and right **pulmonary arteries**, the only arteries in the body that carry deoxygenated blood. However, during pregnancy, the umbilical artery transports deoxygenated blood from the fetus to the placenta to re-oxygenate the fetal blood supply. The left ventricle receives oxygenated blood from the left atrium through the **mitral valve**. The aortic valve remains closed to enable the left ventricle to fill with blood. When filled, the ventricles contract, causing the mitral valve to close and the aortic valve to open. Oxygenated blood is pumped through the aortic valve into the aorta when the left ventricle contracts. The aortic valve directs blood to the rest of the body.

THE CORONARY ARTERIES

The myocardium requires constant oxygen and nutrients to maintain the contractions and relaxations that keep the heart pumping. This blood supply is supported through the **coronary arteries** and veins within the myocardium. The right and left coronary arteries, which branch from the first section of the aorta known as the ascending aorta, located between the left ventricle and the aortic arch, supply blood to a network of capillaries in the myocardium. Deoxygenated blood from the myocardium is transported through **cardiac veins** to the right atrium and drained through the **coronary sinus**. The coronary arteries run along the **coronary sulcus** of the heart's myocardium. Their primary function is to supply blood to the heart, which is crucial for myocardial performance and overall body homeostasis. The four main coronary arteries include the left and right coronary arteries. The left coronary artery extends along the coronary sulcus and supplies the left side of the heart, further branching into the **circumflex artery** and the left anterior descending artery.

The right coronary artery supplies blood to the right ventricle, right atrium, sinoatrial (SA) and atrioventricular (AV) nodes, which regulate heart rhythm. It divides into smaller branches, including the right posterior descending artery and the acute marginal artery (see Figure 14.2).

Anatomy of the Heart

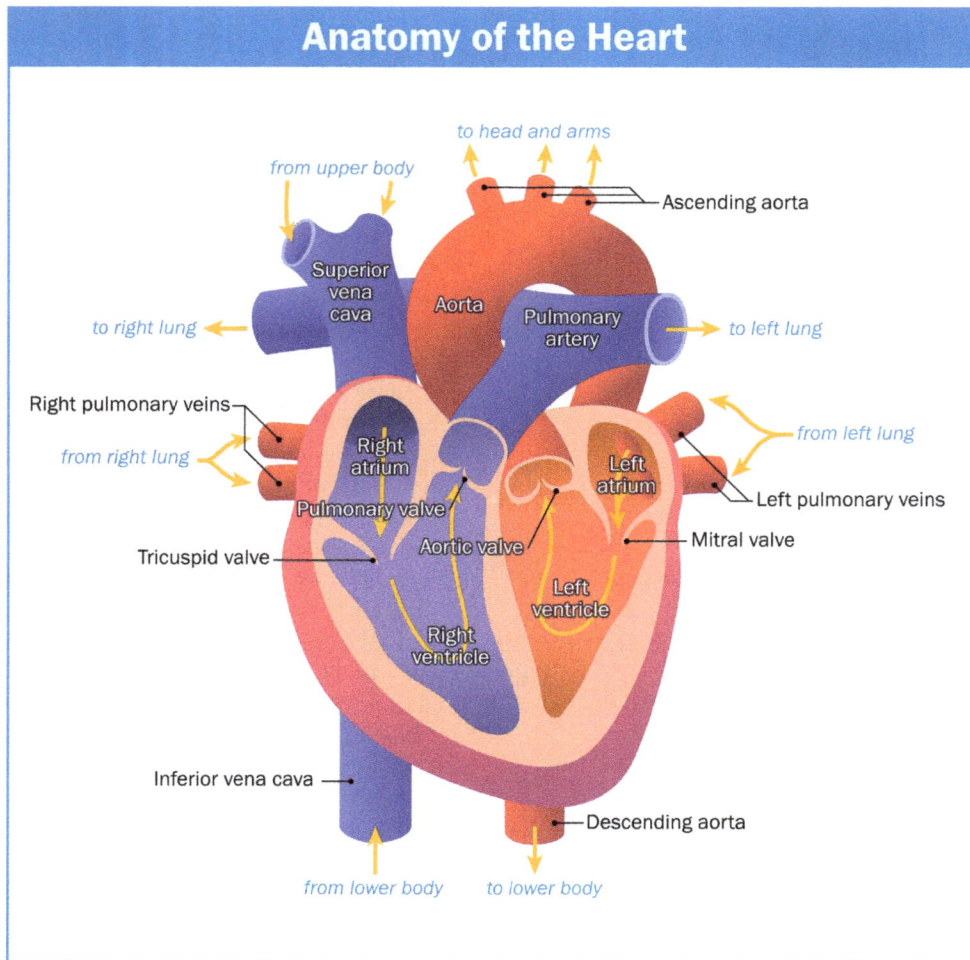

FIGURE 14.1   Structure of the heart.

CORONARY ARTERIES

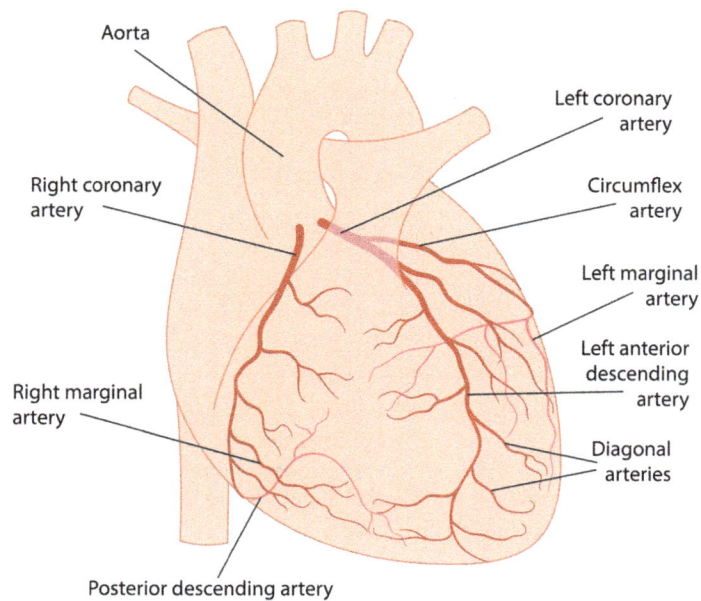

FIGURE 14.2   Coronary arteries.

THE CARDIAC CONDUCTION SYSTEM

The specialized **cardiomyocytes** that make up the conduction system in the human heart initiate the electrical impulse, resulting in the rhythmic and synchronized contraction of the atria and ventricles. An electrical stimulus is generated by the SA node, a tissue located in the right upper atrium. The sinus node regularly produces electrical stimulation, 60–100 times per minute under normal conditions. The atria are activated as the electrical stimulus travels down the conduction pathways, causing the heart's ventricles to contract and pump blood (see Figure 14.3). The atria are stimulated first and contract briefly before the ventricles.

The electrical impulse travels from the SA node to the AV node. The impulses are briefly slowed down, then continue down the conduction pathway via the **bundle of His** into the ventricles. The bundle of His divides into right and left pathways, called bundle branches, to stimulate the right and left ventricles. Typically, at rest, as the electrical impulse moves through the heart, it contracts about 60–100 times a minute, depending on a person's age. Each contraction of the ventricles represents one heartbeat. The atria contract a fraction of a second before the ventricles, allowing their blood to empty into the ventricles before the ventricles contract.

ANGINA PECTORIS

Angina pectoris is a clinical syndrome characterized by precordial discomfort or pressure caused by transient myocardial ischemia without infarction. It is often triggered by exertion or psychological stress. There are four types of angina pectoris:

- **Stable angina:** Pain is predictable in frequency and duration. It is relieved by rest and nitroglycerin.

Heart Conduction System

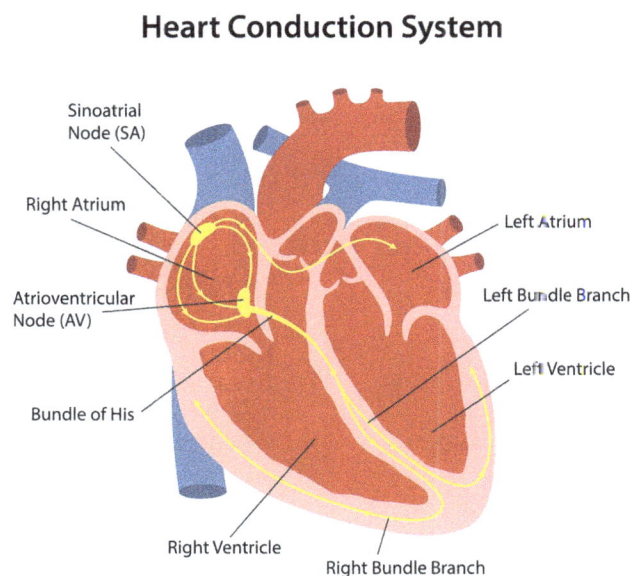

FIGURE 14.3 Cardiac conduction system.

- **Unstable angina:** Pain increases in frequency and duration. It is induced more quickly than stable angina. This indicates worsening CAD that may progress to MI.
- **Prinzmetal's angina:** It is also referred to as *variant angina*. Spasms of the coronary arteries cause pain. It may occur spontaneously and be unrelated to physical exercise or emotional stress.
- **Microvascular angina**: It is caused by impairment of vasodilator reserve, resulting in angina-like chest pain in a person with normal coronary arteries.

ETIOLOGY OF ANGINA PECTORIS

Angina pectoris occurs when the myocardium does not receive enough blood and oxygen, known as ischemia. It can be a symptom of CAD, which develops when plaque accumulates in the coronary arteries that supply blood to the heart. The narrowing or hardening of the arteries, known as atherosclerosis, reduces blood flow to the heart.

PATHOPHYSIOLOGY OF ANGINA PECTORIS

Coronary atherosclerosis typically occurs at sites of turbulence known as vessel bifurcations. If an atherosclerotic plaque ruptures, it may lead to an acute thrombus, resulting in ischemia that can trigger angina pectoris. The effects of acute ischemia are collectively referred to as acute coronary syndromes. A transient, localized increase in vascular tone is known as coronary artery spasm, which significantly narrows the lumen and reduces blood flow, potentially causing symptomatic ischemia or variant angina.

Thrombi may form as a result, leading to infarction or life-threatening arrhythmia. In arteries without atheroma, the baseline coronary artery tone is likely increased. The response to vasoconstricting stimuli may then be heightened. In arteries with atheromas, local hypercontractility may occur. Causes of coronary artery spasms include abnormalities in nitric oxide production, decreased sensitivity to intrinsic vasodilators, increased production of vasoconstrictors, the use of vasoconstricting drugs, and emotional stress.

YOU SHOULD REMEMBER

CAD is a major contributor to cardiovascular morbidity and mortality in the United States, primarily because the initial coronary event is often quickly fatal.

PHARMACOTHERAPY FOR ANGINA PECTORIS

Treatment of CAD involves medications and procedures aimed at reducing ischemia and restoring or improving coronary blood flow. Nitrates, such as nitroglycerin, are commonly administered sublingually, orally, transdermally, or topically in ointment form. Isosorbide may decrease myocardial oxygen consumption. Beta-adrenergic blockers can lessen the heart's workload and oxygen demands by lowering heart rate and peripheral resistance to blood flow. CCBs may help prevent coronary artery spasms. Antiplatelet drugs reduce platelet aggregation and minimize the risk of coronary occlusion. Antilipemic agents can also lower serum cholesterol or triglycerides.

Coronary artery bypass graft (CABG) surgery restores blood flow by bypassing an occluded artery using another vessel, providing a surgical option for certain patients. Angioplasty relieves the blockage for those with partial occlusion and no calcification. Lifestyle modifications are recommended to slow the progression of CAD, including regular exercise, smoking cessation, maintaining a healthy body weight, stress management, and following a low-fat, low-sodium diet.

CLASSIFICATION OF DRUGS FOR ANGINA PECTORIS

Three major anti-ischemia drugs include organic nitrates, β-blockers, angiotensin II receptor blockers, and calcium channel antagonists. Organic nitrates or β-blockers are typically preferred for the initial treatment of angina.

ORGANIC NITRATES

Organic nitrates are potent vasodilators and are the most commonly used antianginal agents during acute events. They selectively dilate epicardial coronary arteries, enhancing collateral flow and inhibiting platelet aggregation. Examples include nitroglycerin, isosorbide dinitrate, and isosorbide mononitrate.

Clinical Indications

Nitroglycerin and long-acting nitrates are commonly used for all anginal symptoms and have proven effective in relieving or preventing myocardial ischemia. They are also utilized to manage hypertension through vasodilation. Nitroglycerin can be administered through various methods, including sublingual and transdermal routes, rectal ointment, and IV infusion.

Mechanism of Action

Organic nitrate esters exert a direct relaxant effect on vascular smooth muscles. The dilation of coronary vessels enhances oxygen supply to the myocardium. Additionally, dilation of peripheral veins and, at higher doses, peripheral arteries decreases preload and afterload, thereby reducing myocardial oxygen consumption.

Adverse Effects

Nitroglycerin's adverse effects include headaches, nausea, dizziness, blurred vision, weakness, flushing, syncope, tachycardia, palpitations, and orthostatic hypotension. Dermal forms can cause local rashes and irritation. Nearly 10% of patients cannot tolerate nitrates due to debilitating headaches or dizziness.

Contraindications

Nitroglycerin is contraindicated for patients with known hypersensitivity to this agent, those who are hypotensive, and children under 12 years of age. Patients with severe anemia or alcohol intoxication should also avoid nitroglycerin.

YOU SHOULD REMEMBER

Sublingual nitroglycerin is administered before acute attacks or to prevent an attack during exertion. Pain relief typically occurs within 1.5–3 minutes. Patients should always carry nitroglycerin tablets or an aerosol spray.

BETA-ADRENERGIC BLOCKERS

Beta-adrenergic blockers inhibit the action of epinephrine and noradrenaline on beta-adrenergic receptors in the body. Several beta-adrenergic blockers are available (see Table 14.1).

Clinical Indications

Beta-adrenergic blockers (β-blockers) are an essential class of drugs used to treat various heart diseases, including angina pectoris, arrhythmias, hypertension, hypertrophic cardiomyopathy, and heart failure. β-Blockers can also help alleviate migraine headaches and glaucoma. As some of the most commonly prescribed medications in the United States, approximately 30 million adults utilize beta-blockers.

Mechanism of Action

Beta-blockers reduce nerve impulses to the heart and blood vessels, minimizing the effects of norepinephrine throughout the body. They cause the heart to beat more slowly and with less force, which lowers cardiac output and blood pressure.

Adverse Effects

Beta-blockers can cause common adverse effects, such as cold fingers or toes, insomnia, nightmares, extreme tiredness, and dizziness. They can also worsen the signs and

TABLE 14.1

Beta-Adrenergic Blockers

Generic Name	Trade Name
Acebutolol	Generic only
Atenolol	Tenormin
Bisoprolol	Zebeta
Carvedilol	Coreg
Esmolol, injectable	Brevibloc
Labetalol	Normodyne
Metoprolol	Toprol XL, Lopressor
Nadolol	Corgard
Nebivolol	Bystolic
Propranolol	Inderal

symptoms of chronic obstructive pulmonary disease and asthma. Additional side effects may include fatigue, brady-cardia, and fluid retention in the extremities.

Contraindications

Beta-blockers can negatively impact several diseases, conditions, and health concerns.

Beta-blockers are contraindicated in peripheral vascular diseases, diabetes mellitus, chronic obstructive pulmonary disease, and asthma. They are also contraindicated in specific arrhythmias, **Raynaud's phenomenon**, and hypoglycemia.

Beta-blockers are classified as pregnancy category C drugs during the first and second trimesters, and as pregnancy category D drugs during the third trimester.

CALCIUM-CHANNEL BLOCKERS

CCBs are a group of medicines commonly prescribed to treat conditions of the heart and blood vessels. They prevent calcium from entering cardiac and peripheral cells, which stops the blood vessels from constricting. The two main subclasses are dihydropyridine and non-dihydropyridine CCBs (see Table 14.2). These categories differ in their pharmacological action. The dihydropyridine CCBs include amlodipine, felodipine, nifedipine, and nicardipine.

Clinical Indications

CCBs are used to treat angina pectoris, coronary spasms, hypertension, supraventricular dysrhythmias, hypertrophic cardiomyopathy, and **pulmonary hypertension**. They are also prescribed for Raynaud's phenomenon, **subarachnoid hemorrhage**, and migraine headaches.

Mechanism of Action

CCBs inhibit the inward movement of calcium by binding to the long-acting voltage-gated calcium channels in the heart and vascular smooth muscle.

TABLE 14.2
Calcium Channel Blockers

Generic Name	Trade Name
Dihydropyridines	
Amlodipine	Norvasc
Clevidipine, injectable	Cleviprex
Felodipine	Plendil
Nicardipine, injectable	Cardene
Nifedipine	Adalat, Procardia
Nimodipine	Nimotop
Nisoldipine	Sular
Nondihydropyridines	
Diltiazem	Cardizem
Verapamil	Calan, Isoptin

Adverse Effects

CCBs may lead **to** severe fatigue, flushing, headaches, nausea, dizziness, leg swelling, arrhythmia, chest pain, and constipation.

Contraindications

Generally, patients with certain types of heart problems or hypotension may not be able to take CCBs. They should be avoided during pregnancy, in cases of liver damage, heart failure, or certain irregular heart rhythms.

Check Your Knowledge

1. What are the indications of CCBs?
2. What are the adverse effects of CCBs?
3. What are the contraindications of CCBs?

MYOCARDIAL INFARCTION

MI involves necrosis of the myocardium resulting from an occlusion of a coronary artery (see Figure 14.4). Approximately 805,000 people experience MIs in the United States annually, and they are fatal in 30%–40% of cases. Due to increased awareness of heart disease prevention methods and improved diagnostic and treatment techniques, the prevalence of MIs is decreasing. *Chronic* MI occurs more than 1 year after the initial presentation of a patient with acute coronary syndrome. *Subacute* MI occurs between 30 days and 1 year after presentation, while *acute* MI occurs less than 30 days after presentation.

The primary cause of MI is a blockage in a coronary artery, usually due to atherosclerosis, which leads to a blood clot that restricts blood flow to the heart muscle. The risk factors include hypertension, smoking, high cholesterol, diabetes, obesity, an unhealthy diet, age, family history, and gender.

The signs and symptoms of a MI typically begin with pain or pressure beneath the sternum, often radiating outward. This pain may extend to the back, jaw, and left arm. The discomfort is generally more intense and prolonged than angina pectoris and may be accompanied by **dyspnea**, fatigue, **diaphoresis**, nausea, vomiting, and syncope. In addition to reinfarction, the potential complications of acute MI include various types of arrhythmias, heart failure, angina, **cardiogenic shock**, embolization, pericarditis, and death. The most prevalent and serious complication associated with MI is arrhythmia. Deaths resulting from ventricular fibrillation can be prevented if cardiopulmonary resuscitation is initiated within 5 minutes of the onset of fibrillation.

PHARMACOTHERAPY FOR MI

Pharmacotherapy for MI aims to reduce morbidity and prevent complications. The primary objectives of emergency department medical therapy include rapid intravenous thrombolysis, timely referral for percutaneous coronary intervention, optimization of oxygenation, reduction of cardiac workload, and pain control using morphine. Pharmacotherapy for MI

Myocardial infarction

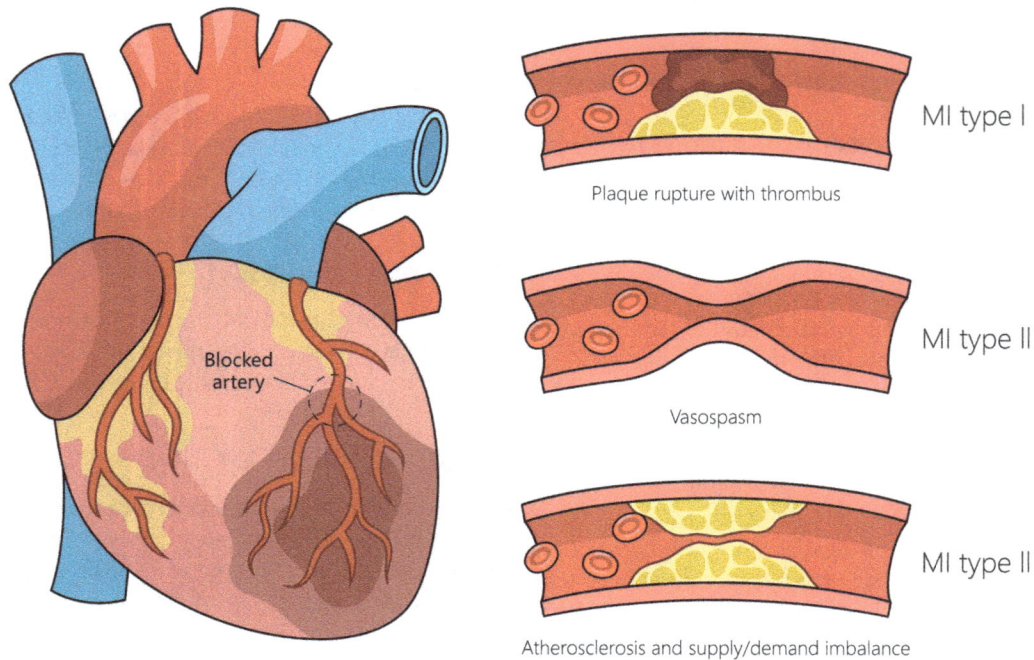

MI type I

Plaque rupture with thrombus

MI type II

Vasospasm

MI type II

Atherosclerosis and supply/demand imbalance

FIGURE 14.4   Myocardial infarction.

encompasses aspirin, thrombolytics, fibrinolytics, nitroglycerin, morphine, beta blockers, and statins.

Thrombolytics or fibrinolytics are a group of medications used to dissolve intravascular clots. They are utilized in the emergency treatment of MI, strokes caused by blood clots, or massive pulmonary embolisms. Some patients who experience MI receive stents in their coronary arteries or undergo CABG, and they are treated with two types of antiplatelet agents simultaneously to prevent blood clotting. Many patients with CAD, including those who have had a heart attack, a stent, or CABG, are prescribed aspirin for the rest of their lives. Examples of thrombolytics include streptokinase (Streptase), tenecteplase (TNKase), reteplase (Retavase Half-Kit), urokinase (Kinlytic), alteplase (Activase), and urokinase (Abbokinase).

Check Your Knowledge

1. What are the descriptions of subacute and chronic MI?
2. What are the contraindications for MI?
3. What is the pharmacotherapy for MI?

CLINICAL CASE STUDIES

Clinical Case Study 1

A 43-year-old man was brought to the emergency department due to severe chest pain that radiated to his left shoulder. The episode lasted for about 15 minutes. His family had no history of CAD, but his father had hypertension and diabetes mellitus. The patient has been smoking for 34 years. Following the performance of the electrocardiogram and cardiac catheterization, he was diagnosed with unstable angina, and dissection of the left anterior descending coronary artery was found.

CRITICAL THINKING QUESTIONS

1. What are the classifications of angina?
2. What are the causes of angina pectoris?
3. What is the treatment for angina pectoris?

Clinical Case Study 2

A 64-year-old woman was transferred to the hospital due to severe chest pain, shortness of breath, sweating, fatigue, nausea, and vomiting. She had a prior diagnosis of hypertension. An EKG showed a heart rate of 140 bpm, sinus rhythm, AV block, and various electrical abnormalities of the heart. Coronary angiography revealed three coronary arteries with 70% blockage. The patient experienced a MI and cardiac arrest. She did not respond to treatment and passed away.

CRITICAL THINKING QUESTIONS

1. What are the causes and risk factors for MI?
2. What are the signs and symptoms of MI?
3. What are the treatments for MI?

FURTHER READING

Allen, B. (2015). *Diagnosis, Pathophysiology, and Treatment of Angina Pectoris*. Foster Academics.

Arampalzis, C., McFadden, E.P., Michalis, L.K., Virmani, R., and Serruys, P.W. (2012). *Coronary Atherosclerosis: Current Management and Treatment*. Informa Healthcare.

Avanzas, P., and Kaski, J.C. (2015). *Pharmacological Treatment of Chronic Stable Angina Pectoris* (Current Cardiovascular Therapy). Springer.

Baliga, R.R., and Eagle, K.A. (2020). *Practical Cardiology: Evaluation and Treatment of Common Cardiovascular Disorders*, 3rd Edition. Springer.

Brown, D.L., and Warriner, D. (2022). *Manual of Cardiac Intensive Care*. Elsevier.

Griffin, B.P. (2018). *Manual of Cardiovascular Medicine*, 5th Edition. Lippincott, Williams, and Wilkins.

Hanna, E.B. (2022). *Practical Cardiovascular Medicine*, 2nd Edition. Wiley Blackwell.

Hawkey, M.C., and Lauck, S.B. (2022). *Valvular Heart Disease — A Guide for Cardiovascular Nurses and Allied Health Professionals*. Springer.

Herring, N., and Paterson, D.J. (2018). *Levick's Introduction to Cardiovascular Physiology*, 6th Edition. CRC Press.

Jevon, P. (2012). *Angina and Heart Attack (The Facts). All The Information You Need Straight from the Experts*. Oxford University Press.

Libby, P., Bonow, R.O., Mann, D.L., Tomaselli, G.F., Bhatt, D.L., and Solomon, S.D. (2022). *Braunwald's Heart Disease — A Textbook of Cardiovascular Medicine*, 12th Edition. Elsevier.

Lilly, L.S. (2020). *Pathophysiology of Heart Disease — An Introduction to Cardiovascular Medicine*, 7th Edition. Lippincott, Williams, and Wilkins.

Miller, D.V., and Revelo, M.P. (2023). *Diagnostic Pathology — Cardiovascular*, 3rd Edition. Elsevier.

U.S. Department of Health and Human Services, Agency for Healthcare Research and Quality. (2013). *Antiplatelet and Anticoagulant Treatments for Unstable Angina/Non-ST Elevation Myocardial Infarction*. AHRC.

15 Drugs Affecting Hyperlipidemia

LEARNING OUTCOMES

After studying the chapter, readers should be able to

1. Review the classifications of lipids and lipoproteins.
2. Explain the role of cholesterol in the human body.
3. Compare the function of low-density lipoprotein (LDL) with high-density lipoprotein (HDL).
4. Summarize the relationship between high cholesterol and atherosclerosis.
5. Identify the most common treatments for hyperlipidemia.
6. Explain the classifications of antidepressants.
7. Discuss the adverse effects of statins.
8. Describe the mechanism of action for bile acid sequestrants (BAS).
9. Explain the significant adverse effects of niacin.
10. Review the newest cholesterol-lowering drugs on the market and their mechanism of action.

OVERVIEW

Hyperlipidemia is characterized by elevated fats or lipids in the blood. "Lipids" generally refer to cholesterol and triglycerides found in the blood. An elevation in low-density lipoproteins, one of the three main subclasses of cholesterol, has been linked to the development of atherosclerosis and the most common type of heart disease, coronary heart disease. Atherosclerosis is the accumulation of plaque, which narrows and stiffens the arteries. This contributes to heart disease, the leading cause of death in the US. Approximately every 34 seconds, one American adult has a coronary event, and roughly every minute, an American adult will die of one type of cardiovascular disease (CVD). This represents a financial burden to society, with the total direct and indirect costs of CVD in 2021 estimated to be $407 billion. For comparison, the estimated cost of all cancers was $201.5 billion. CVD costs more than any other disease state. Because avoiding CVD is desired, hyperlipidemia treatments exist, and these include diet, exercise, and drugs such as statins, BAS, niacin, fibrates, and selective cholesterol absorption inhibitors. Statins are the mainstay of hyperlipidemia treatment.

CLASSIFICATIONS OF LIPIDS AND LIPOPROTEINS

Lipids are a group of organic compounds that are insoluble in water but soluble in organic solvents. They are mainly classified into simple, complex, and derived lipids. Simple lipids are **triglycerides**, esters of fatty acids, and wax esters. The hydrolysis of these lipids gives glycerol and fatty acids.

TABLE 15.1
Classification of Cholesterol Levels

Total Cholesterol

<200 mg/dL	Desirable
200–239 mg/dL	Borderline high
240 mg/dL or higher	High

LDL Cholesterol

<100 mg/dL	Optimal
100–129 mg/dL	Above optimal
130–159 mg/dL	Borderline high
160–189 mg/dL	High
190 mg/dL or higher	Very high

Complex lipids are heterogeneous compounds of steroids, fats, waxes, oils, and other substances. These chemical features are present in many molecules, such as **fatty acids**, **phospholipids**, **sterols**, **sphingolipids**, and **terpenes**.

A lipoprotein's **density** is determined by its protein and lipid content (see Table 15.1). LDL carries 60%–70% of the total blood **cholesterol**. HDL is about 20%–30% of total blood cholesterol. Very-LDL (VLDL) carries approximately 10%–15% of total blood cholesterol and most triglycerides in a fasting state.

Elevated HDL has been linked to removing excess VLDL or LDL from the blood and returning it to the liver or bile for elimination. This process accounts for some of the beneficial effects of high HDL, often called "good cholesterol." When triglycerides are 400 mg/dL, a more complex, specialized laboratory technique is required for accuracy. A fasting lipid panel should be measured in all adults 20 years or older at least once every 5 years. The optimal cholesterol laboratory values for each of the four standard tests in a lipid panel are as follows:

- **Total cholesterol**: Below 200 mg/dL. (Borderline high: 200–239 mg/dL.)
- **HDL cholesterol**: Above 40 mg/dL. (60 mg/dL or above is protective against heart disease.)
- **LDL cholesterol**: Below 100 mg/dL. (For people with diabetes: Below 70 mg/dL.)
- **Triglycerides**: Below 150 mg/dL. (Borderline high: 150–199 mg/dL and High: 200–499 mg/dL.)

FUNCTIONS OF CHOLESTEROL AND LIPOPROTEINS

Cholesterol and triglycerides are the major blood lipids. They are essential for forming cell membranes, hormone production, and providing an energy source of free fatty

DOI: 10.1201/9781003461913-17

acids. Because lipids are not water-soluble, they do not circulate in the blood in free form but are bound to proteins in a macromolecule called a lipoprotein.

The structure of a lipoprotein consists of a lipid core and a surface coat. The lipid core, the interior of a lipoprotein, contains triglycerides and cholesterol esters, both of which are insoluble in water. Cholesterol esters are cholesterol molecules that have been attached to a fatty acid. The surface coat of lipoproteins comprises water-soluble components, including **apolipoprotein B** (Apo-B), phospholipids, and unesterified cholesterol. The phospholipids are oriented so that their water-soluble heads face outward and their fat-soluble tails face inward. Additionally, Apo-B interacts with external water.

Lipids, including triglycerides in food, are broken down into fatty acids and cholesterol by enzymes found in the gut. Fatty acids are then converted to triglycerides and brought back together with cholesterol to make chylomicrons that eventually enter the bloodstream. Intestinal cholesterol absorption is assisted by Niemann-Pick C1-Like 1 protein, which targets the cholesterol-lowering drug ezetimibe. After gaining entry into the circulation, chylomicrons are broken down in muscle and fat cells, where they are stored as energy or used as free fatty acids.

An enzyme breaks down VLDLs into intermediate-density lipoproteins (IDLs). The liver either removes IDLs or further processes them into cholesterol-rich LDLs. LDL particles last longer in circulation than VLDLs and contain about two-thirds of all blood cholesterol. LDLs are recycled into the liver or other peripheral cells via LDL receptors. Cellular cholesterol requirements ultimately regulate this process. When fasting or on a low-fat intake, most cholesterol is created in organs other than the liver. Individual abnormalities may develop in the gene that regulates LDL receptors, proprotein convertase subtilisin-kexin type 9 (PCSK9). This abnormality can cause blood elevations in LDL by increasing the destruction of LDL receptors and decreasing the uptake of LDL into the liver. Lastly, LDL may be excreted into the small intestine fluid and out of the body through the stool.

The liver is primarily responsible for the regulation of cholesterol in the body. The endogenous pathway begins in the liver with the creation of VLDLs. VLDL production is stimulated in the liver when there is an increase in the number of free fatty acids. An enzyme breaks down VLDLs into IDLs. Cellular cholesterol requirements ultimately regulate this process. When fasting or on a low-fat diet, most cholesterol is created in organs other than the liver. Of note, the rate-limiting enzyme for the intracellular creation of cholesterol is 3-hydroxy-3-methylglutaryl coenzyme A (HMG-CoA) reductase, the target of the statin drug class, which includes simvastatin and atorvastatin. A person's genetics may create abnormalities in the gene that regulates LDL receptors, PCSK9. This abnormality can cause blood elevations in LDL by

increasing the destruction of LDL receptors and decreasing the uptake of LDL into the liver. Lastly, LDL may be excreted into the stomach fluid and out of the body through the stool.

BLOOD LIPID PROFILES

A **lipid profile** is a standard blood test that healthcare providers use to monitor and screen for the risk of CVD. A blood lipid profile is used to measure the types and amounts of lipids in the blood, including total cholesterol, HDL cholesterol, LDL cholesterol, and triglycerides. The patient must fast for 12 to 14 hours before the test, consuming only water.

YOU SHOULD REMEMBER

Children can also have high cholesterol to the degree that they may need a lipid profile blood test. High cholesterol in children is linked to heredity, diet, and obesity. In most cases, children with high cholesterol have a parent with elevated cholesterol.

Check Your Knowledge

1. What are the classifications of lipoproteins?
2. What are the laboratory values for optimal cholesterol and triglycerides?
3. Why are cholesterol and triglycerides essential for the body?

BIOSYNTHESIS OF CHOLESTEROL AND EXCRETION

Large amounts of cholesterol are found in the brain, nervous tissues, liver, kidneys, spleen, and skin. The total cholesterol for a man weighing 70 kg is about 140 g. Most of the body's cholesterol is synthesized (about 1 g daily) *de novo*, whereas the average diet provides approximately 0.3 g daily. Because cholesterol is not synthesized in plants, dietary cholesterol is obtained from animal sources such as skeletal muscle, the liver, the brain, and egg yolks.

Slightly less than half of the cholesterol in the body is derived *de novo*: the liver accounts for ~10%; the intestines make up ~15%. The cholesterol biosynthesis pathway involves enzymes in the cytoplasm, **endoplasmic reticulum**, and **peroxisomes**. Because the majority of cholesterol synthesis in the body occurs in extrahepatic tissues, and the only quantitatively significant site for excretion and **catabolism** of cholesterol is the liver, approximately 600–800 mg of cholesterol each day must be transported from peripheral tissues through the plasma compartment to the liver.

Once in the liver, cholesterol is converted into bile salts or secreted as unesterified cholesterol into the **bile**. From there, cholesterol enters the small intestine, is partially reabsorbed, and then is excreted via the feces.

Biliary cholesterol secretion is essential for two major disease complexes: atherosclerotic CVD and cholesterol **gallstone** disease.

YOU SHOULD REMEMBER

Cholesterol homeostasis is associated with multiple human diseases, including CVD, cancer, and neurodegenerative and hepatic diseases.

CHOLESTEROL AND ATHEROSCLEROSIS

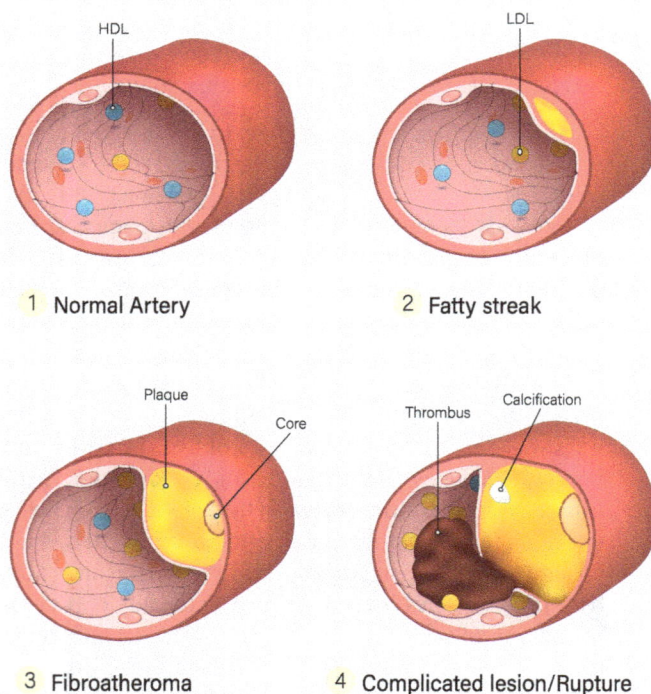

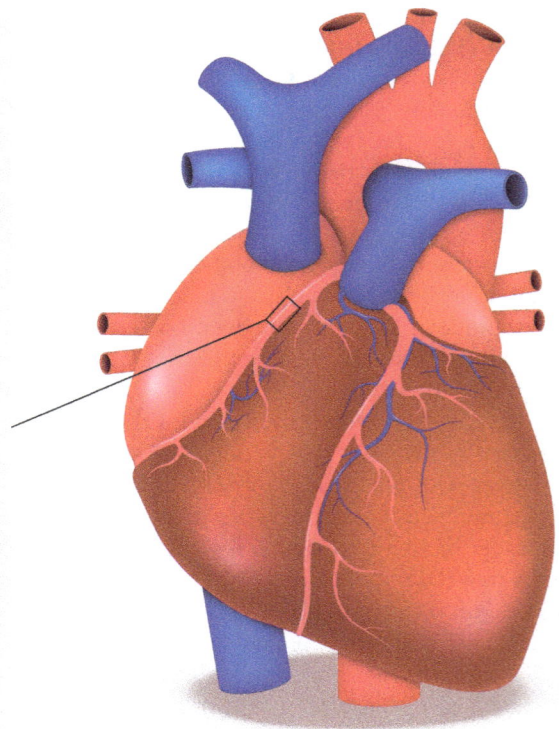

Atherosclerosis is the most common form of **arteriosclerosis** (see Figure 15.1). Atherosclerotic plaques are either stable or unstable. *Stable plaques* can grow slowly over several decades until stenosis or occlusion occurs. *Unstable plaques* are weaker and more likely to erode, develop fissures, or rupture. They can cause acute **thrombosis**, occlusion, and infarction for a long time before causing extreme stenosis. Unstable plaques cause the majority of signs and symptoms. They do not appear to be severe during **angiography**. Therefore, stabilization of unstable plaques may help reduce disease and death. The rupture of plaques is based on activated **macrophages** inside that secrete **cathepsins** and **collagenases**.

PHARMACOTHERAPY

Various medications can lower blood cholesterol. Statins are recommended for most patients to reduce the risk of heart attack or stroke. In most cases, statins continue to provide the most effective lipid-lowering treatment. Other drugs include BAS, PCSK9 inhibitors, fibrates, niacin, cholesterol absorption inhibitors, and omega-3. This chapter discusses them.

STATINS

Statins can help lower LDL and total cholesterol in the blood. They are the most common cholesterol-lowering drugs. Also known as **HMG-CoA reductase inhibitors**, they reduce illness and mortality in those at high risk for CVD. There are various types of statins. Statins are prescribed primarily for their LDL-lowering capabilities, but they can also lower triglycerides and raise HDL. There are seven types of statins (see Table 15.2). Based on their metabolic fates, they differ in their LDL-lowering efficacy and potential for drug and disease interactions. All statins are classified as category X in pregnancy and should be used cautiously in women of childbearing potential because category X drugs are known to cause fetal congenital disabilities.

Atherosclerosis

1 Normal Artery

2 Fatty streak

3 Fibroatheroma

4 Complicated lesion/Rupture

FIGURE 15.1   Atherosclerosis.

TABLE 15.2
Various Types of Statins

Statin	Dose Range
Atorvastatin	Lipitor
Fluvastatin	Lescol
Lovastatin	Mevacor
Pitavastatin	Livalo
Pravastatin	Pravachol
Rosuvastatin	Crestor
Simvastatin	Zocor

Clinical Indications

Statin medications are the mainstay of treatment for hyperlipidemia and **mixed dyslipidemia**.

They are also used for hypertriglyceridemia, atherosclerosis, and the primary prevention of atherosclerotic CVD. They are indicated for adults aged 40–75 years with one or more cardiovascular risk factors such as dyslipidemia, diabetes, hypertension, or smoking.

Mechanism of Action

Statins inhibit the enzyme HMG-CoA reductase. This inhibition occurs early in the hepatic pathway that produces cholesterol and other vital metabolic products. Cholesterol is an intermediate product in pathways involving corticosteroids, sex steroids, vitamin D, and bile acids – metabolic pathways influenced by statins and PCSK9 inhibitors. Decreased cholesterol synthesis results in the upregulation of LDL cholesterol receptors and greater clearance of LDL cholesterol. When PCSK9 is bound to the LDL cholesterol receptor, the internalized receptor is more likely to be degraded. With lower PCSK9 activity due to PCSK9 inhibitors, LDL receptors are more likely to be recycled to the cell surface to clear plasma LDL cholesterol.

Adverse Effects

While statins are generally effective and safe for most individuals, they have been associated with muscle pain, digestive issues, and mental fogginess in some cases. Rarely, they may lead to liver damage. **Rhabdomyolysis** is the most severe adverse effect of statin use, although it seldom occurs (in <0.1% of cases). The most common risk factors for statin-related myopathy include hypothyroidism, polypharmacy, and alcohol misuse. Headache, fatigue, dizziness, and insomnia are frequently reported.

Contraindications

Statins are contraindicated in individuals who have hypersensitivity to the medication.

They are also contraindicated in acute liver failure or decompensated cirrhosis, pregnancy, and lactation.

YOU SHOULD REMEMBER

Grapefruit juice can negatively interact with certain statin medications, potentially increasing the risk of muscle pain (rhabdomyolysis) and liver damage. Consuming grapefruit juice can inhibit the enzyme CYP3A4, which metabolizes some statins, such as simvastatin (Zocor) and lovastatin (Mevacor).

Check Your Knowledge

1. What are the effects of statins on blood lipids?
2. What are the laboratory values for optimal cholesterol and triglycerides?
3. Why are cholesterol and triglycerides essential for the body?

BILE ACID SEQUESTRANTS

BAS are also called anion-exchange resins because they bind bile acid through an anion-exchange reaction. Typically, bile acids emulsify dietary fats to help them be absorbed in the small intestine. Due to decreased absorption of bile acids, the liver increases the production of bile acids from blood cholesterol and thus increases the removal of cholesterol from the circulation. As a class, their main effect is to decrease LDL. There are three BAS available for use: cholestyramine (Questran), colestipol (Colestid), and colesevelam (Welchol). Because BAS's site of action is in the intestines, it can decrease the absorption of multiple drugs when administered simultaneously. It is generally recommended that the co-administration of BAS with other medications should be separated by 1 hour before or 4 hours after other medicines.

Clinical Indications

BAS is FDA-approved to manage hypercholesterolemia. They can be combined with statins or used as monotherapy. BAS is often used as adjuvant therapy to complement exercise and dietary modifications.

Mechanism of Action

The BAS are highly positively charged molecules that bind to the negatively charged bile acids in the intestine, inhibiting their lipid-solubilizing activity and thus blocking cholesterol absorption.

Adverse Effects

Since BAS is not absorbed, it lacks systemic toxicity. Still, GI side effects, including nausea, bloating, constipation, and cramping, often limit their use. Colesevelam has been reported to be more tolerable than the other BAS. Sequestrants increase blood triglycerides, and there are safety concerns.

Contraindications

BAS should not be used in patients with hypersensitivity to active ingredients or components. Cholestyramine use is contraindicated in patients with complete biliary obstruction where bile is not secreted into the intestine. The only medications currently acceptable during pregnancy are BAS since they are not systemically absorbed and, therefore, are not felt to pose a fetal risk.

NIACIN

Niacin (nicotinic acid) describes both nicotinic acid and nicotinamide. However, only nicotinic acid has lipid-lowering properties. Nicotinamide only functions as a vitamin. The mechanism by which niacin lowers blood lipids is not fully understood, but it appears to inhibit the production of lipoproteins and decrease the production of VLDL in the liver. It also reduces blood concentrations of triglycerides and increases HDL. Among all available lipid-lowering agents, niacin appears to have the most favorable effect on the lipid profile. Niacin is available with and without a prescription and is usually inexpensive. The extended-release formulation, Niaspan, requires a prescription. Niacin is typically administered in two to three doses daily, except for Niaspan, which is taken as a single dose at bedtime. Since some niacin preparations are available over the counter, it is essential to counsel patients that their use is associated with many side effects.

Mechanism of Action

Niacin influences lipoprotein metabolism by decreasing triglyceride synthesis via multiple pathways. In adipocytes, it inhibits the lipolysis of triglycerides and retards the mobilization of free fatty acids to the plasma.

Clinical Indications

Niacin oral tablets are indicated as monotherapy or in combination with simvastatin or lovastatin to treat primary hyperlipidemia and mixed dyslipidemia. They can also reduce the risk of nonfatal myocardial infarctions in patients with a history of hyperlipidemia. However, a new analysis featured in *JAMA Network Open*, published by the Journal of the American Medical Association, suggests this old drug offers no benefit for most people.

Adverse Effects

Significant adverse effects of niacin include hot flashes, itching, vomiting, diarrhea, fatigue, and headache. This usually leads to discontinuation of the drug. Most patients develop tolerance after prolonged use. Slowly increasing the niacin dose over weeks is essential to reduce the frequency and severity of skin reactions. Other adverse effects include liver toxicity, **hyperglycemia**, and **hyperuricemia**. Significant liver toxicity is more common with high doses of the short-acting formulation of niacin.

Contraindications

Niacin should be avoided in patients with hypersensitivity, active peptic ulcer disease, active liver disease, or those presenting with unexplained and persistent elevations in hepatic transaminases.

FIBRATES

Fibrates or fibric-acid derivatives have a complex mechanism of action and may vary among the drugs in their class. It is thought that fibrates work in the liver to reduce the formation of VLDL triglycerides. Fibrates are primarily used to decrease triglycerides by up to 35%–50% in people with hypertriglyceridemia but have very modest LDL-lowering effects. Sometimes, LDL rises during this therapy. Elevated triglycerides increase the risk of pancreatitis (inflammation of the pancreas). Two fibrates are currently on the market: gemfibrozil (Lopid) and fenofibrate (Tricor). They are administered once to twice daily and come in tablet and capsule formulations. Fibrates are highly bound to blood albumin (which means they may interact with other medications) and can cause displacement of other protein-bound drugs when administered together.

Clinical Indications

Fibrates help to reduce serum LDL, total cholesterol, triglycerides, and Apo-B and increase HDL. They are to be used as an adjunct to dietary modifications in adults with severe hypertriglyceridemia.

Mechanism of Action

Fibrates reduce hepatic triglycerides by inhibiting hepatic extraction of free fatty acids and, thus, hepatic triglyceride production. These drugs may also lower cholesterol by increasing endothelial lipoprotein lipase activity. After stimulating LDL secretion, they promote receptor-mediated clearance of LDL, causing this catabolism to occur 20% faster than in an untreated patient. This action results in an overall decrease in VLDL.

Adverse Effects

Fibrates are generally well-tolerated, but GI complaints are their most common adverse effects. They also increase the risk of gallstone formation and may increase the risk of muscle toxicity when given together with statins. Notably, the risk of muscle toxicity is reduced with fenofibrate.

Contraindications

Fibrates are contraindicated in patients with known hypersensitivity to the drug class. Fibrates are not to be used if the patient has active liver disease, as they are shown to be hepatotoxic if preexisting liver inflammatory states exist. They are also contraindicated if creatinine clearance is <30 mL/minutes or if the patient has severe renal dysfunction.

Check Your Knowledge

1. What are the contraindications of BAS?
2. What are the adverse effects of niacin?
3. What are the contraindications of fibrates?

CHOLESTEROL ABSORPTION INHIBITORS

Ezetimibe (Zetia) is the only drug in the class of cholesterol absorption inhibitors. It works by preventing a protein that decreases cholesterol absorption in the gut. This inhibition is only for cholesterol, with no impact on the absorption of other drugs or fat-soluble vitamins. A decrease in cholesterol delivery to the liver results in an increase in cholesterol uptake in the bloodstream by the liver and a reduction of systemic cholesterol. Ezetimibe has modest effects on LDL and other cholesterol parameters and is FDA-approved only as an add-on therapy to statins or fibrates. Ezetimibe is primarily used for high-risk patients, such as those with a history of prior myocardial infarction or diabetes, as well as those who cannot tolerate high-intensity statin therapy and do not meet cholesterol goals on lower doses alone. Ezetimibe improves cardiac outcomes in patients who have had a past heart attack. It is well-tolerated and exhibits a side effect profile similar to statins when used in combination therapy.

PROPROTEIN CONVERTASE SUBTILISIN-KEXIN TYPE 9

PCSK9 is an enzyme that regulates cholesterol levels in the blood. It is produced in the liver and breaks down LDL receptors responsible for removing cholesterol from the bloodstream. PCSK9 inhibitors are the latest cholesterol-lowering medications available on the market. They include evolocumab (Repatha) and alirocumab (Praluent). Both drugs are monoclonal antibodies that bind to and inhibit the PCSK9 enzyme responsible for the degradation of LDL receptors in the liver. By increasing LDL receptors in the liver, there is an increase in the removal of cholesterol from the bloodstream. This novel mechanism leads to significantly decreased LDL (up to 70%), more than any other lipid-lowering therapy currently available. Because of their potent LDL-lowering effects, their target population is those who have CVD and need additional lowering of LDL or have a genetic predisposition to high cholesterol. While PCSK9 inhibitors are undoubtedly very effective in lowering LDL levels, it is unknown if they improve long-term cardiovascular outcomes.

OMEGA-3 FATTY ACIDS

Omega-3 fatty acid ethyl esters are a type of healthy fat essential for many bodily functions, including brain cell development, heart health, and eye development. They are derived from chemically changed and purified fish oils.

Omega-3 esters are used in conjunction with dietary changes to help reduce triglycerides. Omega-3 drugs available in the U.S. include omega-3-acid ethyl esters (Lovaza®), icosapent ethyl (Vascepa®), omega-3 carboxylic acids (Epanova®), and omega-3-acid ethyl esters A (Omtryg®).

NONMEDICATION THERAPY

Lifestyle changes are the backbone of treating hyperlipidemia. This includes dietary therapy with low total and saturated fat and reduced cholesterol intake. A registered dietitian may be consulted to help patients change their diet. In addition to making healthy food choices, patients must be encouraged to exercise. A reasonable goal is at least 30 minutes of moderate-intensity exercise at least 5 days a week. All patients should be encouraged to diet and exercise to maintain a healthy weight, limit alcohol use, and avoid tobacco use.

CLINICAL CASE STUDIES

Clinical Case Study 1

A 36-year-old woman developed angina while she was breastfeeding her daughter. Cardiac catheterization revealed a partial blockage of three vessels, which led to coronary artery bypass grafting. The patient's father had previously had the same procedure after he had a myocardial infarction. Therefore, the woman was diagnosed with familial hypercholesterolemia. During follow-up, she was treated with rosuvastatin and ezetimibe (Zetia) daily. She received three coronary stents as well.

CRITICAL THINKING QUESTIONS

1. Which types of drugs are FDA-approved to manage hypercholesterolemia?
2. What are the adverse effects of statins?
3. What is the classification of ezetimibe?

Clinical Case Study 2

A 61-year-old man with a history of hypertension and hyperlipidemia has a record of persistently elevated triglycerides and a family history of atherosclerotic CVD. Additionally, he recently quit smoking after many years. The patient has been taking niacin combined with simvastatin for 6 months, and his condition has improved.

CRITICAL THINKING QUESTIONS

1. What is the mechanism of action of niacin?
2. What are the indications for niacin?
3. What are the adverse effects of niacin?

FURTHER READING

Arpilor, L. (2023). *Health Benefits of Omega-3. Explore the Potential Health Benefits of Omega-3 Fatty Acids Found in Certain Foods and Supplements.* Learn About Their Advantages for Heart and Brain Health. Arpilor.

Banach, M. (2015). *Combination Therapy in Dyslipidemia.* Adis.

Cheng Jiang, X. (2020). *Lipid Transfer in Lipoprotein Metabolism and Cardiovascular Disease* (Advances in Experimental Medicine and Biology, 1276). Springer.

Kassi, E., Randeva, H.S., and Kyrou, I. (2023). *Atherosclerotic and Cardiometabolic Disease: From Molecular Basis to Therapeutic Advances* (International Journal of Molecular Sciences). MDPI.

Keenan, J.M., and Dunn, K.M. (2022). *The Niacin Breakthrough – The Keenan Protocol for Heart Health and Healthy Aging.* Keenan.

Kritchevky, D. (2012). *Lipids, Lipoproteins, and Drugs* (Advances in Experimental Medicine and Biology, 63). Springer.

NetCE, and Lanca, A.J. (2022). *Hyperlipidemias and Atherosclerotic Cardiovascular Disease.* NetCE.

Ollero, M., and Touboul, D. (2020). *Bioactive Lipids and Lipidomics 2018* (International Journal of Molecular Sciences). MDPI.

Reamy, B.V. (2010). *Hyperlipidemia Management for Primary Care: An Evidence-Based Approach.* Springer.

Ridgway, N.D., and McLeod, R.S. (2021). *Biochemistry of Lipids, Lipoproteins and Membranes*, 7th Edition. Elsevier Science.

Sears, W., and Sears, J. (2012). *The Omega-3 Effect.* Little Brown Spark.

Singh, G. (2017). *Role of Fibrates in Abrogated Potentials of Ischemic Preconditioning.* Scholars' Press.

Sniderman, A., and Durrington, P. (2021). *Fast Facts: Hyperlipidemia: Bringing Clarity to Lipid Management*, 6th Edition. S. Karger.

Sorrentino, M.J. (2011). *Hyperlipidemia in Primary Care – A Practical Guide to Risk Reduction.* (Current Clinical Practice). Humana Press.

Yassine, H. (2015). *Lipid Management – From Basics to Clinic.* Springer.

16 Drugs Affecting Hypertension

LEARNING OUTCOMES

After studying this chapter, readers should be able to

1. Review the classifications of hypertension.
2. Describe the causes of hypertension.
3. Summarize the various organs affected by hypertension.
4. Compare prehypertension and resistant hypertension.
5. Discuss the hypertensive crisis and its complications.
6. Explain the main goal in treating hypertension.
7. Describe the contraindications of both angiotensin-converting enzyme (ACE)-I and angiotensin receptor-blocking agents (ARBs).
8. Explain the adverse effects of ARBs.
9. Identify the two direct vasodilators.
10. Describe the trade names of prazosin, terazosin, and carvedilol.

OVERVIEW

Hypertension remains one of the most significant causes of mortality worldwide. It can be prevented through medication and lifestyle modifications. Individuals with hypertension are at greater risk for coronary heart disease, atrial fibrillation, heart failure, stroke, renal damage, peripheral artery disease, and dementia. Primary hypertension may occur without identifiable causes but is influenced by cardiovascular risk factors stemming from environmental and lifestyle changes. Various toxicities, iatrogenic diseases, and congenital conditions contribute to secondary hypertension. The complications of hypertension encompass clinical outcomes of persistent hypertension, leading to cardiovascular disease, atherosclerosis, kidney disease, diabetes mellitus, metabolic syndrome, preeclampsia, erectile dysfunction, and eye disease. Treatment strategies for hypertension include lifestyle modifications – such as a diet rich in fruits, vegetables, and low-fat foods, along with fish that has reduced saturated and total fat content; salt restriction; maintaining an appropriate body weight; regular exercise; moderate alcohol consumption; and smoking cessation – as well as drug therapies. However, these approaches may vary somewhat based on published hypertension treatment guidelines.

CLASSIFICATION

Hypertension refers to chronically elevated blood pressure (BP) in the blood vessels. Patients with high BP can remain asymptomatic for some time. However, hypertension is diagnosed as having a systolic pressure of 130 mm Hg or higher and a diastolic pressure exceeding 80 mm Hg. A person's BP fluctuates throughout the day and is affected by various factors, including caffeine, cigarette smoking, and medications. Hypertension can be categorized as primary, secondary, prehypertension, resistant, or a hypertensive crisis.

PRIMARY AND SECONDARY HYPERTENSION

Primary or essential hypertension is high BP without a single identifiable cause. Instead, it typically results from a combination of factors that arise from environmental and lifestyle changes. **Secondary hypertension, on the other hand,** is caused by various toxicities, **iatrogenic diseases**, and congenital conditions. The complications of hypertension stem from persistently high BP and include cardiovascular disease, atherosclerosis, kidney disease, diabetes mellitus, metabolic syndrome, preeclampsia, erectile dysfunction, and eye disease. BP is categorized into four groups (see Table 16.1).

PREHYPERTENSION

When BP is slightly higher than usual, it indicates prehypertension. A regular reading is 120/80 mmHg. With prehypertension, systolic values can reach as high as 139 mm Hg, or the diastolic pressure can be as high as 89 mm Hg. Approximately 70 million people aged 20 and older have prehypertension, with men being nearly twice as likely to be affected compared to women. This category was added because studies indicated that patients with borderline BP had more risk factors for cardiac disease than nonhypertensive patients. Prehypertension increases the risk of MI by 3.5-fold but does not affect stroke risk.

RESISTANT HYPERTENSION

Resistant hypertension is a condition where BP stays elevated despite the use of three or more antihypertensive medications at their maximum tolerated doses. At least one

TABLE 16.1
Blood Pressure Classifications

Classification	Diastolic BP (mm Hg)
Normal	Less than 80
Prehypertension	Less than 80
Stage 1 hypertension	Or 80 to 89
Stage 2 hypertension	90 or higher

DOI: 10.1201/9781003461913-18

of these medications must be a diuretic, while the others should belong to different classes.

Risk factors associated with resistant hypertension include chronic kidney disease (CKD), older age, left ventricular hypertrophy, African-American race, female sex, and residence in the southeastern United States. CKD may be the most significant factor influencing resistant hypertension. Other factors are modifiable: obesity, excessive dietary salt, and heavy alcohol intake. High salt or alcohol consumption increases the risk of hypertension and interferes with treatment efficacy. Excessive nutritional salt has a particularly harmful impact on patients who are African-American, elderly, or have CKD. Complications of resistant hypertension include heart attack, stroke, kidney failure, heart failure, and death.

HYPERTENSIVE CRISIS

A **hypertensive crisis** occurs when BP exceeds 180/120 mm Hg. If left untreated, it increases the risk of myocardial infarction, stroke, or other severe health conditions. Starting at age 18, monitoring BP at least every 2 years is essential, although some individuals may require more frequent monitoring. A hypertensive crisis is an emergency that demands immediate access to a hospital. This condition can occur in about 1% of patients with underlying hypertension, as chronic hypertension is a highly prevalent condition. Common causes of crisis include nonadherence to antihypertensive medication, lack of a primary care provider, or the use of illicit drugs, such as cocaine.

ETIOLOGY

The cause of primary hypertension is idiopathic, commonly referred to as essential hypertension. It has long been suggested that an increase in salt intake raises the risk of developing hypertension. Approximately 90% of hypertension cases are classified as primary, while the remaining 10% are secondary. Causes of secondary hypertension include obstructive sleep apnea, CKD, **pheochromocytoma**, **hyperthyroidism**, **aldosteronism**, blood vessel abnormalities (coarctation of the aorta), **preeclampsia**, **eclampsia**, and certain medications that may trigger hypertension.

Many drugs can cause secondary hypertension, including birth control pills, ibuprofen, naproxen, and diclofenac. Individuals with uncontrolled hypertension should avoid caffeinated beverages. Nicotine, cocaine, methamphetamine, and other recreational drugs that raise BP should also be avoided. Cocaine leads to a significant increase in BP, particularly if the person is using a beta-blocker.

The risk factors encompass age, family history, race or ethnicity, and unhealthy lifestyle choices, such as a diet high in sodium and low in potassium, inadequate exercise, smoking, and excessive alcohol consumption. Certain diseases, including diabetes and kidney disease, may also contribute as risk factors.

CLINICAL FEATURES

Most patients with hypertension are asymptomatic, which is why it is often called the "silent killer." However, some symptoms of dangerously high BP include headaches, dyspnea, and chest pain. As hypertension progresses, target organs may become damaged. The main organs affected are the heart, brain, kidneys, peripheral blood vessels, and eyes (see Figure 16.1). Patients with very high BP (usually 180/120 mm Hg or higher) may experience severe headaches, dizziness, dyspnea, chest pain, blurred vision, **epistaxis**, anxiety, nausea, and vomiting.

YOU SHOULD REMEMBER

More than one billion adults worldwide have hypertension, with up to 45% of the adult population affected by this condition. The high prevalence of hypertension is consistent across all socio-economic and income strata, and it increases with age, accounting for up to 60% of the population over 60 years of age.

Check Your Knowledge

1. What **is** the etiology of primary and secondary hypertension?
2. What organs are affected by hypertension?

PHARMACOTHERAPY

Hypertension can be treated with lifestyle changes that may reduce or even eliminate the need for drug therapy. These changes include consuming a healthy diet, reducing sodium intake (table salt), exercising adequately, and moderating alcohol consumption, which are believed to positively influence BP. Over 75% of salt intake comes from processed foods, so plans to reduce salt consumption must involve eating out less and consuming fewer packaged foods. A diet containing less than 2.3 g of sodium daily can help prevent and treat hypertension. Natural potassium sources found in fruits and vegetables are recommended. Common medications for hypertension include ACE inhibitors, angiotensin receptor blockers, calcium channel blockers, potassium-sparing diuretics, and other diuretics. A list of commonly used antihypertensive agents can be found in Table 16.2.

ACE INHIBITORS

ACE inhibitors can effectively manage hypertension, heart failure, stroke, and heart attacks. They reduce the activity of the ACE enzyme, which is responsible for producing hormones that regulate BP. This process prevents the body from producing angiotensin II (ATII), which constricts blood vessels and increases BP. Consequently, ACE inhibitors promote vasodilation and lower BP.

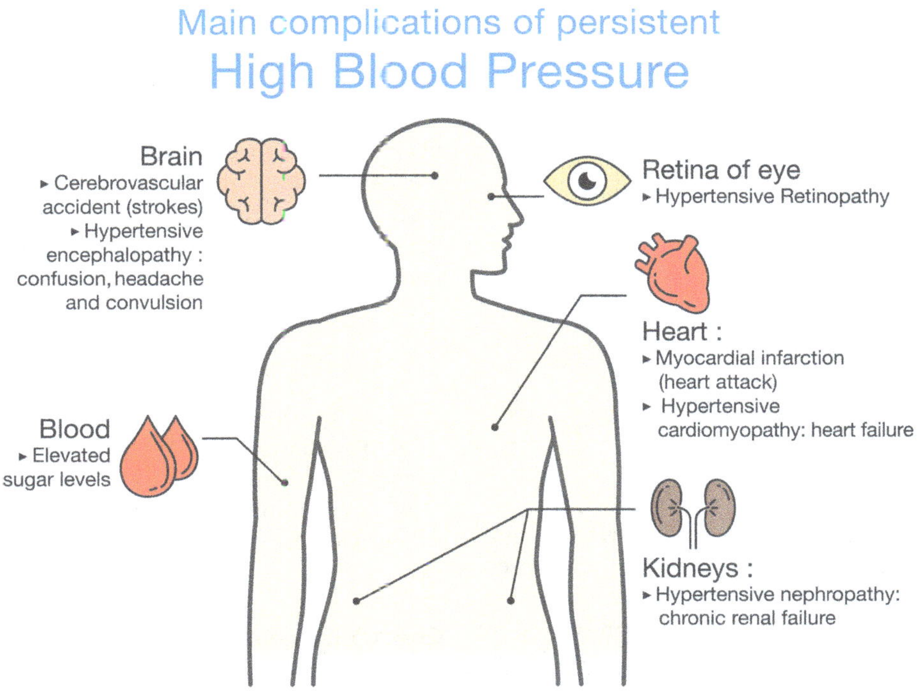

Main complications of persistent
High Blood Pressure

Brain
- ▸ Cerebrovascular accident (strokes)
- ▸ Hypertensive encephalopathy : confusion, headache and convulsion

Retina of eye
- ▸ Hypertensive Retinopathy

Heart :
- ▸ Myocardial infarction (heart attack)
- ▸ Hypertensive cardiomyopathy: heart failure

Blood
- ▸ Elevated sugar levels

Kidneys :
- ▸ Hypertensive nephropathy: chronic renal failure

FIGURE 16.1 Organs commonly affected by hypertension.

TABLE 16.2

Classifications of Commonly Used Antihypertensive Agents

Generic Name	Trade Name
Angiotensin-Converting Enzyme Inhibitors	
Benazepril	Lotensin
Captopril	Capoten
Enalapril	Vasotec
Fosinopril	Monopril
Lisinopril	Zestril
Perindopril	Aceon
Quinapril	Accupril
Ramipril	Altace
Angiotensin Receptor Blockers	
Candesartan	Atacand
Irbesartan	Avapro
Losartan	Cozaar
Telmisartan	Micardis
Valsartan	Diovan
Direct Renin Inhibitor	
Aliskiren	Tekturna

Clinical Manifestations

ACE inhibitors are utilized to treat and manage hypertension, a significant risk factor for coronary disease, heart failure, stroke, and heart attack.

Mechanism of Action

ACE inhibitors block the conversion of angiotensin I (ATI) into the potent vasoconstrictor ATII, which lowers BP. Essentially, they inhibit the renin–angiotensin–aldosterone system by promoting vasodilation and reducing BP.

Adverse Effects

ACE inhibitors are generally well-tolerated; however, adverse effects can occur, such as a dry cough and allergic reactions. More severe adverse effects include kidney damage and angioedema, which may require emergency medical care. Other adverse effects encompass hypotension, hyperkalemia, and hypermagnesemia. ACE inhibitors reduce the production of **aldosterone**, leading to increased potassium levels. One potentially profound side effect that can occur in less than 1% of patients is angioedema. Cholestatic jaundice or hepatitis is another rare but severe adverse effect that can progress to hepatic **necrosis** and, in some cases, death.

Contraindications

ACE inhibitors are contraindicated for patients with a history of **angioedema** or hypersensitivity related to treatment with this agent. Pregnant individuals or those with hereditary angioedema should avoid taking ACE inhibitors. These medications should not be administered during pregnancy. They are classified as category D in pregnancy under the old FDA system because they are known to cause skull hypoplasia, anuria, hypotension, renal failure, lung hypoplasia, skeletal deformations, oligohydramnios, and death.

YOU SHOULD REMEMBER

ACE inhibitors can help prevent the worsening of heart disease and may reduce the risk of stroke or heart attack. They can also aid in treating kidney issues and may even enhance kidney function.

Check Your Knowledge

1. How can lifestyle changes lower BP?
2. What are the classifications of medications used to treat hypertension?
3. What are the contraindications of ACE inhibitors?

ANGIOTENSIN RECEPTOR BLOCKING AGENTS

ARBs treat hypertension, heart failure, and CKD by inhibiting the receptors for ATII. Also referred to as ATII receptor antagonists, ARBs are considered first-line medications for managing hypertension. They are often prescribed for patients who cannot tolerate the side effects of ACE inhibitors, as ARBs tend to have fewer adverse effects.

Clinical Manifestations

ARBs are comparable in efficacy to ACE inhibitors for treating hypertension, heart failure, and kidney disease, and they reduce the risk of heart attacks and strokes. Additionally, ARBs tend to be more expensive than ACE inhibitors since generic substitutes are not available. The routine use of ACE inhibitors and ARBs for hypertension is not recommended. ARBs include losartan, valsartan, and telmisartan.

Mechanism of Action

ARBs block receptors for angiotensin, specifically AT1 receptors in the heart, blood vessels, and kidneys. By blocking the action of ATII, they help lower BP and prevent damage to the heart and kidneys. Renin, secreted from the juxtaglomerular cells of the kidneys, catalyzes the conversion of **angiotensinogen** to ATI in the liver. ATI is converted to ATII by ACE and other non-ACE pathways.

Adverse Effects

ARBs have excellent safety profiles both alone and in combination with other antihypertensive drugs. They exhibit better tolerability compared to ACE inhibitors. Common side effects include headache, dizziness, drowsiness, gastrointestinal disturbances, rash, and fatigue. They can also cause first-dose hypotension and orthostatic hypotension. In patients with renal insufficiency, ARBs may induce hyperkalemia, so potassium levels should be continuously monitored. They can precipitate renal failure in patients with bilateral renal artery stenosis and low, fixed renal blood flow. Cough and angioedema are rare, occurring less frequently than with ACE inhibitors.

Contraindications

ARBs are contraindicated in cases of bilateral **renal artery stenosis**, aortic stenosis, coarctation of the aorta, and during pregnancy. The FDA classifies ARBs as a category D risk for pregnant individuals, and patients who might become pregnant while taking ARBs should recognize the importance of using birth control. There is no evidence supporting the safe use of ARBs during breastfeeding, and the effects of potential exposure to a nursing infant remain unknown.

YOU SHOULD REMEMBER

ARB therapy places the patient at an increased risk of hypotension, renal impairment, and hyperkalemia. Therefore, the patient's BP, renal function, and serum electrolytes should be monitored closely during ARB use.

Check Your Knowledge

1. What is the mechanism of action of ARBs?
2. What are the clinical indications of ARBs?
3. What are the contraindications of ARBs?

CALCIUM CHANNEL BLOCKERS

Calcium channel blockers are medications used to reduce BP. They prevent calcium from entering the cells of the heart and arteries, which causes them to contract more forcefully. By blocking calcium, these medications allow blood vessels to relax and widen. **Chapter 17** discusses calcium channel blockers in detail.

DIRECT VASODILATORS

Hydralazine and minoxidil are **vasodilators** that relax blood vessels, allowing blood to flow more quickly through the body. Both medications directly relax **arterioles** and lower BP. The exact mechanism by which hydralazine induces arterial smooth muscle relaxation remains unknown; however, it influences calcium movement within blood vessels. Calcium is essential for muscle contraction, so disturbances in calcium movement may lead to smooth muscle relaxation in the blood vessels. Hydralazine selectively targets arterioles, while minoxidil can effectively reduce BP in most patients with resistant hypertension, particularly when treatment has failed with multidrug regimens. Due to their distinct side effects, these medications are not utilized as first-line agents. Table 16.3 summarizes some other drugs used to treat hypertension.

COMBINED ALPHA–BETA-BLOCKERS

Combined alpha–beta-blockers, also known as alpha and beta dual-receptor blockers, bind to these receptors,

TABLE 16.3

Direct Vasodilators, Combined Alpha–Beta-Blockers, Alpha-2 Agonists, and Alpha-1 Antagonists

Generic Name	Trade Name
Direct Vasodilator	
Hydralazine	Generic only
Combined Alpha–Beta-Blockers	
Carvedilol	Coreg
Labetalol	Normodyne
Alpha.2 Agonists and Alpha.1 Antagonists	
Clonidine	Catapres
Doxazosin	Cardura
Prazosin	Minipres
Terazosin	Hytrin

preventing their stimulation and allowing blood vessels to dilate. Carvedilol and labetalol are nonselective alpha- and beta-blockers that inhibit alpha-1, beta-1, and beta-2 receptors. Together, these effects reduce cardiac output and systemic BP through vasodilation. They are often used to treat hypertension and heart failure, and they may also help counteract the cardiovascular effects of cocaine use.

COMBINED ALPHA-2 AGONISTS AND ALPHA-1 ANTAGONISTS

Combining alpha-2 agonists and alpha-1 antagonists would generally lower BP due to their opposing effects on the alpha-adrenergic receptors. Alpha-2 agonists reduce sympathetic activity, while alpha-1 antagonists inhibit the vasoconstrictive effects of norepinephrine at alpha-1 receptors, resulting in vasodilation and decreased BP.

When the first-line agents are insufficient or cannot be utilized, alternative agents may be added. Clonidine and methyldopa are central alpha-2 agonists that reduce sympathetic outflow from the brain. Decreased sympathetic activity lowers cardiac output and total peripheral resistance. At higher doses, long-term use leads to sodium and water retention, as well as depression. Reserpine is seldom used as an antihypertensive agent.

BETA-ADRENERGIC BLOCKERS

Beta-adrenergic blockers are medications that lower BP. They inhibit the effects of the hormone epinephrine, leading to hypotension and bradycardia. Additionally, they dilate arteries and veins to enhance blood flow. **Chapter 10** discussed beta-adrenergic blockers.

DIURETICS

Diuretics are medications that help reduce fluid buildup in the body. They are sometimes referred to as "water pills."

Most diuretics assist the kidneys in removing salt and water from the urine, which lowers fluid flow through the veins and arteries. As a result, BP decreases. Diuretics rank among the most commonly used drugs for treating hypertension and edema, or fluid retention. They will be discussed in detail in **Chapter 23**.

CLINICAL CASE STUDIES

Clinical Case Study 1

A 34-year-old Black man with a family history of heart disease visits his primary physician for a checkup. His father had a myocardial infarction at the age of 61. The patient smokes, exercises occasionally, and regularly consumes beer. His BP measures 144/98 mm Hg. He is overweight, his HDL level is low, and his LDL cholesterol is high.

CRITICAL THINKING QUESTIONS

1. What are the types of hypertension?
2. What are the risk factors associated with hypertension?
3. What medications treat hypertension?

Clinical Case Study 2

A 52-year-old African American woman with treatment-resistant hypertension has been taking antihypertensive medications since she was 45 years old. Her hypertension is difficult to manage, and her BP consistently exceeds 150/100 mm Hg. She is currently prescribed valsartan, amlodipine, and extended-release metoprolol. The patient follows a low-sodium diet, rarely consumes alcohol, does not smoke, and walks for 30 minutes each day. Her medical history includes hyperlipidemia, obesity, and diabetes. Her medications comprise ACE inhibitors, angiotensin receptor-blocking agents, and diuretics.

CRITICAL THINKING QUESTIONS

1. What are the risk factors associated with resistant hypertension?
2. What are the adverse effects of ACE inhibitors?
3. What are the contraindications of ARBs?

FURTHER READING

Aeddula, N., Gupta, P., and Ahmed, I. (2023). *Certified Hypertension Specialist: Board and Certification Review.* StatPearls.

Bakris, G.L., Sorrentino, M.J., and Laffin, K.J. (2023). *Hypertension: A Companion to Braunwald's Heart Disease,* 4th Edition. Elsevier.

Berberian, J., Brady, W., and Mattu, A. (2022). *Emergency ECGs—Case-Based Revie and Interpretations.* Emergency Medicine Residents Association.

Hanna, E.B. (2022). *Practical Cardiovascular Medicine,* 2nd Edition. Wiley Blackwell.

Heller, G., and Hendel, R.C. (2022). *Nuclear Cardiology—Practical Applications*, 4th Edition. McGraw Hill.

Herring, N., and Paterson, D.J. (2018). *Levick's Introduction to Cardiovascular Physiology*, 6th Edition. CRC Press.

Houston, M.C., and Bell, L. (2021). *Controlling High Blood Pressure through Nutrition, Nutritional Supplements, Lifestyle, and Drugs*. CRC Press.

Kaplan, N.M., and Victor, R.G. (2014). *Kaplan's Central Hypertension*, 11th Edition. Lippincott, Williams, and Wilkins.

Lerma, E.V., Berns, J.S., and Nissenson, A.R. (2012). *Current Essentials: Nephrology & Hypertension*. McGraw Hill-Lange.

Lerma, E.V., Luther, J.M., and Hiremath, S. (2022). *Hypertension Secrets*, 2nd Edition. Elsevier.

Libby, P., Bonow, R.O., Mann, D.L., Tomaselli, G.F., Bhatt, D.L., and Solomon, S.D. (2020). *Braunwald's Heart Disease—A Textbook of Cardiovascular Medicine*, 12th Edition. Elsevier.

Lilly, L.S. (2020). *Pathophysiology of Heart Disease—An Introduction to Cardiovascular Medicine*, 7th Edition. Lippincott, Williams, and Wilkins.

Nadar, S., and Lip, G.H. (2023). *Hypertension*, 3rd Edition. Oxford University Press.

Rodriguez, K. (2023). *Pulmonary Hypertension: Navigating the Path to Wellness with a Comprehensive Guide to Understanding, Treating, and Overcoming Pulmonary Hypertension for Improved Respiratory Health and Quality of Life*. Rodriguez.

Salvetti, M. (2016). *Resistant Hypertension—Practical Case Studies in Hypertension Management*. Springer.

17 Drugs Affecting Arrhythmia

LEARNING OUTCOMES

After studying the chapter, readers should be able to

1. Summarize the cardiac conducting system.
2. Describe the various types of arrhythmias.
3. Review the drugs used for treating arrhythmias.
4. Describe the contraindication of procainamide.
5. Explain the classes of sodium channel blockers.
6. Describe how beta-adrenergic blockers are used for arrhythmias.
7. Review the indications of amiodarone and common adverse effects.
8. Describe the common calcium channel blocker (CCB) used for arrhythmias.
9. Describe symptoms of digoxin toxicity.
10. Explain the indication of adenosine in the cardiovascular system.

OVERVIEW

Various conditions, such as endocarditis or myocardial infarction (MI), can damage the heart's conduction system and disturb its regular rhythm. These events may lead to arrhythmia, also known as dysrhythmia. These conditions represent abnormal heart rhythms, as the only normal rhythm of the heart is a normal sinus rhythm. With this rhythm, an impulse is generated in the sinoatrial (SA) node, which is conducted through and slowed while passing through the atrioventricular (AV) node. Arrhythmias can be classified as bradycardias or tachycardias. Bradycardias are further divided into sinus bradycardias and heart block. Tachycardias may be supraventricular or ventricular. Antiarrhythmics can prevent and treat irregular heart rhythms. They can reduce symptoms and help avoid life-threatening complications. Their mechanism of action classifies them. The four primary classes of antiarrhythmic medications include class Ia, Ib, Ic, and sodium blockers. Class II drugs are beta-blockers, class III are potassium channel blockers, and class IV drugs include CCBs.

PHYSIOLOGY OF THE CONDUCTING SYSTEM

The conduction system is a network of cells and nodes that generates and conducts electrical impulses to control the heartbeat. Its primary function is to coordinate the heart's rhythmic contractions, achieved by creating and propagating electrical impulses that cause different parts of the heart to expand and contract. The cardiac conduction system is designed for electrical impulse creation and propagation, allowing for initiating impulses in the atrium. The heart's

electrical conduction system includes the **SA node**, the **AV node**, the AV bundle (**bundle of His**), and **Purkinje fibers**.

An **action potential** is a brief electrical event that occurs in a neuron's axon and signals that the neuron is active. It is a rapid sequence of changes in the voltage across the neuron's cell membrane. It is a sudden and transient **depolarization** of the membrane. The rapid rise in potential depolarization in myocytes is an all-or-nothing event initiated by the opening of sodium ion channels within the plasma membrane. **Repolarization** is mediated by the opening of potassium ion channels. Action potentials trigger myocardial contractions, which must be coordinated carefully for the heart to pump effectively.

The conduction of cardiac impulses generates tiny electrical currents in the heart that can be detected beyond its surface. This phenomenon holds significant clinical importance, allowing for the creation of visible records of the heart's electrical activity with an electrocardiograph using electrodes placed on the skin. The electrocardiogram (ECG or EKG) is a graphical representation of the heart's electrical activity and its conduction of impulses (see Figure 17.1). Electrodes from a recording electrocardiograph are attached to the limbs and chest to produce an ECG. Changes in voltage within the cardiac tissue, which represent fluctuations in the heart's electrical activity, are observed as deflections of a line drawn on paper or traced on a video monitor.

ARRHYTHMIA CLASSIFICATIONS

Arrhythmias can be classified by type and location. They can include extra beats, supraventricular tachycardias, ventricular arrhythmias, or bradyarrhythmias, affecting the ventricles or atria. Specific types of arrhythmia may occur in patients with severe heart disease and can cause sudden cardiac arrest. Atrial fibrillation is the most common cardiac abnormality. Arrhythmias can be asymptomatic, **paroxysmal**, or irregular, making it difficult to estimate their true prevalence. Arrhythmias are classified according to their origin, means of transmission, and associated symptoms. The various types include the following:

- **Sinus dysrhythmia** is a variation in heart rate during the breathing cycle. Typically, the rate increases during inspiration and decreases during expiration. The causes of sinus dysrhythmia are unclear. This phenomenon is common in young people and usually does not require treatment.
- **Extrasystoles** occur before the next expected contraction in a series of **cardiac cycles**. They often arise due to a lack of sleep, excessive caffeine or nicotine intake, alcoholism, or heart damage.

Normal Range ECG

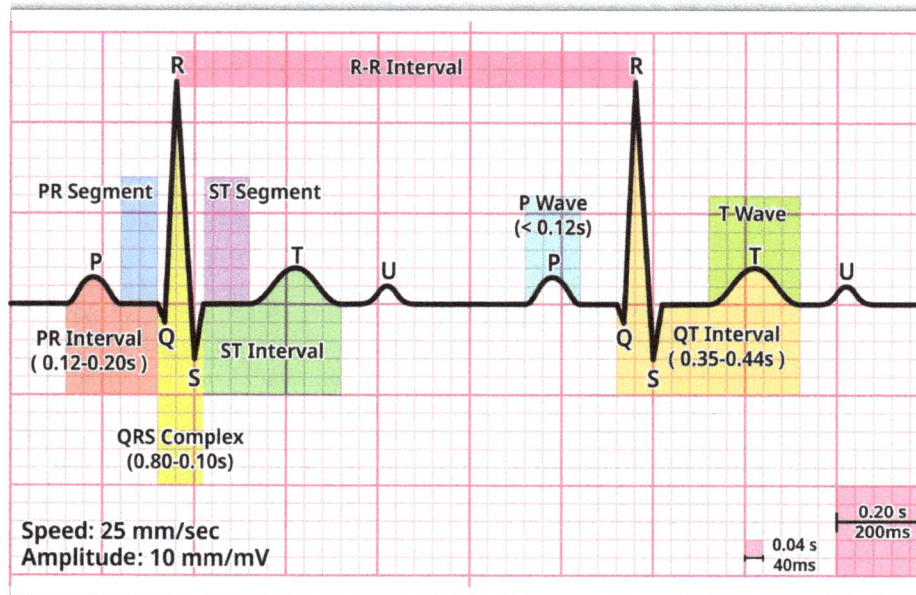

FIGURE 17.1   Normal electrocardiogram.

Ventricular depolarizations that appear earlier than expected are called premature ventricular contractions (PVCs). They present as early, broad QRS complexes without a preceding related P wave on the ECG. PVCs can lead to various conditions, including hypoxemia, electrolyte imbalance, acidosis, stress, ventricular hypertrophy, or drug reactions. Occasional PVCs are not clinically significant in otherwise healthy individuals. However, in patients with heart disease, **cardiac output** may be diminished.

- **Bradycardia** is a slow heart rhythm with fewer than 60 beats/minute. Slight bradycardia is normal during sleep and in conditioned athletes at rest. Abnormal bradycardia may result from improper autonomic nervous control of the heart or damage to the SA node. If the issue is severe, artificial pacemakers can increase the heart rate by replacing the SA node.

- **Tachycardia** occurs when the heart rate exceeds 100 beats/minute. It can be triggered by exercise or stress, alcohol or caffeine consumption, electrolyte imbalances, specific medications, hyperthyroidism, anemia, and smoking or nicotine usage. Abnormal tachycardia may also arise from improper autonomic regulation of the heart, fever, blood loss or shock, and certain drugs.

- **Atrial fibrillation** is an irregular and rapid heartbeat that can be intermittent, long-lasting, or permanent. It is the most common type of treated heart arrhythmia. Symptoms include angina, dyspnea, palpitations, and fatigue. It is the leading cardiac cause of stroke and death. Risk factors for atrial fibrillation include advanced age, hypertension,

underlying heart and lung diseases, congenital heart defects, **rheumatic heart disease**, mitral stenosis, MI, and increased alcohol consumption. This condition can be treated with a beta-blocker (esmolol, atenolol) or a CCB (verapamil or diltiazem). Each drug may be offered depending on the patients' symptoms and overall health. Digoxin can also be prescribed if other medications are unsuitable. Occasionally, **cardioversion** can also be used. The complications of atrial fibrillation include stroke, heart failure, cognitive impairment, hypertension, and syncope.

- **Ventricular fibrillation** is a rapid, irregular heartbeat that can last for a few seconds or much longer. It is the most common life-threatening arrhythmia. Without treatment, the condition can be fatal within minutes. Common causes of VF include angina, MI, heart surgery, family history, certain medications, and electrolyte imbalances. The condition constitutes a medical emergency. Complications of ventricular fibrillation include brain hypoxia, post-defibrillation arrhythmias, injuries from cardiopulmonary resuscitation, and death. Ventricular fibrillation can be treated with **defibrillation** and medications. The drugs used for treatment include amiodarone, lidocaine, sotalol, and mexiletine.

- **Ventricular flutter** is a potentially fatal cardiac arrhythmia that causes the heart's ventricles to beat rapidly, typically between 180 and 300 beats per minute. If left untreated, ventricular flutter can progress to ventricular fibrillation. Figure 17.2 illustrates various types of ventricular tachyarrhythmia.

Monomorphic ventricular tachycardia — Female, 24 years old, clinically diagnosed idiopathic ventricular tachycardia. Ventricular tachycardia originated from the right ventricular outflow tract.

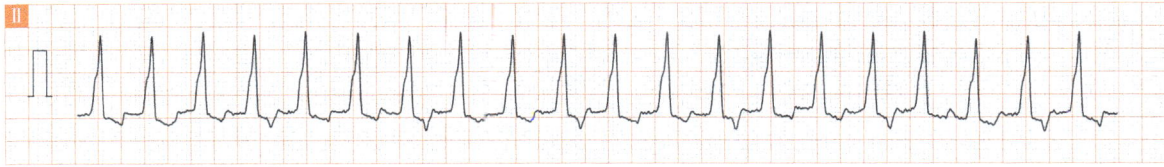

II

Bidirectional ventricular tachycardia — Male, 56 years old, clinically diagnosed aconitine poisoning. He survived and was discharged.

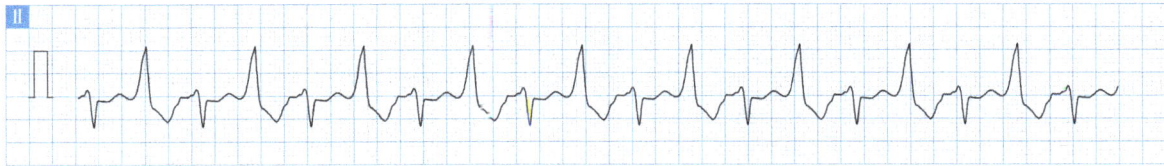

II

Polymorphic ventricular tachycardia — Male, 39 years old, suddenly syncope during treadmill examination. Coronary angiography revealed severe occlusion of the proximal segment of the left anterior descending artery, and coronary stents were implanted. He survived and was discharged.

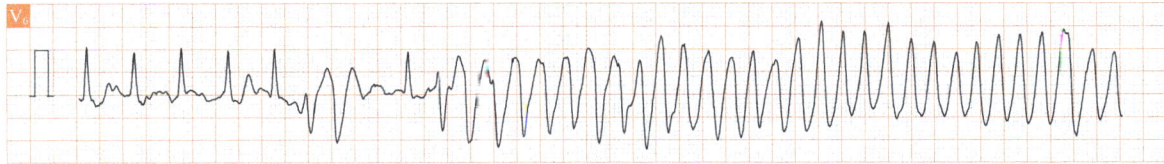

V₆

Torsades de pointes — Male, 3 years old, clinically diagnosed as Jervel-Lange-Nielsen syndrome. He died of ventricular fibrillation.

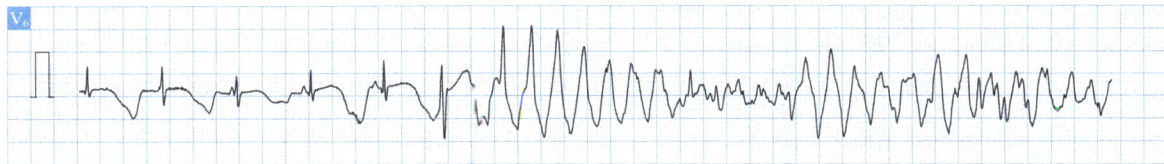

V₆

Ventricular flutter — A 33-year-old female was diagnosed with fulminant myocarditis. She died of ventricular fibrillation.

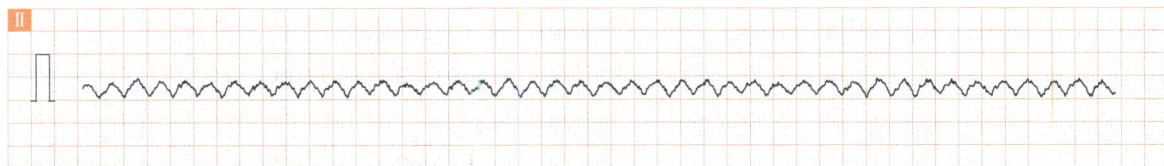

II

Ventricular fibrillation — Male, 60 years old, clinically diagnosed as aortic insufficiency and left heart failure. He died of ventricular fibrillation.

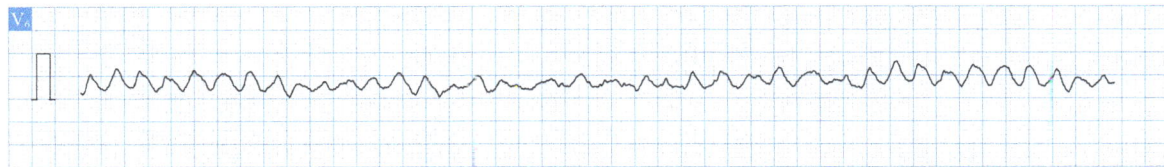

V₆

Ventricular tachyarrhythmia

NO.978 Some ventricular tachycardias are benign and some are malignant.
Note: Morphology and frequency of QRS waves.

ECG
Visual electrocardiogram

FIGURE 17.2 Various types of ventricular tachyarrhythmia.

Check Your Knowledge

1. What is the electrical conduction system?
2. What are the arrhythmia classifications?
3. What are the differences between atrial and ventricular fibrillations?

ANTIARRHYTHMIC AGENTS

Antiarrhythmic agents can prevent and treat abnormal heart rhythms while avoiding life-threatening complications. Some drugs halt irregular, excess electrical impulses, whereas others obstruct abnormally rapid impulses from traveling along heart tissues. Most antiarrhythmics are designed for long-term use. The medications used to treat arrhythmias include sodium channel blockers, beta-blocker antagonists, potassium channel blockers, and calcium channel blockers (see Table 17.1).

CLASS I: SODIUM CHANNEL BLOCKERS

Sodium channel blockers are a class of medications that prevent sodium from passing through cell membranes, thereby slowing electrical impulses in the heart. They are

---

TABLE 17.1

Drugs for Treating Arrhythmias

Generic Names	Trade Names
Class 1A – Sodium Channel Blockers	
Disopyramide	Norpace
Procainamide	Pronestyl
Class 1B – Sodium Channel Blockers	
Lidocaine	Xylocaine
Mexiletine	Mexitil
Phenytoin	Dilantin
Class 1C – Sodium Channel Blockers	
Flecainide	Tambocor
Propafenone	Rythmol
Class II – Beta-adrenergic blockers	
Acebutolol	Sectral
Esmolol	Brevibloc
Propranolol	Inderal, InnoPran XL
Sotalol	Betapace, Sotylize
Class III – Potassium Channel Blockers	
Amiodarone	Pacerone, Cordarone
Dofetilide	Tikosyn
Dronedarone	Multaq
Class IV – Calcium-Channel Blockers	
Diltiazem	Cardizem, Dilacor, Taztia XR
Verapamil	Calan, Isoptin SR, Verelan
Miscellaneous	
Adenosine	Adenocard, Adenoscan
Aigoxin	Lanoxin, Lanoxicaps

---

primarily used to treat heart conditions, such as arrhythmias and long QT syndrome, and can help reduce the risk of stroke or sudden cardiac death. Sodium channel blockers were among the first drugs used for several decades and are divided into three groups based on their cellular electrophysiologic effects. These agents are effective against many arrhythmias, including atrial fibrillation.

The three subgroups of sodium channel blockers include those in class IA, which slow conduction and prolong repolarization (procainamide, disopyramide); those in class IB, which slow conduction and shorten repolarization (lidocaine, mexiletine, phenytoin); and those in class IC, which prolong conduction with little to no effect on repolarization (flecainide, propafenone).

Procainamide

Procainamide is a Class 1A antiarrhythmic agent used to treat life-threatening ventricular arrhythmias. Long-term use of procainamide can induce hypersensitivity reactions, autoantibody formation, and a lupus-like syndrome.

Clinical Indications

Procainamide is indicated for treating ventricular arrhythmias, supraventricular arrhythmias, atrial fibrillation, atrial flutter, and **Wolff–Parkinson–White syndrome**.

Mechanism of Action

Procainamide can bind to fast sodium channels by inhibiting their recovery after repolarization. It can also prolong the action potential and reduce the speed of impulse conduction. Therefore, it inhibits the ionic fluxes required to initiate and conduct impulses, affecting local anesthetic action.

Adverse Effects

The adverse effects of procainamide include bradycardia, hypotension, cardiac toxicity, and blood dyscrasias. Prolongation of the PR interval and QRS complex are among procainamide's most potentially harmful cardiac side effects, which may worsen as serum procainamide concentrations rise. Additionally, procainamide can cause bone marrow toxicity, leading to **pancytopenia**.

Contraindications

Procainamide is contraindicated in patients with complete heart block due to its effects in suppressing ventricular pacemakers, which may lead to **asystole**. It should also be avoided in patients with severe heart failure, **myasthenia gravis**, and peripheral neuropathy.

Lidocaine

Lidocaine is a local anesthetic agent commonly used for local and topical anesthesia. It also has antiarrhythmic and analgesic properties and can be an adjunct to tracheal intubation. Lidocaine belongs to the tertiary amine class of IB antiarrhythmic drugs. It is frequently used in the treatment of ventricular arrhythmias, particularly those associated with acute MI, including ventricular tachycardia, ventricular fibrillation, and multiple ventricular extrasystoles. It is also often effective in treating atrial arrhythmias.

Lidocaine blocks cardiac sodium channels and shortens the action potential. It is used intravenously only for treating arrhythmias. It is not recommended for prophylactic administration to suppress PVCs or to prevent ventricular tachycardia or fibrillation after acute coronary syndrome. The adverse effects of lidocaine include QRS widening, hypotension, shock, and asystole. Lidocaine toxicity is usually dose-dependent, with an increased risk in individuals with renal and hepatic impairments. Lidocaine is also discussed in **Chapter 12**.

Mexiletine

Mexiletine is a class 1B antiarrhythmic with a rapid onset and more pronounced effects at higher heart rates. It inhibits the inward sodium current essential for impulse conduction, thereby decreasing the rate of the action potential. Mexiletine is mainly used to suppress ventricular arrhythmias. Its adverse effects include nausea, paresthesia, seizures, hypotension, sinus bradycardia, bundle branch block, AV heart block, asystole, ventricular fibrillation, cardiovascular collapse, and coma.

Flecainide

Propafenone

Propafenone is a class IC antiarrhythmic agent used to prevent atrial fibrillation and its recurrence. It works directly on the heart tissue and slows the nerve impulses in the heart. It may be used in combination with a beta-blocker or a CCB to cardiovert recent-onset atrial fibrillation in patients without structural heart disease. Adverse effects of propafenone include nausea, vomiting, heartburn, unusual taste, loss of appetite, headache, blurred vision, fatigue, dyspnea, and bradycardia.

Class II: Beta-Adrenergic Blockers

Class II antidysrhythmics, or beta-adrenergic blockers, decrease blood pressure and cause bradycardia. They block the effects of epinephrine, preventing catecholamine-mediated actions on the myocardium. The pharmacologic effects of beta blockers are particularly beneficial after a heart attack. They treat arrhythmias and help avoid complications and sudden cardiac arrest after MI in patients. Although several beta-adrenergic blockers are available, only a few are commonly used as antidysrhythmics. The medications frequently prescribed in the United States include acebutolol, esmolol, propranolol, and sotalol.

Acebutolol

Acebutolol is a cardioselective beta-adrenergic blocker used to treat hypertension and arrhythmias. It blocks specific beta receptors in the blood vessels and the myocardium, inhibiting the action of adrenaline. Common adverse effects of acebutolol include headache, nausea, diarrhea, dizziness, bradycardia, dyspnea, fatigue, muscle aches, and

nightmares. This drug is contraindicated in patients with persistently severe bradycardia, cardiogenic shock, or heart failure.

Esmolol

Propranolol

Propranolol is a short-acting, cardioselective β-blocker and a class II antiarrhythmic agent. It is used alone or in combination with other medications to treat hypertension, atrial fibrillation, thyrotoxicosis, and pheochromocytoma. Propranolol also helps prevent angina and migraine headaches. It works by relaxing blood vessels and slowing the heart rate to improve blood flow and lower blood pressure. The most common adverse effects of propranolol include headaches, dizziness, fatigue, cold fingers or toes, stomach pain, nausea, vomiting, diarrhea, insomnia, and nightmares. Propranolol is contraindicated in patients with asthma, diabetes, hypotension, heart block, uncontrolled heart failure, and Raynaud's disease.

Sotalol

Sotalol is a beta-blocker with antiarrhythmic properties that affect the atria and ventricles. It is used to treat life-threatening ventricular fibrillation or atrial flutter. Sotalol is a competitive inhibitor of the rapid potassium channel. This inhibition lengthens the duration of action potentials and the refractory period in the atria and ventricles. The inhibition of rapid potassium channels increases as the heart rate decreases. Approximately 80%–90% of sotalol is excreted in the urine as unchanged, and a small amount is excreted in the feces as unchanged. The adverse effects of sotalol include chest pain, headache, blurred vision, confusion, fatigue, sweating, skin rash, anxiety, bleeding gums, and edema.

Check Your Knowledge

1. What are the clinical indications of procainamide?
2. What is the classification of mexiletine and propafenone?
3. What are the indications of beta-adrenergic blockers?

Class III: Potassium Channel Blockers

Potassium channel blockers are a class of drugs that inhibit potassium from passing through cell membranes, which can slow electrical impulses in the heart. They are indicated for the treatment of ventricular fibrillation, ventricular tachycardia, hypertension, epilepsy, and multiple sclerosis. Examples of potassium channel blockers include amiodarone, dofetilide, and dronedarone.

Amiodarone

Amiodarone is a class III antiarrhythmic drug. It blocks potassium currents that cause cardiac muscle to repolarize during the third phase of the cardiac action potential.

Clinical Indications

Amiodarone is commonly used to prevent and treat certain types of severe ventricular fibrillation and ventricular tachycardia; however, this drug is also utilized to treat atrial fibrillation.

Mechanism of Action

Amiodarone's mechanism of action includes blocking potassium's ability to repolarize the heart during phase III of the cardiac action potential. This potassium channel-blocking effect increases the duration of the action potential and prolongs the effective refractory period in cardiac muscle, reducing excitability.

Adverse Effects

Adverse effects include blurred vision, photophobia, fatigue, chest pain, dyspnea, headache, dizziness, fainting, chills, liver damage, vomiting, and constipation.

Contraindications

Amiodarone is contraindicated in patients with second- or third-degree heart block who do not have pacemakers. It should be avoided in individuals with Wolff–Parkinson–White syndrome when concurrent atrial fibrillation is present or in patients with cardiogenic shock.

Dofetilide

Dofetilide is a class III antiarrhythmic medication used to treat acute atrial fibrillation or flutter and to prevent recurrence. It is also used to treat paroxysmal supraventricular tachycardias. Dofetilide inhibits potassium channels during phase 3 of the action potential, slowing the efflux of potassium cations out of the myocyte and ultimately slowing the cell's repolarization rate. The adverse effects of dofetilide include dizziness, headaches, chest pain, abdominal pain, nausea, and diarrhea. It is contraindicated in patients with severe renal impairment or congenital or acquired long **QT syndromes**.

> Overdose of sotalol may cause hypotension, bradycardia, congestive heart failure, bronchospasm, and hypoglycemia. Such intentional toxicity may result in a prolonged QT interval, bradycardia, hypotension, cardiac asystole, ventricular tachycardia, and premature ventricular complexes.

Dronedarone

Dronedarone is another class III antiarrhythmic drug that restores normal sinus rhythm in patients with persistent atrial fibrillation and atrial flutter. It manages the condition by controlling the rhythm rate, preventing thromboembolic events, and treating the underlying disease. The adverse effects include nausea, diarrhea, fainting, rashes, dyspnea, and lightheadedness.

YOU SHOULD REMEMBER

Atrial fibrillation is the most common type of arrhythmia, resulting from abnormal electrical activity in the atria. It leads to turbulent and irregular blood flow through the heart chambers, which decreases the heart's effectiveness in pumping blood and increases the likelihood of thrombus formation within the atria. This thrombus can ultimately dislodge and cause a stroke.

Check Your Knowledge

1. What are the indications of potassium channel blockers?
2. What are the adverse effects of amiodarone?
3. What is the description of QT syndromes?

CLASS IV: CCBS

Class IV antidysrhythmic drugs are CCBs. These medications treat heart and blood vessel conditions by blocking calcium from entering muscle cells. Two commonly used drugs for treating dysrhythmias are diltiazem and verapamil.

Diltiazem

Diltiazem increases the supply of blood and oxygen to the heart while reducing its workload. It inhibits the inflow of calcium ions into the myocytes during depolarization, lowering intracellular calcium concentrations through increased smooth muscle relaxation, which leads to arterial vasodilation and decreased blood pressure. Diltiazem is an antiarrhythmic, antihypertensive, and anti-anginal medication used in a variety of clinical situations. The FDA has approved it for atrial arrhythmia, paroxysmal supraventricular tachycardia, pulmonary hypertension, and migraine headaches. Its adverse effects include headaches, bradycardia, flushing, leg edema, dizziness, blurred vision, and stomach pain.

Verapamil

Verapamil may be used alone or in combination with other agents to treat arrhythmia, hypertension, or angina. It inhibits the influx of calcium ions into the myocardium and the coronary artery muscle, resulting in vasodilation. Verapamil also enhances myocardial oxygen delivery. This medication can be administered either orally or intravenously. The adverse effects of verapamil include dizziness, dyspnea, nausea, constipation, **gingival hyperplasia,** bradycardia, hypotension, dyspepsia, and edema.

Check Your Knowledge

1. What are the indications of CCBs?
2. What are the adverse effects of CCBs?
3. What are the contraindications of CCBs?

MISCELLANEOUS DRUGS

The miscellaneous drugs do not belong to a traditional class of antiarrhythmics. They have varied mechanisms of action and uses. The medications in this group include adenosine and digoxin.

DIGOXIN

Digoxin is a cardiotonic glycoside belonging to the digitalis class. Cardiac glycosides, which include digitalis and digoxin, have been used in clinical practice for a long time. This drug is prescribed to treat heart failure, atrial flutter, and atrial fibrillation. It enhances the heart's strength, regulates the heartbeat's rate, and reduces edema in heart failure. Digoxin is utilized to manage congestive heart failure in conjunction with a diuretic. It possesses a very narrow therapeutic window and is a toxic substance known for its cardiotoxic effects. This agent is considered for elderly patients with atrial fibrillation.

Superior therapies with milder adverse effects and better safety profiles, such as beta-blockers and calcium channel blockers, have replaced them. In current practice, they are used as backup drugs when first-line agents are ineffective. Their optimal use is to treat mild to moderate heart failure in adult patients and enhance myocardial contraction.

ADENOSINE

Adenosine is a naturally occurring substance that relaxes and dilates blood vessels and influences the electrical activity of the myocardium. It is utilized in myocardial perfusion **scintigraphy** to treat supraventricular tachycardia and during a heart stress test. Adverse effects of adenosine include facial flushing, dyspnea, chest pain, MI, dizziness, and numbness. Contraindications include known hypersensitivity to adenosine, heart block, asthma, and COPD.

YOU SHOULD REMEMBER

Digoxin toxicity can occur when a patient takes a higher dose or uses it for an extended period. It can lead to fatal cardiac arrhythmias. Symptoms of digoxin toxicity include nausea, vomiting, diarrhea, loss of appetite, yellow or green-tinted vision, blurred vision, blind spots, seeing spots, palpitations, dyspnea, syncope, and atrial tachycardias.

Check Your Knowledge

1. What are the clinical indications of digoxin?
2. What are the symptoms of digoxin overdose?
3. What are the indications of adenosine?

CLINICAL CASE STUDIES

Clinical Case Study 1

A 75-year-old man visited his primary physician for an annual check-up to review his lab blood work from before the visit. His physician performed an EKG in the office and diagnosed him with atrial fibrillation. The patient has never smoked or consumed alcohol. However, a history of hypertension indicates that he may be at risk for arrhythmias. He denies experiencing dyspnea, fatigue, dizziness, palpitations, or chest pain.

CRITICAL THINKING QUESTIONS

1. What is the description of atrial fibrillation?
2. What are the risk factors for atrial fibrillation?
3. What medications are used to treat atrial fibrillation?

Clinical Case Study 2

A 53-year-old woman was transferred to the emergency department with a medical history of hypertension, asthma, and tobacco use. The patient complained of chest pain and dyspnea. She became unresponsive and pulseless. Cardiopulmonary resuscitation was initiated, and she experienced ventricular fibrillation. The patient returned to spontaneous circulation after using the defibrillator and administering amiodarone and lidocaine.

CRITICAL THINKING QUESTIONS

1. What are the primary drugs for treating ventricular fibrillation?
2. What are the mechanisms of action for amiodarone?
3. What are the indications for lidocaine?

FURTHER READING

Bennett, D.H. (2013). *Bennett's Cardiac Arrhythmias – Practical Notes on Interpretation and Treatment*, 8th Edition. Wiley-Blackwell.

Billman, G.E. (2010). *Novel Therapeutic Targets for Antiarrhythmic Drugs*. Wiley.

Breithardt, G., Borggrefe, M., Camm, J.A., and Shenasa, M. (2012). *Antiarrhythmic Drugs – Mechanisms of Antiarrhythmic and Proarrhythmic Actions*. Springer.

Garcia, T.B., and Garcia, D.J. (2019). *Arrhythmia Recognition – The Art of Interpretation*, 2nd Edition. Jones & Bartlett Learning.

Ho, R.T. (2022). *Electrophysiology of Arrhythmias – Practical Images for Diagnosis and Ablation*, 2nd Edition. Wolters Kluwer.

Huff, J. (2022). *ECG Workout – Exercises in Arrhythmia Interpretation*, 8th Edition. Wolters Kluwer.

Kass, R.E., and Clancy, C.E. (2010). *Basis and Treatment of Cardiac Arrhythmias* (Handbook of Experimental Pharmacology, 171). Springer.

Kowey, P.R., and Naccarelli, G.V. (2010). *Advances in Antiarrhythmic Drug Therapy* – An Issue of Cardiac Electrophysiology Clinics (Volumes 2–3) (The Clinics – Internal Medicine). Saunders.

Kowey, P., Piccini, J.P., Naccarelli, G., and Reiffel, J.A. (2017). *Cardiac Arrhythmias, Pacing and Sudden Death.* Springer.

Martinez-Rubio, A., Tamargo, J., and Dan, G.A. (2020). *Antiarrhythmic Drugs* (Current Cardiovascular Therapy). Springer.

Murray, D. (2023). *A Clinical Approach to Antiarrhythmic Drugs.* American Medical Publishers.

NEDU. (2021). *EKG/ECG Interpretation Made Easy – An Illustrated Study Guide for Students to Easily Learn How to Read and Interpret ECG Strips.* NEDU LLC.

Stone, B. (2023). *Critical Care Medications – Pharmacology of Common Anti-Arrhythmic Medications Used in Critical Care – A Study Guide and Resource Book for Nurses, Physicians and More.* Stone.

Walraven, G. (2016). *Basic Arrhythmias with 12-Lead EKGs,* 8th Edition. Pearson.

Wang, P., and Al-Ahmad, A. (2011). *Advances in Antiarrhythmic Drug Therapy – An Issue of Cardiac Electrophysiology* (The Clinics – Internal Medicine Book 3). Saunders.

18 Drugs Affecting Coagulation Disorders

LEARNING OUTCOMES

After studying the chapter, readers should be able to

1. Summarize the various coagulation disorders.
2. Identify the most common cause of pulmonary embolism (PE).
3. Explain the complications of atrial fibrillation (AFib).
4. Describe the mechanism of action of warfarin.
5. Summarize the indications for using low-molecular-weight heparin (LMWH) agents.
6. Review the adverse effects of LMWH agents.
7. Explain the newest direct thrombin inhibitors (DTIs) and their contraindications.
8. Review the clinical indications of LMWH agents.
9. Explain the adverse effects of factor Xa inhibitors.
10. Describe the indications of antiplatelet drugs.

OVERVIEW

Anticoagulants are drugs that prevent blood from clotting. They can break down existing clots or prevent them from forming. These medications help prevent life-threatening conditions such as PEs, strokes, and heart attacks. Anticoagulation therapy prevents and treats blood clots associated with various diseases. Thrombolytics can break up clots, enhancing blood flow to the heart, brain, lungs, and other organs. Anticoagulant use primarily aims to prevent venous thromboembolic events or blood clots. Deep vein thrombosis (DVT) and PE are common disorders. It is estimated that over 100,000 patients die from coagulation problems. AFib is the most prevalent heart arrhythmia associated with coagulopathies and is recognized as the most significant independent risk factor for stroke. Anticoagulants come in various forms, including intravenous injections or oral medications. Heparin has two types: unfractionated heparin (UFH), LMWH, and fondaparinux. Thrombin inhibitors work by binding to thrombin, preventing it from facilitating the clotting process. Warfarin is a vitamin K antagonist, meaning it blocks the use of vitamin K and is administered orally.

COAGULATION DISORDERS

Coagulation disorders can be inherited or acquired conditions. They include clotting factor deficiencies and **von Willebrand disease**. Acquired coagulation disorders develop after birth due to another condition or injury rather than being inherited. They may result from changes in the platelets, blood vessels, or coagulation factors and can lead to a tendency to bleed or clot. Antiphospholipid syndrome is the most common acquired clotting disorder and is classified as an **autoimmune** condition. Higher levels of antiphospholipid syndrome antibodies in the blood raise the risk of blood clots. Disseminated intravascular coagulation is a rare but serious condition that causes abnormal blood clotting. It can be life-threatening and requires immediate treatment. Acquired coagulation conditions typically result from surgery, trauma, medications, or a medical condition that increases the risk of developing blood clots.

Thrombosis occurs when blood clots and obstructs veins or arteries (see Figure 18.1). Signs and symptoms include pain and swelling in one leg, chest pain, or numbness on one side of the body. Complications of thrombosis can be life-threatening, such as PE, stroke, AFib, or heart attack.

DEEP VEIN THROMBOSIS

DVT occurs when a clot forms in the deep veins, usually developing in the leg, thigh, or pelvis. DVT can lead to severe disability, illness, and even death. When caught early, it is preventable and treatable. If the blood clot breaks off from the vein wall, it may travel through the bloodstream and lodge in a lung, resulting in PE. Thrombosis is widespread and is the underlying cause of one in four deaths worldwide.

Thrombosis symptoms depend on the size of the clot, its location (where it forms or becomes lodged), and the complications it causes. Blockages are more likely to occur in areas with small blood vessels, particularly in the lungs, brain, legs, and arms. The most common symptoms of DVT include localized swelling, pain, and skin redness.

STROKE

A stroke is a life-threatening condition that occurs when part of the brain does not receive enough blood flow. It is also known as a cerebrovascular accident. This typically happens due to a blocked artery or bleeding in the brain. Such an event leads to brain necrosis from a lack of oxygen. Strokes are common worldwide and rank as the second leading cause of death. In the United States, stroke is the fifth leading cause of death. Additionally, strokes are a significant cause of disability globally.

The risk factors for stroke include hypertension, smoking, obesity, physical inactivity, genetics, age, sex, race and ethnicity, stress, and sleep apnea. Symptoms include numbness in the face, arm, or leg, particularly on one side of the body. Patients may experience double vision, loss of balance, or lack of coordination. Other warning signs include drowsiness, severe headache, and nausea or vomiting. Treating a stroke depends on various factors. For ischemic strokes, the top priority is restoring circulation to

DOI: 10.1201/9781003461913-20

Thrombosis

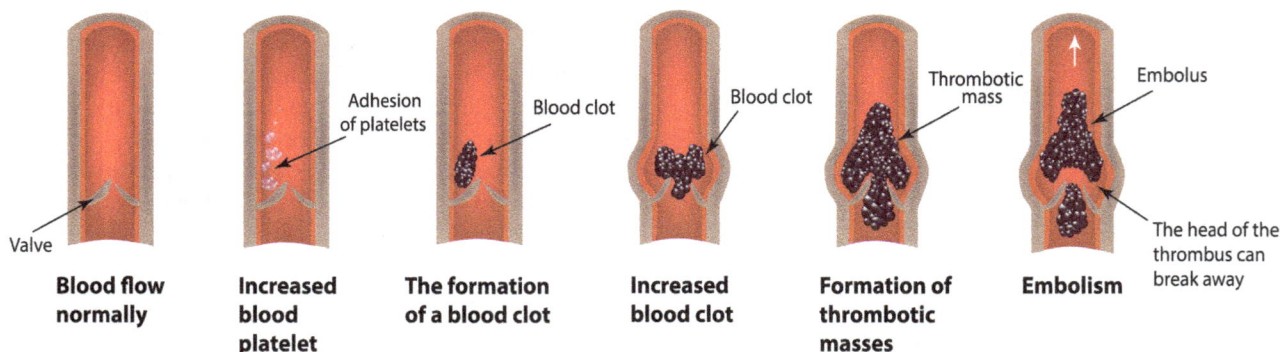

FIGURE 18.1   Thrombosis formation.

the affected brain areas. If this occurs quickly enough, it can prevent permanent damage or limit a stroke's severity. Restoring circulation typically involves thrombolytics but may also require a catheterization procedure.

PULMONARY EMBOLISM

A PE is a blood clot that blocks and stops blood flow to an artery in the lung. In most cases, the blood clot originates in a deep vein in the leg and travels to the lungs. Rarely, the clot forms due to DVT. Pulmonary emboli may result from an injury to the blood vessel wall, slowed blood flow, or an abnormality in coagulation. However, DVT is the most common cause of PE. **Fat emboli**, amniotic fluid, air bubbles, and upper body DVT can also lead to pulmonary emboli. Risk factors for PE include a family history of blood clotting disorders, surgery or injury, prolonged bed rest, long-distance travel, paralysis, older age, obesity, cigarette smoking, cancer treatment, and the use of oral contraceptives.

Symptoms of PE include fever, dyspnea, chest pain, tachypnea, sweating, wheezing, coughing, cyanosis, and fainting. Treatment options encompass anticoagulants, fibrinolytic therapy, a vena cava filter, pulmonary embolectomy, and percutaneous thrombectomy. In severe cases, anticoagulants include heparin, warfarin, and newer oral anticoagulants such as apixaban, rivaroxaban, dabigatran, and thrombolytics.

ATRIAL FIBRILLATION

AFib can lead to blood clots in the heart, increasing the risk of stroke, heart failure, and other heart-related complications. It is estimated that approximately 12 million people in the U.S. will have AFib by 2030. Individuals of European descent are more likely to develop AFib than those of African American descent. Risk factors for AFib include advancing age, hypertension, obesity, hyperthyroidism, diabetes, chronic kidney disease, heart failure, and smoking. Symptoms may include palpitations, dyspnea, fatigue, chest pain, and dizziness, although some individuals may experience no symptoms. Treatments include beta blockers, anticoagulants (such as apixaban, dabigatran, rivaroxaban, or edoxaban), and calcium channel blockers.

Check Your Knowledge

1. What are some examples of inherited or acquired conditions?
2. What are the causes of DVT?
3. What are the signs and symptoms of thrombosis in the brain and lungs?

ANTICOAGULANTS

Anticoagulants, or blood thinners, prevent or reduce blood clotting. They are administered to patients at high risk of developing blood clots to lower the chances of severe conditions such as strokes and heart attacks. These medications are categorized into four primary types: warfarin, heparins, DTIs, and **factor Xa inhibitors**. The three main anticoagulant medications include vitamin K antagonists, direct oral anticoagulants, and LMWHs. Table 18.1 lists the anticoagulants used in the US.

WARFARIN

Warfarin is an oral anticoagulant used to treat and prevent blood clots. It is the most widely prescribed outpatient

TABLE 18.1
Anticoagulants

Generic Name	Brand Name
Warfarin	Coumadin, Jantoven
Rivaroxaban	Xarelto
Apixaban	Eliquis
Enoxaparin	Lovenox
Dalteparin	Fragmin
Tinzaparin	Innohep
Fondaparinux	Arixtra
Dabigatran Etexilate	Pradaxa
Bivalirudin	Angiomax
Desirudin	Iprivask
Argatroban	Argatroban, Acova

anticoagulant. Warfarin works by inhibiting the synthesis of vitamin K-dependent clotting factors, reducing the blood's clotting ability.

Clinical Indications

Warfarin is prescribed to prevent and treat venous thromboembolism, PE, thromboembolism associated with AFib, thromboembolism related to cardiac valve replacement, and thromboembolic events following myocardial infarction.

Mechanism of Action

Warfarin inhibits the synthesis of biologically active forms of vitamin K-dependent clotting factors II, VII, IX, and X.

Adverse Effects

The adverse effects of warfarin include bleeding gums, hematuria, **melena**, unusual bruising, gastrointestinal bleeding, and **hemarthrosis.** Other adverse effects include headache, nausea, vomiting, diarrhea, flatulence, abdominal pain, and alterations in the sense of taste. Warfarin can cross the placental barrier and may cause spontaneous abortion or **stillbirth**.
Bleeding in the fetus.

Contraindications

Warfarin is contraindicated for patients with hypersensitivity to the drug, hemorrhagic tendencies, and recent eye or brain surgery. Elderly patients are more vulnerable to bleeding complications due to falls, drug interactions, cognitive conditions, and their living situations. Warfarin should be avoided during the first trimester and close to delivery.

YOU SHOULD REMEMBER

Vitamin K_1 is the only effective antidote for long-term management, but it takes several hours to reverse the anticoagulation effect. Oral vitamin K_1 has excellent bioavailability and is rapidly absorbed, so it is recommended for use without severe or life-threatening bleeding.

Check Your Knowledge

1. What are the indications of warfarin?
2. What are the contraindications of warfarin?
3. What are the adverse effects of warfarin?

UNFRACTIONATED HEPARIN

UFH is a naturally occurring glycosaminoglycan medication released from **mast cells** in the blood. It is an anticoagulant that enhances the activity of antithrombin. LMWHs differ from UFH in several ways, including their average molecular weight, the requirement for only once or twice-daily dosing, and the lack of need to monitor the **activated partial prothrombin time** (aPTT).

UFH is associated with a higher risk of bleeding and long-term osteoporosis. It is more specific than LMWH for thrombin. Furthermore, the effects of UFH can typically be reversed using protamine sulfate, such as heparin, bivalirudin, desirudin, and argatroban.

LOW MOLECULAR WEIGHT HEPARINS

LMWHs are a new class of anticoagulants derived from UFH. They offer several advantages over UFH, including increased use for specific thromboembolic indications. Dosing adjustments are advised for patients with kidney impairment and obesity. LMWHs can be administered intravenously and subcutaneously. Dalteparin has been approved for long-term secondary prevention of VTE.

Clinical Indications

LMWHs prevent venous thromboembolic disease (VTE) during acute or elective hospital admissions and treat DVT and PE. They are also prescribed for unstable angina and venous thromboembolism in pregnancy.

Mechanism of Action

LMWHs act by inhibiting the coagulation cascade pathway and facilitating the conversion of fibrinogen into fibrin through thrombin activity (see Figure 18.2). The LMWHs inhibit coagulation by activating antithrombin III (ATIII), which binds to and inhibits factor Xa.

Adverse Effects

The primary adverse effects of LMWH include bleeding, osteoporosis, spontaneous fractures, and hypersensitivity reactions. Possible side effects associated with dalteparin use may include injection site reactions such as irritation, pain, and bruising.

Contraindications

LMWHs are contraindicated in patients with peptic ulcer disease, severe hypertension, recent stroke, or recent eye surgery. Since LMWHs are self-administered, it is essential to consider dosing in cases of chronic kidney disease, where there is a risk of accumulation and, consequently, a higher likelihood of problematic bleeding.

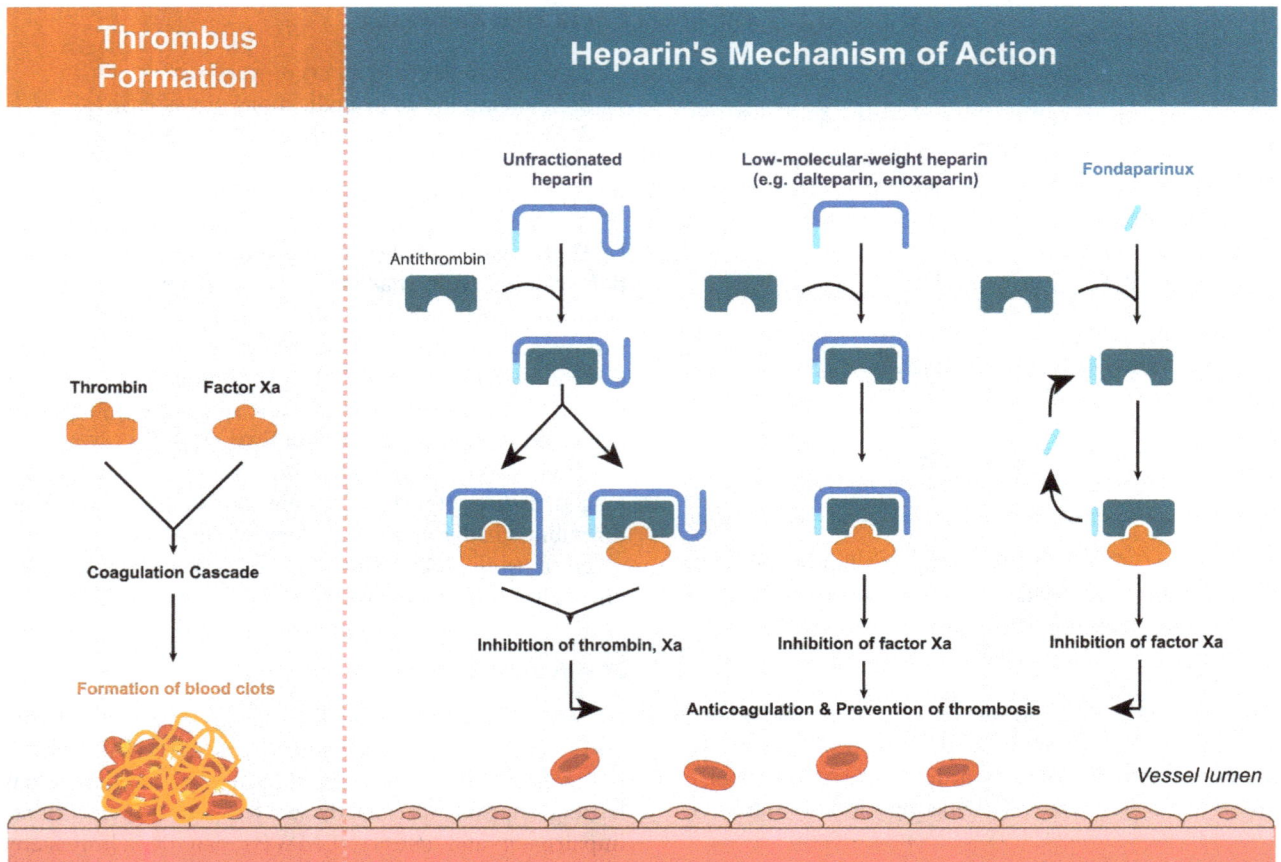

FIGURE 18.2   The mechanism of action of heparins.

YOU SHOULD REMEMBER

LMWHs are a new class of anticoagulants that offer pharmacokinetic and biological advantages over UFH. These advantages are clinically evident in (1) greater convenience due to the ability to administer LMWH via subcutaneous injection without laboratory monitoring, and (2) the associated cost reduction stemming from shorter hospital stays.

Check Your Knowledge

1. What are the differences between UFH and LMWHs?
2. What are the contraindications of LMWHs?
3. What are the adverse effects of LMWHs?

Direct Thrombin Inhibitors

Four parenteral DTIs of thrombin activity – lepirudin, desirudin, bivalirudin, and argatroban – are currently FDA-approved in the United States. The oral dabigatran etexilate is the most promising.

Clinical Indications

DTIs are used for the treatment or prevention of acute coronary syndrome, AFib, and VTE. They represent a novel class of drugs developed as an effective alternative for anticoagulation, particularly in patients suffering from heparin-induced **thrombocytopenia**, those requiring dialysis, and individuals undergoing cardiopulmonary bypass, as well as for managing thromboembolic disorders.

Mechanism of Action

DTIs bind directly to thrombin and do not require a cofactor like antithrombin to exert their effect. They can inhibit both soluble thrombin and fibrin-bound thrombin.

Adverse Effects

Common adverse effects of DTIs include life-threatening bleeding, chest pain, cardiac arrest, bruising, indigestion, abdominal pain, diarrhea, and hypersensitivity. Other adverse effects may include headache, dizziness, fever, swelling, arm and leg pain, and itching.

Contraindications

DTIs are contraindicated in severe renal impairment or hemodialysis, hypersensitivity, impaired hemostasis, active pathologic bleeding, and mechanical prosthetic heart valves.

Check Your Knowledge

1. What are the clinical indications of DTIs?
2. What are the mechanisms of action of DTIs?
3. What are the adverse effects of DTIs?

Factor Xa Inhibitors

Factor Xa inhibitors are newer anticoagulants that offer several advantages over older ones. The most significant benefit is that patients taking them do not require frequent blood tests. They are used to prevent or treat blood clots. Depending on the situation, individuals can take them for days or months. Fondaparinux is the only injectable factor Xa inhibitor available and is approved by the FDA for various indications. However, its unique application is for patients allergic to UFH, LMWH, or HIT. Patients can use this product at home to prevent or treat VTE.

Clinical Indications

Fondaparinux prevents DVT in high-risk patients following hip fracture surgery, hip or knee replacement surgery, or abdominal surgery. Combined with warfarin, it actively treats DVT with or without PE. Additionally, it is used to treat DVT and PE while preventing recurrent DVT and PE.

Mechanism of Action

Fondaparinux exhibits antithrombotic activity due to the ATIII-mediated selective inhibition of factor Xa. By selectively binding to ATIII, fondaparinux activates ATIII's natural ability to neutralize factor Xa. This neutralization interrupts the blood coagulation cascade, inhibiting thrombin formation and thrombus development.

Adverse Effects

The adverse effects of fondaparinux include mild injection site reactions, anemia, skin rash, edema, nausea, vomiting, insomnia, constipation, and fever.

Contraindications

All currently available factor Xa inhibitors are contraindicated in patients with active bleeding.

Check Your Knowledge

1. What are the benefits of factor Xa inhibitors compared to older anticoagulants?
2. What are the indications of fondaparinux?
3. What are the adverse effects of fondaparinux?

ANTIPLATELET DRUGS

Antiplatelet drugs prevent platelets from clumping together and decrease the body's ability to form blood clots. They are used to treat and may help prevent heart attacks and strokes. Spontaneous bleeding can occur with a platelet count under 10,000/µL, and surgical bleeding can occur with counts below 50,000/µL. Aspirin is the most commonly used antiplatelet drug. The others include clopidogrel (Plavix), ticagrelor (Brilinta), prasugrel (Effient), dipyridamole/aspirin (Aggrenox), and eptifibatide (Integrilin).

Clinical Indications

Antiplatelet drugs lower the risk of stroke, MI, or blood clots. Clopidogrel is more effective than aspirin in improving cardiovascular outcomes for patients with cardiovascular disease and offers extra benefits for patients with acute coronary syndromes who are already taking aspirin.

Mechanism of Action

Antiplatelet agents interfere with platelet binding. They slow down clotting, reduce fibrin formation, and prevent clots from forming and growing.

Adverse Effects

Clopidogrel is considered a safer alternative to other antiplatelet drugs due to a lower incidence of hematologic adverse effects. However, purpura is associated with the most significant adverse hematologic effects of clopidogrel, including neutropenia and idiopathic immune thrombocytopenia.

Contraindications

Antiplatelet drugs should be avoided in patients with intracranial hemorrhage, peptic ulcers, liver disease, or recent surgery within 72 hours.

Check Your Knowledge

1. What are some examples of antiplatelet drugs?
2. What is the advantage of clopidogrel?
3. What are the contraindications of antiplatelet drugs?

THROMBOLYTICS

Thrombolytics, or fibrinolytics, break up and dissolve blood clots. Currently available thrombolytic agents include the following:

- **Alteplase (tPA):** This is often the primary choice for strokes, PEs, and cardiovascular cases. It seldom causes allergic reactions.
- **Anistreplase:** It can work throughout the body because it targets all plasminogens, not just those attached to fibrin.
- **Prourokinase:** Researchers are testing this new drug, which must be converted to urokinase to be effective.
- **Reteplase:** This acts more quickly than other thrombolytics.
- **Streptokinase** is the most commonly used drug worldwide, though it is less frequently utilized in the US. It is cheaper than other options, but many individuals experience allergic reactions. Similar to anistreplase, it can have effects on the entire body.
- **Tenecteplase:** This is efficient and a common choice among many providers in North America and Europe. It also carries a lower risk of bleeding.
- **Urokinase:** Healthcare providers may choose this option for peripheral vascular clots in the legs and for catheters with clots. Many providers outside the US prefer it because it is less expensive than other thrombolytic drugs.

CLINICAL INDICATIONS

Thrombolytic therapy is indicated to prevent or dissolve the formation of new blood clots. It is used for acute DVT, PE, myocardial infarction, stroke, acute peripheral arterial occlusion, and intracardiac thrombus formation.

MECHANISM OF ACTION

Thrombolytic agents convert plasminogen to plasmin, an enzyme that breaks down fibrin, forming the backbone of blood clots. Exogenous plasminogen activators dissolve a thrombus and affect circulating plasminogen, decreasing levels of both plasminogen and fibrin.

ADVERSE EFFECTS

The adverse effects of thrombolytic medications include severe bleeding in the brain, hypertension, kidney disease, and damage to blood vessels. The main risk associated with thrombolytic therapy is internal bleeding.

CONTRAINDICATIONS

Thrombolytic drugs should be avoided in patients with intracranial bleeding, aortic dissection, brain tumors, head injuries, or recent surgeries.

Check Your Knowledge

1. What are some examples of thrombolytics?
2. What are the mechanisms of action of thrombolytics?
3. What are the adverse effects of thrombolytics?

CLINICAL CASE STUDIES

Clinical Case Study 1

A 38-year-old woman with a 20-year history of taking contraceptive pills and smoking presented to the emergency department with severe pain and swelling in her right leg that had persisted for 2 weeks. Over the past 3 days, she experienced dyspnea, wheezing, and cyanosis, along with tenderness in the right calf. A duplex scan confirmed acute DVT in the right posterior tibial region. CT pulmonary angiography was performed immediately, revealing extensive unilateral pulmonary emboli. Fondaparinux was administered right away.

CRITICAL THINKING QUESTIONS

1. What are the risk factors for PE?
2. What are the indications for fondaparinux?
3. What are the potential adverse effects of fondaparinux?

Clinical Case Study 2

A 79-year-old man was brought to the emergency room, complaining of dyspnea, palpitations, fatigue, chest pain, and dizziness. His medical history included hypertension and mild aortic stenosis. His EKG revealed AFib, and his blood pressure was 150/100. He was placed on rivaroxaban (Xarelto), one of the newer anticoagulants (DTIs).

CRITICAL THINKING QUESTIONS

1. What are the complications associated with AFib?
2. What are the clinical indications for DTIs?
3. What are the adverse effects of DTIs?

FURTHER READING

DeLoughery, T.G. (2019). *Hemostasis and Thrombosis*, 4th Edition. Springer.

Dorgalaleh, A. (2023). *Congenital Bleeding Disorders – Diagnosis and Management*, 2nd Edition. Springer.

Favaloro, E.J., and Lippi, G. (2017). *Hemostasis and Thrombosis – Methods and Protocols* (Methods in Molecular Biology, 1646). Humana Press.

Goodnight, Jr., S.H., and Hathway, W.E. (2000). *Disorders of Hemostasis and Thrombosis: A Clinical Guide*, 2nd Edition. McGraw-Hill/Medical.

Green, D., and Kwaan, H.C. (2009). *Coagulation in Cancer* (Cancer Treatment and Research, 148). Springer.

Hudnall, S.D. (2023). *Hematology – A Pathophysiologic Approach* (Physiology Series/Physiology Monograph), 2nd Edition. Mosby/Elsevier.

Ibrahim, A., El-Sanhoty, H., and El-Sayed, S. (2012). *Coagulopathy in Chronic Liver Diseases – Impact of Liver Diseases in Coagulation Disorders*. Lap Lambert Academic Publishing.

Kamat, D.M., and Frei-Jones, M. (2021). *Benign Hematologic Disorders in Children – A Clinical Guide*. Springer.

Kaushansky, K., and Levi, M. (2017). *Williams Hematology – Hemostasis and Thrombosis*. McGraw-Hill.

Kitchen, S., Olson, J.D., and Preston, F.E. (2013). *Quality in Laboratory Hemostasis and Thrombosis*, 2nd Edition. Wiley-Blackwell.

Laposata, M. (2011). *Coagulation Disorders: Quality in Laboratory Diagnosis* (Diagnostic Standards of Care). Demos Medical.

Lichtin, A., and Bartholomew, J. (2014). *The Coagulation Consult – A Case-Based Guide*. Springer.

Marder, V.J., Aird, W.C., Bennett, J.S., Schulman, S., and White, III, G.C. (2012). *Hemostasis and Thrombosis – Basic Principles and Clinical Practice*. Wolters Kluwer/Lippincott, Williams, and Wilkins.

Saleh, A. (2020). *Drugs for Coagulation Disorders*. Saleh.

Taloyan, A.M., and Bankiewicz, D.S. (2012). *Coagulation: Kinetics, Structure Formation and Disorders* (Recent Advances in Hematology Research). Nova Biomedical.

Wahed, A., Quesada, A., and Dasgupta, A. (2019). *Hematology and Coagulation: A Comprehensive Review for Board Preparation, Certification and Clinical Practice*, 2nd Edition. Elsevier.

19 Anti-Anemic Drugs

LEARNING OUTCOMES

After studying the chapter, readers should be able to

1. Compare hematopoiesis and hematopoietic growth factors.
2. Explain the clinical indications of epoetin alfa.
3. Review the adverse effects of epoetin alfa.
4. Summarize the role of iron in the body.
5. Describe the signs and symptoms of iron deficiency.
6. Explain the mechanism of action of ferrous sulfate.
7. Summarize the contraindications of ferrous sulfate.
8. Describe the causes of pernicious anemia.
9. Discuss the clinical indications of cyanocobalamin.
10. Describe the contraindications of cyanocobalamin.

OVERVIEW

Anemia is a common blood disorder characterized by abnormally low levels of red blood cells (RBCs). It is critical to health as it contains the protein hemoglobin (Hb), which carries oxygen from the lungs to the body tissues. Several types of anemia exist, each with different causes and treatments. Anemia may be classified as iron deficiency anemia, vitamin B_{12} deficiency anemia, folate deficiency anemia, hemolytic anemia, sickle cell anemia, and aplastic anemia. Some types of anemia may be a symptom of another disease due to bleeding or significant blood loss, destruction of RBCs, poor nutrition, certain medications, and other specific diseases. Nutritional deficiency-related anemia can be prevented by consuming a diet rich in iron and vitamin B_{12} or supplementing these minerals and vitamins when dietary intake is insufficient. Most other types of anemia cannot be prevented, and treatment depends on the severity and cause of the anemia. Severe anemia may require a blood transfusion. Anemia caused by drug therapy can be treated by discontinuing the medication. In contrast, inherited hemolytic anemia may require a splenectomy.

HEMATOPOIESIS

Hematopoiesis refers to the production of blood cells. The body continually generates new blood cells to replace old ones, ensuring a steady blood supply. Before birth, **during** the first few months, the liver and spleen are responsible for making blood cells. After birth, production primarily takes place in the **bone marrow**. Three types of hematopoiesis include leukopoiesis, erythropoiesis, and thrombopoiesis (see Figure 19.1).

Leukocytes are found in blood and lymph tissue. They are part of the body's immune system and help fight infections, protecting the body from pathogens. Leukocytes are also called white blood cells. They are colorless but can appear light purple to pink when examined under a microscope and stained with dye. Leukocytes are tiny cells with a round shape and a nucleus. The five types of leukocytes include neutrophils, basophils, eosinophils, monocytes, and lymphocytes.

The mature **erythrocytes** (RBCs) have a biconcave discoid shape and lack a nucleus. This design allows for the flexibility needed to navigate the cardiovascular system. It provides an increased surface area, which supports sufficient gas exchange and enables the cell to perform its function. Erythrocytes produced through erythropoiesis contain the protein Hb, which carries oxygen from the lungs to all body tissues and transports carbon dioxide from the body to the lungs. Due to their importance in oxygenating the tissues, blood contains more erythrocytes than any other type of blood cell. Various conditions and disorders may affect RBCs, including anemias, which will be discussed later in this chapter.

Thrombocytes (platelets) are colorless cell fragments in the blood that clump together when they detect damaged blood vessels. They seal off injured tissue, preventing blood loss. They are produced through thrombopoiesis in the bone marrow. A normal platelet count ranges from 150,000 to 450,000 platelets per microliter of blood. Having more than 450,000 platelets is a condition called **thrombocytosis**; having fewer than 150,000 is known as thrombocytopenia.

Check Your Knowledge

1. What is hematopoiesis?
2. What causes the three types of hematopoiesis?
3. What are the features of erythrocytes?

HEMATOPOIETIC GROWTH FACTORS

Hematopoietic growth factors are glycoproteins with critical regulatory functions in blood cell differentiation, **proliferation**, and maturation. They produce blood cells in the bone marrow, making them useful for various clinical conditions. Colony-stimulating factors are used to support patients following bone marrow transplantation and chemotherapy. They are also indicated in the treatment of infectious diseases. Erythropoietin (EPO) is widely used for **aplastic anemia** and renal failure. **Thrombopoietin** (TPO) is a thrombocyte growth factor, mimicking its platelet enhancers.

DOI: 10.1201/9781003461913-21

Hematopoiesis

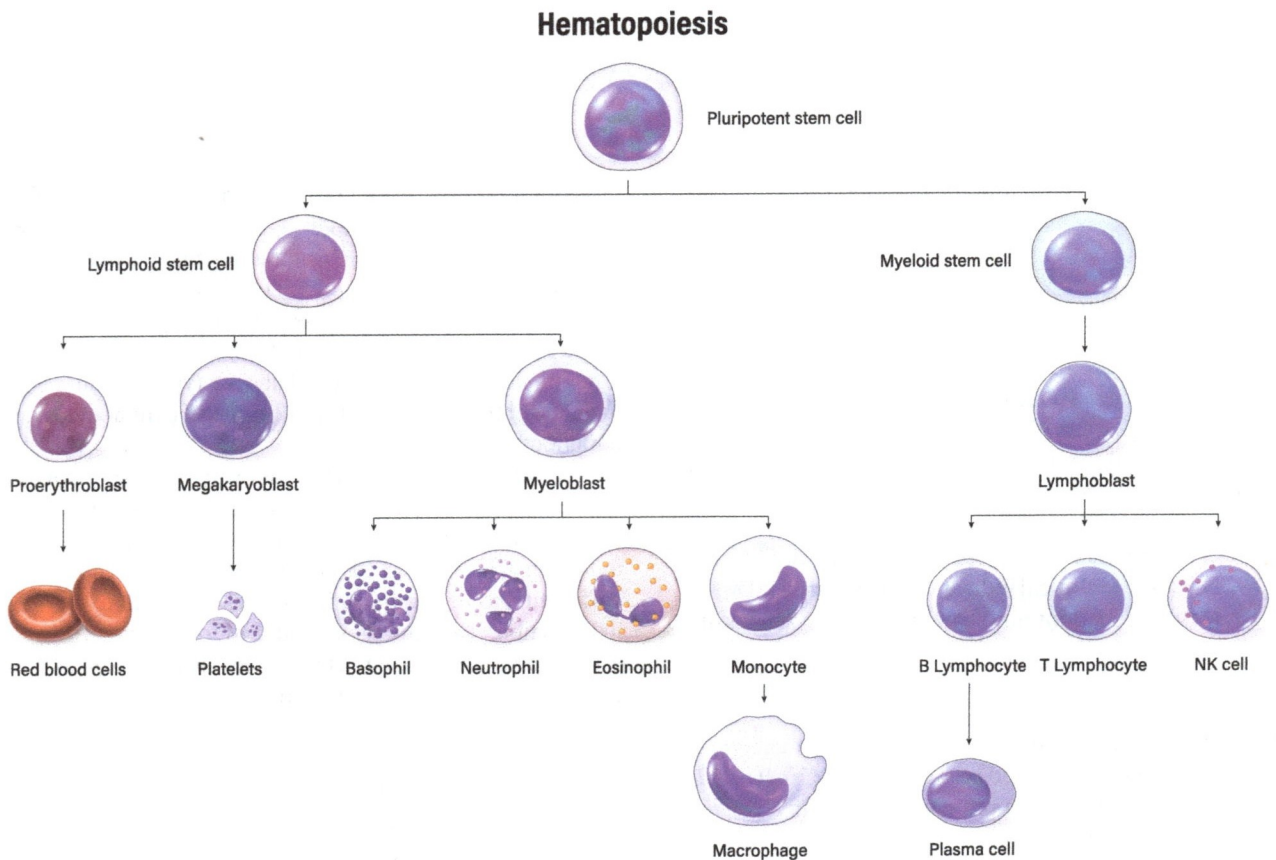

FIGURE 19.1 Hematopoiesis.

TABLE 19.1

Hematopoietic Growth Factors

Generic Name	Trade Name
Erythropoietic Growth Factors	
Darbepoetin alfa	Aranesp
Epoetin alfa (man-made EPO)	Epogen, Procrit
Colony-Stimulating Factors	
Filgrastim	Granix, Neupogen,
Pegfilgrastim	Zarxio
Sargramostim	Neulasta
	Leukine
Platelet Enhancers	
Eltrombopag	Promacta
Oprelvekin	Neumega
Romiplostim	Nplate

Other hematopoietic growth factors include **fibroblasts**, T lymphocytes, monocytes, and **endothelial cells**. Fibroblasts are crucial cellular sources of hematopoietic growth factors, except for EPO and TPO. The adult kidneys primarily produce EPO, while TPO is synthesized in the liver and kidneys. Table 19.1 lists hematopoietic growth factors.

Epoetin alfa is a medication used to treat aplastic anemia. It belongs to the class of drugs that stimulate erythropoiesis. Epoetin can be administered either intravenously or subcutaneously.

CLINICAL INDICATIONS

Epoetin alfa injection treats anemia in patients with chronic kidney disease, whether or not they are on dialysis. This medication may also prevent or treat anemia caused by surgery or by drugs such as zidovudine used for HIV or cancer.

MECHANISM OF ACTION

Epoetin alfa signals the bone marrow to produce more RBCs. This medication closely resembles the natural substance in the body (EPO) that helps prevent anemia.

ADVERSE EFFECTS

The adverse effects of epoetin alfa may include severe headaches, myocardial infarction, stroke, thrombosis, seizures, fatigue, hyperglycemia, hypokalemia, and hypertension. Common adverse effects comprise nausea, vomiting, fever, chills, cough, dyspnea, itching, weight loss, and insomnia.

CONTRAINDICATIONS

Epoetin alfa is contraindicated for patients with known hypersensitivity, uncontrolled hypertension, stroke, myocardial infarction, heart failure, and a history of seizures.

YOU SHOULD REMEMBER

Thrombocytopenia symptoms include easy bruising and frequent bleeding from the gums, nose, or gastrointestinal tract. Specific medications, genetic factors, cancers, chemotherapy, and renal diseases may cause it.

Check Your Knowledge

1. What is a description of hematopoietic growth factors?
2. What are the trade names of epoetin alfa and eltrombopag?
3. What are the contraindications of epoetin?

ANEMIAS

Anemia is a blood disorder that occurs when the body cannot produce enough healthy RBCs or when the RBCs do not function correctly. This condition reduces the blood's ability to carry oxygen to the body's tissues, potentially leading to several signs and symptoms.

Patients may experience tiredness, dyspnea, dizziness, headaches, or arrhythmia. In this condition, Hb, hematocrit, or RBC count decreases. The patients are at risk for anemia due to poor diet, infections, chronic diseases, or intestinal disorders. Pregnant women, children, older adults, and patients with chronic medical conditions are particularly at risk. Various types of anemia include iron, vitamin B_{12}, or folate deficiencies, as well as aplastic, hemolytic, and sickle cell anemias. This chapter will focus on iron and vitamin B_{12} deficiencies.

IRON DEFICIENCY ANEMIA

Iron deficiency anemia is a common condition caused by inadequate iron levels. This deficiency reduces the number of RBCs and Hb, both essential for transporting oxygen to the body's tissues. Approximately 70% of iron is stored as ferritin and hemosiderin in the bone marrow, spleen, and liver. The causes of iron deficiency anemia include blood loss, inadequate dietary iron intake, impaired iron absorption, and pregnancy. Signs and symptoms of iron deficiency anemia include headaches, palpitations, irritability, dyspnea, fatigue, tachycardia, **glossitis**, splenomegaly, and **pica**. Individuals should increase their iron intake through dietary sources, iron supplements, or blood transfusions to prevent iron deficiency anemia. Ferrous sulfate is an iron supplement used to treat this condition and may be

necessary if the diet does not provide sufficient iron. It is available in both tablet and liquid forms.

CLINICAL INDICATIONS

Ferrous sulfate treats or prevents anemia, blood loss, nutritional deficiencies, iron malabsorption, and pregnancy-related issues. Iron supplements may be administered orally, intramuscularly, or intravenously. Ferrous sulfate is available in tablet, drop, and elixir forms. Patients should take ferrous sulfate on an empty stomach, at least 1 hour before or 2 hours after a meal. Antacids and antibiotics should be avoided within 2 hours before or after ferrous sulfate.

MECHANISM OF ACTION

The mechanism of action of ferrous sulfate is to bind with **porphyrin** and globin chains to form Hb, which is essential for oxygen delivery from the lungs to other tissues. In the case of iron deficiency, Hb cannot be synthesized, leading to **microcytic anemia** characterized by the formation of small RBCs with insufficient Hb.

ADVERSE EFFECTS

The adverse effects of ferrous sulfate include heartburn, stomach pain, loss of appetite, constipation, nausea, vomiting, diarrhea, darkening of the stool, and black teeth staining.

CONTRAINDICATIONS

Ferrous sulfate should be avoided in patients with hereditary **hemochromatosis**, **hemosiderosis,** or a history of **hemolytic anemia**.

YOU SHOULD REMEMBER

The Z-track method should be employed for intramuscular iron injection to prevent the leakage of irritating and discoloring medications (such as iron dextran) into the subcutaneous tissue.

Check Your Knowledge

1. What are the signs and symptoms of anemia?
2. What are the various types of anemia?
3. What are the contraindications of ferrous sulfate?
4. What are the adverse effects of ferrous sulfate?

PERNICIOUS ANEMIA

Pernicious anemia is a rare and complex disease with a clear autoimmune basis. The anemia is megaloblastic and results from a deficiency of vitamin B12 due to a lack of **intrinsic factor**. This factor is secreted by the parietal cells in the stomach, binds to B12, and facilitates its transport to the terminal ileum for absorption. It affects individuals of all ages worldwide, primarily those over 60. The signs and symptoms include fatigue, pale skin, headaches, depression, dementia, peripheral neuropathy, and weight loss. Treatment for pernicious anemia primarily involves lifelong vitamin B12 supplementation, often through intramuscular injections, to restore and maintain normal B12 levels. Initially, higher doses of B12 may be administered to rapidly correct the deficiency, followed by maintenance doses to prevent recurrence. Occasionally, high-dose oral B12 supplements may be used as an alternative. Vitamin B12 is also known as cobalamin. Cyanocobalamin is a synthetic compound of vitamin B12.

CLINICAL INDICATIONS

Cobalamin can be administered orally or intramuscularly, depending on the patient's cause, presentation, and needs. A patient with severe cobalamin deficiency is initially treated via the intramuscular route. When the deficiency is less severe, the oral formulation is used. Intramuscular cobalamin remains available in two forms: cyanocobalamin and hydroxocobalamin. Cobalamin is utilized to treat pernicious anemia, malnutrition, *Helicobacter pylori* infection, gastric atrophy, stomach cancer, pancreatic insufficiency, and **gastrectomy**. Vitamin B$_{12}$ is crucial for DNA synthesis and energy production in the blood, particularly in erythroid progenitor cells.

MECHANISM OF ACTION

Cyanocobalamin binds to intrinsic factor, and absorption occurs in the ileum at the end of the small intestine. When administered parenterally, cobalamin bypasses the intestinal barrier, absorbs quickly through diffusion, and enters the bloodstream.

ADVERSE EFFECTS

The adverse effects of cyanocobalamin include anaphylaxis, dyspnea, peripheral vascular thrombosis, optic nerve atrophy, weight gain, **pulmonary edema**, and congestive heart failure. Other adverse effects consist of leg cramps, irregular heartbeats, numbness, hypokalemia, fatigue, and joint pain.

CONTRAINDICATIONS

Cyanocobalamin is contraindicated for patients with allergies, optic neuropathy, gout, and hypokalemia.

YOU SHOULD REMEMBER

Vitamin B$_{12}$ deficiency can lead to paresthesia and subacute combined nerve degeneration. It may also result in glossitis, changes in hair and nails, and skin hyperpigmentation.

Check Your Knowledge

1. What are the characteristics of pernicious anemia?
2. What is the mechanism of action of cobalamin?
3. What are the adverse effects of cobalamin?

CLINICAL CASE STUDIES

Clinical Case Study 1

A 41-year-old woman delivered her eighth child 1 week ago. She was diagnosed with iron deficiency anemia and prescribed oral iron supplementation. However, after 6 weeks, there was no significant improvement. Therefore, intravenous ferrous sulfate was administered, leading to substantial progress.

CRITICAL THINKING QUESTIONS

1. What are the causes and symptoms of iron deficiency anemia?
2. What are the adverse effects of ferrous sulfate?
3. What are the contraindications of ferrous sulfate?

Clinical Case Study 2

A 71-year-old man visited his physician, complaining of fatigue, pale skin, headaches, depression, peripheral neuropathy, dementia, and weight loss. His medical record indicated that he had undergone a partial gastrectomy due to a complication from a peptic ulcer. His laboratory data revealed extremely low vitamin B12 levels, leading to a diagnosis of pernicious anemia.

CRITICAL THINKING QUESTIONS

1. What are the causes and symptoms of pernicious anemia?
2. What are the mechanisms of action of cyanocobalamin?
3. What are the adverse effects of cyanocobalamin?

FURTHER READING

Al Mosawi, A. (2019). *Hypoparathyroidism, Vitiligo, Poliosis, and Macrocytic Anemia Syndrome*. Lap Lambert Academic Publishing.

Arpilor, L. (2023). *Pernicious Anemia Symptoms—Learn About the Symptoms of Pernicious Anemia, From Fatigue to Pale Skin. Discover Potential Signs and the Implications of Vitamin B$_{12}$ Deficiency*. Arpilor.

Bunn, H.F., and Aster, J.C. (2016). *Pathophysiology of Blood Disorders*, 2nd Edition. McGraw-Hill/Medical.

Eugenia, S. (2023). *Iron Deficiency Anemia—Comprehensive Insights, Interventions, and Future Frontiers* (Medical Care and Health). Eugenia.

Grayson, T. (2021). *Understanding Iron Deficiency Anemia—A Unique Guide to Understanding the Symptoms, Diagnosis, Treatment of Iron Deficiency Anemia with Medical and Alternative Forms of Treatment*. Grayson.

Hooper, M. (2012). *Pernicious Anemia—The Forgotten Disease—The Causes and Consequences of Vitamin B_{12} Deficiency.* Hammersmith Health Books.

Hooper, M. (2015). *What You Need to Know About Pernicious Anemia and Vitamin B_{12} Deficiency.* Hammersmith Health Books.

Jeon, K.W. (2011). *International Review of Cell and Molecular Biology.* Academic Press.

Jean, R. (2022). *Blood Disorders—The Complete Guide on Blood Disorders Causes, Symptoms, Treatment and Remedies for Your Complete Wellness.* Jean.

Maheshwari, N. (2018). *Clinical Pathology, Hematology, and Blood Banking*, 3rd Edition. Jaypee.

McCandless, D.W. (2010). *Metabolic Encephalopathy.* Springer.

Meselson, A. (2013). *The Complete Guide on Anemia—Learn Anemia Symptoms, Causes, and Treatments.* Meselson.

Pratt, M. (2022). *Anemia—Clinical Aspects.* American Medical Publishers.

20 Drugs Affecting the Endocrine System

LEARNING OUTCOMES

After studying this chapter, readers should be able to

1. Summarize the main hormones of the pituitary gland.
2. Explain the effects of growth hormones (GHs) in the body.
3. Review the impact of thyroid hormones on body systems.
4. Explain the first-line drugs for the treatment of hypothyroidism.
5. Discuss the contraindications of levothyroxine.
6. Explain the primary functions of the three layers of the adrenal cortex.
7. Describe the adrenal medulla hormones.
8. Summarize the clinical indications of corticosteroids.
9. Explain the adverse effects of spironolactone.
10. Explain the indications for estrogens.
11. Discuss the clinical indications for oxytocin.
12. Review the contraindications of testosterone.

OVERVIEW

The hypothalamus coordinates the endocrine system, releasing and inhibiting hormones that act on the pituitary gland to influence the functions of the thyroid, adrenal, and reproductive organs. The thyroid gland secretes thyroxine and triiodothyronine hormones through a negative feedback process. High or low levels of thyroid-stimulating and free thyroid hormones characterize thyroid dysfunction. The adrenal glands secrete several hormones, with the primary hormones of the adrenal cortex being corticosteroids and mineralocorticoids. Corticosteroids are among the most widely prescribed drug classes worldwide, serving both endocrine and nonendocrine indications. Aldosterone regulates the body's salt and water balance by increasing sodium and water retention while promoting the excretion of potassium by the kidneys. The medulla secretes epinephrine and norepinephrine. The female ovaries are responsible for the development and maturation of ova and the secretion of estrogen and progesterone. At the same time, the testes secrete testosterone for the development and maturation of sperm and the sexual differentiation of the gonads. Estrogen and progesterone replacement therapy are often used to manage and treat menopausal symptoms. Testosterone is the treatment of choice for testosterone deficiency and hypogonadism. Various medications and hormones are used to treat endocrine disorders or conditions.

PITUITARY GLAND

The **pituitary gland** is a pea-sized structure located within the **sphenoid bone** of the skull. It secretes several hormones and is connected to the hypothalamus by the **infundibulum**. Composed of both neural and glandular tissue, the anterior lobe produces hormones that regulate the activity of other endocrine glands. However, it is controlled by the hypothalamus, which stimulates the anterior lobe to release specific hormones. Four of the six anterior pituitary hormones are referred to as **trophic hormones**. The anterior pituitary hormones include:

- GH, known as **somatotropin**, primarily targets bones and skeletal muscles. Its effects are mainly regulated by **insulin-like growth factors** produced by the liver, skeletal muscle, and other tissues. GH stimulates the breakdown of stored triglycerides in adipocytes, releasing fatty acids into the bloodstream. In the liver, GH promotes the breakdown of **glycogen** reserves by liver cells, releasing glucose into the bloodstream.
- **Thyroid-stimulating hormone** (TSH), or **thyrotropin**, is a trophic hormone that stimulates normal development and secretion from the thyroid gland. It is released following the hypothalamic–pituitary–target endocrine organ feedback loop – thyrotrophs of the anterior pituitary release TSH, triggered by thyrotropin-releasing hormone (TRH).
- **Adrenocorticotropic hormone** (ACTH), also known as **corticotropin**, is secreted by the *corticotrophs* of the anterior pituitary. It stimulates the adrenal cortex to release corticosteroid hormones. ACTH release has a daily rhythm, peaking in the morning shortly before awakening. ACTH specifically targets the cells that secrete corticosteroids, which influence glucose metabolism.
- **Follicle-stimulating hormone** (FSH) and **luteinizing hormone** (LH), also known as **gonadotropins,** regulate the functions of the ovaries in females and the testes in males. FSH stimulates the production of sperm and eggs, while LH promotes the production of gonadal hormones. In females, these hormones facilitate the maturation of the ovarian follicle. LH then triggers **ovulation** and the release of ovarian hormones. In males, this gonadotropin is called **interstitial cell-stimulating hormone** because it stimulates testosterone production. Testosterone is also classified as an **androgen**.

DOI: 10.1201/9781003461913-22

PITUITARY GLAND

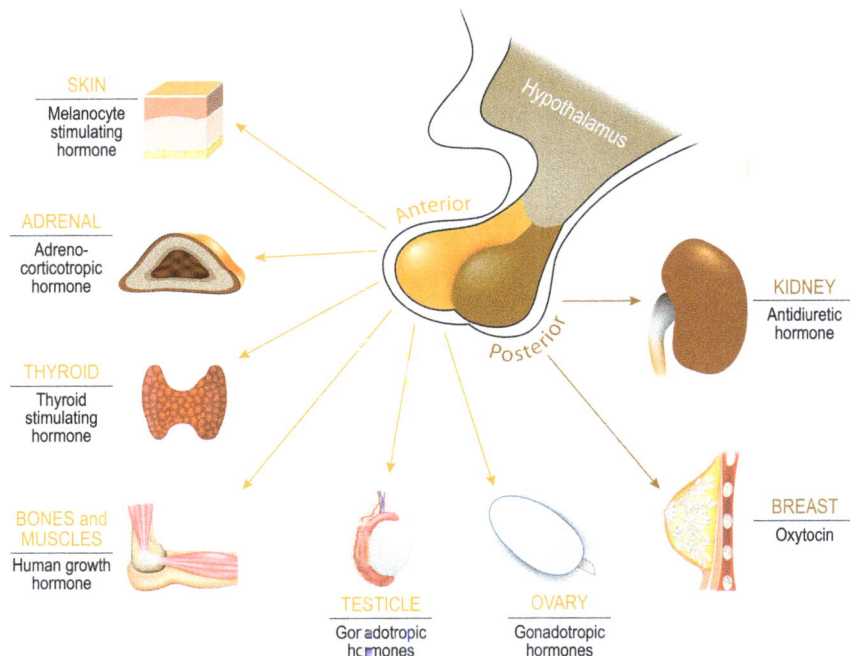

SKIN
Melanocyte stimulating hormone

ADRENAL
Adreno-corticotropic hormone

THYROID
Thyroid stimulating hormone

BONES and MUSCLES
Human growth hormone

TESTICLE
Gonadotropic hormones

OVARY
Gonadotropic hormones

KIDNEY
Antidiuretic hormone

BREAST
Oxytocin

Hypothalamus
Anterior
Posterior

FIGURE 20.1 The effects of pituitary hormones.

- **Prolactin** (PRL) is produced by lactotrophs. PRL stimulates milk production in females in humans, but its role in males remains unclear. It is regulated by the **prolactin-inhibiting hormone**, now known as **dopamine**. In females, prolactin levels fluctuate with estrogen blood levels. Prolactin is also referred to as **mammotropin**. Prolactin, estrogens, progesterone, glucocorticoids, and pancreatic hormones work together to prepare the mammary glands for milk secretion.

The hypothalamus releases two neurohormones: antidiuretic hormone (ADH) and oxytocin. ADH, also known as **vasopressin**, regulates water balance and helps prevent dehydration and water overload. Thus, its primary function is to reduce the amount of water lost by the kidneys. **Osmoreceptors** monitor the concentrations of solutes and water in the blood. When blood levels of ADH are high, it induces vasoconstriction of most visceral blood vessels. In cases of severe blood loss, large amounts of ADH are released, leading to an increase in blood pressure.

Oxytocin is a potent stimulant of uterine contractions, primarily released during childbirth and lactation. The stretching of the uterus and cervix triggers its release. Oxytocin also promotes milk ejection due to positive feedback mechanisms like uterine contractions. In males, it stimulates smooth muscle contractions in the ductus deferens and prostate gland, which may be essential for producing reproductive secretions. In females, oxytocin may induce smooth muscle contractions in the uterus and vagina, facilitating the transport of sperm toward the uterine tubes. The effects of anterior and posterior pituitary hormones on various body parts are illustrated in Figure 20.1.

YOU SHOULD REMEMBER

Consuming alcoholic beverages inhibits ADH secretion and causes polyuria. Its release is also inhibited by drinking excessive amounts of water.

Check Your Knowledge

1. What are the functions of the pituitary gland?
2. What are the functions of trophic hormones?
3. What is the hypothalamus's role in the pituitary gland?

GH DEFICIENCY AND DIABETES INSIPIDUS

GH deficiency (GHD) occurs when the pituitary gland fails to produce sufficient GH to stimulate the body's growth.

DWARFISM

Pituitary **dwarfism** occurs in children with an abnormally short stature and normal body proportions. Congenital GHD can manifest at birth or develop later. In adults, GHD results from damage to the pituitary gland or hypothalamus caused by a tumor, surgery, or radiation. It is a rare condition and is not life-threatening. **Achondroplasia** is primarily recognized as the leading cause of dwarfism.

Dwarfism can be treated with the synthetic hormone vosoritide (Voxzogo), which promotes height growth in patients. Children typically receive daily injections for several years until they achieve their maximum adult height, often within their family's average adult range.

Clinical Indications

Vosoritide is indicated to increase linear growth in pediatric patients with achondroplasia who are 5 years of age or older and have open **epiphyses**. This indication has received accelerated approval based on an improvement in annualized growth velocity.

Mechanism of Action

Vosoritide binds to a target receptor known as natriuretic peptide receptor type B, diminishing the activity of fibroblast growth factor receptor 3. Subsequently, vosoritide boosts cartilage cell growth, enhancing bone growth.

Adverse Effects

The adverse effects of vosoritide include insomnia, hypotension, dizziness, confusion, cough, chills, fever, fatigue, blurred vision, headaches, sore throat, muscle pain, loss of appetite, nausea, diarrhea, and vomiting.

Contraindications

Vosoritide should not be administered to patients with active cancer or those seriously ill due to complications from open-heart or abdominal surgery.

DIABETES INSIPIDUS

Diabetes insipidus is a rare chronic condition caused by a deficiency of ADH. The primary symptoms include **polydipsia** and **polyuria**. A patient can produce nearly 20 L of urine daily. Damage to the hypothalamus or the pituitary gland can lead to diabetes insipidus. Risk factors consist of a family history of the disorder and certain medications, such as diuretics, hypercalcemia, and hypokalemia.

Clinical Indications

Desmopressin is utilized for diabetes insipidus as a replacement therapy for ADH and for managing temporary polyuria and polydipsia following head trauma or surgery.

Mechanism of Action

Desmopressin is a selective vasopressin receptor agonist in the kidney's collecting ducts. This G-protein-coupled receptor is activated, leading to increased water permeability. This activity reduces urine volume and raises osmolality.

Adverse Effects

Although adverse effects from desmopressin are rare, they include confusion, convulsions, decreased urine output, dizziness, arrhythmia, headache, increased thirst, muscle pain or cramps, nausea or vomiting, shortness of breath, edema (on the face, hands, or ankles), fatigue, weight gain, delusions, dementia, and diarrhea.

Contraindications

Desmopressin is contraindicated for patients with moderate to severe renal impairment. It is also contraindicated for patients with hyponatremia or a history of hyponatremia.

Check Your Knowledge

1. What is the cause of pituitary dwarfism?
2. What is the mechanism of action for vosoritide?
3. What are the potential adverse effects of vosoritide?

THYROID GLAND

The thyroid gland is the largest endocrine gland in the anterior neck and is shaped like a butterfly. It controls metabolism, **gluconeogenesis**, protein synthesis, **glycogenolysis**, **lipogenesis**, and thermogenesis. The parathyroid glands are embedded in the posterior surface of the thyroid gland's lateral lobes, aiding in calcium homeostasis.

TSH regulates the thyroid gland. The hypothalamus detects low plasma thyroid hormone concentrations through negative feedback and releases TRH into the hypophyseal portal system. The TRH subsequently stimulates the release of TSH into the systemic circulation, which liberates the hormones *triiodothyronine* (T_3) and *thyroxine* (T_4). Figure 20.2 **illustrate**s the structures that release the various thyroid hormones.

The thyroid contains **parafollicular cells**, also known as C cells. These cells produce **calcitonin,** which primarily lowers calcium levels in the blood and inhibits osteoclasts' activity. Osteoclasts break down bone and release calcium into the bloodstream.

Table 20.1 lists the significant effects of thyroid hormone throughout the body.

HYPOTHYROIDISM AND PHARMACOTHERAPY

Hypothyroidism decreases the levels of free thyroid hormones in circulation. Resistance to hormone action may also contribute to this condition. The thyroid gland can be permanently lost or destroyed. Although it occurs at all ages, it is most common among older adults and women. While it is typically easy to diagnose in younger adults, it can be subtle and manifest atypically in older adults.

Cretinism occurs when congenital hypothyroidism remains untreated for extended periods and represents a severe form of hypothyroidism. It develops in regions of the world with significant iodine deficiency. Neonates with congenital hypothyroidism are often asymptomatic at birth and

THYROID HORMONES

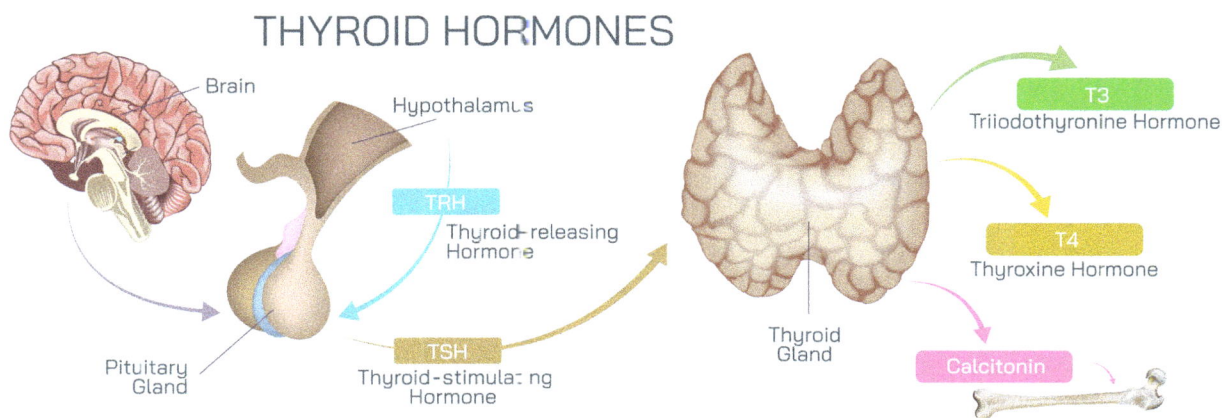

FIGURE 20.2 Structures involved with thyroid hormone release.

TABLE 20.1

Thyroid Hormone and Body System

Body System	Effects
Cardiovascular	Promotes normal functioning of the heart
Nervous	Promotes normal development of the nervous system in the fetus and infant; promotes normal adult nervous system function; enhances effects of the sympathetic nervous system
Musculoskeletal	Promotes normal muscular development and function; promotes normal growth and maturation of the skeleton
Gastrointestinal	Promotes normal development of the nervous system in the fetus and infant; promotes normal adult nervous system function; enhances effects of the sympathetic nervous system
Integumentary	Promotes normal hydration and secretory activity of the skin
Reproductive	Promotes normal female reproductive ability and lactation
Respiratory	Promotes normal oxygen use and basal metabolic rate

are identified through newborn blood screening. However, signs and symptoms may include thickened skin, an enlarged tongue, a protruding abdomen, an umbilical hernia, stunted growth, intellectual disabilities, and neurological impairments such as hearing and speech defects. Treatment with levothyroxine (L-T4) should commence immediately after the diagnosis of congenital hypothyroidism.

Myxedema is a form of acquired hypothyroidism in adults. It affects women significantly more often than men, with most patients being middle-aged. Symptoms include slowed mental and physical activity, leading to apathy. Patients may experience slower speech and intellectual functions, hair and tooth loss, and changes in movement, sensation, consciousness, and intellect.

Levothyroxine (T4) is the preferred treatment for hypothyroidism. Several brand names are available in the United States, including Synthroid®, Levoxyl®, Unithroid®, and Tirosint®.

Clinical Indications

Levothyroxine is primarily used for replacement or supplemental therapy in congenital or acquired hypothyroidism of any etiology, except for transient hypothyroidism during the recovery phase of subacute thyroiditis. Primary hypothyroidism may result from functional deficiency, primary atrophy, partial or total congenital absence of the thyroid gland, or from the effects of surgery, radiation, or drugs, with or without the presence of goiter. Levothyroxine is also used in the treatment or prevention of various types of **euthyroid goiters,** such as thyroid nodules and Hashimoto's thyroiditis. Other indications include being an adjunct to surgery and radioiodine therapy.

Mechanism of Action

L-thyroxine (T4) binds to thyroid receptor proteins attached to DNA. The receptor complex activates gene transcription and the synthesis of messenger RNA and cytoplasmic proteins.

Adverse Effects

The adverse effects of levothyroxine include headaches, rashes, dyspnea, tachycardia, tremors, irregular menstruation, diarrhea, mood changes, and heat intolerance. It can also lead to bone loss and fractures.

Contraindications

Levothyroxine is contraindicated in patients with thyrotoxicosis and acute myocardial infarction. It should also be avoided in patients with uncorrected adrenal insufficiency, as thyroid hormones may precipitate an acute adrenal crisis by increasing the metabolic clearance of glucocorticoids. Finally, it is contraindicated in patients with hypersensitivity to any of the inactive ingredients.

YOU SHOULD REMEMBER

Patients who take thyroid hormones must continue them for the rest of their lives, especially following thyroidectomy or radioactive iodine treatment.

Check Your Knowledge

1. What are the differences between cretinism and myxedema?
2. What is the indication for levothyroxine?
3. What are the side effects of levothyroxine?

HYPERTHYROIDISM AND PHARMACOTHERAPY

Hyperthyroidism occurs when the thyroid gland produces excessive thyroid hormone, speeding up metabolism and impacting many bodily functions. Conversely, **thyrotoxicosis** arises when the thyroid gland produces too much thyroid hormone. **Graves' disease** is an autoimmune disorder that can lead to hyperthyroidism. The immune system attacks the thyroid gland, producing more thyroid hormones than necessary, thereby increasing the activity of many body functions. Secondary hyperthyroidism may occur due to conditions outside the thyroid, such as a TSH-secreting pituitary tumor.

Treatment for hyperthyroidism depends on its underlying cause. Antihyperthyroidism drugs include propylthiouracil (PTU) and methimazole (Tapazole®). Radioactive iodine treatment can destroy thyroid tissue and reduce hormone levels. In some cases, subtotal **thyroidectomy** may be required.

Clinical Indications

PTU is indicated for patients with hyperthyroidism prior to thyroidectomy or radioactive iodine therapy. It is also used to treat thyroid storms and thyrotoxicosis crises. Additionally, PTU is utilized before thyroid surgery or radioactive iodine treatment in patients who have previously been treated with ineffective medications (e.g., methimazole).

Mechanism of Action

PTU inhibits the enzyme thyroid peroxidase, which usually converts iodide into an iodine molecule and incorporates that molecule into the amino acid **tyrosine**. Consequently, it inhibits the conversion of T4 to T3.

Adverse Effects

The common adverse effects of PTU include hepatotoxicity and acute liver failure, acute kidney injury, loss of taste, stomach pain, nausea, vomiting, headache, paresthesia, drowsiness, vertigo, lymphadenopathy, splenomegaly, leukopenia, aplastic anemia, agranulocytosis, hemorrhage, hypoprothrombinemia, and interstitial pneumonitis.

Contraindications

PTU should be avoided in patients with a history of hypersensitivity to the drug. Caution is advised in cases of myelosuppression, hepatic damage, and pediatric patients.

Check Your Knowledge

1. What are the treatments for hyperthyroidism?
2. What are the contraindications of PTU?
3. What are the adverse effects of PTU?

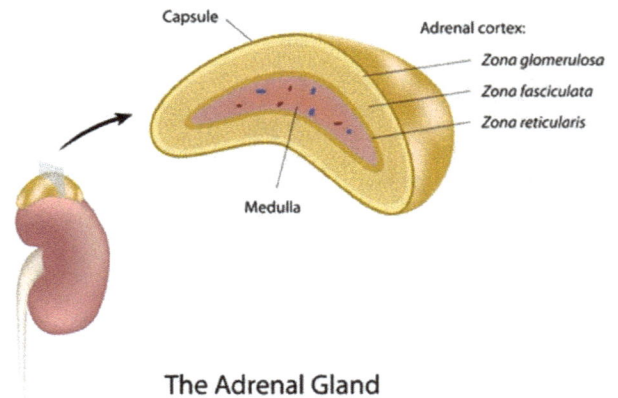

FIGURE 20.3   The adrenal cortex and medulla.

ADRENAL GLANDS

The adrenal glands are triangular and located on top of the kidneys. The cortex is the outer portion, while the medulla is the inner part. They consist of different embryological tissues with distinct structures and functions. The adrenal cortex comprises regular endocrine tissue, whereas the medulla comprises neurosecretory tissue. Glandular function occurs in the outer adrenal cortex, and the inner medulla functions as part of the sympathetic nervous system (see Figure 20.3).

ADRENAL CORTEX

The adrenal cortex comprises three zones: the zona glomerulosa, the zona fasciculata, and the zona reticularis. The cells in the outer zone secrete mineralocorticoids, while those in the middle zone secrete glucocorticoids. The inner zone is responsible for secreting gonadocorticoids.

- **Zona glomerulosa** releases mineralocorticoids that help balance minerals and water in the blood. **Aldosterone,** the most potent mineralocorticoid, reduces sodium excretion and eliminates potassium ions from the body.
- **Zona fasciculata** produces **glucocorticoids** that regulate glucose metabolism and assist the body **in resisting** stress. The most abundant glucocorticoids are *cortisol (hydrocortisone), cortisone,* and *corticosterone.*
- **Zona reticularis** forms gonadocorticoids, which refer to sex hormones. Under normal conditions, the adrenal cortex secretes small amounts of androgens. Usually, not enough androgen is produced to give women masculine characteristics, but it is sufficient to influence the appearance of pubic and axillary hair in both genders.

ADRENAL MEDULLA

The medulla contains tissue similar to that of the **sympathetic ganglia**. It produces the catecholamines **epinephrine**

TABLE 20.2
Adrenal Hormones

Zone	Hormone	Functions
Zona glomerulosa (adrenal cortex)	Mineralocorticoids (aldosterone)	Increase renal absorption of sodium ions and water and increase the loss of potassium in urine
Zona fasciculate (adrenal cortex)	Glucocorticoids (cortisol and corticosterone)	Promote liver formation of glycogen and glucose; promote peripheral use of lipids; anti-inflammatory effects
Zona reticularis (adrenal cortex)	Androgens	Encourages bone and muscle growth, and blood formation in children
Adrenal medulla	Epinephrine and norepinephrine	Increases cardiac activity, blood pressure, breakdown of glycogen, blood glucose levels

and **norepinephrine**, which are essential for managing stress. Norepinephrine significantly influences blood pressure and peripheral vasoconstriction, while epinephrine promotes bronchial dilation and increases blood flow to the heart and skeletal muscles. Table 20.2 lists the adrenal hormones and their functions.

HYPERADRENALISM

Hyperaldosteronism is a chronic condition characterized by excess aldosterone secretion. Aldosterone regulates blood pressure and electrolyte balance by promoting sodium retention and potassium excretion in the urine. Hyperaldosteronism may be classified as primary or secondary. Increased production of aldosterone may lead to primary hyperaldosteronism (**Conn's syndrome**). The signs and symptoms associated with Conn's syndrome include hypertension, hypokalemia, headache, fatigue, muscle cramps, palpitations, anxiety, depression, and memory difficulties. The treatment of hyperaldosteronism depends on its underlying cause. If a tumor is responsible, adrenalectomy is required. However, the primary goal is to manage hypertension. The most common medications for treating Conn's syndrome are aldosterone antagonists.

Cushing's syndrome is characterized by prolonged exposure to high cortisol levels. The body produces too much cortisol, leading to various physical and psychological symptoms. Cushing's syndrome may result from the administration of exogenous cortisol, excessive secretion of ACTH, or an adrenal adenoma. Features include a buffalo hump, a moon-shaped face, weight gain, **hirsutism**, fatigue, hypertension, decreased libido, menstrual irregularity, and **truncal obesity**. Treatment depends on the cause. Patients may need surgery to remove tumors or the adrenal glands, radiation, chemotherapy, and certain hormone-inhibiting drugs.

Hypersecretion of androgens can result from virilizing tumors. Virilizing syndrome, also known as virilization, is a condition that causes a female to develop characteristics associated with male hormones or for a newborn to exhibit signs of male hormone exposure at birth. Symptoms of virilization include hirsutism, baldness, acne, deepening of the voice, increased muscularity, and heightened sex drive. In women, the uterus shrinks, the clitoris enlarges, the breasts become smaller, and normal menstruation ceases.

HYPOADRENALISM

Hypoaldosteronism is a condition marked by a decrease in aldosterone. Symptoms include hyponatremia, hyperkalemia, and metabolic acidosis. Other symptoms include nausea, fatigue, arrhythmia, palpitations, and hypotension.

Addison's disease is a rare endocrine disorder in which the adrenal glands fail to produce sufficient amounts of the hormones cortisol and aldosterone. Without adequate cortisol and aldosterone, the body may experience various symptoms, including fatigue, weakness, hypotension, skin pigmentation, and electrolyte imbalances. Addison's disease can result from an ACTH deficiency or the **atrophy** of the adrenal cortex. Other potential causes include genetics, tuberculosis, and decreased cortisol levels. This condition is treated with medication that replaces corticosteroids and aldosterone.

YOU SHOULD REMEMBER

Addison's crisis is an acute adrenal failure that can lead to life-threatening shock. Symptoms include hypotension, fatigue, confusion, delirium, vomiting, diarrhea, and dehydration.

Check Your Knowledge

1. What is the structure of the adrenal glands?
2. What are the causes of Cushing's syndrome?
3. What are the causes of Addison's disease?

PHARMACOTHERAPY

The primary medications for treating Addison's disease are corticosteroids and aldosterone. Treatment entails lifelong hormone replacement therapy to sustain normal hormone levels.

CORTICOSTEROID

Corticosteroids have several effects on the body, allowing them to treat various conditions and disorders. They can reduce inflammation, suppress overactive immune responses, and help with hormonal imbalances. They may be administered locally or systemically to target the exact location of a problem. Table 20.3 lists various types of corticosteroids.

TABLE 20.3

Corticosteroids

Generic Name	Trade Name
Betamethasone	Celestone, Diprolene
Budesonide	Entocort EC, Rhinocort
Dexamethasone	Baycadron Elixer, Decadron
Hydrocortisone	Cortef, Solu – Cortef, Hydrocort
Methylprednisolone	Depo.Medrol
Prednisone	Deltasone
Triamcinolone	Aristospan, Kenalog
Prednisolone	Orapred, Prelone

Clinical Indications

Corticosteroids are among the most widely prescribed drug classes worldwide, with both endocrine and nonendocrine indications. They serve as replacement therapy for adrenal insufficiency and have anti-inflammatory and immunosuppressive effects. Additionally, corticosteroids are employed to alleviate symptoms of Addison's disease.

Mechanism of Action

The mechanism of action of corticosteroids involves both genomic and nongenomic effects. The glucocorticoid receptor mediates the genomic action, resulting in most anti-inflammatory and immunosuppressive effects. This receptor is located in the cell nucleus and inhibits gene expression and translation in inflammatory leukocytes and epithelium.

Adverse Effects

The adverse effects of corticosteroids depend on the dose, type of steroid, and length of treatment. They are more common at higher dosages and with chronic use. Common adverse effects include increased appetite, weight gain, edema, a "puffy" face, acne, fatigue, and mood changes. Other adverse effects may include agitation, insomnia, diabetes, osteoporosis, hypertension, gastritis, cataracts, and glaucoma.

Contraindications

Corticosteroids are contraindicated in patients who have hypersensitivity to any component of the formulation, systemic fungal infection, varicella infection, herpes simplex keratitis, concurrent administration of live vaccines, diabetes mellitus, glaucoma, osteoporosis, hypertension, and joint infection. They should also be avoided in patients with peptic ulcer disease and congestive heart failure.

ALDOSTERONE ANTAGONISTS

Aldosterone antagonists, also known as potassium-sparing diuretics, prevent the kidneys from absorbing too much salt and help avoid hypokalemia. For a more detailed discussion, see **Chapter 23**.

FEMALE REPRODUCTIVE HORMONE

Gonadotropin-releasing hormone regulation relies on precisely coordinating internal and external signals. A rapid increase in plasma FSH and LH triggers the enhanced release of gonadotropin-releasing hormone. FSH promotes estrogen production, while LH facilitates progesterone release.

Estrogen hormones support the reproductive activities of females, including the development of the **ovum**, the formation and maintenance of the **corpus luteum** to sustain a fertilized ovum, the upkeep of pregnancy, and the preparation of the breasts for lactation. Table 20.4 lists selected estrogens and progestins.

ESTROGENS

Estrogen is involved in both the male and female reproductive systems. The primary source of estrogen, ovarian 17 beta-estradiol, is generally converted to estrone and estriol. In postmenopausal women, the primary sources of estrogen are nonovarian, including the adrenal cortex, adipose tissue, skin, kidney, and brain.

Clinical Indications

The clinical indications of estrogen include primary ovarian insufficiency and female **hypogonadism**. It is used to treat menopausal symptoms such as hot flashes, night sweats, vaginal dryness, itching, burning, and **dyspareunia**. Typically, women with an intact uterus receive a combination of estrogen and progesterone therapy, while those who have had hysterectomies receive estrogen alone.

TABLE 20.4

Estrogens, Progestins, and Estrogen–Progestin Combinations

Generic Name	Trade Name
Estrogens	
Estradiol	Alora, Estrace, Estraderm,
Estradiol valerate	Delestrogen
Estrogen, conjugated	Cenestin, Enjuvia, Premarin
Estropipate	Ogen
Progestins	
Medroxyprogesterone	Depo-Provera, Provera, MPA,
Progesterone	Depo-SubQ
	Crinone, Endometrin, Prochieve,
	Progestin
Estrogen–Progestin Combinations	
Conjugated estrogens with medroxyprogesterone	Premphase, Prempro
Estradiol with norgestimate	MonoNessa, Ortho-Cyclen.28
Ethinyl estradiol with norethindrone acetate	Activella

Mechanism of Action

Estrogen receptors regulate transcriptional processes. Their mechanism of action involves estrogen binding to receptors in the nucleus, after which the receptors dimerize and bind to specific response elements located within the genes. However, estrogen receptors can also regulate gene expression without directly binding to DNA.

Adverse Effects

The adverse effects of estrogens include headaches, breast tenderness, diarrhea, vaginal discharge, chills, dizziness, chest pain, fatigue, and dark urine.

Contraindications

Estrogen should be avoided in patients with a history of any form of breast or uterine cancer, deep vein thrombosis, pulmonary embolism, myocardial infarction, or stroke. These contraindications do not apply to estrogen used transvaginally, such as in creams or suppositories.

PROGESTINS

Progestin is a synthetic progestogen, a class of hormones with many uses, including birth control, treating gynecological conditions, and managing cancer.

Clinical Indications

Progestins can effectively treat symptoms of menopause, such as hot flashes and vaginal dryness. They are also utilized in various birth control methods, including pills, injections, intrauterine devices, and implants. Additionally, progestins are indicated for the treatment of breast, kidney, and uterine cancers.

Mechanism of Action

Progestin binds to and activates its nuclear receptor, which is integral to the signaling stimuli that maintain the endometrium during preparation for pregnancy. Additionally, it may prevent pregnancy by inhibiting FSH, which typically induces ovulation.

Adverse Effects

The adverse effects of progestin include bleeding, headaches, abdominal cramping, nausea, breast tenderness, and increased vaginal discharge or decreased libido. Nausea can be minimized by taking the medication at night before bed. Smoking cigarettes while using oral contraceptives can significantly elevate the risk of severe adverse effects. Cigarette smoking also heightens the risk of serious cardiovascular adverse effects from oral contraceptive use. The risk increases with age and with heavy smoking.

Contraindications

Progestin should not be used during pregnancy, unexplained vaginal bleeding, breast cancer, endometrial cancer, migraines with auras, hypertension, ischemic heart disease, cirrhosis, malignant hepatoma, and hepatocellular adenoma.

Check Your Knowledge

1. What are the clinical indications of estrogen and progestin?
2. What are the contraindications of estrogen and progestin?
3. What are the adverse effects of estrogen and progestin?

MENSTRUAL CYCLE

The menstrual cycle is a natural process that involves hormonal and physical changes in the uterus and ovaries. It occurs from the first day of one period to the first day of the next. The average menstrual cycle lasts 28 days, but it can vary from person to person. Estrogen and progesterone are responsible for the changes in the **endometrium**, cervix, and vagina, and they also regulate the secretion of FSH and LH. The cycle consists of two phases: the follicular phase, which lasts 14 days before ovulation. This phase, also called the proliferative phase, is dominated by estrogen.

The luteal phase is the 14 days following ovulation. This phase, also called the secretory phase, is dominated by progesterone. The endometrium slows its proliferation and decreases in thickness. The menstrual cycle includes the development of an ovarian follicle and its **oocyte**, ovulation, the preparation of the reproductive tract to receive the fertilized ovum, and the shedding of the endometrial lining if **fertilization** does not occur. The hormonal changes and events of a 28-day menstrual cycle are shown in Figure 20.4.

UTERINE STIMULANTS

Uterine stimulants induce contractions of the uterus. They are primarily used to initiate labor, reduce postpartum bleeding, and facilitate abortion. Oxytocin and prostaglandin are available options. However, they carry the risk of uterine hyperstimulation and rupture. Oxytocin is a natural hormone that helps facilitate childbirth.

CLINICAL INDICATIONS

Oxytocin is indicated both antepartum and postpartum. During the **antepartum** period, exogenous oxytocin is used to strengthen uterine contractions to support the successful vaginal delivery of the fetus. In the postpartum period, it may induce contractions in the third stage of labor to control postpartum bleeding. Additionally, oxytocin is recommended for mothers with **preeclampsia**, maternal diabetes, and premature rupture of the membranes.

MECHANISM OF ACTION

Oxytocin's mechanism of action involves binding to a G protein-coupled receptor on the cell surface. This activation triggers intracellular signaling pathways that lead to calcium mobilization and smooth muscle contractions.

Menstrual cycle

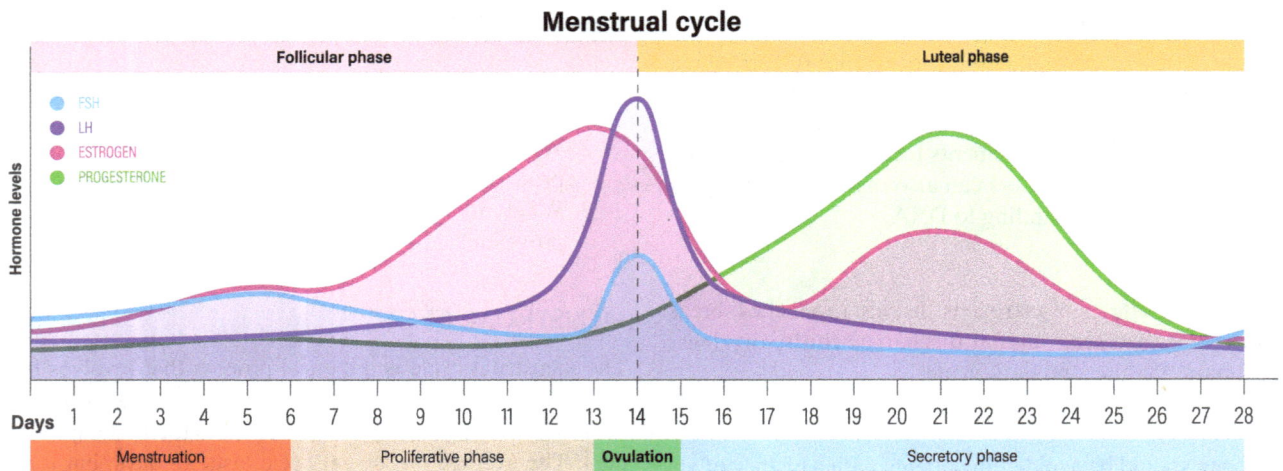

FIGURE 20.4   The hormonal changes and events of the menstrual cycle.

During labor, oxytocin stimulates uterine contractions that aid in delivery. It also helps squeeze milk into the **lactiferous ducts.**

ADVERSE EFFECTS

The adverse effects of oxytocin include hypertension, dysrhythmia, blurred vision, nausea, vomiting, severe allergic reactions, confusion, bleeding after labor, and rupture of the uterus.

CONTRAINDICATIONS

Oxytocin is contraindicated in conditions such as prematurity, **placenta previa**, fetal distress, previous Caesarean section, **hydramnios**, and any condition that could potentially lead to uterine rupture.

YOU SHOULD REMEMBER

Oxytocin is best known for its contractile activity on uterine smooth muscle, and it is still widely used to reduce uterine bleeding after the delivery of the fetus. However, when given in a rapid bolus, oxytocin can cause hypotension and tachycardia. Therefore, it is recommended to administer a dose of five international units of oxytocin via slow intravenous injection instead of a rapid bolus injection.

UTERINE RELAXANTS

Uterine relaxants, also known as **tocolytics**, are medications that decrease the activity of the uterine muscles. They are used to stop premature labor and provide the fetal lungs time to mature. Table 20.5 lists various uterine stimulants and relaxants.

TABLE 20.5

Uterine Stimulants and Relaxants

Generic Name	Trade Name
Stimulants (Oxytocics)	
Oxytocin	Pitocin
Ergot Alkaloid	
Methylergonovine	Methergine
Prostaglandins	
Carboprost	Hemabate
Dinoprostone	Cervidil, Prepidil, Prostin E2
Relaxants (Tocolytics)	
Magnesium sulfate	(Generic only)
Nifedipine	Adalat, Procardia
Terbutaline	Brethine

MALE REPRODUCTIVE HORMONES

The male reproductive system refers to the organs involved in sexual function and the production of children. These organs are both external and internal. Together, they store and ejaculate sperm. The male reproductive system also produces hormones such as testosterone, which play a key role in male development. The hypothalamic–pituitary–gonadal axis plays a significant role in sexual maturity, sperm production, and the development of secondary sex characteristics (see Figure 20.5). The anterior pituitary secretes LH and FSH, which act on membrane receptors in the Leydig and Sertoli cells of the testes, respectively.

TESTOSTERONE

Testosterone is a hormone primarily produced in the testicles by **Leydig cells**. It is crucial in maintaining sperm production, sex drive, muscle strength and mass, bone

Male Hypothalamic-Pituitary-Gonadal Axis

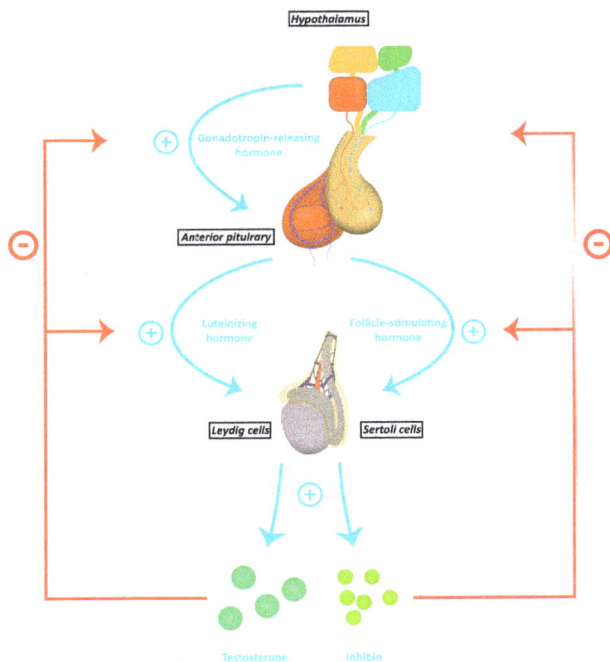

FIGURE 20.5   Male hypothalamic–pituitary–gonadal axis.

TABLE 20.6

Various Forms of Testosterone

Generic Name	Trade Name
Fluoxymesterone	(Generic only)
Methyltestosterone	Android, Methitest, Testred
Nandrolone	(Generic only)
Oxandrolone	Oxandrin
Oxymetholone	Anadrol.50
Testosterone	Striant, Androderm, AndroGel, others
Testosterone cypionate	Depo.Testosterone
Testosterone enanthate	Delatestryl, Xyosted
Testosterone undecanoate	Aveed, Jatenzo

density, fat distribution, facial and body hair, and red blood cell production. Table 20.6 lists various forms of testosterone.

Clinical Indications

Testosterone replacement therapy is indicated for the treatment of patients with hypogonadism, **gender dysphoria**, sexual dysfunction in men, and hot flashes in women. Potential benefits include increased libido and energy levels, bone density, strength, and cardioprotective effects.

Testosterone therapy is often used to treat symptoms of hormone deficiency in postmenopausal women. It can be administered orally, buccally, sublingually, intranasally, transdermally, vaginally, and rectally. It can also be given through intramuscular or subcutaneous injections or implants.

Mechanism of Action

Testosterone binds to androgen receptors. Once bound, the receptor–hormone complex enters the cell nucleus and binds to DNA. This attachment alters DNA transcription and protein synthesis, leading to increased testosterone production.

Adverse Effects

The adverse effects of testosterone in men include acne, hypertension, peripheral edema, hepatic dysfunction, dysuria, depression, chest pain, and breast swelling. However, the adverse effects of testosterone in women are uncommon if levels are maintained within the female physiological range. The most common effects are weight gain, hirsutism, and acne.

Contraindications

Testosterone is contraindicated for patients with myocardial infarction, stroke, heart failure, prostate or breast cancers, renal disease, liver failure, and untreated obstructive sleep apnea.

YOU SHOULD REMEMBER

Testosterone replacement therapy has been used in millions of men worldwide to treat diminished libido and erectile dysfunction and to improve strength and physical function.

CLINICAL CASE STUDIES

Clinical Case 1

An 18-month-old girl was diagnosed with congenital hypothyroidism. Her growth was slower than the average rate for her age. Other symptoms included an enlarged tongue, lethargy, dry skin, thickened skin, and an umbilical hernia. She received levothyroxine.

CRITICAL THINKING QUESTIONS

1. What are the signs and symptoms of congenital hypothyroidism?
2. What are the adverse effects of levothyroxine?
3. What are the contraindications of levothyroxine?

Clinical Case Study 2

A 47-year-old man was admitted to a local hospital with symptoms of fatigue, weakness, hypotension, skin pigmentation, and electrolyte imbalances. His medical history revealed that he had had tuberculosis for several years and had received treatment. Blood tests showed very low levels of cortisol and aldosterone hormones. He was diagnosed with Addison's disease.

CRITICAL THINKING QUESTIONS

1. What are the causes of Addison's disease?
2. What are the adverse effects of cortisol?
3. What are the contraindications of corticosteroids?

FURTHER READING

Alfred, M.J. (2022). *The Human Growth Hormone and Its Corresponding Effects on Your Body*. Alfred.

Beck-Peccoz, P. (2004). *Syndromes of Hormone Resistance on the Hypothalamic-Pituitary-Thyroid Axis* (Endocrine Updates, Book 22). Springer.

Beck-Peccoz Edison, D. (2021). *Understanding Hyperthyroidism – A Unique Guide to Understanding the Symptoms, Diagnosis, Treatment of Hyperthyroidism with Medical and Alternative Forms of Treatment*. Edison.

Bennie, A., MacLaren, K., and O'Keeffe, C. (2021). *Our Guide to Understanding Premature Ovarian Insufficiency*. Daisy Network/Grosvenor House Publishing Limited.

Cohlen, B.J., Van Santbrink, E., and Laven, J.S.E. (2016). *Ovulation Induction – Evidence-Based Guidelines for Daily Practice – Reproductive Medicine and Assisted Reproductive Techniques Series*. CRC Press.

Doster, M.R., Petrock, I.R., and Petrock, E.R. (2024). *Muscle Through Menopause – Kinda Vintage, Kinda Savage – Real Conversation, Facts and Solutions to Take Control of the Menopause Dilemma through Muscular Strength and Nutrition*. Doster Petrock.

Elliott, O. (2023). *Achieving Hormone Balance – A Science-Backed Blueprint to Unlock Natural Remedies, Manage Menopause, and Thrive with More Energy and Joy*. Shuo Luo.

Fritz, M.A., and Speroff, L. (2010). *Clinical Gynecologic Endocrinology and Infertility*, 8th Edition. Wolters Kluwer/ Lippincott, Williams, and Wilkins.

Hacknew, A.C. (2023). *Sex Hormones, Exercise and Woman – Scientific and Clinical Aspects*, 2nd Edition. Springer.

Hoffman, A.B. (2013). *Sex Hormones – Development, Regulation and Disorders* (Endocrinology Research and Clinical Developments). Nova Biomedical.

Hohner, M. (2023). *A Practical Treatise on Disorders of the Sexual Function in the Male and Female*. Legare Street Press.

Holt, E.H., Lupsa, B., Leo, G.S., Bassyouri, H., and Peery, H.E. (2021). *Goodman's Basic Medical Endocrinology*, 5th Edition. Elsevier.

Kuehn, F.S., and Lozada, M.P. (2010). *Thyroid Hormones – Functions, Related Diseases and Uses* (Endocrinology Research and Clinical Development). Nova Science Publishers, Inc.

Leng, G. (2018). *The Heart of the Brain – The Hypothalamus and Its Hormones*. The MIT Press.

Litwack, G. (2019). *Hormonal Signaling in Biology and Medicine* (Comprehensive Modern Endocrinology). Academic Press.

Litwack, G. (2022). *Hormones*, 4th Edition. Academic Press.

Pitt-Rivers, R., Tata, J.R., Callow, R.K., Campbell, P.N., and Datta, S.P. (2013). *The Thyroid Hormones*. Pergamon Press.

Radovick, S., and Misra, M. (2018). *Pediatric Endocrinology – A Practical Clinical Guide*, 3rd Edition. Springer.

Samson, S.L., and Ioachimescu, A.G. (2022). *Pituitary Disorders throughout the Life Cycle – A Case-Based Guide*. Springer.

Sofferman, R.A., and Ahuja, A.T. (2012). *Ultrasound of the Thyroid and Parathyroid Glands*. Springer.

21 Antidiabetic Drugs

LEARNING OUTCOMES

After studying this chapter, readers should be able to

1. Review the regulation of blood glucose.
2. Summarize the characteristics of different types of diabetes.
3. Describe the significant complications of diabetes.
4. Discuss the function of glucagon in the body.
5. Review the contraindications of glucagon administration.
6. Summarize the various types of insulin.
7. Identify the adverse effects of Humulin R U-500.
8. Review oral diabetes medications and their classifications.
9. Discuss the contraindications of thiazolidinediones (TZDs).
10. Describe the adverse effects of metformin.

OVERVIEW

Diabetes is a common condition that affects people of all ages. There are several forms of diabetes, with Type 2 being the most common. Prediabetes is the stage that precedes type 2 diabetes (T2D), in which blood glucose levels are higher than usual but not high enough for an official diagnosis. Combining treatment strategies can help patients manage the condition, enabling them to lead a healthy life and prevent complications. Diabetes occurs when the body is unable to properly regulate blood glucose levels. Patients with type 1 diabetes (T1D) do not produce enough insulin, while those with T2D cannot use insulin effectively. Gestational diabetes develops in some individuals during pregnancy and usually resolves after childbirth. However, those who experience gestational diabetes face a higher risk of developing T2D later in life. The risk factors for T2D include being overweight or obese, being inactive, and having a family history of diabetes. Individuals with T1D generally do not produce sufficient insulin, while those with T2D struggle to utilize their insulin correctly.

REGULATION OF BLOOD GLUCOSE

In blood glucose regulation, the hormone insulin plays a key role. When blood sugar rises, insulin signals the liver, muscles, and other cells to store the excess glucose. Some is stored as body fat, while others are converted into **glycogen** in the liver and muscles. As blood glucose levels decrease, the pancreas releases the hormone **glucagon**. This hormone travels through the bloodstream to the liver, where it triggers the breakdown of glycogen into glucose. The glucose enters the bloodstream, causing glucose levels to return to normal. Glucose is also derived from the breakdown of fats and proteins through gluconeogenesis. Insulin and glucagon are produced in the pancreatic islets of Langerhans (see Figure 21.1). The pancreatic cells include alpha, beta,

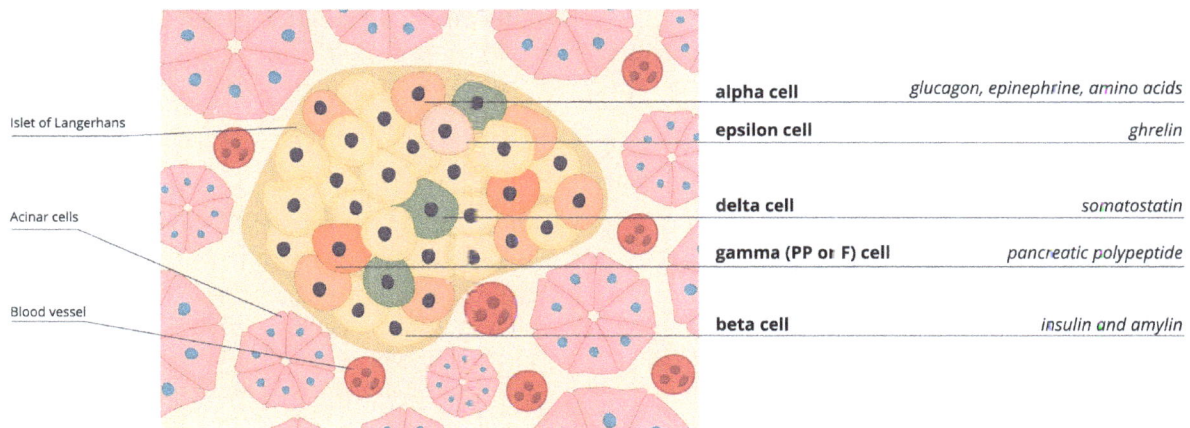

alpha cell	glucagon, epinephrine, amino acids
epsilon cell	ghrelin
delta cell	somatostatin
gamma (PP or F) cell	pancreatic polypeptide
beta cell	insulin and amylin

Islet of Langerhans

Acinar cells

Blood vessel

Islets of Langerhans, endocrine cells: secretion

FIGURE 21.1 Pancreatic cells.

DOI: 10.1201/9781003461913-23

delta, epsilon, and gamma. Insulin is secreted by beta cells, while glucagon is secreted by alpha cells. A balance exists between these two hormones depending on the body's metabolic state, with insulin having higher concentrations during energy-rich states and glucagon during fasting. Insulin, glucagon, and other hormones also regulate glucose transport in and out of cells.

Glucose plays a significant role in all body organs. However, specific organs are crucial in regulating glucose. Somatostatin from the delta cells of the pancreas also decreases blood glucose levels. The liver maintains blood glucose levels. Glycogen is primarily found in the liver and skeletal muscles. Glucose is released from glycogen under the influence of glucagon and fasting conditions, raising blood glucose levels. Glucocorticoids and thyroxine also increase blood glucose levels. This sugar is the brain's primary energy source and requires a steady supply of glucose to function correctly. **Hypoglycemia** is much more dangerous to the brain and can cause permanent injury.

DIABETES MELLITUS

Diabetes mellitus (DM) is a condition in which the body cannot regulate blood glucose levels. As of 2023, over 537 million adults worldwide have diabetes, and this number is expected to rise to 643 million by 2045. There are two main types of diabetes: type 1 and type 2. T2D is the most common form, accounting for almost 90% of all cases, while approximately 5%–10% are T1D. Patients with T1D do not produce enough insulin, whereas those with T2D cannot effectively use it. Consequently, excess glucose can damage various organs in the body. T1D is generally caused by an autoimmune attack against the beta cells, resulting in a lack of insulin. The other less common types include gestational (GDM) and neonatal diabetes.

The underlying causes of diabetes are complex and vary depending on the type. In T1D, the body's immune system mistakenly attacks and destroys the insulin-producing beta cells in the pancreas. Specific genes increase the risk of developing T1D. The underlying causes of T2D include genetics, environmental factors, obesity, and lifestyle choices. T2D typically begins with a body cell's resistance to insulin and beta cell failure. GDM may be caused by β-cell dysfunction and chronic insulin resistance during pregnancy.

CLINICAL MANIFESTATIONS

The signs and symptoms of DM include polyphagia, polydipsia, polyuria, blurred vision, urinary tract infections, skin infections, weight loss, fatigue, and weakness. Additionally, hyperglycemia and electrolyte imbalance may be present. The onset of symptoms in adult-onset diabetes is more variable than in children. It can be challenging to distinguish T1D from T2D.

TABLE 21.1

Hemoglobin A1c Levels with Estimated Average Blood Glucose

Hemoglobin A1c (Percentage)	Blood Glucose (mg/dL)
5	97
6	111
7	154
8	183
9	212
10	240
11	269
12	298

DIAGNOSIS

Diagnosis is often based on clinical manifestations and is generally straightforward. Laboratory tests for diagnosing diabetes include fasting blood sugar tests, glucose tolerance tests, **hemoglobin** A1c tests, random blood sugar tests, and urine tests. Since most erythrocytes live for 3–4 months, hemoglobin A1c can be measured. When tested, it is expressed as a percentage of glycated hemoglobin. This level reflects the average blood glucose concentration over that period. The correlation between A1c levels and average blood glucose is shown in Table 21.1.

COMPLICATIONS

DM can lead to many severe complications that impact the body in various ways, including the nerves, eyes, kidneys, and blood vessels. These complications can be disabling or even life-threatening. Significant complications of DM include **diabetic ketoacidosis**, retinopathy, macular edema, coronary artery disease, heart failure, cardiomyopathy, stroke, foot ulcers, **amputations**, nephropathy, neuropathy, and hearing loss.

YOU SHOULD REMEMBER

Hypoglycemia is common in patients with diabetes; however, it can also occur in people without diabetes. Immediate treatment for hypoglycemia involves consuming carbohydrates. If left untreated, hypoglycemia can be dangerous and life-threatening. Glucagon can also be used to treat severe hypoglycemia.

Check Your Knowledge

1. How is blood glucose regulated in the body?
2. What are the various types of DM?
3. What is the concept of hemoglobin A1c?

GLUCAGON

Hypoglycemia is a common and sometimes life-threatening adverse event associated with insulin, sulfonylurea, and meglitinide therapies. For patients who are disoriented or unconscious, treatment with injectable glucagon is recommended.

Clinical Indications

Glucagon injection is an emergency medication used to treat severe hypoglycemia in diabetic patients who are treated with insulin and are either unconscious or unable to consume oral foods or beverages containing sugar.

Mechanism of Action

Glucagon binds to specific receptors on liver cells, initiating a signaling cascade that results in the breakdown of glycogen stores and glucose production through gluconeogenesis. Consequently, it elevates blood sugar levels. Glucagon serves as a counterregulatory hormone to insulin by enhancing hepatic glucose output.

Adverse Effects

The adverse effects of glucagon administration include headache, nausea, hypertension, allergic reaction, dyspnea, and diabetes. Additionally, it may cause blurred vision, cold sweats, confusion, dizziness, anxiety, and coma.

Contraindications

Glucagon administration is contraindicated in patients who have hypersensitivity to the medication. It should be avoided in neonates or children with inadequate glycogen stores. Glucagon is not advisable for patients with a known lactose allergy.

Check Your Knowledge

1. What is the role of glucagon?
2. What are the contraindications for glucagon?
3. What are the adverse effects of glucagon?

INSULIN THERAPY

Insulin therapy replaces the insulin hormone that the body usually produces. Patients with T1D must administer insulin daily, as it is a crucial component of diabetes treatment. Insulin therapy helps maintain blood sugar control and prevents complications associated with diabetes. It functions similarly to the hormone insulin that the body typically makes. Patients with T2D must use insulin when other treatments and medications fail to manage blood sugar levels. The primary types of insulin therapy include long-, intermediate-, and short-acting insulins (see Table 21.2).

TABLE 21.2

Insulin Categories

Long-Acting	Intermediate-Acting	Short-Acting and Ultra-Rapid
Insulin degludec U.100 (Tresiba) or U.200 (Tresiba)	Human insulin NPH (Humulin N, Novolin N)	Insulin Aspart U.100 (Novolog, Fiasp)
Insulin detemir (Levemir)	Human insulin regular (Humulin R U.500)	Insulin glulisine U.100 (Apidra)
Insulin Glargine U.100 (Lantus, Basaglar) or U.300 (Toujeo)		Human insulin regular (Humulin R U.100, Novolin R U.100)
		Inhaled insulin (Afrezza)
		Insulin Lispro U.100 (Humalog, Admelog) or U.200 (Humalog)

SHORT-ACTING AND ULTRA-RAPID-ACTING INSULIN

Short-acting insulins quickly lower blood sugar levels. They are typically injected 30 minutes before meals. Insulin aspart, lispro U-100, and insulin glulisine are rapid-acting insulins. Admelog is the first rapid-acting follow-up to biological insulin. The pens can deliver up to 60 units in a single injection. Reusable pens with pen cartridges are also available. These reusable pens are divided into half units, which benefits patients on very low insulin doses. These insulins are commonly used in insulin pumps.

The newest insulin is aspart, an ultra-rapid-acting insulin. Its onset is less than 2.5 minutes, compared to the 15 minutes observed with rapid-acting insulins. Fiasp is available in both an insulin pen and a vial. This type of insulin may be preferred in insulin pumps and is particularly effective for quickly lowering high blood glucose.

INTERMEDIATE-ACTING INSULINS

Intermediate-acting insulins provide a slower and longer-lasting effect than short-acting insulins. They typically cover the body's insulin needs between meals and overnight. They include *regular insulin (Humulin R and Novolin R)* and *neutral protamine Hagedorn (NPH)*. The newer rapid-acting and long-acting insulins have mostly replaced intermediate-acting insulins. Regular insulins are used as bolus insulin but have largely been supplanted by rapid-acting insulins due to their faster onset and offset, which helps reduce the chances of hypoglycemia. Regular insulins take longer to become effective and should be administered 30 minutes before meals.

LONG-ACTING INSULINS

Long-acting insulins provide a steady, prolonged effect on blood sugar levels. They are typically injected subcutaneously once or twice a day. These include insulin glargine U-100 (Lantus, Basaglar), insulin glargine U-300 (Toujeo), insulin detemir (Levemir), and insulin degludec U-100, U-200 (Tresiba). Generally, most long-acting insulins can be administered once daily. The exceptions are insulin glargine U-100 and insulin detemir because they have shorter durations of action than the others; in some patients, they may act for less than 24 hours, making twice-daily dosing more effective for glucose control.

INHALED INSULIN

Afrezza is an inhaled, rapid-acting insulin powder used to manage blood sugar levels in adults with type 1 or 2 diabetes. It is the only FDA-approved inhaled insulin medication available. Afrezza serves as an alternative for patients looking to reduce the burden of injections. It is administered at the start of a meal and is inhaled through a small device known as an inhaler.

The insulin powder is rapidly absorbed into the lungs, with an onset and offset quicker than injectable rapid-acting insulin. Afrezza should be avoided in patients with COPD or those who smoke. The most common adverse effects of Afrezza include cough, throat irritation, and chest tightness. More severe adverse effects encompass hypoglycemia, allergic reactions, and asthma. The inhaler should be replaced every 2 weeks.

PRE-MIXED INSULINS

Pre-mixed insulins combine short- and long-acting insulins, administered via injection, and are frequently prescribed for patients with T2D. For example,

Novolog Mix 70/30 contains 70% insulin as part of protamine and 30% insulin alone. It can be taken every 12 hours instead of four times daily. However, such insulin formulations are less flexible. When a dose is adjusted because it was pre-mixed, the intermediate-acting and mealtime insulin must be adjusted simultaneously. Achieving tight glycemic control is often challenging without significant hypoglycemia.

U-500 INSULIN

Humulin U-500 is a concentrated form of insulin. Historically, severe insulin resistance has been characterized by patients requiring a total daily insulin dose exceeding 200 units daily. With the increasing prevalence of diabetes and obesity in the United States, severe insulin resistance has become more common, heightening the demand for U-500.

Clinical Indications

Humulin R U-500 is used to treat patients with diabetes who require higher doses than those typically available in U-100. It is intended exclusively for individuals with severe insulin resistance. U-500 insulin is commonly prescribed for T2D, though insulin resistance can also develop in T1D. The safe and effective use of U-500 demands careful monitoring to optimize glycemic control while minimizing the risk of hypoglycemia.

[H3] Mechanism of Action

Humulin U-500 decreases blood glucose by stimulating peripheral fat and skeletal muscle glucose uptake. Therefore, it inhibits hepatic glucose production.

Adverse Effects

The adverse effects of Humulin R U-500 include hypoglycemia, chills, sweating, chest pain, allergic reactions, rash, and injection site reactions. Other side effects include anxiety, blurred vision, cough, flushing, dizziness, tachycardia, polydipsia, polyphagia, **oliguria**, weight gain, and seizures.

Contraindications

Humulin R U-500 should be avoided in patients with hypoglycemia or those who are hypersensitive to it. Insulin dosage adjustments should be made cautiously and only under medical supervision.

YOU SHOULD REMEMBER

Humulin R U-500 is five times more concentrated than standard U-100 insulin. It is designed for adults and children who require more than 200 units of insulin daily to manage their blood glucose levels. It is available in a disposable U-500 pen or a U-500 vial, which is used with a specific BD U-500 insulin syringe.

Check Your Knowledge

1. What are the different classifications of insulin?
2. What are the clinical indications for U-500 insulin?
3. What are the side effects of U-500 insulin?

DRUG THERAPY FOR T2D

There are several types of drugs used to treat T2D, including oral diabetes medications that help patients with T2D manage their blood sugar levels. Some patients may need to take both oral medications and insulin injections. The noninsulin medication classes are listed in Table 21.3.

TABLE 21.3

Non-Insulin Medication Classes

Class	Generic Name	Brand Name
Meglitinides	Nateglinide	Starlix
	Repaglinide	Prandin
Thiazolidinediones	Pioglitazone	Actos
	Rosiglitazone	Avandia
Sulfonylureas	Glimepiride	Amaryl
	Glipizide	Glucotrol
	Glyburide	Micronase, Glynase
Biguanide	Metformin	Glucophage, Riomet
Alpha glucosidase inhibitors	Acarbose	Precose
	Miglitol	Glyset
	Voglibose	Volix

MEGLITINIDES

Meglitinides are a class of oral antidiabetic medications used to improve blood sugar control in T2D. These medications include nateglinide and repaglinide, which can increase insulin secretion from the pancreas. While meglitinides are relatively inexpensive, they are less effective than sulfonylureas.

Clinical Indications

Meglitinides are primarily used to enhance blood glucose control in patients with T2D, in conjunction with diet and exercise. They stimulate the pancreas to release more insulin, particularly after meals. They are unsuitable for T1D or diabetic ketoacidosis.

Mechanism of Action

Meglitinides bind to receptors on the pancreas's beta cells and stimulate insulin release. Additionally, they attach to an adenosine triphosphate (ATP)-dependent potassium channel on beta cells. Since they promote insulin secretion, they benefit only patients with some beta-cell function.

Adverse Effects

The adverse effects of meglitinides include diarrhea, nausea, vomiting, stomach upset, myalgia, arthralgia, weight gain, dizziness, hypoglycemia, and headaches.

Contraindications

Meglitinides are contraindicated for patients with T1DM, diabetic ketoacidosis, and severe liver damage.

THIAZOLIDINEDIONES

TZDs are a new class of oral hypoglycemic agents that enhance insulin sensitivity. Previously, the only oral agent used to treat T2D was metformin. TZDs improve insulin sensitivity in muscle and fat cells. Pioglitazone and rosiglitazone are TZDs that are administered once daily.

Clinical Indications

TZDs are commonly used to treat type 2 DM. These drugs offer several therapeutic effects, including reducing insulin resistance, lowering hyperglycemia, providing anti-inflammatory effects, and managing hypertension.

Mechanism of Action

The precise mechanism of action of TZDs is still unknown. However, much of their antidiabetic action is believed to be mediated by binding to the adipogenic transcription factor, peroxisome proliferator-activated receptor-γ, in white adipose tissue.

Adverse Effects

The primary adverse effects of TZDs include weight gain, **pedal edema**, bone loss, fractures, renal and liver toxicity, and congestive heart failure. They carry a black box warning that indicates an increased risk of ischemic cardiovascular events.

Contraindications

TZDs are contraindicated in patients with heart failure, liver disease, and T1D. They should be avoided in patients with allergies to this medication.

SULFONYLUREAS

Sulfonylureas are a class of drugs used to manage T2D. They lower blood glucose levels by stimulating the release of insulin from the pancreas. Drugs in this class include glimepiride, glyburide, and glipizide, all of which work by prompting the pancreas to release more insulin.

Clinical Indications

Sulfonylureas are employed to manage hyperglycemia in individuals with T2D who are unable to attain adequate control through diet and exercise modifications alone.

Mechanism of Action

The sulfonylureas stimulate insulin secretion from the pancreas by blocking the ATP-dependent K^+ current in pancreatic β cells. These drugs begin their effects by binding to and inhibiting an ATP-sensitive K^+ channel.

Adverse Effects

The adverse effects of sulfonylureas include hypoglycemia, palpitations, shaking, nausea, diaphoresis, dizziness, confusion, fatigue, headache, seizures, and coma.

Contraindications

Sulfonylureas should be avoided in patients hypersensitive to these drugs. They are also contraindicated in individuals with T1D or severe renal or hepatic failure. Patients experiencing ketoacidosis should receive insulin rather than an oral antihyperglycemic agent. Additionally, sulfonylureas should be avoided during pregnancy and lactation.

BIGUANIDES

Biguanides are a class of drugs used to treat T2D. Examples of biguanides include metformin and phenformin. Metformin, the most common biguanide, is available as Glucophage, Riomet, and Fortamet. Phenformin has been discontinued due to a comparatively high incidence of lactic acidosis.

Clinical Indications

Metformin is used to treat T2D and **gestational diabetes**. It is also indicated for the prevention of T2D in patients at high risk of developing the condition.

Mechanism of Action

Metformin lowers blood glucose levels by reducing hepatic glucose production, decreasing glucose absorption, enhancing peripheral glucose uptake and utilization, and increasing insulin sensitivity.

Adverse Effects

The adverse effects of metformin include tachycardia, dizziness, a metallic taste, peripheral neuropathy, nausea, diarrhea, bloating, and stomach pain.

Contraindications

Metformin should be avoided in patients who have experienced allergic reactions, uncontrolled diabetes, congestive heart failure, metabolic acidosis, renal impairment, or liver dysfunction.

Check Your Knowledge

1. What are the classifications of drug therapies for T2D?
2. What are the contraindications of meglitinides?
3. What are the adverse effects of TZDs?

ALPHA-GLUCOSIDASE INHIBITORS

Alpha-glucosidase inhibitors (AGIs) are relatively inexpensive oral medications that manage T2D by slowing the digestion of carbohydrates in the small intestine. Examples of AGIs are acarbose, miglitol, and voglibose.

Clinical Indications

AGIs are utilized to treat T2D and may help prevent or delay the onset of the disease. They are taken three times daily, with each dose taken upon the first bite of each main meal.

Mechanism of Action

AGIs work by inhibiting the action of alpha-glucosidase, an enzyme in the small intestine that breaks down carbohydrates. This inhibition slows digestion and absorption, lowering blood sugar levels after eating.

Adverse Effects

The adverse effects of AGIs include abdominal pain, bloating, diarrhea, itching, and skin rash. It is advisable to start with a lower dose and gradually increase it. If hypoglycemia is not treated, severe symptoms may include seizures, confusion, and unconsciousness.

Contraindications

AGIs are contraindicated for patients who are hypersensitive to the drug or any of its components. Patients should avoid taking them if they have inflammatory bowel disease, colonic ulceration, partial intestinal obstruction, or conditions that may worsen with increased intestinal gas.

NEW DIABETES DRUGS FOR WEIGHT LOSS

The FDA has approved several new diabetes drugs that can promote weight loss. Losing 5% to 10% of body weight through diet and exercise has been shown to reduce cardiovascular disease risk in obese or overweight adults. These drugs include semaglutide (Ozempic) and tirzepatide (Mounjaro).

OZEMPIC

Ozempic is specifically approved for treating T2D, controlling blood sugar, and offering cardiovascular benefits. By helping to regulate blood sugar levels, Ozempic allows adults with T2D who also have known heart disease to reduce their risk of heart attack, stroke, or other major cardiovascular events. Ozempic, semaglutide, is effective for weight loss, according to research. The same company manufactures both Ozempic and Wegovy. They contain the same active ingredient, semaglutide. However, Wegovy contains higher doses and is designed for weight loss, while Ozempic has lower doses of semaglutide and was developed specifically for T2D patients.

The adverse effects of semaglutide include nausea, vomiting, diarrhea, stomach pain, constipation, heartburn, burping, bloating, loss of appetite, runny nose or sore throat, headache, dizziness, tiredness, and low blood sugar (in patients with T2D). Some of the most severe side effects and potential adverse reactions of long-term Ozempic use include acute gallbladder disease, acute kidney injury, pancreatitis, and thyroid cancer.

MOUNJARO

Mounjaro treats T2D and is effective for weight loss. It mimics hormones that regulate appetite and blood sugar levels. However, the FDA has not yet approved this drug for weight loss.

CLINICAL CASE STUDIES

Clinical Case Study 1

A 5-year-old boy was transferred to the emergency room because he was unconscious. His mother reported that he had dysuria, no appetite, polyuria, polydipsia, and weight loss over the past few weeks. There was no history of diabetes in his family. His blood glucose level was extremely high, and the urine test was also positive for glucose. He was diagnosed with type 1 DM.

CRITICAL THINKING QUESTIONS

1. What are the signs and symptoms of DM?
2. What are the primary types of insulin?
3. What are the adverse effects of Humulin R U-500?

Clinical Case Study 2

A 62-year-old woman with a 13-year history of T2D weighs 218 pounds, and her A1C has never fallen below 12%. She has been taking oral antidiabetic medications, including meglitinides, sulfonylureas, and metformin. The patient was advised to lose at least 35 pounds.

CRITICAL THINKING QUESTIONS

1. What are the clinical indications and adverse effects of meglitinides?
2. What are the adverse effects and contraindications of sulfonylureas?
3. What does metformin (biguanides) indicate?

FURTHER READING

Chawla, R. and Jaggi, S. (2019). *Novel Insights on SGLT-2 Inhibitors*. Jaypee Brothers Medical Publishers, Ltd.

DeFronzo, R.A., Ferrannini, E., Zimmet, P., and Alberti, G. (2015). *International Textbook of Diabetes Mellitus*, 2 Volume Set, 4th Edition. Wiley-Blackwell.

Ghani, U. (2019). *Alpha-Glucosidase Inhibitors — Clinically Promising Candidates for Anti-diabetic Drug Discovery*. Elsevier.

Grau, M. (2023). *Pathological Mechanisms in Diabetes*. Mdpi AG.

Khardori, R., and Gossain, V.V. (2021). *Drugs in Diabetes*. Jaypee Brothers Medical Publishers, Ltd.

Leahy, J.L., and Cefalu, W.T. (2013). *Insulin Therapy — An Issue of Endocrinology and Metabolism Clinics (Internal Medicine Book 41)*. Saunders.

Mohan, V., and Unnikrishnan, R. (2012). *World Clinics — Diabetology — Type 2 Diabetes Mellitus*. Jaypee Brothers Medical Publishers, Ltd.

Nandaniya, R., Ram, P., and Pada, R. (2015). *Anti-Diabetic Drugs*. Lap Lambert Academic Publishing.

Poretsky, L. (2017). *Principles of Diabetes Mellitus*, 3rd Edition. Springer Reference.

Strachan, M.W.J., and Frier, B.M. (2013). *Insulin Therapy — A Pocket Guide*. Springer.

Thomas, N. (2018). *A Practical Guide to Diabetes Mellitus*, 8th Edition. Jaypee Brothers Medical Publishers Ltd.

Todkar, A., Thite, D., and Suryawanshi, S. (2023). *Diabetes Mellitus — Pathophysiology of Diabetes*. Lap Lambert Academic Publishing.

Umpierrez, G.E. (2014). *Therapy for Diabetes Mellitus and Related Disorders*, 6th Edition. American Diabetes Association.

White, J.R. (2023). *The 2022—2023 Guide to Medications for the Treatment of Diabetes Mellitus*. American Di.

22 Drugs Affecting the Respiratory System

LEARNING OUTCOMES

After studying the chapter, readers should be able to

1. Review the respiratory tract functions.
2. Describe the cause of allergic rhinitis and chronic obstructive pulmonary disease (COPD).
3. Summarize the adverse effects of fexofenadine.
4. Describe the clinical indications of fluticasone.
5. Determine the adverse effects of fluticasone.
6. Describe the adverse effects of omalizumab.
7. Assess the contraindications of steroid inhalers.
8. Describe mast cell stabilizers and their mechanism of action.
9. Explain the inhaler drugs and their classification.
10. Describe the adverse effects of mucolytics.

OVERVIEW

The respiratory system is a network of organs that facilitates the exchange of gases between the body and the atmosphere. It primarily takes in oxygen and releases carbon dioxide. The body must produce sufficient energy to sustain life, making oxygen consumption and carbon dioxide production indispensable. The respiratory system enables oxygen to enter the body and carbon dioxide to exit. Examples of respiratory disorders include allergic rhinitis, asthma, chronic obstructive pulmonary disease, cystic fibrosis (CF), lung cancer, and pulmonary hypertension. If left untreated, lung disease can lead to health complications, bothersome symptoms, and life-threatening conditions. Respiratory medications address various breathing issues, including allergic rhinitis, asthma, chronic bronchitis, and chronic obstructive pulmonary disease. These medications can come in many forms, including oral tablets, liquids, injections, and inhalations. Inhalations are particularly effective as they deliver the medication directly to the lungs. Over-the-counter analgesics, nasal decongestants, and antihistamines effectively treat cold symptoms in adults.

THE RESPIRATORY TRACT FUNCTIONS

The primary function of the respiratory tract is to facilitate gas exchange. Through breathing, it takes in oxygen and releases carbon dioxide from the body while filtering, warming, and humidifying the air as it enters. The respiratory tract also performs functions such as air movement, sound production, and protection against foreign particles through mucus and cilia that line the airway.

The respiratory tract consists of the upper and lower airways. The upper airways begin at the nares and mouth and

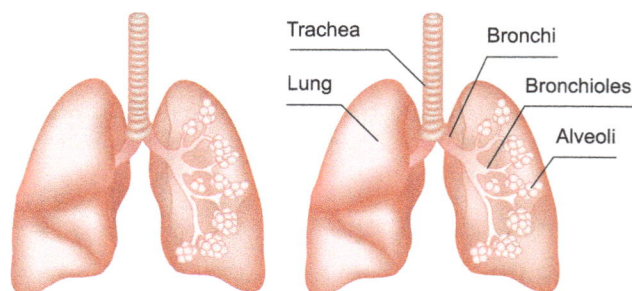

FIGURE 22.1 Physiology of the lower respiratory tract.

end at the larynx. The nasal cavity and nasopharynx have a large surface area and a curving airflow path conducive to filtering, warming, and humidifying air. The lower airways include the trachea, bronchi, bronchioles, alveolar ducts, and alveoli (see Figure 22.1).

The trachea divides at its lower end into two **primary bronchi**. **T**he right bronchus is wider, shorter, and more vertical than the left, which explains why aspirated foreign objects often enter the right bronchus. Ciliated mucosa lines the bronchi. Each primary bronchus enters the lung on its respective side and divides into smaller branches called **secondary bronchi**. The secondary bronchi continue to branch, forming **tertiary bronchi** and small bronchioles. The trachea, the two primary bronchi, and their many branches create the **bronchial tree**.

The terminal bronchioles branch into respiratory bronchioles, characterized by thin, gas-exchanging walls. The respiratory bronchioles continue into alveolar ducts, which end in one or more alveolar sacs. Approximately 300 million alveoli are present in the lungs. These tiny, balloon-shaped air sacs are located at the end of the bronchioles, the branch-like tubes in the lungs. They are enveloped by capillary networks, fulfilling the lungs' primary and vital function: gas exchange between air and blood. The alveoli are the site where oxygen and carbon dioxide are exchanged during breathing. Oxygen inhaled from the air passes through the alveoli and into the blood, then travels to the body's tissues. Carbon dioxide from the body's tissues travels in the blood and passes through the alveoli to be exhaled.

RESPIRATORY DISORDERS

Respiratory diseases affect the airways and other lung structures. They can manifest as acute or chronic conditions. Acute respiratory infections encompass a range of conditions, including the common cold, pharyngitis, otitis media, epiglottitis, laryngitis, bronchitis, and pneumonia.

DOI: 10.1201/9781003461913-24

Chronic respiratory diseases (CRDs) are long-term conditions that impact the lungs and airways. The most common CRDs include allergic rhinitis, asthma, COPD, occupational lung diseases, and pulmonary hypertension. In addition to tobacco smoke, other risk factors include air pollution, occupational chemicals, dust, and a history of frequent lower respiratory infections during childhood. CRDs are not curable; however, various treatments can help open the air passages, control symptoms, and improve daily life for people living with these conditions.

The aim is to reduce morbidity, disability, and premature mortality related to CRDs, specifically asthma and chronic obstructive pulmonary disease. In this chapter, we focus on pharmacotherapy for certain CRDs.

Allergic Rhinitis

Allergic rhinitis, or **hay fever**, is an allergic reaction that causes sneezing, congestion, nasal itching, and eye watering. Although the condition can be severe, relief can be achieved through allergy shots, medications, and lifestyle modifications. It is triggered by exposure to **allergens** such as insects, mold, pet dander, and pollen.

The early-phase reaction is marked by mast cell degranulation. This phase is associated with the rapid onset of acute nasal symptoms within minutes. These symptoms may arise from the release of histamine, primarily from mast cells in the nasal mucosa. The release of histamine, along with the effects of other chemical substances such as **leukotrienes** and **eicosanoids**, can enhance vascular permeability, leading to the development of edema.

The late-phase reaction involves recruiting basophils, neutrophils, monocytes, eosinophils, and T-lymphocytes and releasing mediators that influence the inflammatory response. It also requires tissue remodeling, additional tissue edema, and nasal congestion. Due to mucosal inflammation, the tissues react more intensely to allergen exposure. Late-phase reactions and modifications of tissue responsiveness contribute to bronchial hyperresponsiveness.

Asthma

Asthma is a chronic pulmonary disease that inflames and obstructs the airways. It is the most common disorder in childhood and affects people of all ages, making breathing difficult. A combination of environmental and genetic factors causes asthma, which occurs when the bronchi and bronchioles swell and narrow, increasing their sensitivity. The most critical factors for developing asthma include a history of allergic rhinitis, **conjunctivitis**, dermatitis, or viral respiratory infections in early childhood.

Asthma manifests as inflammation and obstruction of the airways. Inflammation is the primary issue, causing the airways to swell and react to various inhaled stimuli. The obstruction results from the contraction of the smooth muscles in the airway. Signs and symptoms include dyspnea, **wheezing**, chest tightness, **hypoxia**, and cough. Increased mucus production also leads to narrowing of the airways.

Status asthmaticus is the most severe form of asthma and constitutes a true medical emergency. If left untreated, patients may experience respiratory distress and possibly death.

Chronic Obstructive Pulmonary Disease

COPD is a chronic lung condition that makes it difficult to breathe by restricting airflow into and out of the lungs. It permanently damages lung function and includes emphysema and chronic bronchitis. According to the WHO, COPD is the third leading cause of death worldwide, with more than 65 million people suffering from this condition and over 3 million dying each year globally. COPD can be prevented by avoiding smoking, reducing exposure to air pollution, and receiving vaccines against infections. However, smoking is one of the main risk factors for developing COPD; people who have never smoked may also develop the disease. Other risk factors include genetics and aging.

The signs and symptoms of COPD include cough, wheezing, dyspnea on exertion, fatigue, weight loss, and disability. Mucus plugs lead to air trapping and narrowed airways. During inspiration, the airways are opened, allowing gases to flow past the obstruction. However, during expiration, decreased elastic recoil of the bronchial walls results in airway collapse, preventing normal expiratory airflow.

Treatment includes quitting smoking, using supplemental oxygen, and taking medications. Various medications are available for COPD, including corticosteroids, bronchodilators, and antibiotics, which can be administered alone or in combination.

Cystic Fibrosis

CF is a genetic disease that leads to thick mucus buildup in the lungs and digestive tract. This mucus can cause breathing and digestion problems and increase the risk of infections. CF is a life-threatening condition; however, advancements in treatment and care have significantly improved life expectancy and quality of life. Approximately 40,000 individuals are affected by CF in the United States. The signs and symptoms include wheezing, dyspnea, frequent pneumonia or bronchitis, persistent coughing, poor growth or weight gain, salty-tasting skin, difficulty with bowel movements, and **clubbing of fingers**.

In most cases, CF can be diagnosed during childhood through newborn screening, sweat tests, genetic tests, chest X-rays, lung function tests, and sputum cultures. There is no cure for CF, but treatment can ease symptoms, reduce complications, and improve the quality of life. Close monitoring and early, aggressive intervention are recommended to slow the progression of CF, which can lead to a longer life. Medications include mucolytics, expectorants, inhaled medications, and, in some cases, antibiotics, which will be discussed in this chapter.

Check Your Knowledge

1. What factors could trigger allergic rhinitis?
2. What is the manifestation of asthma?
3. What causes COPD?
4. What are the signs and symptoms of CF?

PHARMACOTHERAPY OF RESPIRATORY DISORDERS

Various medications are utilized to treat respiratory disorders. Most patients with allergic rhinitis, asthma, COPD, CF, and the common cold require treatments to manage their symptoms.

ANTIHISTAMINES

Antihistamines are a class of drugs commonly used to treat allergy symptoms. They help address conditions caused by the body's immune system producing too much histamine. People with allergic reactions to pollen and other allergens most often use antihistamines. Table 22.1 lists the selective antihistamines.

Fexofenadine is a nondrowsy second-generation antihistamine that alleviates allergic symptoms. It is an over-the-counter medication approved for use in both children and adults.

Clinical Indications

Fexofenadine prevents and treats allergic symptoms such as sneezing, a runny nose, eye redness, and hives. This agent also addresses seasonal allergies and chronic idiopathic urticaria. Fexofenadine is approved for use in children and adults, with a minimum age requirement of 6 months.

Mechanism of Action

Fexofenadine is a second-generation H1 receptor inhibitor that selectively antagonizes H1 receptors on the surfaces of cells in various organs. It also influences inflammatory mediators.

Adverse Effects

Adverse effects of fexofenadine include allergic reactions, **epistaxis**, headaches, cough, **aphonia**, vomiting, diarrhea, drowsiness, dry mouth, dizziness, back pain, chills, **leukopenia**, and rashes.

Contraindications

Fexofenadine must be avoided in patients who have hypersensitivity, renal failure, hypertension, coronary artery disease, and arrhythmias.

INTRANASAL CORTICOSTEROIDS

Intranasal corticosteroids are anti-inflammatory drugs sprayed into the nose to treat various conditions, including allergic rhinitis, sinusitis, and nasal polyps. They are the first-line treatment for allergic rhinitis and are more effective than oral antihistamines. Table 22.2 lists intranasal corticosteroids, and fluticasone is an example of these drugs, which are discussed in detail here.

Fluticasone is a steroid medication used to treat various conditions. It can be administered via nasal, oral, or topical routes and is available over-the-counter or by prescription.

Clinical Indications

Fluticasone nasal spray is indicated for allergic rhinitis and hay fever symptoms, including sneezing, a runny nose, itchy or watery eyes, and nasal obstruction. It can also treat chronic nasal congestion, hay fever, COPD, asthma, and nasal polyps.

Mechanism of Action

Fluticasone has a direct local effect of vasoconstriction and anti-inflammatory activity. It decreases inflammation in the airways and nose by reducing the number of inflammatory cells, including mast cells, eosinophils, and monocytes. Additionally, fluticasone increases beta-2 receptors in airway muscle and reduces mucus gland secretions.

Adverse Effects

The adverse effects of intranasal fluticasone include cough, fever, sore throat, skin rash, bloody nose, fatigue, dyspnea,

---

TABLE 22.1

OTC Antihistamine Combinations

Generic Name	Trade Name
Brompheniramine	Dimetapp children's cold and allergy
Cetirizine	Zyrtec.D, Allertec
Chlorpheniramine	Actifed cold and allergy tablets, Sudafed PE sinus and allergy tablets, Triaminic cold/allergy, tylenol allergy multi-symptoms gels
Clemastine	Tavist allergy tablets
Diphenhydramine	Benadryl allergy/cold caplets, Sudafed PE nighttime cold, Tylenol PM Gelcaps
Doxylamine	Unisom, Advil PM, NyQuil
Fexofenadine	Allegra-D
Loratadine	Claritin-D, Alavert
Triprolidine	Actifed Plus

TABLE 22.2

Corticosteroids Drugs for Allergic Rhinitis

Generic Name	Trade Name
Corticosteroids	
Beclomethasone	Qnasl, Qvar
Budesonide	Rhinocort Aqua
Ciclesonide	Omnaris, Zetonna
Flunisolide	Nasalide, Nasarel
Fluticasone	Flonase, Veramyst
Mometasone	Nasonex, Sinuva
Triamcinolone	Nasacort AQ

headache, myalgia, and stuffy nose. Less common adverse effects include blurred vision, glaucoma, cataracts, changes in vision, and blindness.

Contraindications

Fluticasone is contraindicated for patients with hypersensitivity to the medication. It should be avoided in individuals experiencing status asthmaticus and acute bronchospasm.

MAST CELL STABILIZERS

Mast cell stabilizers are medications that prevent or treat allergic disorders and are involved in type I allergic reactions. This class of drugs includes cromolyn sodium, lodoxamide, nedocromil sodium, and pemirolast potassium. Cromolyn sodium serves as an alternative therapy for mild persistent asthma and allergic rhinitis.

Clinical Indications

Cromolyn oral inhalation prevents mild to moderate bronchial asthma and bronchospasm, and serves as an adjunctive treatment for allergic rhinitis and systemic **mastocytosis**.

Mechanism of Action

Cromolyn sodium inhibits the release of inflammatory mediators such as histamine and leukotrienes from sensitized mast cells, contributing to allergies. It operates at the cell surface, blocking degranulation and potentially worsening allergic reactions. It differs from antihistamine medications in that it reduces the action of histamines released from mast cells.

Adverse Effects

Cromolyn sodium has adverse effects, including coughing, headache, facial edema, dysphagia, nausea, abdominal pain, diarrhea, myalgia, wheezing, and itching. Severe adverse effects include flushing, palpitations, premature ventricular contractions, **hypoesthesia**, and changes in behavior.

Contraindications

Hypersensitivity to cromolyn products or formulation components is an absolute contraindication for using cromolyn sodium. This drug is rarely used due to its dosing frequency and the adverse effects of coughing and wheezing, which may be mistaken for asthma symptoms. Theophylline relaxes smooth muscle, thereby dilating the airways.

YOU SHOULD REMEMBER

Cromolyn sodium may be used cautiously during pregnancy because of its low systemic absorption. It is regarded as an alternative therapy to inhaled corticosteroids (ICS) for managing mild persistent asthma during pregnancy.

Check Your Knowledge

1. What is the mechanism of action of fexofenadine?
2. What are the contraindications for fluticasone?
3. What are the indications for cromolyn sodium?

LEUKOTRIENE MODIFIERS

Leukotriene modifiers (receptor antagonists) are medications that manage allergic rhinitis, COPD, and asthma by blocking the action of leukotrienes. They are generally not used as the first line of treatment. Examples include montelukast, zafirlukast, and zileuton.

Clinical Indications

Montelukast is commonly used to relieve nasal congestion and inflammation linked to allergic rhinitis and to address narrowing and blockage of the airways in asthma.

Mechanism of Action

Montelukast inhibits the action of leukotriene receptors in the lungs, reducing inflammation and promoting relaxation of smooth muscle.

Adverse Effects

The adverse effects of montelukast include headaches, rhinorrhea, sore throat, otitis media, sinusitis, fever, fatigue, epistaxis, conjunctivitis, myalgia, arthralgia, and hepatotoxicity. Additional adverse effects of montelukast include agitation, anxiety, aggressiveness, insomnia, depression, memory loss, seizures, and suicidal thoughts.

Contraindications

Montelukast is contraindicated in patients with a history of hypersensitivity to the drug or its components and those with phenylketonuria since the chewable tablet contains phenylalanine.

IMMUNOMODULATORS

Immunomodulators are substances that can stimulate or suppress the immune response. They may include drugs, vaccines, or other agents specifically targeting parts of the immune system. They enhance infection resistance by boosting the immune system's baseline response and increasing effectiveness. Omalizumab (Xolair®) is the first available immunomodulator therapy for children with asthma. Other classes of immunomodulators include bevacizumab (Fasenra®), reslizumab (Cinqair®), sarilumab (Dupixent®), and tezepelumab-ekko (Tezspire®).

Clinical Indications

Omalizumab treats moderate to severe asthma that is not controlled by other medications. It is also used to treat chronic rhinosinusitis with nasal polyps in patients who have tried nasal corticosteroids but require additional relief.

Mechanism of Action

Omalizumab inhibits the binding of IgE to the high-affinity IgE receptor on the surface of mast cells and basophils. This action effectively lowers immunoglobulin levels and prevents interaction with the high-affinity receptor, primarily in eosinophils, basophils, and mast cells.

Adverse Effects

Although omalizumab is generally regarded as having an excellent safety profile, there are significant concerns regarding its potential adverse effects. These include systemic reactions, possible effects on malignancy, risk of parasitic diseases, and immunological effects, including **serum sickness**. Anaphylaxis, a potential adverse effect, necessitates extra monitoring by the provider. Patients should have access to prefilled auto-injectors containing epinephrine for self-administration. These are known as *EpiPen auto-injectors*, as Figure 22.2 shows. Other adverse effects associated with omalizumab include localized injection-site reactions and headaches.

Contraindications

Omalizumab has contraindications, including a severe hypersensitivity reaction, and reported cases of anaphylaxis have occurred. Due to the risk of anaphylaxis, patients must be closely observed for an appropriate period after omalizumab administration.

YOU SHOULD REMEMBER

Immunoglobulin E is traditionally recognized as the primary antibody that fights parasitic diseases. Consequently, there are concerns that omalizumab therapy may elevate patients' risk of helminth infections.

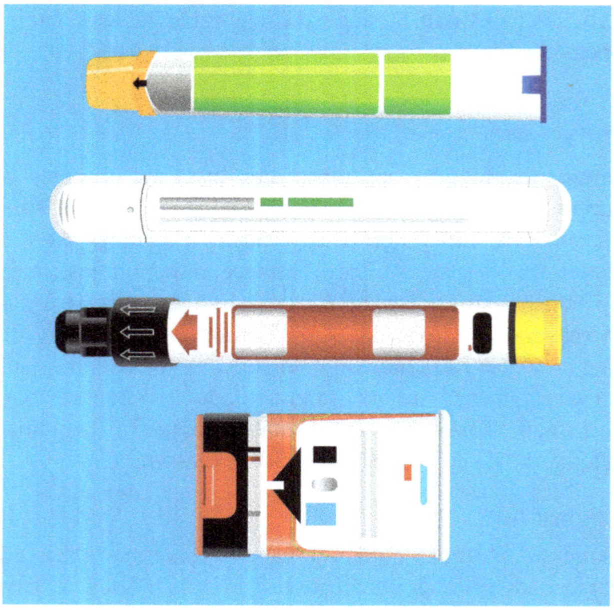

FIGURE 22.2 EpiPen auto-injectors.

Check Your Knowledge

1. What is the definition of immunomodulators?
2. What are the indications for omalizumab?
3. What are the side effects of omalizumab?

INHALATION DRUGS

Inhaled drugs are medicines breathed directly into the lungs to treat chronic diseases such as asthma and COPD. They can be delivered using various devices, including metered-dose inhalers (MDIs), nebulizers, and dry powder inhalers. Bronchodilators and corticosteroids are common inhaled medications. MDIs are pocket-sized, hand-held drug delivery devices that utilize the energy of compressed propellant to generate aerosols (see Figure 22.3). These specified-dose devices provide a precise dose of the active drug in each puff.

A nebulizer compressor system delivers a large drug dose efficiently and directly to the lungs. Hand-held inhalers and spacers are equally effective as nebulizers in controlling routine symptoms. Nebulization treatments typically last at least 5 minutes but may extend over 20 minutes. The nebulizer is bulkier, more expensive, and demands more maintenance. *Nebulizers* are commonly used to treat COPD, asthma, and CF. They can administer corticosteroids or bronchodilators. A *small-volume nebulizer* mixes the desired medication into a fine mist that can be inhaled easily into the lungs, allowing the medication to take effect quickly. An example of a nebulizer is shown in Figure 22.4.

A dry powdered inhaler (DPI) is an asthma treatment option for older children and teens. Using a DPI allows medicine to reach deep into the lungs. Unlike other inhalers, which deliver a puff of medicine, these inhalers contain the drug as a dry powder. The advantages of using a DPI over an MDI include being less time-consuming. It is important to advise patients not to shake the inhaler before use.

FIGURE 22.3 Metered-dose inhaler (MDI).

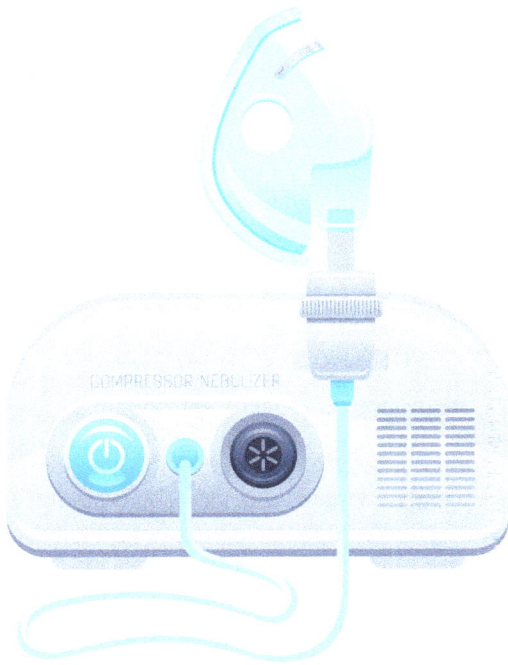

FIGURE 22.4   A nebulizer.

Check Your Knowledge

1. What are some examples of bronchodilators?
2. What devices are used for administering inhaled drugs?
3. What are the indications for nebulizers?

INHALED CORTICOSTEROIDS

Several anti-inflammatory drugs, including corticosteroids, bronchodilators, and leukotriene modifiers, can help reduce lung inflammation and control asthma symptoms. Therefore, the potent anti-inflammatory actions of ICS make these medications the therapy of choice for all types of asthma. Common types of corticosteroids include beclomethasone dipropionate, budesonide, fluticasone propionate, mometasone furoate, and triamcinolone acetonide. Relatively newer inhaled steroid molecules include ciclesonide and flunisolide.

Clinical Indications

ICS are the most effective asthma controllers. They decrease the number and severity of asthma attacks when used regularly every day. However, they do not relieve an asthma attack that has already started. Most asthma patients can be treated and managed with inhaled steroids, but some remain uncontrolled despite adequate asthma therapy. They are also used to treat COPD.

Mechanism of Action

ICS work in COPD by entering airway cells and binding to glucocorticoid receptors, which then suppress the transcription of pro-inflammatory genes, leading to a reduction in the production of inflammatory mediators and decreasing the recruitment of inflammatory cells, such as eosinophils and T-lymphocytes, into the airways.

Contraindications

ICS should not be used in patients with acute tuberculosis, untreated systemic fungal, bacterial, viral, or parasitic infections, or ocular herpes simplex.

Adverse Effects

The adverse effects of ICS include **oral thrush**, dysphonia, cataracts, decreased growth in children, osteoporosis, and an increased risk of infections, particularly at high doses or with long-term use.

BRONCHODILATORS

Bronchodilators help relieve asthma and COPD symptoms by relaxing the smooth muscles of the airways and clearing mucus from the lungs. They are available in long- and short-acting forms. Patients may use them as inhalers, nebulizers, or tablets. The three most widely used bronchodilators include β2-agonists, anticholinergics, and theophylline. Some examples of β2-agonists are albuterol, salmeterol, formoterol, and vilanterol. Examples of anticholinergics include ipratropium, tiotropium, umeclidinium, and aclidinium.

Theophylline is only available as a long-acting oral tablet. Albuterol, also known as salbutamol, is approved by the FDA for treating and preventing acute or severe bronchospasm in patients with reversible COPD.

Clinical Indications

Albuterol is used to treat or prevent acute and severe bronchospasm in patients with asthma, bronchitis, emphysema, and other lung diseases. It is also used to prevent exercise-induced bronchospasm.

Mechanism of Action

Albuterol acts on β2-adrenergic receptors, inducing relaxation of bronchial smooth muscle and inhibiting the release of immediate hypersensitivity mediators, particularly those from mast cells. Although it also affects β1-adrenergic receptors, this impact is minimal and has little effect on heart rate. Anticholinergic agents are competitive antagonists of the neurotransmitter acetylcholine at receptor sites in the cholinergic system.

Contraindications

Albuterol is contraindicated in patients with a history of hypersensitivity to it. Rare cases of hypersensitivity reactions, including **urticaria**, rash, and **angioedema**, have been reported after the use of albuterol sulfate. Caution is required when using anticholinergics, especially in older adults.

TABLE 22.3
Nasal Decongestants

Generic Name	Trade Name
Naphazoline	Privine
Oxymetazoline	Afrin 12 hour, Neo-Synephrine 12 hour
Phenylephrine	Afrin, Neo-Synephrine
Pseudoephedrine	Sudafed
Tetrahydrozoline	Tyzine
Xylometazoline	(Generic only)

Adverse Effects

The adverse effects of albuterol include immediate hypersensitivity reactions, bronchospasm, hypokalemia, and dysrhythmia. These effects are associated with anticholinergic agents and encompass upper respiratory tract infections, sinusitis, headache, dry mouth, dizziness, and bronchitis.

NASAL DECONGESTANTS

Nasal decongestants are medicines that help relieve stuffy, congested noses. They come in various forms, including pills, liquids, drops, and nasal sprays. While many individuals can safely use nasal decongestants, they are not suitable for everyone. Table 22.3 lists nasal decongestants. Pseudoephedrine relieves nasal or sinus congestion caused by allergies, hay fever, and the common cold by constricting the capillaries in the nasal cavities.

Clinical Indications

Pseudoephedrine alleviates nasal or sinus congestion due to the common cold, hay fever, sinusitis, and other respiratory allergies. It is also utilized to relieve ear congestion resulting from otitis media. Some of these formulations are available only by prescription.

Mechanism of Action

Pseudoephedrine is a sympathomimetic with a mixed mechanism of action involving both direct and indirect effects. It directly stimulates beta-adrenergic receptors while also indirectly stimulating alpha-adrenergic receptors by promoting the release of endogenous norepinephrine from the granules of neurons.

Adverse Effects

The adverse effects of pseudoephedrine include insomnia, headaches, tremors, fatigue, hypertension, tachycardia, dysuria, nausea or vomiting, anxiety, and dyspnea.

An overdose of pseudoephedrine may cause seizures, hallucinations, tachycardia, hypertension, dyspnea, circulatory collapse, xerostomia, and coma.

Contraindications

Contraindications of pseudoephedrine include hypersensitivity to the drug and treatment with monoamine oxidase inhibitors (MAOIs) either currently or within the last 2 weeks. It should be used with caution in patients with hypertension, coronary artery disease, liver or renal impairment, benign prostatic hyperplasia, diabetes mellitus, hyperthyroidism, **narrow-angle glaucoma**, and mental agitation. Pseudoephedrine should not be used during pregnancy, lactation, or in children under 2 years of age. The extended-release form of the drug should not be used until the age of 12 years.

YOU SHOULD REMEMBER

Pharmacotherapy for cold symptoms in adults includes over-the-counter analgesics, zinc, nasal decongestants with or without antihistamines, and ipratropium for cough. Acetylcysteine, nasal saline irrigation, and intranasal ipratropium are the only established safe and effective treatments for children. Over-the-counter cold medications should not be used in children under four years of age.

Check Your Knowledge

1. What are the clinical indications for pseudoephedrine?
2. What are the contraindications of pseudoephedrine?
3. What are the effects of pseudoephedrine?

ANTITUSSIVES

Antitussive agents reduce coughing. The most common antitussive products for adults include products that may contain codeine or dextromethorphan. Codeine requires a prescription, whereas dextromethorphan is available over the counter. Table 22.4 lists antitussive drugs.

TABLE 22.4
Antitussives

Generic Name	Trade Name
Opioid Antitussives	
Codeine	Hycodan, Hydromet
Hydrocodone/homatropine	
Nonopioid Antitussives	
Benzonatate	Tessalon
Dextromethorphan	Delsym, Robitussin DM, others
Expectorant	
Guaifenesin	Mucinex, Robitussin, others
Mucolytics	
Acetylcysteine	Mucomyst
Dornase alfa	Pulmozyme

Clinical Indications

Dextromethorphan solution often alleviates coughs caused by the common cold, flu, or other conditions. However, it does not cure the underlying cause of the cough or expedite recovery.

Mechanism of Action

The primary mechanism of action of dextromethorphan as a cough suppressant is unknown. However, it may act on the brainstem, where the pulmonary vagal afferent fibers synapse in the central nervous system.

Adverse Effects

Adverse effects of dextromethorphan are rare; however, the most common symptoms include nausea, vomiting, gastrointestinal discomfort, drowsiness, **nystagmus**, and dizziness. More severe adverse effects associated with dextromethorphan include serotonin syndrome.

Contraindications

Dextromethorphan is contraindicated for patients with known hypersensitivity and those experiencing an idiosyncratic reaction upon administration. It should not be given to patients taking selective serotonin reuptake inhibitors due to the risk of **serotonin syndrome**. Co-administration of dextromethorphan with MAOIs is not recommended, nor is it advised for 2 weeks after discontinuing MAOI medications.

YOU SHOULD REMEMBER

Serotonin syndrome is a potentially life-threatening condition that occurs when the body has excessive serotonin. Dextromethorphan can trigger serotonin syndrome, particularly when taken with other medications that raise serotonin levels.

EXPECTORANTS

Expectorants help to loosen and thin mucus in the airways, making it easier to cough, discharge, or expel from the respiratory tract. They treat the symptoms of respiratory tract infections, such as bronchitis, pneumonia, and CF. Two expectorant ingredients are available in the United States: guaifenesin and potassium iodide. Guaifenesin is the most commonly used expectorant in many cold, cough, and flu medications, including Mucinex® and Robitussin®. Potential adverse effects of guaifenesin include dizziness, fatigue, headache, nausea, vomiting, and constipation.

Potassium iodide is believed to function as an expectorant by increasing secretions in the respiratory tract, thereby reducing mucus viscosity. The oral solution treats symptomatic chronic pulmonary diseases, including bronchial asthma, bronchitis, and pulmonary emphysema, where thick mucus complicates the condition. Potassium iodide is contraindicated in patients with a known allergy to iodides. When administered to a pregnant woman, potassium iodide can cause fetal harm, abnormal thyroid function, and goiter. Due to the potential for fetal goiter development, if the drug is used during pregnancy or if the patient becomes pregnant during therapy, inform the patient of the associated risks.

MUCOLYTIC DRUGS

Mucolytic drugs break down mucus to help clear it from the lungs, making coughing easier. Examples of mucolytics include acetylcysteine (Mucomyst), bromhexine (Mucolite), and dornase alfa (Pulmozyme). Acetylcysteine also possesses anti-inflammatory and antioxidant properties. These drugs are available in oral, intravenous (IV), and nebulizer forms.

Mechanism of Action

Mucolytic drugs decrease the viscosity of mucus in the airways. They break down the chemical bonds within the secretions, making them thinner and easier to expel.

Clinical Indications

Mucolytic drugs are prescribed to treat chronic lung or breathing disorders that lead to thick mucus production. They also treat CF, asthma, bronchiectasis, COPD, and emphysema.

Adverse Effects

The adverse effects of mucolytics depend on the type and formulation of the drug. Over-the-counter mucolytics, such as bromhexine, may cause mild side effects like headache, drowsiness, and dizziness. In contrast, acetylcysteine has common side effects that include nausea, vomiting, diarrhea, sore throat, loss of voice, and chest pain.

Contraindications

Mucolytics are generally safe when used as prescribed; however, they should not be used in children under six. They should be used with caution in patients who are pregnant or breastfeeding.

CLINICAL CASE STUDIES

Clinical Case Study 1

A 53-year-old woman visited her family physician with a common cold that had lasted for 3 days. Her medical history revealed chronic bronchitis and a smoking history. She also had a history of asthma. Every morning, she coughed up sputum. Her physician prescribed pseudoephedrine, dextromethorphan, and acetaminophen.

CRITICAL THINKING QUESTIONS

1. What is the clinical indication for pseudoephedrine?
2. What is the mechanism of action for pseudoephedrine?
3. What are the contraindications for pseudoephedrine?

Clinical Case Study 2

A teenager with a history of bronchial asthma was brought to the emergency department with a 2-day history of progressive dyspnea and wheezing following a common cold. He was using a regular metered-dose inhaler of albuterol, taking one puff four times daily. He reported coughing and wheezing but had no fever, runny nose, or upper respiratory tract symptoms.

CRITICAL THINKING QUESTIONS

1. What is the mechanism of action for albuterol?
2. What are the side effects of albuterol?
3. What are the contraindications of albuterol?

FURTHER READING

Des Jardins, T., and Burton, G.G. (2019). *Clinical Manifestations and Assessment of Respiratory Disease*, 8th Edition. Mosby.

Duggins Davis, S., Rosenfeld, M., and Chmiel, J. (2020). *Cystic Fibrosis — A Multi-Organ System Approach* (Respiratory Medicine). Humana Press.

Evans, S.H. (2020). *The Sinus Solution — The Ultimate Guide to Getting Permanent Relief from Chronic Sinusitis*. Evans.

Global Initiative for Chronic Obstructive Lung Disease. (2022). *Pocket Guide to COPD Diagnosis, Management and Prevention — A Guide for Health Care Professionals*. GICOLD.

Juan, C. (2024). *Allergic Rhinitis –- A Comprehensive Guide to the Understanding, Treatment, and Recovery from Allergic Rhinitis*. Juan.

Lockey, R.F., and Ledford, D.K. (2020). *Allergens and Allergen Immunotherapy — Subcutaneous, Sublingual, and Oral*, 6th Edition. CRC Press.

Louis, D. (2020). *Allergic Rhinitis — The Complete Guide on Everything You Need to Know About Allergic Rhinitis, Causes, Cure, Prevention, and Management*. Louis.

Medwiz Healthcare Communications. (2023). *Allergic Rhinitis — Comprehensive Management and Emerging Innovations*. Medwiz Healthcare Communications.

Melan, A. (2022). *Cystic Fibrosis (CF) — A Manual for Those Who Suffer from Cystic Fibrosis and the Search for a Health Medication Really the Most Effective*. Melan.

Orenstein, D.M., Spahr, J.E., and Weinger, D.J. (2011). *Cystic Fibrosis — A Guide for Patient and Family*, 4th Edition. Wolters Kluwer/Lippincott, Williams, and Wilkins.

Pereira, C. (2021). *Corticosteroids — A Paradigmatic Drug Class*. IntechOpen.

Popov, N. (2022). *Common Cold — A Complete Guide on Treatment of the Common Cold and 10 Tips to Prevent It*. Popov.

Ryan, F. (2020). *Virusphere — From Common Colds to Ebola Epidemics — Why We Need the Viruses that Plague Us*. Prometheus.

Thomson-Smith, L.D. (2013). *Allergic Rhinitis — The Causes and Consequences*. Facebook Publishing.

Zanasi, A., Fontana, G.A., and Mutolo, D. (2020). *Cough — Pathophysiology, Diagnosis and Treatment*. Springer.

Zinserling, V. (2021). *Infectious Pathology of the Respiratory Tract*. Springer.

23 Drugs Affecting the Urinary System

LEARNING OUTCOMES

After studying this chapter, readers should be able to

1. Review the components of the nephron.
2. Describe the causes of chronic kidney disease (CKD).
3. Summarize the classifications of diuretics.
4. Explain the adverse effects of thiazides and loop diuretics.
5. Discuss the clinical indications of all subclasses of diuretics.
6. Describe the contraindications of loop diuretics.
7. Summarize the mechanism of action of potassium-sparing diuretics.
8. Describe the indications of osmotic diuretics.
9. Explain the mechanism of action of carbonic anhydrase inhibitors (CAIs).
10. Review the contraindications of CAIs.

OVERVIEW

The urinary system filters blood and produces urine as a waste by-product. Its organs include the kidneys, renal pelvis, ureters, bladder, and urethra. The kidneys also regulate blood pressure and produce erythropoietin, a hormone that controls the production of red blood cells in the bone marrow. They regulate the acid–base balance and electrolytes. Diuretics increase the rate of urine flow and sodium excretion to adjust the volume and composition of body fluids. They are used to treat hypertension, heart failure, edema, and cirrhosis of the liver. Diuretics have several significant categories and physicochemical properties. They affect renal hemodynamics and their mechanism of action. Typically, they consist of five categories: thiazide diuretics, loop diuretics, potassium-sparing diuretics, osmotic diuretics, and CAIs. Diuretics promote sodium and water excretion by inhibiting sodium reabsorption in the renal tubules. Each class of diuretic can affect different parts of the nephron.

PHYSIOLOGY OF A NEPHRON

The kidney contains over 1 million functioning units called **nephron**s. Approximately 200 L of fluid are filtered from blood flow daily, allowing for the removal of metabolic waste products, toxins, and excess ions while retaining essential substances in the blood. The kidney regulates **plasma osmolarity** by controlling the concentration of plasma **solutes** and **electrolytes**. It also maintains acid–base balance and produces erythropoietin. The kidney also produces renin for blood pressure regulation and converts vitamin D into its active form.

Each nephron consists of the **glomerulus**, **Bowman's capsule**, **proximal convoluted tubule** (PCT), **loop of Henle**, **distal convoluted tubule** (DCT), and **collecting duct** (see Figure 23.1). The glomerulus filters blood, removing cells and large proteins. From the PCT, non-reabsorbed filtrates move to the loop of Henle to be reabsorbed. The loop is divided into descending and ascending limbs. Sodium, potassium, and chloride are reabsorbed in the thick segment of the ascending limb, while magnesium and calcium ions are reabsorbed in the thin segment of the ascending limb. Water is not reabsorbed in the ascending limb.

The DCT is a short nephron segment between the **macula densa** and the collecting duct. It plays a vital role in regulating electrolytes and fluid volume in the body (see Figure 23.2). The **collecting tubule** is where the final reabsorption stage occurs, and it is essential for maintaining fluid–electrolyte balance. In response to the antidiuretic hormone, the collecting duct reabsorbs water, while the hormone aldosterone stimulates it to reabsorb sodium.

The kidney regulates plasma osmolarity by modulating the concentration of water, solutes, and electrolytes. Erythropoietin stimulates the production of red blood cells and renin, which helps regulate blood pressure. The kidney also converts vitamin D into its active form. There are three steps in urine formation: glomerular filtration, tubular reabsorption, and tubular secretion.

Glomerular filtration is the process of filtering waste and excess fluid from the blood to produce urine. The glomeruli are tiny filters in the kidneys that carry out this function. Glomerular filtration is a passive process in which **hydrostatic pressure** pushes fluid and solutes through a membrane. **Osmotic pressure** also significantly impacts the **glomerular filtration rate**.

Tubular reabsorption is the second step in urine formation. In this step, the kidneys remove water and solutes from the filtrate and return them to the bloodstream. The kidneys also reabsorb glucose, amino acids, and electrolytes. The PCT reabsorbs water by osmosis, driven by solute reabsorption. It also reabsorbs lipid-soluble solutes via passive diffusion due to the concentration gradient created by water reabsorption. Urea reabsorption also occurs in the PCT through passive paracellular diffusion driven by a chemical gradient.

From the PCT, the non-reabsorbed filtrates move on to the loop of Henle, which contains a descending and an ascending limb. The descending limb reabsorbs water through osmosis, while solutes are not reabsorbed here. In contrast, the ascending limb reabsorbs sodium, calcium, magnesium, potassium, and chloride but does not reabsorb water.

DOI: 10.1201/9781003461913-25

Nephron Structure

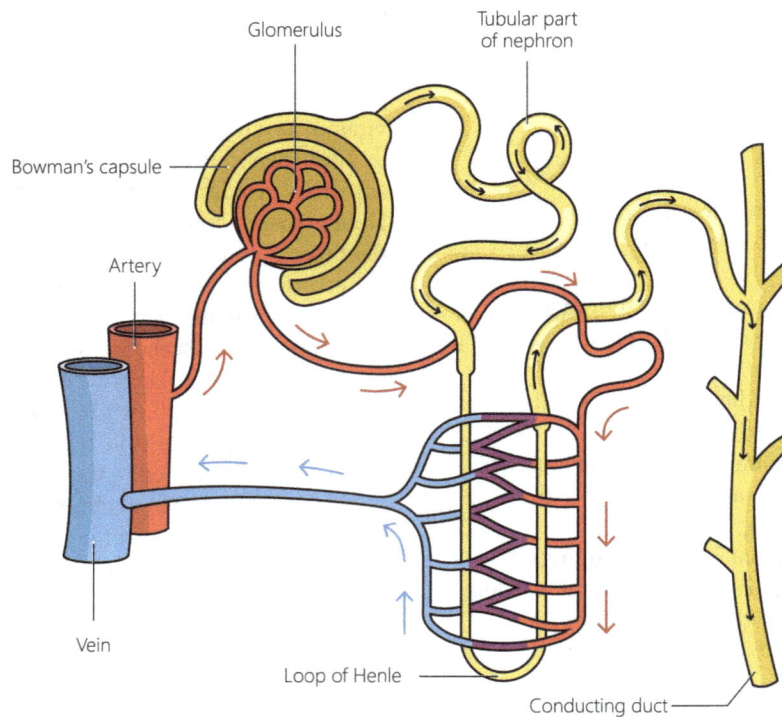

FIGURE 23.1   The structure of the nephron.

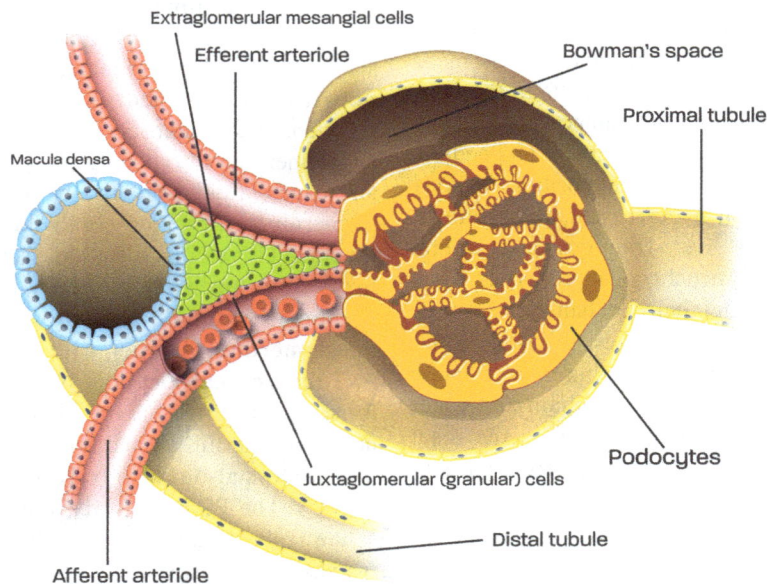

FIGURE 23.2   The distal convoluted tubule, macula densa, and surrounding structures.

The DCT regulates sodium chloride reabsorption through aldosterone, and calcium reabsorption occurs via passive uptake controlled by parathyroid hormone. The collecting duct is crucial for fluid–electrolyte balance, specifically reabsorbing sodium and water in response to antidiuretic hormone and aldosterone.

Tubular secretion is the process of moving waste products from the blood into the renal tubular lumen, or urine-collecting duct. The kidneys reabsorb valuable substances back into the bloodstream. Tubular secretion occurs in the distal and collecting tubules of the nephron. It removes substances too large to be filtered, such as

FIGURE 23.3   Polycystic kidney disease.

antibiotics and toxins, or substances in excess in the blood, like hydrogen and potassium ions.

Check Your Knowledge

1. What is the structure of nephrons?
2. What are the functions of the kidneys?
3. What are the three steps in urine formation?

CHRONIC KIDNEY DISEASE

CKD is a devastating disorder caused by various disease pathways that irreversibly alter the kidney's structure and function over months or years. Some examples of CKD include glomerulonephritis, polycystic kidney disease, and nephrotic syndrome.

Glomerulonephritis is characterized by inflammation of the glomeruli. The kidneys gradually lose their ability to remove waste and excess fluid from the blood, resulting in the production of urine. Glomerulonephritis can occur independently or as part of another disease, such as lupus or diabetes. Severe or prolonged inflammation related to glomerulonephritis can damage the kidneys. Signs and symptoms include hematuria, proteinuria, hypertension, edema, **oliguria**, muscle cramps, nausea, and vomiting.

Nephritic syndrome is a group of symptoms that can occur due to glomerulonephritis. It is primarily characterized by hematuria, hypertension, proteinuria, and oliguria. **Polycystic kidney disease** is a genetic disorder marked by the formation of numerous cysts in the kidneys (see Figure 23.3). These cysts gradually enlarge, leading to progressive kidney damage and eventual kidney failure. It may be asymptomatic in the early stages. Over time, the kidneys cannot filter waste and fluid out of the blood. Therefore, symptoms arise, which include itching, edema, muscle cramps, oliguria, dyspnea, insomnia, weight loss, loss of appetite, nausea, vomiting, and foamy urine. There's no cure for CKD, but treatment can help relieve the symptoms and prevent it from worsening. The main treatments are lifestyle changes, medication, dialysis, and a kidney transplant.

There is no specific medicine for CKD, but it can help manage many of the issues that lead to the condition and the complications that may arise from it.

The complications of CKD contribute to high morbidity and mortality, as well as a poor quality of life. Some of these complications include congestive heart failure, hypertension, volume overload, anemia, mineral bone disorder, electrolyte imbalances, and acid–base abnormalities. They may require specific management approaches, such as diuretics for heart failure and hypertension, or erythropoiesis-stimulating medications to correct anemia. In this chapter, our focus will be on the discussion of diuretics.

The five stages of CKD indicate how effectively the kidneys are functioning. Kidney disease progresses over time. In the early stages (Stages 1–3), the kidneys can still filter waste from the blood. In the later stages (Stages 4–5), they must work harder to filter the blood, and they may eventually stop functioning altogether.

Check Your Knowledge

1. What are the causes of CKD?
2. What signs and symptoms indicate glomerulonephritis?
3. What are the characteristics of nephritic syndrome?

DIURETICS

Diuretics are medications that reduce fluid accumulation in the body. Most diuretics eliminate salt and water through urine, effectively lowering blood pressure. They are classified based on their sites of action within the nephron, chemical structure, and diuretic potency. Diuretics can be categorized into five subclasses: thiazide and thiazide-like, loop, potassium-sparing, osmotic diuretics, and CAIs. They are typically used in combination with another antihypertensive agent and are more effective at reducing hypertension than either drug alone.

THIAZIDE AND THIAZIDE-LIKE DIURETICS

Thiazides and thiazide-like diuretics are moderately potent and are often combined with loop diuretics to achieve a more powerful effect. The most commonly used thiazides are chlorothiazide and hydrochlorothiazide, while the most commonly used thiazide-like diuretics are metolazone, indapamide, and chlorthalidone. Thiazide and thiazide-like diuretics are considered safe and effective for treating high blood pressure and edema (see Table 23.1).

Clinical Indications

Thiazide and thiazide-like diuretics are first-line therapies for hypertension. They are also used to treat edema, heart failure, idiopathic hypercalciuria, pulmonary edema, and hepatic cirrhosis. Diuretics, along with salt restriction, are recommended as the first-line therapy for ascites due to liver cirrhosis.

Mechanism of Action

The mechanism of action for thiazide and thiazide-like diuretics inhibits sodium-chloride reabsorption in the DCT to reduce extracellular fluid and decrease blood volume.

Adverse Effects

Electrolyte imbalances are the most common adverse effects associated with chronic thiazide therapy for hypertension. They occur most often with longer-acting drugs, such as chlorthalidone and metolazone, but a low-dose strategy can mitigate this to some extent. Thiazide and thiazide-like diuretics may cause glucose intolerance, hypokalemia, metabolic alkalosis, hyponatremia, hypercalcemia, hypomagnesemia, hyperuricemia, hypertriglyceridemia, and hypercholesterolemia.

Contraindications

Thiazide and thiazide-like diuretics should be avoided in patients with renal failure, hypotension, hepatic coma, **gout**, or allergies to this medication. They are also contraindicated in patients with Addison's disease, hyponatremia, and hypokalemia.

LOOP DIURETICS

Loop diuretics are the most potent, effectively decreasing extracellular fluid, cardiac output, and blood pressure. They are commonly used to treat edematous conditions and congestive heart failure. However, their use for hypertension is less prevalent (see Table 23.2).

Clinical Indications

Loop diuretics are the preferred option for treating hypertension, heart failure, nephrotic syndrome, and pulmonary edema. A thiazide or amiloride may be added to enhance the response in patients with heart failure who do not respond adequately to loop diuretics. However, high doses of loop diuretics can lead to ototoxicity.

Mechanism of Action

Loop diuretics, including furosemide and bumetanide, function by inhibiting the reabsorption of sodium, potassium, and chloride ions in the thick ascending limb of the loop of Henle within the kidneys.

TABLE 23.1
Thiazide and Thiazide-Like Diuretics

Generic Names	Brand Names
Hydrochlorothiazide	HCTZ, Hydrodiuril
Chlorthalidone	Hygroton
Metolazone	Zaroxolyn
Indapamide	Lozol
Chlorothiazide	Thalitone

TABLE 23.2
Loop Diuretics

Generic Names	Brand Names
Furosemide	Lasix
Ethacrynic acid	Edecrin
Bumetanide	Bumex
Torsemide	Demadex

Adverse Effects

The adverse effects of loop diuretics include anaphylaxis, fever, headache, vertigo, dizziness, restlessness, **orthostatic hypotension**, and syncope. They may also cause severe electrolyte imbalances such as hyponatremia, hypochloremia, hypokalemia, azotemia, metabolic alkalosis, dehydration, and hyperuricemia. Other adverse reactions include skin rash, **tinnitus**, ototoxicity, photosensitivity, and **interstitial nephritis**. Thrombocytopenia, aplastic anemia, hemolytic anemia, leukopenia, agranulocytosis, **pneumonitis**, pulmonary edema, blurred vision, abdominal pain, anorexia, diarrhea, **urticaria**, jaundice, pancreatitis, hepatic coma, and **impotence** may also occur.

Contraindications

Loop diuretics should be avoided in patients with severe hypersensitivity, hyponatremia, hypokalemia, azotemia, hypotension, **anuria**, and hepatic coma. They are also contraindicated when fluid depletion is anticipated, such as during surgery.

YOU SHOULD REMEMBER

Loop diuretics are named for their site of action in the thick ascending limb of the loop of Henle, where they inhibit Na^+ and Cl^- reabsorption, as 25% of filtered Na^+ is reabsorbed in this segment.

Check Your Knowledge

1. What are the clinical uses of thiazides and thiazide-like diuretics?
2. What are the contraindications of loop diuretics?
3. What are the side effects of loop diuretics?

POTASSIUM-SPARING DIURETICS

Potassium-sparing diuretics increase urine output while maintaining potassium levels in the body. They are classified as aldosterone antagonists and encourage the retention of sodium and water. These medications are not as potent as thiazides or loop diuretics, and they are often combined with a thiazide diuretic to reduce patient potassium loss. Table 23.3 lists potassium-sparing diuretics.

Clinical Indications

Spironolactone is indicated for managing and treating hypertension and heart failure, as well as other conditions beyond cardiovascular disease. It is also used to diagnose and treat hyperaldosteronism. Additionally, this medication may be employed to treat edema in patients with congestive heart failure, liver cirrhosis, or nephrotic syndrome.

Mechanism of Action

Spironolactone specifically works by competitively blocking the action mediated by aldosterone receptors. The effect of this blockade is that sodium reabsorption and water retention do not occur, leading to increased potassium retention.

Adverse Effects

The adverse effects of spironolactone include allergic reactions, seizures, dehydration, hyperkalemia, hyponatremia, headache, bleeding gums, chest pain, cough, dyspnea, loss of appetite, vomiting, diarrhea, dizziness, loss of consciousness, arrhythmias, and hematuria. Additionally, spironolactone may cause menstrual irregularities, loss of libido, **gynecomastia,** and impotence.

Contraindications

Spironolactone should be avoided in patients with hypersensitivity to it, renal failure, fluid imbalances, hyponatremia, hyperkalemia, anuria, gout, liver disease, and Addison's disease.

Osmotic Diuretics

Osmotic diuretics inhibit solute and water reabsorption by altering the osmotic driving forces along the nephron. These drugs promote sodium ion and water loss through the nephron by excreting non-reabsorbable filtrate. Examples include mannitol, isosorbide, and glycerin. Table 23.4 lists osmotic diuretics.

Clinical Indications

Osmotic diuretics are utilized when increased water excretion with minimal sodium excretion is necessary. They are the first choice for treating **cerebral edema**. Mannitol's primary indication is for the treatment of acute angle-closure glaucoma. Glycerin helps lower intraocular pressure and reduces superficial **corneal edema**. These drugs can also treat congenital glaucoma and **keratoplasty** following **retinal detachment** repair.

TABLE 23.3
Potassium-Sparing Diuretics

Generic Names	Brand Names
Spironolactone	Midamor Aldactone, Carospir
Eplerenone	Inspra
Amiloride	Midamor
Triamterene	Dyrenium

TABLE 23.4
Osmotic Diuretics

Generic Names	Brand Names
Mannitol	Osmitrol
Isosorbide	Imdur
Glycerin	None

Mechanism of Action

Osmotic diuretics primarily inhibit the reabsorption of water and sodium chloride in the PCT, descending loop of Henle, and collecting duct, which subsequently affects the reabsorption of urea and uric acid. Mannitol reduces tubular water reabsorption and increases sodium and chloride excretion by raising the osmolarity of the glomerular filtrate.

Adverse Effects

The adverse effects of osmotic diuretics include chills, nausea, vomiting, fever, dizziness, headache, **rhinorrhea**, seizures, chest pain, hypotension, **phlebitis**, fluid or electrolyte imbalances, acidosis, polydipsia, blurred vision, urinary retention, and skin rashes. Another adverse effect of osmotic diuretics is acute **closed-angle glaucoma**.

Contraindications

Osmotic diuretics should be avoided in patients with kidney failure, anuria, or heart failure. Mannitol and urea are contraindicated in individuals with cranial bleeding. Glycerin must be avoided in patients with known hypersensitivity or diabetes mellitus. It is also contraindicated for those with hypervolemia and cardiac failure.

Carbonic Anhydrase Inhibitors

CAIs are still widely used systemic antiglaucoma drugs.

Carbonic anhydrases catalyze the conversion of carbon dioxide to bicarbonate. CAIs are provided in Table 23.5.

Clinical Indications

CAIs are indicated for the treatment of **open-angle glaucoma**, intracranial hypertension, congestive heart failure, edema, **periodic idiopathic paralysis**, high-altitude sickness, and epilepsy. Acetazolamide can be utilized to manage edema that is resistant to other diuretics. However, CAIs are significantly less potent as diuretics than others.

Mechanism of Action

CAIs inhibit the enzyme carbonic anhydrase in the proximal tubule cells, which prevents the reabsorption of bicarbonate ions (HCO_3^-) from the urine. This leads to increased bicarbonate excretion and a resultant mild metabolic acidosis; ultimately, it causes increased urine output due to the associated loss of sodium and water and the bicarbonate ions.

Adverse Effects

The common adverse effects of CAIs include nausea, vomiting, diarrhea, black stools, abdominal pain, and fatigue. Other unwanted effects are **paresthesia**, tinnitus, decreased libido, and taste alterations. Additionally, Stevens–Johnson syndrome, aplastic anemia, and severe hepatic necrosis must also be considered as their adverse effects.

Contraindications

CAIs are contraindicated in patients with known drug hypersensitivity, liver and kidney failure, type 2 diabetes mellitus, emphysema, gout, Addison's disease, pregnancy, and breastfeeding. Acetazolamide should be avoided in individuals with hypokalemia or hyponatremia. Patients experiencing hyperchloremic acidosis should not use acetazolamide.

Check Your Knowledge

1. What are the clinical indications for spironolactone?
2. What are the adverse effects of spironolactone?
3. What are the indications for osmotic diuretics?
4. What are the adverse effects of osmotic diuretics?

CLINICAL CASE STUDIES

Clinical Case Study 1

A 77-year-old woman with a history of CKD and hypertension takes three antihypertensive medications. However, her blood pressure remains at 186/112 mm Hg. The patient complains of dyspnea and swelling in her legs. She has been taking furosemide every 12 hours for several months. During her visit, hydrochlorothiazide is prescribed to improve hypertension control. A few days later, she was transferred to the emergency department with symptoms of syncope, fatigue, anorexia, and dizziness. Her blood pressure was 88/56 mmHg, and she had lost 9 pounds in the past week.

CRITICAL THINKING QUESTIONS

1. What are the clinical indications for loop diuretics?
2. What are the adverse effects of loop diuretics?
3. What are the contraindications of loop diuretics?

Clinical Case Study 2

A 63-year-old man with a history of hypertension visited his ophthalmologist, complaining of severe pain in his right eye, blurred vision, headache, and nausea. The patient was diagnosed with acute angle-closure glaucoma and prescribed mannitol, a CAI.

TABLE 23.5

Carbonic Anhydrase Inhibitors

Generic Names	Brand Names
Acetazolamide	Diamox
Brinzolamide	Azopt
Dorzolamide	Trusopt
Diclofenamide	Cambia, Voltaren
Methazolamide	Neptazane
Ethoxolamide	Cardrase
Zonisamide	Zonegran

CRITICAL THINKING QUESTIONS

1. What are the clinical indications of osmotic diuretics?
2. What is the mechanism of action for osmotic diuretics?
3. What are the contraindications of osmotic diuretics?

FURTHER READING

Aiden, R. (2022). *Diuretics Complete Guide—Ultimate Guide: Understanding the Purpose, Benefits, Downsides, Interactions, Dosages & Many More*. Aiden.

Chegwidden, W.R., and Carter, N.D. (2021). *The Carbonic Anhydrases—Current and Emerging Therapeutic Targets*. Springer.

Earlstein, F. (2018). *Glaucoma Explained—Glaucoma Facts, Diagnosis, Symptoms, Treatment, Causes, Effects, Alternative Medicines, Therapeutic Methods, History, Home Remedies, and More!* Pack & Post Plus, LLC.

Eaton, D.C., and Pooler, J.P. (2023). *Vander's Renal Physiology*, 10th Edition. McGraw-Hill/Lange.

Fadem, S.Z. (2022). *Staying Healthy with Kidney Disease—A Complete Guide for Patients*. Springer.

Gilbert, S.J., Weiner, D.E., Bomback, A.S., Perazella, M.A., and Rifkin, D.E. (2022). *National Kidney Foundation's Primer on Kidney Diseases*, 8th Edition. Elsevier.

Hewitt, J., and Gabata, M. (2011). *Nephrotic Syndrome—Causes, Tests, and Treatments*. CreateSpace Independent Publishing Platform.

Johnson, R.J., Floege, J., and Tonelli, M. (2023). *Comprehensive Clinical Nephrology*, 7th Edition. Elsevier.

Kalogeropoulos, A.P., Skopicki, H., Butler, J., and Murali, S. (2022). *Heart Failure — An Essential Clinical Guide*. CRC Press.

Koeppen, B.M., and Stanton, B.A. (2018). *Renal Physiology—Physiology Series*, 6th Edition. Mosby/Elsevier.

Koyner, J.L., Topf, J.M., and Lerma, E.V. (2021). *Handbook of Critical Care Nephrology*. Wolters Kluwer.

Landoni, G., Pisano, A., Zangrillo, A., and Bellomo, R. (2016). *Reducing Mortality in Acute Kidney Injury*. Springer.

Saleh, A. (2020). *Diuretic Therapy and Drugs for Kidney Failure*. Saleh.

Wilcox, C.S., Choi, M., Segal, M.S., Chen, L., and Williams, W.W. (2022). *Handbook of Nephrology and Hypertension*, 7th Edition. Wolters Kluwer.

Wouters, L. (2013). *Diuretics—Pharmacology, Therapeutic Uses and Adverse Side Effects*. Nova Medical.

Zuther, J.E. (2022). *It's Not Just a Swelling! Lymphedema—Causes, Prevention, Treatment, Self-Management*. Prolymph Consultants.

24 Drugs Affecting the Digestive System

LEARNING OUTCOMES

After studying the chapter, readers should be able to

1. Describe the function of the digestive tract.
2. Identify the hormones that are released from the digestive tract.
3. Review the accessory organs of the digestive system.
4. Explain the common causes of peptic ulcer disease (PUD).
5. Summarize the drugs used to treat peptic ulcers.
6. Describe the clinical indications of H2-blockers.
7. Discuss the treatments of *H. pylori*.
8. Describe the features of inflammatory bowel disease (IBD).
9. Review the characteristics of ulcerative colitis and Crohn's disease.
10. List the medications that are used to treat IBD.
11. Describe the drugs used to treat diarrhea.
12. Explain the adverse effects of diphenoxylate with atropine.

OVERVIEW

The gastrointestinal (GI) system begins in the mouth and ends in the anus. The primary function of the GI system is food digestion and absorption. Food breakdown starts in the mouth with chewing and continues in the stomach, where food is mixed with acid, mucus, enzymes, and other substances. From the stomach, the fluid and partially digested food pass into the small intestine, where biochemicals and enzymes secreted by the liver, exocrine pancreas, and small intestinal epithelium break it down into absorbable components of proteins, carbohydrates, and fats. Peptic ulcer disease (PUD) involves erosions in the GI tract due to mucosal damage, often caused by pepsin and gastric acid secretion. The stomach and duodenum are the most common sites for *H. pylori* ulcers. Medications and surgery are options for treating PUD. Typically, medications are the first line of treatment for ulcers; however, surgery may be indicated in certain patients with complications. The lower GI tract includes most of the small intestine (jejunum and ileum). The primary function of the large intestine is to digest, absorb nutrients, and eliminate waste products. Extensive reabsorption of water and electrolytes occurs in the right proximal colon.

PHYSIOLOGY OF THE DIGESTIVE SYSTEM

This system consists of the GI tract and accessory organs of digestion, such as the liver, gallbladder, and exocrine pancreas (see Figure 24.1). The upper digestive tract extends from the mouth to the stomach. The mouth, also known as the oral cavity, receives food through ingestion, breaking it into small particles. The digestive process begins in the mouth during chewing. The salivary glands secrete saliva, which moistens food to help move a bolus more quickly through the esophagus into the stomach. Saliva contains amylase that begins to break down starches in the food. The surface of the oral mucosa is covered by squamous stratified epithelium. This is an avascular and semipermeable tissue whose thickness and degree of keratinization vary

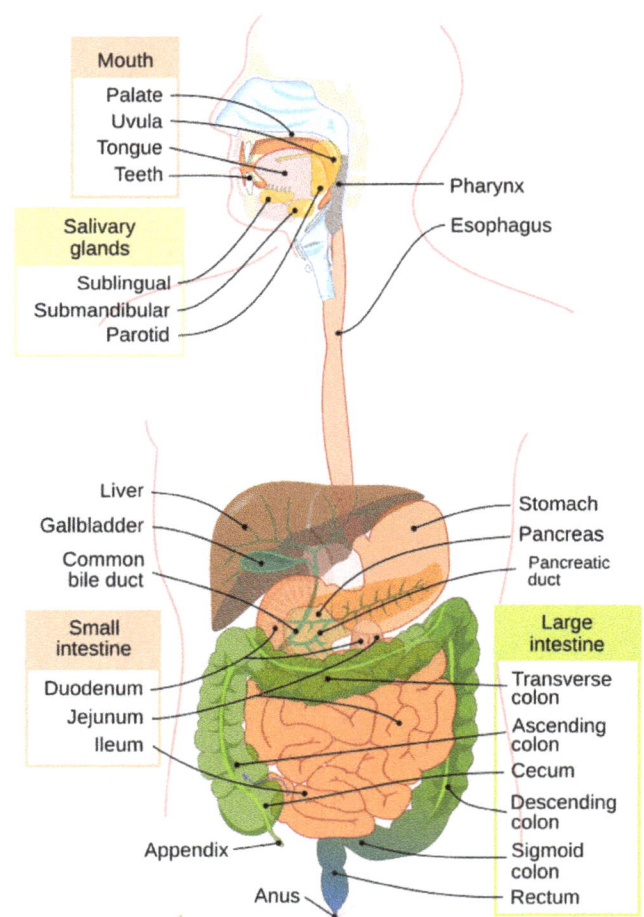

FIGURE 24.1 The digestive system.

DOI: 10.1201/9781003461913-26

HUMAN STOMACH

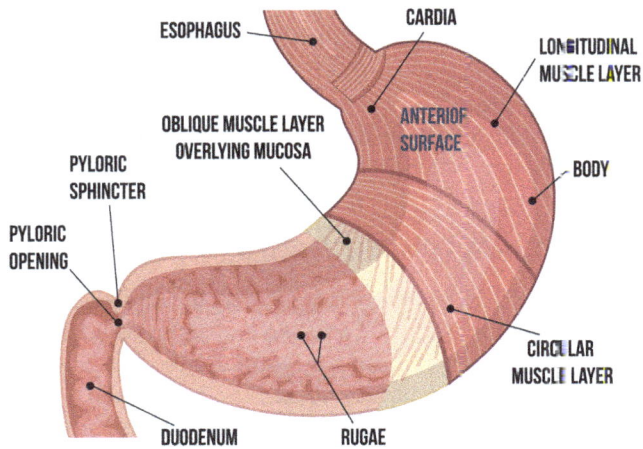

FIGURE 24.2 The layers of smooth muscle in the stomach.

Large intestine
(sections of the colon)

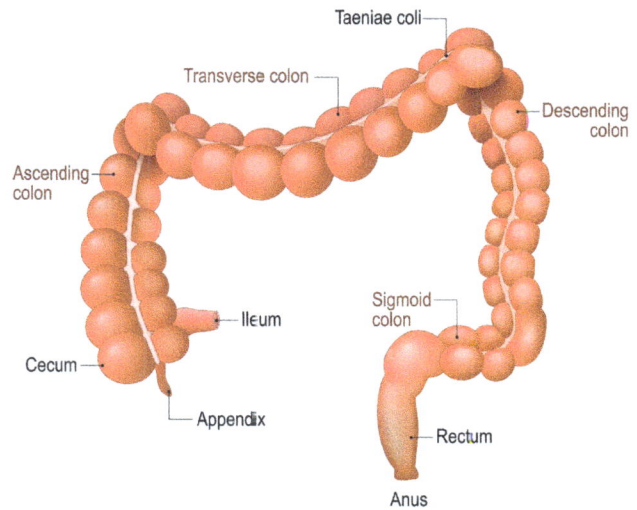

FIGURE 24.3 The large intestine.

according to the location in the oral cavity and the area's functional and mechanical requirements. The pharynx and esophagus facilitate the passage of material from the mouth and throat to the stomach.

The stomach comprises three muscle layers: an inner oblique layer, a middle circular layer, and an outer longitudinal layer (see Figure 24.2). It mixes food, churns it, and pushes the contents into the small intestine. The gastric mucosa contains various cells and glands in its inner layer, including parietal, chief, endocrine, and enteroendocrine cells. Parietal cells secrete hydrochloric acid and **intrinsic factors**, lowering the stomach's pH and activating **pepsin**. Chief cells secrete pepsinogen, stimulated by cholinergic activity from the vagus nerve.

Gastric enteroendocrine cells (**G cells**) in the stomach and small intestine secrete **gastrin**, a hormone that stimulates the stomach to release gastric acid. Additionally, **ghrelin** is another hormone produced by the stomach. The stomach can absorb most acidic and weakly alkaline drugs, including salicylic acid, aspirin, thiopental, secobarbital, and antipyrine.

The small intestine is approximately 20–26 feet long. It consists of the duodenum, the jejunum, and the ileum. The duodenum is the first part attached to the stomach and is about 10 inches long. The jejunum is approximately 8 feet long, **while** the ileum measures about 12 feet. The small intestine breaks down large food molecules into absorbable components, including carbohydrates, proteins, and fats. The nutrients pass through the small intestinal epithelium into blood and lymphatic vessels, carrying them to the liver via the hepatic portal circulation for further processing.

Villi are located in the small intestine, with microvilli increasing the surface area. Mucus-secreting **goblet cells** are found in both the small and large intestines. The duodenum and jejunum are responsible for the secretion of **cholecystokinin**, secretin, gastric inhibitory peptide, and **motilin**. Cholecystokinin regulates the ingestion, digestion,

and absorption of nutrients. The primary absorption of nearly all drugs occurs in the small intestine through oral administration.

The first portion of the large intestine is the **cecum**. The colon includes the ascending, transverse, descending, and sigmoid sections (see Figure 24.3). The cecum joins the ileum, and the **ileocecal valve** allows material to pass into the large intestine. The large intestine absorbs water and electrolytes, producing vitamins B, K, and biotin. The rectum holds feces in place until defecation occurs. Drug absorption in the lower rectum is transported directly to systemic blood circulation. However, drug absorption in the colon passes to the liver through the portal system.

Check Your Knowledge

1. Which organs are considered part of the digestive system?
2. What chemical substances are released by the stomach?
3. What are the functions of the villi in the small intestine?

ACCESSORY ORGANS OF THE DIGESTIVE SYSTEM

The accessory organs of the digestive system include the teeth, tongue, salivary glands, liver, gallbladder, and pancreas. The organs with essential pharmacological functions are the liver and pancreas. The liver is the body's largest organ and occupies most of the right side of the diaphragm. It is one of the body's most vital organs, performing functions critical to healthy living. The liver's metabolic functions require a large volume of blood. The liver receives blood from both arterial and venous sources. The hepatic artery branches from the celiac artery and supplies oxygenated blood, while the hepatic portal vein receives deoxygenated blood from the inferior and superior mesenteric veins and the splenic vein. The hepatic portal vein, which carries 70% of the blood supply to the liver, is rich in nutrients absorbed from the digestive tract. Figure 24.4 illustrates the internal anatomy of the liver, including the hepatic portal system.

The liver's primary functions include detoxifying various substances, secreting bile, breaking down and removing old red blood cells, recycling iron from hemoglobin, metabolizing foods and drugs, storing fat-soluble vitamins, and producing clotting factors and albumin.

Hepatocytes produce bile, which is essential for the intestinal emulsification and absorption of fats. The common bile duct and pancreatic duct converge at the ampulla of Vater, which discharges into the duodenum. To reduce their toxicity or biological activity, the liver modifies exogenous and endogenous chemicals, including drugs, foreign molecules, and hormones. Cytochrome P450 enzymes play a crucial role in metabolizing many medications, participating in over 90% of enzymatic reactions. Cytochrome P450 enzymes can be inhibited or induced by drugs, leading to clinically significant drug–drug interactions that may result in unanticipated adverse reactions or therapeutic failures.

The pancreas comprises **acinar cells** that secrete enzymes and a network of ducts that release alkaline fluids, which play critical roles in digestion. The acinar units, which make up most of the pancreatic tissue, secrete pancreatic juice. This digestive fluid is primarily water but contains sodium bicarbonate and various digestive enzymes. Therefore, the exocrine part of the pancreas plays a vital role in digestion. The endocrine functions of the pancreas were discussed in **Chapter 21**.

Check Your Knowledge

1. What organs are classified as accessory organs of the digestive system?
2. What are the liver's functions?
3. Which enzyme in the liver is essential for drug–drug interactions?

DIGESTIVE SYSTEM DISORDERS

Digestive system disorders, which affect the digestive tract, can range from mild to severe and can be problematic for patients. This chapter discusses common digestive diseases, including gastroesophageal reflux disease (GERD), PUD, and IBD (**ulcerative colitis** and **Crohn's disease**), as well as symptoms such as diarrhea, nausea, vomiting, and constipation, which their pharmacotherapy will cover.

GASTROESOPHAGEAL REFLUX DISEASE

GERD is a condition that develops when stomach contents flow retrograde into the esophagus. It can present as non-erosive reflux disease or erosive esophagitis. GERD is one of the most commonly diagnosed digestive disorders in the United States. This condition is caused by multiple mechanisms that can be intrinsic, structural, or both, leading to the disruption of the esophagogastric junction barrier and exposing the esophagus to acidic gastric contents. Clinically, GERD typically manifests with symptoms of heartburn and regurgitation. Based on endoscopic and histopathologic appearance, GERD is classified into three

Internal anatomy of human liver

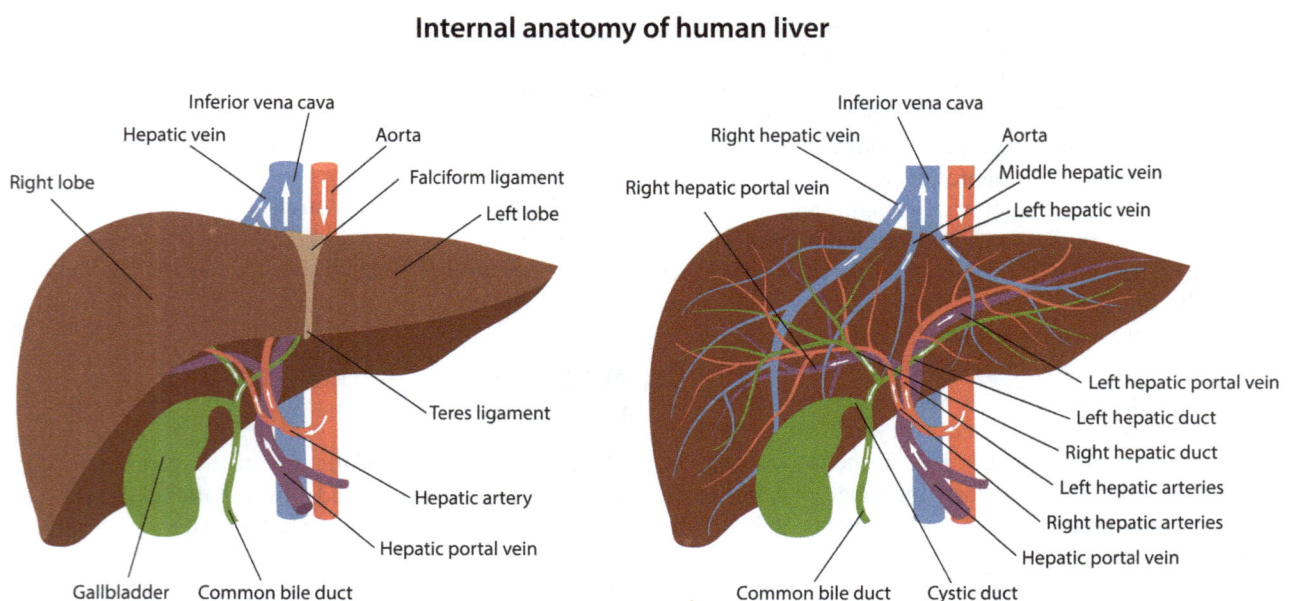

FIGURE 24.4   The internal anatomy of the liver, including the hepatic portal system.

phenotypes: non-erosive reflux disease, erosive esophagitis, and Barrett's esophagus. Over the years, the mainstay in managing GERD has been lifestyle modifications and proton pump inhibitors.

Peptic Ulcer Disease

PUD is characterized by discontinuity in the inner lining of the GI tract caused by gastric acid secretion or pepsin. It extends into the muscularis layer of the gastric epithelium. PUD usually occurs in the stomach and proximal duodenum but may also involve the lower esophagus, distal duodenum, or jejunum. PUD consists of stomach and duodenal ulcers. Peptic refers to the fact that acid and pepsin are necessary to produce the break in the mucosa, known as an ulcer. Patients whose parietal cells are continuously stimulated to secrete acid by gastrin from a gastrinoma develop ulcers. Individuals with pernicious anemia rarely develop ulcers due to achlorhydria.

The two most common causes of peptic ulcers are *Helicobacter pylori* (*H. pylori*) infection and the long-term use of nonsteroidal anti-inflammatory drugs. Risk factors for peptic ulcers include corticosteroid use, cigarette smoking, chemotherapy, radiation therapy, potassium chloride, and cocaine use.

The signs and symptoms of peptic ulcers include stomach pain, nausea, burping, bloating, dark blood in the stool, and weight loss. The patient describes a burning pain in the epigastric region and chest that radiates toward her back. The pain worsens after taking aspirin or drinking coffee and is alleviated by antacids. The patient has been diagnosed with PUD.

Helicobacter pylori Infection

Helicobacter pylori (*H. pylori*) infection occurs when the bacteria infect the stomach. It typically begins during childhood and is a common cause of peptic ulcers. This infection may affect more than half of the global population. Most individuals are unaware that they have an *H. pylori* infection because they do not experience any symptoms. However, testing may yield a positive result for the disease if patients develop signs and symptoms of a peptic ulcer. Signs and symptoms of an *H. pylori* infection are usually related to gastritis or a peptic ulcer. Complications associated with *H. pylori* infection include peptic ulcers, gastritis, and stomach cancer.

Treatment for peptic ulcers involves eliminating the *H. pylori* germ if necessary. It may also include reducing or stopping the use of nonsteroidal anti-inflammatory drugs and taking medications such as antibiotics, proton pump inhibitors (PPIs), and histamine (H-2) blockers.

Inflammatory Bowel Disease

IBD comprises a group of chronic, relapsing conditions of unknown origin, including Crohn's disease and ulcerative colitis. **Crohn's disease** is known as granulomatous colitis,

ileocolitis, and regional enteritis. It is an idiopathic inflammatory disorder that can affect any part of the GI tract, from the mouth to the anus. Typically, the distal small intestine and proximal large colon are impacted. Risk factors include family history, smoking, Jewish ethnicity, urban living conditions, age under 40 years, and, to a lesser extent, female gender. The pathogenesis may be associated with an overly aggressive response to normal flora bacteria in genetically predisposed individuals.

Ulcerative colitis can cause ulceration of the colonic mucosa. It extends proximally from the rectum into the colon. Lesions appear in susceptible people between the ages of 20 and 40. Risk factors include a family history of ulcerative colitis and Jewish descent. The disease is more prevalent in Caucasian and Northern European populations. For unknown reasons, it is less common among smokers. Dietary, infectious, genetic, and immunologic factors may all be causative. The inflammation may be due to commensal or pathogenic enteric microorganisms, increased mucosal adherence and invasion, and persistent T-cell activation. Typical chronic lesions are granulomas, with projections of inflamed tissue surrounded by ulcerated areas.

Check Your Knowledge

1. What causes PUD?
2. Which disorders are classified as IBD?
3. What are the risk factors for ulcerative colitis?

Diarrhea

Diarrhea is an abnormal increase in the fluidity and frequency of bowel movements. The small intestine receives about 9 L of fluid daily, most of which is reabsorbed; only about 1 L reaches the colon. If the small and large intestines do not reabsorb enough fluid, diarrhea may occur. Periodic diarrhea is self-limiting and does not usually require medications. Sometimes, diarrhea signifies the body's defense mechanism to eliminate pathogens and toxins quickly. However, if prolonged or severe, diarrhea can lead to significant loss of body fluids, necessitating pharmacotherapy. Prolonged diarrhea may result in imbalances in fluid, acid–base, or electrolytes.

Common causes of diarrhea include infections, IBD, adverse reactions to medications, food poisoning, food allergies, and malabsorption. Viral and bacterial infections are the most prevalent causes. Other, less common reasons for diarrhea include damage from radiation treatments or tumors that produce excessive hormones. If diarrhea is not treated, patients are at risk for dehydration. Severe dehydration can lead to organ damage, shock, fainting, or coma.

Nausea and Vomiting

Nausea is the unpleasant feeling that triggers the desire to vomit, while vomiting involves the contraction of abdominal and diaphragmatic muscles, leading to the expulsion of stomach contents. Common causes of nausea and vomiting in patients with severe illness include stomach flu, bowel obstruction, infections, medications, radiation, chemotherapy, meningitis, brain trauma or tumors, **vestibulitis**, surgery, dehydration, migraines, **morning sickness**, and food poisoning.

CONSTIPATION

Constipation is a bowel dysfunction that occurs when patients experience infrequent or difficult bowel movements. It can result in hard, dry, or rough stools. Other symptoms include abdominal pain and bloating. Various factors, including inadequate fiber intake, insufficient fluid consumption, a sedentary lifestyle, and changes in diet or routine, can contribute to constipation. Medical conditions such as irritable bowel syndrome, hypothyroidism, Parkinson's disease, spinal cord injuries, and intestinal obstructions can contribute to this issue. Additionally, medications like opioids, antihistamines, calcium channel blockers, and iron supplements may also cause constipation. Stress, anxiety, pregnancy, aging, surgery, and travel are other potential causes.

PHARMACOTHERAPY OF PEPTIC ULCER

Medications are the first line of treatment for PUD. However, surgery may be necessary for certain patients with complications. Medications include PPIs, H2-receptor antagonists, and antacids. Examples of PPIs are omeprazole, esomeprazole, lansoprazole, pantoprazole, rabeprazole, and dexlansoprazole.

PROTON PUMP INHIBITORS

PPIs are the first line of treatment for PUD. They effectively reduce stomach acid to promote healing of the ulcer. Examples of PPIs are listed in Table 24.1.

Clinical Indications

Omeprazole treats certain conditions characterized by excessive stomach acid. It addresses dyspepsia, gastric and duodenal ulcers, erosive esophagitis, and GERD. In some cases, it is prescribed in combination with antibiotics (e.g., amoxicillin and clarithromycin) to treat ulcers associated with

infections caused by the *H. pylori* bacterium. Omeprazole is also utilized for managing Zollinger–Ellison syndrome, a disorder in which the stomach produces an excess of acid. This medication is available in several dosage forms, including tablets, capsules, and powders.

Mechanism of Action

Omeprazole reduces the stomach's acid production and irreversibly blocks an enzyme called H+/K+ ATPase, which controls acid production. This enzyme, also known as the proton pump, is found in the parietal cells of the stomach wall. PPIs inhibit meal-stimulated acid production. Dosing PPIs twice daily allows more pumps to be inhibited, leading to more rapid and complete gastric acid suppression.

Adverse Effects

The adverse effects are typically minimal. The most common include headaches, stomach pain, and constipation.

Contraindications

Omeprazole should be avoided in patients with a history of allergy to the medication or in individuals with current and past health conditions. It is contraindicated in patients with renal or liver disease. It must be used cautiously during pregnancy or breastfeeding.

H2-RECEPTOR ANTAGONISTS

H2 receptor antagonist drugs, also known as H2 blockers, reduce stomach acid production. They are available by prescription and over-the-counter. H2 blockers work more quickly than PPIs, providing relief in as little as 15–30 minutes. PPIs may take longer to take effect, but they offer longer-lasting relief, making them ideal for those who suffer from frequent heartburn. They are listed in Table 24.2.

Clinical Indications

H2 receptor blockers are short-term treatments for stomach and duodenal ulcers, GERD, and Zollinger–Ellison disease.

Mechanism of Action

The H2-receptor blockers bind to histamine type 2 receptors on gastric parietal cells, thereby disrupting the pathways of gastric acid production and secretion. They exert minimal or no influence on histamine type 1 receptors, can be inhibited by typical antihistamines that address allergic reactions, and have a negligible effect on gastric acid production.

TABLE 24.1

Proton Pump Inhibitors

Generic Name	Trade Name
Dexlansoprazole	Dexilant, Kapidex
Esomeprazole	Nexium
Lansoprazole	Prevacid
Omeprazole	Prilosec
Pantoprazole	Protonix
Rabeprazole	AcipHex

TABLE 24.2

H2-Receptor Antagonists

Generic Name	Trade Name
Cimetidine	Tagamet
Famotidine	Pepcid
Nizatidine	Axid

Adverse Effects

The adverse effects of H2-receptor blockers include head-ache, dizziness, chest pain, diarrhea or constipation, fatigue, and myalgia. In some patients, hallucinations and confusion may occur.

Contraindications

Histamine H2-receptor antagonists are contraindicated in patients with known allergies to these products. They should also be used cautiously during pregnancy or lacta-tion, in individuals with renal or liver impairments, and in cases of prolonged use.

YOU SHOULD REMEMBER

Zantac was removed from the market in April 2020 in the U.S. due to the potential risk of increased can-cer, including pharyngeal, esophageal, stomach, liver, pancreatic, and colorectal cancers.

Check Your Knowledge

1. What is the first line of medications for treating peptic ulcers?
2. What are the adverse effects of omeprazole?
3. What are some examples of H2-receptor antagonists?

ANTACIDS

Antacids are medications used to treat heartburn and indi-gestion by reducing stomach acid. They neutralize the acid in the stomach by inhibiting pepsin. Antacids contain vari-ous compounds, including calcium, magnesium, and alumi-num. Table 24.3 lists some examples of antacids.

Clinical Indications

Currently, the use of antacids is limited to alleviating mild, intermittent GERD accompanied by heartburn.

TABLE 24.3
Antacids

Generic Name	Trade Name
Aluminum hydroxide	AlternaGEL, others
Bismuth subsalicylate	Pepto-Bismol
Calcium carbonate	Titralac, Tums
Calcium carbonate and magnesium hydroxide	Mylanta Supreme, Rolaids
Magaldrate	Riopan
Magnesium hydroxide	Milk of Magnesia
Magnesium trisilicate and aluminum hydroxide	Gaviscon
Magnesium hydroxide and aluminum hydroxide with simethicone	Mylanta, Maalox Plus
Sodium bicarbonate	Alka-Seltzer, baking soda

Mechanism of Action

Aluminum hydroxide neutralizes gastric acid, inhib-its pepsin, and binds bile acids. It also enhances mucosal defense and suppresses *H. pylori*. Antacids typically con-tain aluminum or magnesium and provide more immediate relief for dyspepsia symptoms. However, they can interact with many drugs, potentially decreasing the absorption of certain medications.

Adverse Effects

The adverse effects of aluminum hydroxide include nausea, vomiting, aluminum intoxication, a chalky taste, osteope-nia, hypomagnesemia, osteomalacia, hypophosphatemia, constipation, fecal impaction, milk–alkali syndrome, and stomach cramps.

Contraindications

Aluminum hydroxide should be avoided in patients with known hypersensitivity to the drug. It is also contraindi-cated in patients with heart failure, renal calculus, renal failure, or those undergoing dialysis. This medication should be avoided in patients with achlorhydria and severe diarrhea.

YOU SHOULD REMEMBER

Antacids containing aluminum salts are considered safe for aspiration prophylaxis in pregnant women and women in labor. Although the information regarding the use of aluminum-containing antacids in breastfeeding females has not been studied, it is known that aluminum is endogenous to breast milk.

PHARMACOTHERAPY OF *HELICOBACTER PYLORI*

H. pylori infections are typically treated with a combination of medications. Antibiotics such as amoxicillin, clarithro-mycin, and metronidazole are frequently used. Additionally, proton pump inhibitors, bismuth subsalicylate, and histamine H-2 blockers are included in the treatment. At least 4 weeks after therapy, repeat testing for *H. pylori* is conducted. If the tests indicate that the infection persists, further treatment with a different combination of antibiotics may be necessary.

PHARMACOTHERAPY OF IBD

Treating IBD aims to reduce the inflammation that triggers symptoms. In ideal cases, this may lead to symptom relief, long-term remission, and a lower risk of complications. IBD treatment typically involves either medications or surgery. Table 24.4 lists pharmacotherapy for IBD.

Sulfasalazine reduces inflammation, pain, and swelling. Adults and children older than six can take it. It is usually

TABLE 24.4

Pharmacotherapy for Inflammatory Bowel Disease

Generic Name	Trade Name
Balsalazide	Colazal, Giazo
Mesalamine	Asacol, Canasa, Lialda, others
Olsalazine	Dipentum
Sulfasalazine	Azulfidine

administered as an oral tablet that must be swallowed whole with a glass of water. The typical dose is two 500 mg tablets twice daily. The medication is also available as an oral liquid or as a rectal suppository.

Clinical Indications

Sulfasalazine is used to treat and prevent acute attacks of mild to moderate ulcerative colitis. It reduces inflammation and other symptoms of the disease. Additionally, sulfasalazine is used to treat Crohn's disease and **rheumatoid arthritis**.

Mechanism of Action

Sulfasalazine's mechanism of action is not fully understood but is thought to involve reducing the activity of the immune system and acting as an anti-inflammatory. This medication may also inhibit the cystine–glutamate antiporter and sepiapterin reductase.

Adverse Effects

Generally, sulfasalazine is well tolerated. Its adverse effects include headache, tinnitus, fatigue, insomnia, mouth sores, loss of appetite, gastric upset, nausea, vomiting, and diarrhea. Severe adverse effects include sun sensitivity, hearing loss, mood changes, hematuria, cold sweats, and blurred vision.

Contraindications

Sulfasalazine is contraindicated for patients with allergies, **porphyria**, heart, lung, kidney, or liver disease, and glucose-6-phosphate dehydrogenase deficiency.

Check Your Knowledge

1. What are the indications for sulfasalazine?
2. What are the adverse effects of sulfasalazine?
3. What are the adverse effects of sulfasalazine?

PHARMACOTHERAPY OF DIARRHEA

In most cases, patients can safely treat acute diarrhea with over-the-counter medicines such as loperamide and bismuth subsalicylate. Occasionally, opiates and other miscellaneous antidiarrheals, anticholinergics, and adsorbents may be indicated, including kaolin, pectin, cholestyramine resin, and *Lactobacillus acidophilus*. Table 24.5 lists various antidiarrheals.

TABLE 24.5

Antidiarrheals

Generic Name	Trade Name
Opioids	
Difenoxin with atropine	Motofen
Diphenoxylate with atropine	Lomotil
Loperamide	Imodium
Opium tincture	Paregoric
Miscellaneous Antidiarrheals	
Bismuth subsalicylate	Kaopectate, Pepto-Bismol
Octreotide	Sandostatin
Telotristat	Xermelo

Clinical Indications

To treat severe diarrhea, a combination of diphenoxylate and atropine, along with other supportive measures such as fluids and electrolyte replacement, is used. Diphenoxylate slows down the movement of the intestines, which helps stop diarrhea.

Mechanism of Action

Diphenoxylate is an opioid agonist that primarily targets presynaptic opioid receptors. It inhibits acetylcholine's release, reducing motility and decreasing colon contractions. The drug also decreases the epithelial secretion of fluid and electrolytes. Atropine acts as a competitive inhibitor of acetylcholine receptors, helping to prevent misuse of the diphenoxylate portion of the formulation.

Adverse Effects

The adverse effects of diphenoxylate with atropine include headaches, dizziness, respiratory depression, flushing, confusion, tachycardia, hyperthermia, urinary retention, dryness of skin and mouth, vomiting, and abdominal pain.

Contraindications

Contraindications of diphenoxylate with atropine include pediatric patients under 6 years of age due to the risks of respiratory and CNS depression, patients with known hypersensitivity to this medication, and obstructive jaundice.

PHARMACOTHERAPY OF NAUSEA AND VOMITING

Many drugs can alleviate nausea and vomiting, including anticholinergics, antihistamines, benzodiazepines, cannabinoids, corticosteroids, neurokinin receptor antagonists, phenothiazines, phenothiazine-like drugs, and serotonin ($5\text{-}HT_3$) receptor antagonists. Table 24.6 lists the various antiemetics.

Clinical Indications

Metoclopramide alleviates symptoms of nausea, vomiting, heartburn, a sense of fullness after meals, and loss of appetite. It is also utilized to treat heartburn in patients with GERD.

TABLE 24.6
Antiemetics

Generic Name	Trade Name
Anticholinergics-Antihistamines	
Dimenhydrinate	Dramamine
Diphenhydramine	Benadryl
Doxylamine and pyridoxine	Diclegis
Hydroxyzine	Atarax, Vistaril
Meclizine	Antivert, Bonine
Scopolamine	Hyoscine, Transderm-Scop
Benzodiazepines	
lorazepam	Ativan
Cannabinoids	
Dronabinol	Marinol, Syndros
Nabilone	Cesamet
Corticosteroids	
Dexamethasone	Decadron
Methylprednisolone	Medrol, Solu-Medrol, others
Neurokinin Receptor Antagonists	
Aprepitant	Emend Capsules
Fosaprepitant	Emend for Injection
Phenothiazine and Phenothiazine-like Drugs	
Metoclopramide	Reglan
Perphenazine	(Generic only)
Prochlorperazine	Compazine, others
Promethazine	Phenergan, others
Trimethobenzamide	Tigan
Serotonin (5-HT$_3$) Receptor Antagonists	
Dolasetron	Anzemet
Granisetron	Sancuso, Sustol
Ondansetron	Zofran, Zuplenz
Palonosetron	Aloxi

Mechanism of Action

Metoclopramide enhances GI motility in three ways: by inhibiting presynaptic and postsynaptic D_2 receptors, stimulating presynaptic excitatory $5\text{-}HT_4$ receptors, and antagonizing presynaptic inhibition of muscarinic receptors.

Adverse Effects

The adverse effects of metoclopramide include anorexia, diarrhea, dark urine, fatigue, drowsiness, depression, chills, fever, seizures, confusion, hypertension, decreased sexual ability, breast pain, muscle pain, and reduced libido.

Contraindications

Metoclopramide should be avoided in patients with known allergies, GI bleeding, kidney or liver issues, obstructions, or perforations. It is also contraindicated in patients with bradycardia or adrenal gland tumors.

PHARMACOTHERAPY OF CONSTIPATION

Constipation is often treated with lifestyle and dietary changes, but it can also be managed with medication or other therapies. Pharmacotherapy for constipation includes laxatives and cathartics. Laxatives soften hard stools or stimulate the movement of the intestines, while cathartics accelerate defecation. Subtypes of these medications include bulk-forming agents, saline, stimulants, stool softeners, herbal products, opioid antagonists, and miscellaneous laxatives. Table 24.7 lists laxatives and cathartics.

Clinical Indications

Psyllium mucilloid is the most commonly used medication for constipation. It is a bulk-forming laxative that absorbs excess water while stimulating regular bowel movements. It helps relieve constipation, irritable bowel syndrome, and hemorrhoids. Patients must drink enough fluids to prevent intestinal obstruction when using psyllium mucilloid.

Mechanism of Action

Psyllium mucilloid seeds are high in dietary fiber. They form a gel-like mass that acts as a mild laxative when mixed with water. This gel moves through the digestive system and softens the stool by increasing its water content.

Adverse Effects

The adverse effects of psyllium mucilloid include dysphasia, abdominal pain, rectal pain, nausea, vomiting, skin rash, itching, and dyspnea.

TABLE 24.7
Laxatives and Cathartics

Generic Name	Trade Name
Bulk-Forming	
Calcium polycarbophil	Equalactin, FiberCon, others
Methylcellulose	Citrucel (generic only)
Psyllium mucilloid	Metamucil, Naturacil, others
Saline (Osmotic)	
Magnesium hydroxide	Milk of Magnesia
Polyethylene glycol	MiraLAX
Sodium biphosphate	Fleet Phospho.Soda
Stimulants	
Bisacodyl	Correctol, Dulcolax, others
Castor oil	Emulsoil, Neoloid
Stool Softeners	
Docusate	Colace, DulcoEase
Herbal Products	
Castor oil	(Generic only)
Senna	Ex-Lax, Senokot, others
Opioid Antagonists	
Alvimopan	Entereg
Methylnaltrexone	Relistor
Naloxegol	Movantik

Contraindications

Psyllium mucilloid should be avoided in patients with bowel obstructions, spasms, or dysphagia. It is also contraindicated in patients with **atresia** or GI tract obstruction. This medication should not be administered to those who have heart disease, hypertension, kidney disease, diabetes mellitus, rectal bleeding, intestinal blockage, or difficulty swallowing.

YOU SHOULD REMEMBER

Psyllium can worsen constipation if not taken appropriately. It is essential to instruct patients who are receiving psyllium ingredients to drink a good amount of fluids to avoid the development of bowel obstruction, especially in the long-term use of such laxatives.

CLINICAL CASE STUDIES

Clinical Case Study 1

A 45-year-old man visits his primary physician due to stomach pain, nausea, burping, bloating, dark blood in his stool, and weight loss. The patient describes burning pain in the epigastric area and chest that radiates toward his back. The pain worsens after taking aspirin or drinking coffee and is relieved by antacids. The patient is diagnosed with PUD.

CRITICAL THINKING QUESTIONS

1. What are the initial lines of medications used to treat peptic ulcers?
2. What are the mechanisms of action for omeprazole?
3. What are the adverse effects of omeprazole?

Clinical Case Study 2

A 32-year-old woman was brought to the emergency department with abdominal pain, bloody diarrhea, fatigue, and weight loss. Her medical history reveals that she has ulcerative colitis. She was given an IV solution and oral medication. After 8 hours, she was discharged and continued to follow a clear liquid diet along with her medication (sulfasalazine).

CRITICAL THINKING QUESTIONS

1. What causes ulcerative colitis?
2. What medications are used to treat IBD?
3. What are sulfasalazine's contraindications?

FURTHER READING

Ananthakrishnan, A.N., Xavier, R.J., and Podolsky, D.K. (2017). *Inflammatory Bowel Diseases — A Clinician's Guide*. Wiley Blackwell.

Bowen, R. (2023). *Inflammatory Bowel Disease — Diagnosis and Therapeutics*. American Medical Publishers.

Cohen, R.D. (2017). *Inflammatory Bowel Disease — Diagnosis and Therapeutics*, 3rd Edition (Clinical Gastroenterology). Humana Press.

Donald, H. (2022). *H. pylori Treatment Manual — The Definitive Step-by-Step Guide on the Best Treatment and Cure for Helicobacter Pylori Infection*. Donald.

Ehret, A., and Lust, J. (2021). *Definite Cure of Chronic Constipation Also Overcoming Constipation Naturally*. Benedict Lust Publications.

Farraye, F.A., and Kane, S.V. (2022). *Mayo Clinic on Crohn's Disease and Ulcerative Colitis — Strategies to Manage IBD and Take Charge of Your Life*. Mayo Clinic Press.

Feldman, M., Friedman, L.S., and Brandt, L.J. (2020). *Sleisenger and Fordtran's Gastrointestinal and Liver Disease — 2 Volume Set — Pathophysiology, Diagnosis, Management*, 11th Edition. Elsevier.

Fleisher, D.R. (2014). *Management of Functional Gastrointestinal Disorders in Children — Biopsychosocial Concepts for Clinical Practice*. Springer.

Hall, I. (2022). *Gastrointestinal Tract Disorders — A Clinician's Guide*. State Academic Press.

Lacy, B.E., Crowell, M.D., and DiBaise, J.K. (2015). *Functional and Motility Disorders of the Gastrointestinal Tract — A Case Study Approach*. Springer.

Morrison, C.M. (2023). *Peptic Ulcer — A Book That Will Give You a Better Understanding of What Peptic Ulcer Is, Its Symptoms, Treatment, Management, Choice of Diet, Myths & Facts About It, and More!* Morrison.

Natalie, R. (2022). *Antacids — The Complete Guide on Everything You Need Know About Antacids*. Natalie.

Parker, B. (2015). *Proton Pump Inhibitors (PPIs) — Prevalence of Use, Effectiveness and Implications for Clinicians* (Pharmacology-Research, Safety Testing and Regulation). Nova Science Publishers, Inc.

Rajapakse, R. (2021). *Inflammatory Bowel Disease — Pathogenesis, Diagnosis and Management* (Clinical Gastroenterology). Humana Press.

Said, H.M. (2018). *Physiology of the Gastrointestinal Tract*, 6th Edition. Academic Press.

Sandberg-Lewis, S. (2017). *Functional Gastroenterology — Assessing and Addressing the Causes of Functional Gastrointestinal Disorders*, 2nd Edition. Sandberg-Lewis.

Targan, S.R., Shanahan, F., and Karp, L.C. (2011). *Inflammatory Bowel Disease — Translating Basic Science into Clinical Practice*. Wiley-Blackwell.

Unigastro Profesor. (2019). *Digestive Diseases 2019 to 2022 Edition*. Editorial Edra.

Wingate Todd, T. (2023). *The Clinical Anatomy of the Gastrointestinal Tract*. Legare Street Press.

25 Drugs Affecting the Skeletal System

LEARNING OUTCOMES

After studying the chapter, readers should be able to

1. Summarize the pathophysiology of rheumatoid arthritis (RA).
2. Describe the primary goals of treatments for RA.
3. Review the contraindications of methotrexate (MTX).
4. List drugs for relieving the pain of RA and osteoarthritis (OA).
5. Describe the primary effects of OA.
6. Explain the clinical indications for adalimumab.
7. Explain the first-line agents used to treat gout.
8. Describe the clinical indications of allopurinol.
9. Discuss the mechanism of action for colchicine.
10. Assess the efficacy and safety of allopurinol.

OVERVIEW

Many disorders can affect the skeletal system, including bone diseases and arthritis. These can weaken bones, leading to fractures or deformities. Common bone and joint diseases include RA, OA, ankylosing spondylitis, osteoporosis, and gout. These disorders impact the bones, joints, and ligaments. RA is a chronic autoimmune disease of the joints, causing inflammation and pain. It occurs more frequently in females than in males. Treatment options include medication and physical therapy. OA is a progressive, degenerative joint disease and is the most common type of arthritis, especially among older individuals. Patients often experience joint pain and stiffness. Risk factors include aging, obesity, injury or surgery, and a family history of OA. Gout is a metabolic disorder characterized by joint inflammation, primarily caused by the accumulation of needle-shaped uric acid crystals in the joints. Gout has no cure; however, it can be managed with medication and self-management strategies.

PHYSIOLOGY OF THE SKELETAL SYSTEM

The skeletal system's functions include providing structural support for the body, enabling movement through muscle attachments, protecting vital organs, storing minerals such as calcium, and producing blood cells within the red bone marrow. The skeletal system comprises bones and cartilage, which are connected by ligaments to form a framework for the rest of the body tissues. There are two parts to the skeleton: the axial and appendicular skeletons. The axial skeleton includes the skull, vertebral column, and ribcage, while the appendicular skeleton consists of the upper and lower limbs, pelvic girdle, and shoulder girdle.

The bones serve as the primary reservoirs for minerals in the body. They house approximately 99% of the body's calcium, 85% of the body's phosphate, and 50% of the body's magnesium. Bones play a crucial role in maintaining mineral homeostasis in the blood. The minerals stored in the bones are released according to the body's demands, and their levels are maintained and regulated by hormones, including parathyroid hormone.

Bone marrow, classified as red or yellow, is present in bones. Figure 25.1 illustrates the structure of a long bone. Red bone marrow produces blood cells, including erythrocytes, leukocytes, and platelets. Yellow bone marrow serves as a potential energy reserve for the body, composed mainly of adipose cells that store triglycerides in the blood.

DISORDERS OF BONES AND JOINTS

Disorders of the bones and joints include OA (degenerative joint disease), RA (inflammatory joint disease), **osteoporosis** (weakened bones), gout (a painful joint disorder caused by uric acid accumulation), **Paget's disease** (bone deformities), **osteomalacia** (soft bones due to vitamin D deficiency), **osteogenesis imperfecta** (brittle bone disease), **scoliosis** (spinal curvature), and fibromyalgia (muscle and joint pain with widespread tenderness).

RA is common, affecting more than 18 million people worldwide. About 70% of these patients are women. OA is another prevalent type of arthritis that impacts millions of people globally. **Gout** is becoming increasingly common around the world; it increases with age and affects men more than women. Bone and joint disorders can lead to pain, stiffness, and impaired mobility in the affected areas. It is estimated that 75 million people have osteoporosis worldwide, including high-risk patients who have already fractured. This chapter will discuss RA, OA, and gout.

RHEUMATOID ARTHRITIS

RA is a progressive autoimmune disorder that leads to joint inflammation, pain, swelling, stiffness, and loss of function. It is the most common form of arthritis. In RA, the immune system attacks its tissues, resulting in significant inflammation in the joints and other organs. Over time, RA progressively damages the joints and **cartilage**. It may also affect the blood vessels, heart, lungs, and skin. Consequently, the clinical symptoms are similar to those of systemic autoimmune disorders, such as **systemic lupus erythematosus**. This disease is more prevalent in women, with about 75% of RA patients

DOI: 10.1201/9781003461913-27

Structure of a Long bone. Humerus. Tube.

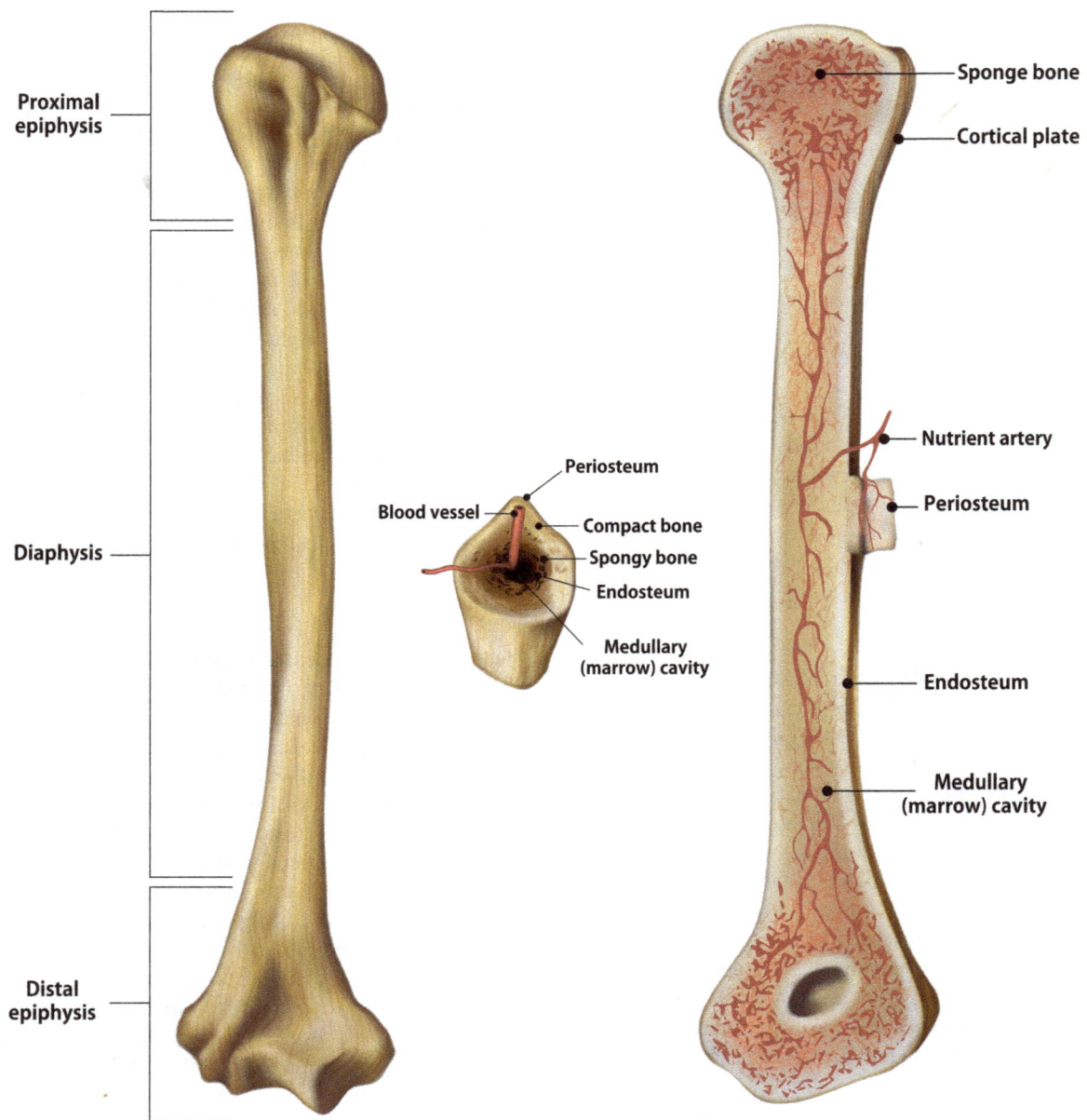

Proximal epiphysis

Diaphysis

Distal epiphysis

Sponge bone

Cortical plate

Nutrient artery

Periosteum

Endosteum

Medullary (marrow) cavity

Periosteum

Blood vessel

Compact bone

Spongy bone

Endosteum

Medullary (marrow) cavity

FIGURE 25.1   Bone structure.

being female. Symptoms typically begin between the ages of 30 and 50 but can occur at any age. RA is a chronic condition, and there is currently no cure; however, many treatments and management strategies are available to alleviate symptoms. Early diagnosis and appropriate treatments can ease symptoms and help prevent joint damage or disability.

RA causes joint deformities, starting in the hands and wrists. These deformities also impact the supporting ligaments, tendons, and muscles, leading to severe pain. Occasionally, additional **synovial joints** are affected. RA is challenging to diagnose in its early stages because the initial signs and symptoms resemble those of many other diseases. Figure 25.2 shows a hand affected by RA.

PHARMACOTHERAPY OF RA

Pharmacotherapy's primary goals are to reduce pain, manage inflammation, and enhance physical abilities. Therapy typically begins with nonsteroidal anti-inflammatory drugs (NSAIDs), which alleviate pain and inflammation. However, aspirin should be avoided for long-term treatment due to adverse effects related to gastrointestinal bleeding and **platelet aggregation**. Acetaminophen can reduce pain and fever, but it does not address inflammation. Corticosteroids are also administered at the lowest effective daily doses.

A disease-modifying antirheumatic drug (DMARD) is typically the first line of treatment for RA. Common

Rheumatoid Arthritis in Hand

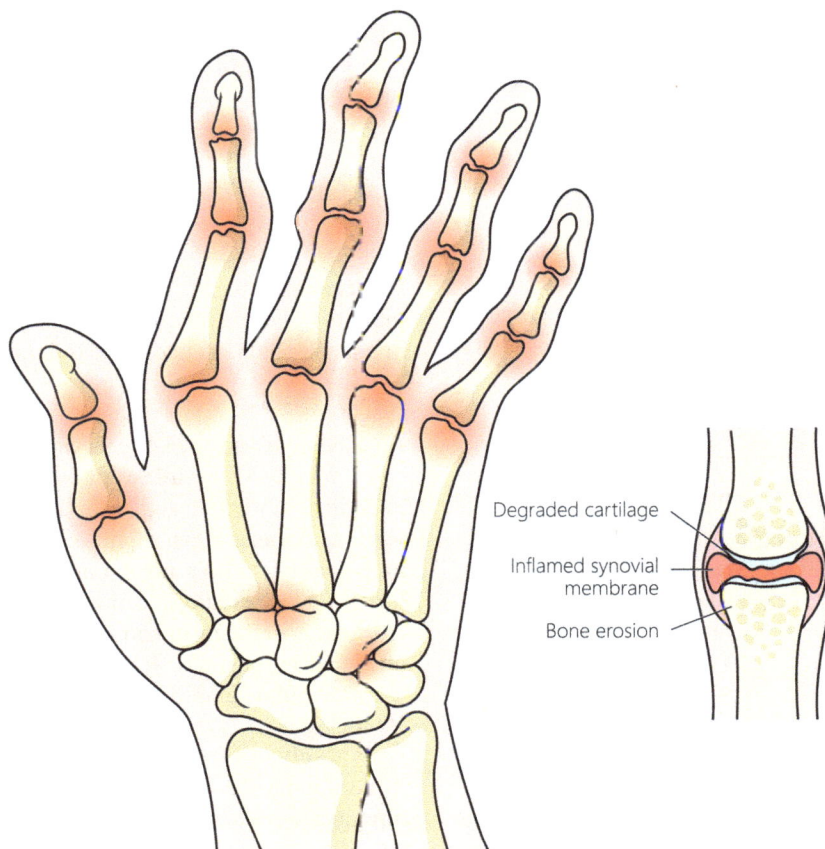

FIGURE 25.2 A hand affected by rheumatoid arthritis.

DMARDs include MTX, leflunomide, hydroxychloroquine, and sulfasalazine. If DMARDs alone do not control RA inflammation, a biological drug such as adalimumab, etanercept, tocilizumab, or abatacept is used.

DMARDs modify immune and inflammatory responses, reduce mortality rates, improve symptoms, and enhance quality of life. They are classified as *nonbiologic, biologic (antitumor necrosis factor* (anti-TNF) or TNF), and biologic (non-anti-TNF). Table 25.1 lists the DMARDs used for RA.

MTX is the first-line therapy for RA. However, hydroxychloroquine and sulfasalazine are also used for initial treatment. Nonbiologic DMARDS such as azathioprine and leflunomide have more adverse effects than the other DMARDs but can be combined with biologic therapies.

Methotrexate

MTX is the most common antimetabolite medication for RA. It reduces the immune system's activity, helping prevent further permanent damage.

Clinical Indications

MTX is indicated for the treatment of moderate to severe active RA in adults, either as a standalone therapy or in combination with biologics. It can also be used to treat patients with JIA. MTX is often used as a first-line treatment for RA because it reduces inflammation and slows disease progression. Additionally, MTX is employed off-label for various cancers and autoimmune disorders, although these uses are secondary to its primary role in RA management. MTX is also effective for patients who have received a donated organ because of its anti-inflammatory and immunomodulatory activity. Additionally, this drug can be combined with anti-TNF medications for treating patients with ulcerative colitis, breast carcinoma, specific lung cancer, epidermal tumors of the head and neck, and ovarian carcinoma.

Mechanism of Action

MTX inhibits dihydrofolate reductase, an enzyme essential for folate metabolism and nucleic acid synthesis. This inhibition disrupts the proliferation of inflammatory cells, thereby reducing inflammation and slowing disease progression in RA.

Adverse Effects

Even a low dose of MTX can cause adverse effects. This agent may cause severe, life-threatening adverse effects. Common adverse effects of MTX include feeling sick,

TABLE 25.1

Disease-Modifying Antirheumatic Drugs (DMARDs) Used for Rheumatoid Arthritis

Generic Name	Trade Name
Nonbiologic DMARDs	
Azathioprine	Azasan, Imuran
Cyclosporine a	Neoral, Sandimmune
Hydroxychloroquine	Plaquenil
Leflunomide	Arava
Methotrexate	Otrexup, Rheumatrex, Trexall
Sulfasalazine	Azulfidine
Biologic DMARDs (Anti-TNF Drugs)	
Adalimumab	Humira
Certolizumab pegol	Cimzia
Etancercept	Enbrel
Golimumab	Simponi
Infliximab	Remicade
Biologic DMARDs (Non-Anti-TNF Drugs)	
Abatacept	Orencia
Anakinra	Kineret
Rituximab	Rituxan
Sarilumab	Kevzara
Tocilizumab	Actemra
tofacitinib	Xeljanz

headaches, fever, mouth ulcers, nausea, vomiting, diarrhea, dyspnea, rashes, liver damage, hair loss, and fatigue.

Contraindications

MTX should be avoided in patients with hypersensitivity reactions to this drug. It is also contraindicated in patients with severe liver disease, renal failure, HIV/AIDS, and blood dyscrasias. MTX has black box warnings for potentially causing severe liver toxicity and bone marrow suppression. This agent is a category X drug and is contraindicated during pregnancy.

YOU SHOULD REMEMBER

Gold therapy is used for RA; gold salts help control the disease and relieve joint pain and stiffness. It is considered toxic and has many side effects, but it can be effective for some patients.

Check Your Knowledge

1. What are the primary goals of treatment for RA?
2. What is the first treatment for RA?
3. What are the contraindications of MTX?

ADALIMUMAB

Adalimumab is a high-affinity recombinant monoclonal antibody that inhibits TNF-alpha (TNF-α). It treats various autoimmune conditions, such as RA, ankylosing spondylitis, psoriasis, psoriatic arthritis, Crohn's disease, and ulcerative colitis.

Clinical Indications

Adalimumab is indicated as monotherapy or in combination with MTX for moderate to severe active RA in adults. It is also used for JIA, **psoriatic arthritis**, and **ankylosing spondylitis**.

Mechanism of Action

Adalimumab blocks TNF's inflammatory activity and inhibits inflammation. High levels of TNF are often found in patients with RA and other inflammatory disorders.

Adverse Effects

Adalimumab's adverse effects usually include injection-site pain, increased **creatine phosphate**, upper respiratory infection, headache, and rash. This drug has a black box warning because it can suppress the immune system, potentially leading to infections. Latent infections such as hepatitis B and tuberculosis can become reactivated. The TNF blockers have been linked to malignancies such as leukemias and lymphomas. Rarely, fatal cases of hepatosplenic T-cell lymphoma have occurred.

Contraindications

Adalimumab contraindications are rare. However, due to its immunosuppressive effects, patients must be monitored for infections or malignancies.

Check Your Knowledge

1. What is the drug category of adalimumab?
2. What are the clinical indications of adalimumab?
3. What are the adverse effects of adalimumab?

OSTEOARTHRITIS

OA is a degenerative joint disease that causes inflammation and pain. It affects the cartilage at the ends of bones. OA is the most common form of arthritis, affecting millions worldwide. It occurs when the cartilage cushions wear down over time. The disease often affects a few joints but can be generalized. In less than 10% of cases, OA can be seen in younger patients who have some risk factors, such as joint deformity, joint injury, **hemochromatosis**, diabetes mellitus, **ochronosis**, and significant obesity. The knees and hands are more commonly affected in females, and the hips in males. Signs and symptoms of OA include loss of flexibility, stiffness, pain, tenderness, bone spurs, and joint edema.

OA primarily affects the knees, hips, lumbar and cervical vertebrae, **interphalangeal joints**, and first **tarsometatarsal joints**. **Heberden nodes**, manifestations of **osteophytes** at the distal interphalangeal joints, can occur. The wrists, elbows, and shoulders are often spared.

PHARMACOTHERAPY OF OA

OA cannot be reversed, but treatments can reduce pain and help patients move better. OA treatment depends on the severity of the condition, age, general health, and symptoms. Treatment aims to alleviate pain and stiffness, as well as enhance joint mobility. Treatments include medications, lifestyle changes, surgery, and physical therapy. Patients may benefit from medication for mild symptoms. Acetaminophen has been shown to help some patients with OA who have mild to moderate pain. However, taking more than the recommended dose can cause liver damage. Systemic NSAIDs are the most common medications for relieving pain and inflammation in OA, and NSAID creams are also used as topical treatments. Corticosteroid injections into the joint might alleviate pain for a few weeks.

Opioids, like codeine, can relieve severe pain but should only be used when all other treatment options have failed. Duloxetine is used for patients with chronic pain. Nonpharmacologic measures to treat OA include weight loss, exercise, braces, and joint replacement surgery. Physical activity can strengthen the muscles surrounding the joint and enhance overall fitness. Transcutaneous **electrical nerve stimulation** uses a low-voltage electrical current to relieve pain. This device relieves short-term pain for some patients with knee and hip OA.

YOU SHOULD REMEMBER

OA is caused by wear and tear on joint cartilage. At the same time, RA is an autoimmune disease in which the body's immune system attacks the joint lining, resulting in inflammation and pain in multiple joints simultaneously.

GOUT

Gout is a painful form of arthritis caused by **hyperuricemia**. It is a painful form of arthritis characterized by sharp crystals that can form in the joints (usually the big toe) but can also affect other joints like the knees, ankles, hands, and wrists. Sudden pain and edema in the toe are the most common signs and symptoms. Increased uric acid biosynthesis is due to unknown causes. Patients may produce excessive uric acid due to an enzyme deficiency or decreased excretion.

Hyperuricemia may appear after 20–30 years and is more common in males. Gout usually develops in middle age, but females typically develop it after menopause. Excessive alcohol use, obesity, and diuretics like thiazides can lead to toxicity and cause gout. **Tophus** formation is the primary manifestation of advanced gout (see Figure 25.3). It is a prerequisite for monosodium urate crystal formation. In most patients, urate underexcretion occurs via renal and GI mechanisms.

GOUT

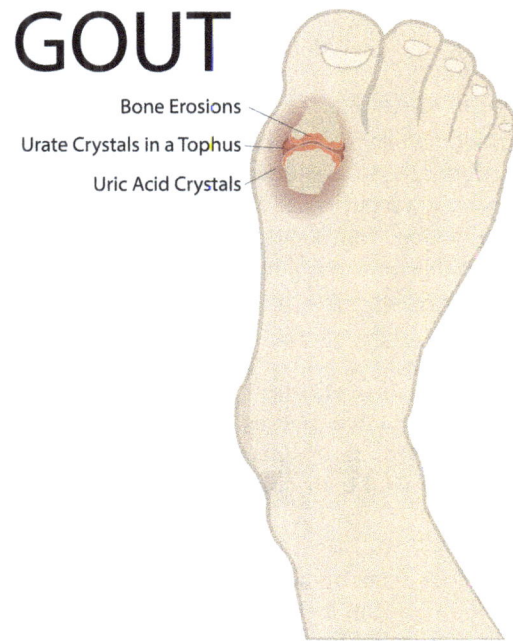

- Bone Erosions
- Urate Crystals in a Tophus
- Uric Acid Crystals

FIGURE 25.3   A tophus indicating advanced gout.

TABLE 25.2

Pharmacotherapies Used for Gout

Generic Name	Trade Name
Allopurinol	Aloprim, Zyloprim
Colchicine	Colcrys
Febuxostat	Uloric
Lesinurad	Zurampic
Pegloticase	Krystexxa
Probenecid	Probalan

PHARMACOTHERAPY OF GOUT

Acute gouty arthritis is treated with NSAIDs, corticosteroids, and colchicine. Preventing gout attacks and treating chronic gout requires the use of a xanthine oxidase inhibitor, a uricosuric agent, or a combination of these agents, along with lifestyle modifications. Allopurinol is a first-line agent for gout prevention in patients requiring urate-lowering therapy. Pegloticase, a specialty drug, should be reserved for patients whose treatment has failed. Table 25.2 lists the various pharmacotherapies used for gout.

ALLOPURINOL

Allopurinol, a xanthine oxidase inhibitor, is considered one of the most effective urate-lowering drugs and is frequently used to prevent gout attacks.

Clinical Indications

Allopurinol prevents gout by lowering high blood uric acid levels, regardless of the cause. It is also used to avoid or lower excess uric acid levels caused by cancer medicines or in patients with kidney stones.

Mechanism of Action

Allopurinol inhibits the enzyme xanthine oxidase, which is responsible for the final step in the production of uric acid. It effectively lowers uric acid levels in the body by preventing its formation from xanthine, essentially acting as an inhibitor of this enzyme. This makes it helpful in treating conditions like gout, where high uric acid levels are present.

Adverse Effects

The adverse effects of allopurinol include headache, drowsiness, malaise, **vertigo**, nausea, vomiting, diarrhea, cataracts, **retinopathy**, thrombocytopenia, **Stevens–Johnson syndrome**, and fatal **toxic epidermal necrolysis**.

Contraindications

Allopurinol is contraindicated in patients with known allergies to this agent. It should also be avoided in patients with liver and renal failure and a history of peptic ulcers. Allopurinol is contraindicated for those who have been diagnosed with bone marrow depression or are pregnant.

COLCHICINE

Colchicine is an oral, plant-based prescription alkaloid derived from the dried seeds of the autumn crocus or meadow saffron that may be used to treat and prevent gout flares in adults. It is an alkaloid used to treat acute gout flares due to its potent anti-inflammatory and analgesic effects. It is also a natural product for **Behcet's disease** and **familial Mediterranean fever** (FMF). Due to its narrow therapeutic index and links to deaths in the USA, injectable colchicine was removed from the market, and now only the safe oral form is available.

Clinical Indications

Colchicine prevents or treats gout attacks caused by too much uric acid in the blood. It can also treat FMF. To relieve symptoms, it decreases inflammation and reduces the buildup of uric acid in the joints. The brand name of this medication is Colcrys®. This medication can reduce the risks of myocardial infarction and stroke.

Mechanism of Action

Colchicine has primarily anti-inflammatory properties. It inhibits microtubule synthesis and prevents neutrophil activation, degranulation, and migration, mediating some gout symptoms. The mechanism of action of colchicine in the treatment of FMF is unknown. It may interfere with the intracellular assembly of the inflammasome complex present in neutrophils and monocytes, which mediates the activation of interleukin-1-beta.

Adverse Effects

The most significant adverse effects of colchicine include abdominal pain, anorexia, nausea, vomiting, and diarrhea.

The agent may also cause aplastic anemia, agranulocytosis, leukopenia, or thrombocytopenia. Other adverse effects are mental confusion, **azotemia**, liver impairment, oliguria, proteinuria, and hematuria.

Contraindications

Colchicine is contraindicated in patients with a known hypersensitivity to the drug and in patients with kidney or hepatic impairment. It should be used with caution in pregnant people and those with GI, hepatic, or cardiac impairment. Patients with blood dyscrasias must not receive colchicine due to the risk of myelosuppression, leukopenia, thrombocytopenia, and aplastic anemia. The drug should also be used with caution in older adults, debilitated patients, and those with early signs of GI, kidney, liver, or cardiac disease.

YOU SHOULD REMEMBER

Since gout is associated with obesity and weight gain, weight loss is one good way to manage and control it. However, some methods are better than others. For instance, a keto diet is not recommended for those with gout, as ketosis can elevate uric acid levels in the body.

Check Your Knowledge

1. What is the pathophysiology of gout?
2. What are the treatments for acute gouty arthritis?
3. What are the contraindications of colchicine?
4. What are the adverse effects of allopurinol and colchicine?

CLINICAL CASE STUDIES

Clinical Case Study 1

A 68-year-old man has had RA for the past 3 years. He complains of increased joint pain and inflammation in his hands, and he is unable to perform his daily activities. The patient has been taking naproxen for symptomatic relief and is currently on MTX. He was evaluated for worsening RA and started on adalimumab.

CRITICAL THINKING QUESTIONS

1. What are the clinical indications of adalimumab?
2. What are the adverse effects of adalimumab?
3. What are the contraindications of adalimumab?

Clinical Case Study 2

A 42-year-old woman visits a family practitioner complaining of pain, tenderness, edema, and stiffness in her left knee.

She is overweight, and her mother has been diagnosed with OA. An X-ray of her knee shows narrowing of the joint space, mild effusion, and osteophyte formations. The patient received a diagnosis of OA and was treated with corticosteroid injections alongside oral and topical NSAIDs.

CRITICAL THINKING QUESTIONS

1. What is the pathophysiology of OA?
2. What are some risk factors for OA?
3. What are the treatments for cases of OA?

FURTHER READING

Atkinson, K. (2020). *Rheumatoid Arthritis, Juvenile Idiopathic Arthritis, Ankylosing Spondylitis, Psoriatic Arthritis, and Gout – Targeted Biological Treatments*, Book 3. Atkinson.

Baring Garrod, A. (2023). *A Treatise on Gout and Rheumatic Gout* (Rheumatoid Arthritis). Legare Street Press.

Bozeman, W. (2024). *Stem Cells for Osteoarthritis*. RELATED Media.

Hook, M. (2024). *Rheumatoid Arthritis Handbook — A Complete Guide Providing Insights into Understanding, Preventing, Managing, and Treating RA*. Hook.

Jenkins, J. (2011). *Rheumatoid Arthritis*, 3rd Edition. How To Books.

Koonce, R.C. (2019). *Non-Surgical Treatment Options for Hip Osteoarthritis — An Informative Guide for Patients*. OrthoSkool.

Newcombe, D.S., and Robinson, D.R. (2013). *Gout — Basic Science and Clinical Practice*. Springer.

O'Neill Young, K. (2017). *Rheumatoid Arthritis Unmasked — 10 Dangers of RA*. RA Patient Insights, LLC.

Peterson, L.S. (2020). *Mayo Clinic Guide to Arthritis — Managing Joint Pain for an Active Life*. Mayo Clinic Press.

Pizzo, G.M. (2023). *Rheumatoid Arthritis — A Comprehensive Guide to Understanding, Preventing, Managing, and Treating RA*. Pizzo.

Poole, E.K. (2022). *A Guide on Osteoarthritis — How to Regain Your Life, What You Need to Know About the Treatment and Management of Osteoarthritis*. Poole.

Shlotzhauer, T.L. (2014). *Living with Rheumatoid Arthritis*. Johns Hopkins Press.

Terkeltaub, R., and Edwards, N.L. (2016). *Gout — Diagnosis and Management of Gouty Arthritis and Hyperuricemia*, 4th Edition. Professional Communications, Inc.

Westlake, S. (2016). *Hip Osteoarthritis CAN be Cured — Treating OA with Physical Therapy — A Self Help Guide*. CreateSpace Independent Publishing Platform.

Wilde, P. (2016). *The Physiology of Gout, Rheumatism and Arthritis as a Guide to Accurate Diagnosis and Efficient Treatment*. Forgotten Books.

26 Muscle Relaxants

LEARNING OUTCOMES

After studying this chapter, readers should be able to

1. Describe the characteristics of skeletal muscles.
2. Explain the risk factors for muscle spasms.
3. List examples of medications that can cause muscle spasms.
4. Discuss diseases that cause muscle spasticity.
5. Review the various types of muscle relaxants.
6. Summarize the centrally acting muscle relaxants.
7. Identify the clinical indications of baclofen.
8. Describe the indications of dantrolene.
9. Review the contraindications of dantrolene.
10. Explain the contraindications of botulinum toxins (BTXs).

OVERVIEW

The skeletal muscle system, including muscles, bones, tendons, and ligaments, is vital to the musculoskeletal system. It is responsible for movement, body position, stabilizing joints, and generating heat. The skeletal muscles are voluntary, and the central nervous system controls them. A whole skeletal muscle is considered an organ of the muscular system. Each organ or muscle consists of skeletal muscle tissue, connective tissue, nerve tissue, and blood or vascular tissue. Skeletal muscles vary in size, shape, and arrangement of fibers. There are more than 600 muscles in the body. Skeletal muscle disorders are a group of conditions that can affect the skeletal muscles, for example, muscle spasms, muscular dystrophy, fibromyalgia, myasthenia gravis, and amyotrophic lateral sclerosis. Approximately 1.71 billion people have musculoskeletal conditions worldwide that can cause disability. Low back pain is the single leading cause of disability. Muscle disorders may cause weakness or paralysis in the presence of an intact nervous system. Muscle spasms commonly occur, and muscle relaxants are frequently prescribed.

SKELETAL MUSCLE SYSTEM

Skeletal muscles comprise 40% of body weight. There are more than 600 muscles in the body. They attach to the bones for various movements and functions. Skeletal muscles are voluntary, meaning the central nervous system controls them. They produce movement, maintain posture, stabilize joints, and generate heat. Therefore, they perform essential functions in the body. Transmission of impulses from motor nerves to muscle cells occurs across neuromuscular junctions (see Figure 26.1). At this point of contact between a motor neuron and a muscle fiber, the nervous system's electrical impulses are converted into muscle action.

Musculoskeletal system diseases affect the body's bones, muscles, joints, tendons, ligaments, and cartilage. This chapter will discuss only muscle spasms and their pharmacotherapy. The body's bones, joints, tendons, ligaments, and cartilage were discussed in **Chapter 25**.

SKELETAL MUSCLE DISORDERS

Many muscular diseases include muscle spasms, amyotrophic lateral sclerosis, **myasthenia gravis**, myositis, **muscular dystrophy**, neuropathy, and **rhabdomyolysis**. Approximately 1.72 billion people worldwide suffer from musculoskeletal conditions, with low back pain being the primary contributor, affecting 570 million individuals globally. About 50%–60% of adults in the U.S. experience muscle cramps, with the prevalence increasing with age. This chapter will focus on muscle spasms and the pharmacotherapy of muscle relaxants.

MUSCLE SPASMS

Muscle spasms or cramps cause painful contractions. A muscle cramp is a sudden, unexpected contraction of one or more muscles. Generally, the cramp can last from seconds to a few minutes due to idiopathic or known causes in healthy subjects or the presence of diseases. They may be *tonic* or *clonic*. A *tonic* spasm is a single, prolonged contraction, while clonic spasms are multiple, rapidly repeated contractions.

Muscle cramps typically occur in the legs, particularly in the calf muscles, and can last from a few seconds to several minutes. However, the most common type of spasms and pain is crippling back spasms, which can occur due to injuries, overuse, poor posture, **degenerative disc disease**, **herniated disc**, dehydration, stress, and anxiety.

Severe sprains or strains are a common cause of debilitating back spasms. Repetitive movements or heavy lifting can also cause them. Sitting or standing with improper form can contribute to back spasms. A bulging or herniated disc can irritate a spinal nerve root, causing inflammation and spasms. Exercising or engaging in physical activities in a hot environment can also cause muscle spasms.

After the cramp eases, the area might be sore for hours or days. Muscle spasms can be **idiopathic** or result from muscle fatigue, dehydration, or electrolyte depletion. The spasms can occur in people who have had a stroke, **cerebral palsy**, multiple sclerosis, and **amyotrophic lateral sclerosis**. The risk factors for muscle spasms include age, obesity, pregnancy, and extreme sweating. Drugs that may cause muscle spasms after several months of taking them include the following:

DOI: 10.1201/9781003461913-28

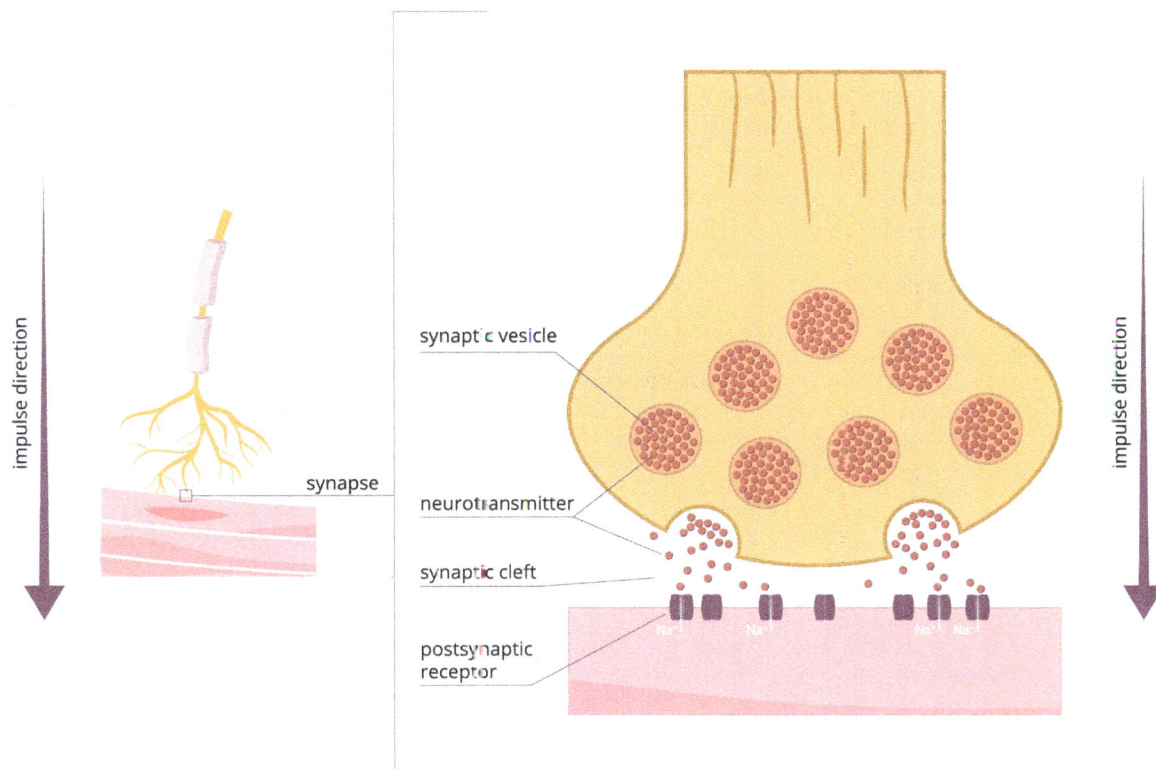

Neuromuscular junction

FIGURE 26.1 Neuromuscular junctions.

- **Antihyperlipidemics**: statins, such as rosuvastatin (Crestor), atorvastatin (Lipitor), and simvastatin (Zocor); excessive niacin (Vitamin B3)
- **Antineoplastics**: cisplatin (Platinol), imatinib (Gleevec), vincristine (Oncovin)
- **Calcium channel blockers**: nifedipine (Procardia)
- **Antifungals**: fluconazole (Diflucan), ketoconazole (Nizoral)
- **Diuretics**: furosemide (Lasix), hydrochlorothiazide (Microzide)
- **Proton pump inhibitors**: esomeprazole (Nexium), omeprazole (Prilosec)
- **Cholinesterase inhibitors**: donepezil (Aricept)
- **Adrenergic agonists**: albuterol (Proventil), terbutaline (Brethine)
- **Immunosuppressants**: leflunomide (Arava), corticosteroids such as prednisone (Deltasone)

Statins are the most common drugs that can cause muscle spasms. They may lead to rhabdomyolysis, resulting in extreme disability. Most muscle spasms and strains respond well to rest, physical therapy, and the short-term use of non-steroidal anti-inflammatory drugs (NSAIDs) or non-opioid analgesics. Additionally, muscle relaxants, massage, and stretching can be beneficial. Muscle spasms range from mild, uncomfortable twitches to intense, severe pain. However, steps can be taken to prevent or treat them.

YOU SHOULD REMEMBER

A muscle cramp can occur after overworking a muscle, straining it, losing body fluids through sweat, or simply maintaining a position for an extended period. Often, however, the exact cause remains unknown.

Check Your Knowledge

1. What are the characteristics of tonic and clonic spasms?
2. What are the risk factors for muscle spasms?
3. What medications can cause muscle spasms?

SKELETAL MUSCLE RELAXANTS

Skeletal muscle relaxants decrease muscle tension and spasms while preventing muscle movement during surgery. They also alleviate painful muscle spasms and stiffness associated with acute musculoskeletal conditions, such as spasms, sprains, strains, and injuries. Skeletal muscle relaxants are frequently used alongside rest and physical therapy, especially when the spasms are linked to conditions like cerebral palsy, multiple sclerosis, or spinal cord injuries.

There are several types of muscle relaxers, each working in different ways. Nonpharmacological treatments include immobilization, massage, application of heat or cold, and hydrotherapy. Local anesthesia may also influence the relaxation of specific muscle groups. Many conditions causing spasticity may require long-term therapy with muscle relaxants. Muscle relaxants are categorized into centrally *acting* and *directly acting* skeletal muscle relaxants.

CENTRALLY ACTING MUSCLE RELAXANTS

Centrally acting muscle relaxants exert CNS effects that alleviate muscle spasms and may induce sedation. They can be categorized into spasmolytic and antispasmodic medications. These agents may be combined with analgesics and anti-inflammatory drugs. A list of centrally acting muscle relaxants is provided in Table 26.1.

Clinical Indications

Baclofen is used to help relax specific muscles in the body. It relieves muscle spasms, cramping, and tightness caused by multiple sclerosis or specific spine injuries.

Mechanism of Action

Baclofen acts as an agonist at the β subunit of gamma-aminobutyric acid. It is thought to decrease the release of excitatory neurotransmitters in presynaptic neurons and enhance inhibitory neuronal signals in postsynaptic neurons, resulting in relief from spasticity.

TABLE 26.1
Centrally Acting Muscle Relaxants

Generic Drugs	Trade Name
Baclofen	Lioresal
Carisoprodol	Soma
Chlorzoxazone	Lorzone
Cyclobenzaprine	Amrix
Deutetrabenazine	Austedo
Metaxalone	Skelaxin
Methocarbamol	Robaxin
Orphenadrine	Norflex
Tetrabenazine	Xenazine
Tizanidine	Zanaflex

Adverse Effects

The most common adverse effects associated with baclofen administration include nausea, hypotension, confusion, vertigo, muscle fatigue, tremor, insomnia, urinary retention, and impotence. Abrupt discontinuation of oral baclofen therapy may cause seizures and hallucinations. Gradual dose reduction is recommended to prevent withdrawal symptoms.

Contraindications

Baclofen should be avoided in patients who have a hypersensitivity to the drug. It is contraindicated for patients with a history of stroke, Parkinson's disease, or skeletal muscle spasms related to rheumatoid disorders.

Check Your Knowledge

1. What causes skeletal muscle spasms?
2. What are the mechanisms by which centrally acting muscle relaxants operate?
3. What are the contraindications of baclofen?

YOU SHOULD REMEMBER

The FDA requires a boxed warning for intrathecal baclofen. This drug may cause a hypermetabolic state resulting in hyperpyrexia, impaired mental status, muscle rigidity, and severe rebound spasticity, which could potentially lead to rhabdomyolysis and multiorgan system failure.

DIRECT-ACTING MUSCLE RELAXANTS

Direct-acting muscle relaxants prevent the release of calcium ions from muscle cells, thereby relieving muscle spasticity. They include dantrolene (Dantrium) and various BTXs. Dantrolene acts directly on skeletal muscle as a calcium-release blocker within skeletal muscle cells. BTXs affect nerves and cause muscle paralysis. The skeletal muscle relaxants include abobotulinumtoxinA (Dysport), incobotulinumtoxinA (Xeomin), onabotulinumtoxinA (Botox Cosmetic), and rimabotulinumtoxinB (Myobloc). They are the most poisonous substances known, requiring extreme caution.

Clinical Indications

Dantrolene is used to relieve muscle spasms and stiffness, which may be caused by multiple sclerosis, cerebral palsy, and stroke. It is also used to treat and prevent malignant hyperthermia, which causes a life-threatening reaction to certain anesthetic drugs. BTX (Botox) is used to manage and treat both therapeutic and cosmetic purposes. It is used to treat spastic disorders, **hemifacial spasms**, chronic migraine, cervical dystonia, **strabismus**, **blepharospasm**, and detrusor hyperactivity.

Mechanism of Action

Dantrolene directly relaxes spastic muscles via interference with calcium ion release from the **sarcoplasmic reticulum**. This action decreases muscle force and relaxes the muscle.

BTX inhibits the release of acetylcholine at the neuromuscular junction, paralyzing muscles. This process is reversible, so the effects of BTX are temporary.

Adverse Effects

Dantrolene's adverse effects include seizures, allergic reactions, dyspnea, decreased inspiratory capacity, dizziness, muscle fatigue, nausea, and diarrhea. Its oral form contains a black box warning about the potential for hepatotoxicity, including overt hepatitis. BTX's adverse effects are generally moderate and self-limited. However, headache, infection, **ectropion**, and xerophthalmia may occur.

Contraindications

Oral dantrolene is contraindicated in patients with liver cirrhosis, nonalcoholic steatohepatitis, and hepatitis B or C infections. However, no contraindications exist for using IV dantrolene to treat malignant hyperthermia. BTX injections are generally safe, but specific contraindications include an allergy to BTX. It should be avoided in myasthenia gravis, neuropathies, psychological conditions, during pregnancy, or lactation.

Check Your Knowledge

1. What are the clinical indications for dantrolene?
2. What is dantrolene's mechanism of action?
3. What are the adverse effects of dantrolene?
4. What are the indications for BTX?

YOU SHOULD REMEMBER

Malignant hyperthermia is a severe reaction to specific anesthetic drugs. This reaction typically includes dangerously high fever, muscle spasms, and tachycardia. Without prompt treatment, complications can be fatal.

NONSTEROIDAL ANTI-INFLAMMATORY DRUGS

Nonsteroidal anti-inflammatory drugs (NSAIDs), such as aspirin, ibuprofen, and naproxen, can relieve muscle pain and reduce inflammation. They are usually used for minor to moderate pain caused by muscle overexertion. NSAIDs can also relieve nerve pressure and enhance mobility. For a more detailed discussion of NSAIDs, see **Chapter 27**.

CLINICAL CASE STUDIES

Clinical Case Study 1

A 53-year-old office worker was brought to the emergency department with severe lower back pain that worsens with prolonged sitting, bending, and lifting. The pain radiates down his left leg. His medical history revealed a previous back injury from a lifting incident 2 years ago, and his physical examination indicated decreased lumbar range of motion, tenderness in specific areas of the lower back, and positive neurological signs suggesting nerve root irritation. An X-ray shows no disc herniation or bone fractures. The patient received a prescription for baclofen to be taken three times a day for 10 days and was discharged.

CRITICAL THINKING QUESTIONS

1. What are the indications for baclofen?
2. What are the adverse effects of baclofen?
3. What are the contraindications of baclofen?

Clinical Case Study 2

A 42-year-old woman with multiple sclerosis had been experiencing muscle spasticity and fatigue for the past 3 years. Her signs and symptoms had progressed significantly in recent months, leading to increased difficulty walking. The patient began a course of gait training using a walker, physiotherapy, and medication with dantrolene.

CRITICAL THINKING QUESTIONS

1. What are the indications for dantrolene?
2. What are the adverse effects of dantrolene?
3. What are the contraindications for dantrolene?

FURTHER READING

Bakheit, M. (2007). *Botulinum Toxin Treatment of Muscle Spasticity*. AuthorHouse.

Barnes, M.P., and Johnson, G.R. (2008). *Upper Motor Neurone Syndrome and Spasticity: Clinical Management and Neurophysiology*, 2nd Edition. Cambridge University Press.

Bavikatte, G., Mackarel, D., Shippen, C., Roberts, R., Mahendran, S., Mackarel, S., and Subramanya, S. (2018). *Spasticity Early & Ongoing Management: Common Patterns, Botulinum Toxin Treatment, Ultrasound Localization, Post-Treatment Exercises*. The Walton Centre NHS Foundation.

Bedi, J., Kumar Malhotra, K., and Mehta, N. (2017). *Effect of an Adjuvant on the Properties of Muscle Relaxant*. Lap Lambert Academic Publishing.

Brashear, A. (2015). *Spasticity: Diagnosis and Management*, 2nd Edition. DemosMedical.

Buvanendran, A., Diwan, S., and Deer, T. (2011). *Intrathecal Drug Delivery for Pain and Spasticity: Interventional and Neuromodulatory Techniques for Pain Management*. Elsevier.

Foldes, F.F., and Adriani, J. (2013). *Muscle Relaxants in Anesthesiology: American Lecture Series*, Number 294. Literary Licensing, LLC.

Ghai, A., and Aggarwal, R. (2016). *Ultrasound Guidance in Obturator Nerve Block in Lower Limb Spasticity*. Lap Lambert Academic Publishing.

Jann, M.W., Penzak, S.R., and Cohen, L.J. (2016). *Applied Clinical Pharmacokinetics and Pharmacodynamics of Psychopharmacological Agents*. Adis.

Jil, R.A. (2020). *Sciatica Treatments: Chiropractic Care, Antibiotics, Corticosteroids, Analgesics, Epidural Steroid Injections, Muscle Relaxants, Physical Therapy, Surgery, Massage, Other Therapies and Options*. CreateSpace Independent Publishing Platform.

Pandyan, A.D., Hermens, H.J., and Conway, B.A. (2018). *Neurological Rehabilitation: Spatisticity and Contractures in Clinical Practice and Research* (Rehabilitation Science in Practice Series). CRC Press.

Pottenger, F.M. (2012). *Muscle Spasm and Degeneration in Intrathoracic Inflammations: Their Importance as Diagnostic Aids and Their Influence in Producing and Altering the Well Established Physical Signs, also a Consideration of Their Part in the Causation of Changes in the Body*. Ulan Press.

Rosales, R.L., and Dressler, D. (2016). *Botulinum Toxin Therapy Manual for Dystonia and Spasticity*. InTechOpen.

Sindou, M., Georgoulis, G., and Mertens, P. (2014). *Neurosurgery for Spasticity: A Practical Guide for Treating Children and Adults*. Springer.

Stevenson, V.L., and Jarrett, L. (2018). *Spasticity Management: A Practical Multidisciplinary Guide*, 2nd Edition. CRC Press.

Winchell Magoun, H., Rhines, R., and Pitts, R.F. (2013). *Spasticity: The Stretch Reflex and Extrapyramidal Systems*. Literary Licensing, LLC.

Part III

Drugs Affecting the Cardiovascular System

27 Acetaminophen and NSAIDS

LEARNING OUTCOMES

After studying the chapter, readers should be able to

1. Compare the differences between acetaminophen and NSAIDs.
2. Review the mechanism of action of acetaminophen.
3. Describe the adverse effects of acetaminophen.
4. Explain the contraindications of acetaminophen.
5. Identify the common adverse effects of aspirin.
6. Summarize the contraindications of ibuprofen.
7. Review the mechanism of action of celecoxib.
8. Discuss the nonselective COX inhibitors.

OVERVIEW

Acetaminophen and nonsteroidal anti-inflammatory drugs (NSAIDs) are the preferred analgesics for mild to moderate pain and are typically used before prescribing opioid analgesics. Acetaminophen is both an analgesic and an antipyretic. It is also known as *N*-acetyl-para-aminophenol, a nonopioid analgesic and antipyretic agent used to treat pain and fever. NSAIDs are commonly utilized to manage mild and moderate acute and chronic pain. Individuals who take NSAIDs (excluding aspirin), such as ibuprofen, may have an increased risk of experiencing a heart attack or stroke compared to those who do not use them. These events may occur unexpectedly and can result in death. The risk may be heightened for patients who use NSAIDs for an extended period or at higher doses. NSAIDs like ibuprofen should be avoided in patients who have recently had a heart attack, hypertension, hypercholesterolemia, or stroke. While NSAIDs can be used as monotherapy, their efficacy is greater in combination with opioids for pain management. One of the primary disadvantages of NSAIDs is the risk of gastrointestinal side effects, particularly GI bleeding. This complication can occur more significantly in patients with ulcerative disease, those on higher doses, the elderly, and with concomitant use of corticosteroids, anticoagulants, or other NSAIDs.

ACETAMINOPHEN

Acetaminophen is an antipyretic analgesic used to treat mild to moderate pain and reduce fever. It is the first-line treatment for acute pain and fever. Most cases of liver injury are associated with using acetaminophen at higher doses than the recommended maximum daily limits.

Clinical Indications

Acetaminophen is one of the most widely used over-the-counter analgesics and antipyretics. It is utilized to treat pain and fever. Effective pain management is critical. Acetaminophen can be used for mild to moderate pain or in conjunction with an opioid analgesic for severe pain. Additionally, acetaminophen can treat common colds and the flu. Injectable acetaminophen is available for managing mild to moderate pain in patients over 2 years old.

Mechanism of Action

Acetaminophen's mechanism of action remains unclear but is believed to involve activating descending serotonergic inhibitory pathways in the CNS and other nociceptive systems.

Adverse Effects

Acetaminophen's common adverse effects include hypersensitivity reactions, skin rash, nephrotoxicity, elevated **blood urea nitrogen** and **creatinine levels**, anemia, leukopenia, neutropenia, and pancytopenia. It may also lead to **hyperammonemia**, hyperchloremia, hyperuricemia, increased serum glucose, and elevated bilirubin and alkaline phosphatase levels. Higher doses of this agent can cause impairment and liver failure. Other adverse effects include headache, nausea, vomiting, constipation, **pruritus**, and abdominal pain.

Contraindications

Contraindications for acetaminophen include hypersensitivity to the drug and severe liver impairment. Acetaminophen should be avoided during alcohol consumption, as its use can be associated with liver failure, which may occasionally necessitate **liver transplants** or result in fatalities.

YOU SHOULD REMEMBER

It emphasizes the importance of preventing dosing errors, especially when administering acetaminophen to children. Additionally, it emphasizes the importance of ensuring that the total daily dose of acetaminophen does not exceed the recommended maximum when considering all medications that contain acetaminophen.

DOI: 10.1201/9781003461913-30

NONSTEROIDAL ANTI-INFLAMMATORY DRUGS

Nonsteroidal anti-inflammatory drugs (NSAIDs) are used as antipyretic, anti-inflammatory, and analgesic agents. These effects make NSAIDs helpful in treating muscle pain, pyrexia, dysmenorrhea, arthritic pain, **gout**, and migraines. They are also used as opioid-sparing agents in certain acute trauma cases. Common NSAIDs include aspirin, ibuprofen, naproxen, diclofenac, celecoxib, and meloxicam. They are available in both over-the-counter and prescription options. They are some of the most widely used therapeutics. Their anti-inflammatory, antipyretic, and analgesic activities make them a preferred treatment for various inflammatory diseases (see Table 27.1).

ASPIRIN

Aspirin, a nonsteroidal anti-inflammatory drug from the salicylate group, has been used for over 100 years. It inhibits the production of specific natural substances that cause fever, pain, swelling, and blood clots. Furthermore, aspirin can be found with other medications, such as antacids, pain relievers, and cough and cold remedies. Aspirin also lowers the risk of heart attack, stroke, or blood clots.

Clinical Indications

Aspirin is used to treat pain, fever, inflammation, and migraines. It also reduces the risk of major adverse cardiovascular events. Aspirin protects against atherothrombosis while increasing the risk of significant bleeding.

Mechanism of Action

Aspirin is a cyclooxygenase-1 (COX-1) inhibitor and alters the enzymatic activity of cyclooxygenase-2 (COX-2). Unlike other NSAIDs, such as ibuprofen, which bind reversibly to this enzyme, aspirin binds irreversibly.

Adverse Effects

Aspirin reduces the risk of major adverse cardiovascular events in patients with diabetes who do not have cardiovascular disease. However, it may lead to an increased risk of bleeding and gastrointestinal complications. The most common side effect of aspirin is gastrointestinal bleeding.

TABLE 27.1
NSAID Dosing

Generic Name	Trade Name
Aspirin	Bayer Aspirin, many others
Celecoxib	Celebrex
Diclofenac	Cambia, Voltaren
Ibuprofen	Advil, Motrin
Indomethacin	Indocin
Ketorolac	Toradol
Naproxen	Aleve

Contraindications

Individuals allergic to ibuprofen should avoid taking aspirin due to cross-reactivity. Patients with asthma should exercise caution if they have known bronchospasm associated with NSAIDs. Aspirin raises the risk of gastrointestinal bleeding in patients with existing peptic ulcer disease or gastritis.

> **YOU SHOULD REMEMBER**
>
> Reye syndrome is a rare but fatal condition. It is a type of encephalopathy resulting from fatty changes in an otherwise healthy liver. This condition occurs after a viral upper respiratory tract infection in children treated with aspirin to reduce fever. The incidence has significantly decreased due to increased awareness and the preference for acetaminophen over aspirin in managing fever in children.

IBUPROFEN

Ibuprofen is a nonsteroidal anti-inflammatory drug used to treat mild to moderate pain and alleviate symptoms of **osteoarthritis**, rheumatoid arthritis, or juvenile arthritis. Common brand names include Advil®, Midol®, and Motrin®. This medication is available over the counter and in prescription strength.

Clinical Indications

Ibuprofen is approved to treat inflammatory and rheumatoid disorders. It also relieves minor aches and pains from headaches, myalgia, arthritis, **menstrual periods**, the common cold, toothaches, and backaches. It has become one of the most commonly used medications worldwide.

Mechanism of Action

Ibuprofen primarily inhibits the precursors of prostaglandins. The cyclooxygenase pathway plays a significant role in its intended uses.

Adverse Effects

Ibuprofen increases the risk of gastrointestinal complications, including bleeding, perforation, and ulceration. This risk is further elevated in older patients and those with pre-existing gastrointestinal conditions.

Contraindications

Ibuprofen is contraindicated for patients with a known history of hypersensitivity or allergic reactions to the drug itself, other NSAIDs, or aspirin. It raises the risk of serious cardiovascular events such as myocardial infarction and stroke. Due to the potential for Reye syndrome, it should be avoided in children with viral upper respiratory infections.

CELECOXIB

Celecoxib is a nonsteroidal anti-inflammatory drug that treats mild to moderate pain and alleviates arthritis symptoms, dysmenorrhea, acute pain, and migraine.

Clinical Indications

The mechanism of celecoxib's action is due to the selective inhibition of cyclooxygenase-2 (COX-2), which synthesizes prostaglandins, a key component in the pain and inflammation pathway. This pharmacological activity is responsible for celecoxib's analgesic, anti-inflammatory, and antipyretic effects. Additionally, celecoxib exerts a weak inhibitory effect on COX-1, impacting platelet function less than aspirin.

Mechanism of Action

The mechanism of action involves the selective inhibition of cyclooxygenase-2 (COX-2), which is responsible for prostaglandin synthesis, a crucial component of the pain and inflammation pathway. This pharmacologic activity provides celecoxib with analgesic, anti-inflammatory, and antipyretic effects. Additionally, celecoxib weakly inhibits COX-1, exerting a lesser impact on platelet function than aspirin.

Adverse Effects

Celecoxib is generally regarded as a safe and well-tolerated medication with risks similar to those of most other NSAIDs. Common side effects include dyspepsia, nausea, vomiting, abdominal pain, and diarrhea. Severe side effects of celecoxib may include gastrointestinal ulcer perforation, bleeding, thromboembolism, stroke, and heart attack. Additionally, celecoxib can exacerbate hypertension, cause fluid retention in patients with congestive heart failure, and lead to renal toxicity, liver toxicity, anaphylactic reactions, and Stevens–Johnson syndrome.

Contraindications

Celecoxib is contraindicated in patients who have demonstrated severe allergic reactions to the sulfonamide group. It is also contraindicated in patients with ischemic heart disease, stroke, heart failure, active gastrointestinal ulceration, bleeding, or inflammatory bowel disease.

As with all NSAIDs, celecoxib should not be taken after 29 weeks of pregnancy because of the risk of patent ductus arteriosus.

CLINICAL CASE STUDIES

Clinical Case Study 1

A 77-year-old man has frequently used acetaminophen for musculoskeletal pain while also consuming alcohol. He developed nausea, vomiting, jaundice, abdominal pain, ascites, and confusion. Liver function tests indicated severe liver damage, and a CT scan confirmed this with visual proof. The patient was placed on the list for liver transplantation.

CRITICAL THINKING QUESTIONS

1. What are the indications for acetaminophen?
2. What are the adverse effects of acetaminophen?
3. What are the contraindications for acetaminophen?

Clinical Case Study 2

A 54-year-old woman visits her primary care physician. She has been diagnosed with osteoarthritis and has been taking ibuprofen for nearly 8 months. She reports experiencing abdominal pain and noticing dark blood in her stool.

CRITICAL THINKING QUESTIONS

1. What is the mechanism behind the action of ibuprofen?
2. What are the adverse effects of ibuprofen?
3. What are the contraindications of ibuprofen?

FURTHER READING

Arpilor, L. (2023). *Signs of Liver Damage – Understand the Signs of Liver Damage, from Jaundice to Abdominal Pain. Learn About Potential Indicators and the Implications for Liver Health*. Arpilor.

Ayoub, S.S., Flower, R.J., and Seed, M.P. (2010). *Cyclooxygenases – Methods and Protocols* (Methods in Molecular Biology, 644). Humana Press.

Beom Kim, K., Movahed, R., Malhotra, R.K., and Stanley, J.L. (2021). *Management of Obstructive Sleep Apnea – An Evidence-Based, Multidisciplinary Textbook*. Springer.

Burra, P. (2022). *Textbook of Liver Transplantation – A Multidisciplinary Approach*. Springer.

Javaherian, A., and Latifpour, P. (2012). *Acetaminophen – Properties, Clinical Uses and Adverse Effects* (Pharmacology– Research, Safety Testing and Regulation – Public Health in the 21st Century). Nova Biomedical.

Kaplan, S. (2021). *A Comprehensive Guide to Non-steroidal Anti-inflammatory Drugs*. Nova Science Publishers Inc.

Lanas, A. (2016). *NSAIDs and Aspirin – Recent Advances and Implications for Clinical Management*. Springer.

Oluynin, Y. (2022). *Osteoarthritis – Current Issues of Treatment*. Our Knowledge Publishing.

Peterson, L.S. (2020). *Mayo Clinic Guide to Arthritis – Managing Joint Pain for an Active Life*. Mayo Clinic Press.

Rajendram, R., Preedy, V.R., Patel, V.B., and Martin, C.R. (2021). *Treatments, Mechanisms, and Adverse Reactions of Anesthetics and Analgesics*. Academic Press.

Rice, R. (2022). *Tylenol for Pain and Fever Management – Acetaminophen for Rapid Relief of Pain and Fever Reducer*. Rice.

Riley, M. (2021). *The Essential Guide to Acetaminophen – Usage, Precautions, Interactions and Side Effects*. Interactive Media Licensing.

Saraf, S. (2019). *NSAIDs (Nonsteroidal Anti-Inflammatory Drugs) – An Overview*. BSP Books.

Sinatra, R.S., Jahr, J.S., and Watkins-Pitchford, J.M. (2010). *The Essence of Analgesia and Analgesics*. Cambridge University Press.

Terkeltaub, R., and Edwards, N.L. (2016). *Gout – Diagnosis and Management of Gouty Arthritis and Hyperuricemia*, 4th Edition. Professional Communications, Inc.

Tsagareli, M.G., and Tisklauri, N. (2012). *Behavioral Study of 'Non-Opioid Tolerance'* (Pain and its Origins, Diagnosis and Treatments – Pain Management Research and Technology). Nova Biomedical.

Wesson, D.E. (2016). *Metabolic Acidosis – A Guide to Clinical Assessment and Management*. Springer.

28 Opioid Analgesics

LEARNING OUTCOMES

After studying the chapter, readers should be able to

1. Summarize the classifications of pain.
2. Review pain processing and substance P.
3. Distinguish between acute and chronic pain and nociceptive somatic pain.
4. Identify the classifications of opioids.
5. Explain the trade names of fentanyl, hydrocodone, and oxycodone.
6. Review opioid analgesics and their indications.
7. Identify the common adverse effects of opioids.
8. Distinguish narcotic antagonists and the typical reactions of naltrexone.
9. Describe the effects of buprenorphine.
10. Assess the potential risks of methadone.

OVERVIEW

Pain is an unpleasant sensory and emotional experience. It can be mild or severe, encompassing steady, throbbing, stabbing, aching, burning, pinching, pricking, tingling, or stinging sensations. Pain may be classified as acute or chronic. It serves a vital function in the nervous system by warning of potential bodily injury. Additionally, pain can be influenced by psychological factors. Chronic pain can lead to complications beyond physical symptoms, with patients often experiencing depression, anxiety, and insomnia. Treatments for pain include medications, physical therapy, psychological therapies, and alternative therapies. Nonsteroidal anti-inflammatory drugs are most effective for mild to moderate pain associated with inflammation. They are commonly used for arthritis and pain resulting from muscle sprains, strains, back and neck injuries, or menstrual cramps. Acetaminophen is generally the first recommendation for mild to moderate pain. Narcotic medications are prescribed for severe pain, with opioid analgesics now among the most commonly prescribed drugs for patients in the United States. Effective pain management strategies are essential for treating chronic pain disorders.

PAIN PROCESSING

Pain is a signal in the nervous system that causes an unpleasant feeling, including prickling, tingling, stinging, burning, or aching sensations. It may be sharp or dull, intermittent, or constant. Pain can indicate actual or potential tissue damage. Its signals are transmitted from the peripheral nervous system to the central nervous system, where they are perceived. Environmental, emotional, spiritual, cultural, and cognitive factors often modify this experience. Pain can be classified as acute or chronic and can be nociceptive or neuropathic.

Nociceptive pain is the most common type, resulting from tissue damage caused by an injury outside the nervous system. Neuropathic pain can be caused by brain damage, such as a stroke or a tumor. Peripheral neuropathic pain results from damage to the spinal cord, which can often be limited to damage to the upper and lower extremities.

Pain processing involves four stages: **transduction**, transmission, **modulation**, and **perception**. The first step in this process is transduction, where a painful stimulus is received by pain receptors located near the source of the pain. This triggers the release of chemicals that generate an action potential, which is conducted from nerve fibers to the spinal cord. The second step is transmission, occurring when the nerve signal reaches the spinal cord. In this phase, excitatory chemicals, such as glutamate and **substance P**, are released, contributing to slower neurotransmission by the C fibers. The destruction of substance P fibers leads to analgesia. The third stage, modulation, involves alterations of the pain signals along the transmission pathway. During this stage, pain is understood in various intensities, and the signal is then sent to the brain via ascending spinal cord pathways. The fourth stage, perception, refers to the conscious awareness of pain impulses.

ACUTE PAIN

Acute pain is considered short-term pain lasting less than 6 months. It is a complex, unpleasant experience with emotional, cognitive, and sensory features that occurs in response to tissue trauma. Acute pain occurs suddenly after an injury and typically resolves as the underlying cause heals. It often happens after bone fractures, surgery, burns, cuts, and dental procedures.

CHRONIC PAIN

Chronic pain lasts longer than 6 months and is not associated with a disease process. It can persist even after the injury or illness has healed. Pain signals remain active in the nervous system for weeks, months, or even years. Chronic pain may be linked to conditions such as headaches, cancers, arthritis, and **fibromyalgia**. The most common chronic pain is persistent low back pain, followed by **myofascial pain syndromes**.

Chronic pain is common, affecting more than 100 million Americans, and has a significant financial impact. This pain can lead to decreased work productivity and increased utilization of "sick days," potentially resulting in permanent disability. Chronic pain is most prevalent among

DOI: 10.1201/9781003461913-31

adults living in poverty and those who did not complete high school. The age-adjusted prevalence is much higher in unemployed women. These consequences can cost society 300 billion dollars annually, emphasizing the need for effective treatment of chronic pain and a focus on improving function in these patients.

YOU SHOULD REMEMBER

Women are more likely to experience chronic pain than men. They are more prone to chronic pain conditions such as migraine headaches, osteoarthritis, fibromyalgia, chronic fatigue syndrome, and rheumatoid arthritis. Additionally, women suffer from gender-specific chronic pain conditions like endometriosis.

Check Your Knowledge

1. What are the definitions of acute and chronic pain?
2. What are the four stages of pain processing?
3. What are some examples of chronic pain?

OPIOIDS

Opioids are used to treat moderate to severe pain. They are ordered after an injury, surgery, or for chronic pain conditions like back pain and osteoarthritis. They are classified as natural, semisynthetic, and synthetic drugs.

- **Natural (opiates):** occurring in plants such as the *opium poppy*, containing base chemical compounds that include codeine and morphine (see Figure 28.1).
- **Semisynthetic**: provided in laboratories from natural opiates; examples include heroin, hydromorphone, hydrocodone, and oxycodone.
- **Synthetic drugs** include fentanyl, methadone, dextropropoxyphene, levorphanol, and tramadol.

Opioids are also classified by their mechanisms of action, which include agonists and antagonists (Figure 28.2).

OPIOID ANALGESICS

Opioid analgesics directly work on opioid receptors, causing pain relief. Some examples include morphine, hydrocodone, and oxycodone. Generic and trade names of opioid analgesics are listed in Table 28.1. They are typically not the first step in managing acute and chronic pain due to the potential for addiction and abuse. In the current **opioid crisis**, prescribing recommendations for opioids have become more restricted. Opioid analgesics bind to opioid receptors and mimic the actions of endogenous opioids (naturally made by the body). Opioid receptors are divided into three classes: mu (μ), delta (δ), and kappa (κ). Most opioid analgesics used in treatment bind to the μ-opioid receptor as agonists, which results in a reduction of pain signals. Opioid analgesics have many risks.

FIGURE 28.1 Opium poppy.

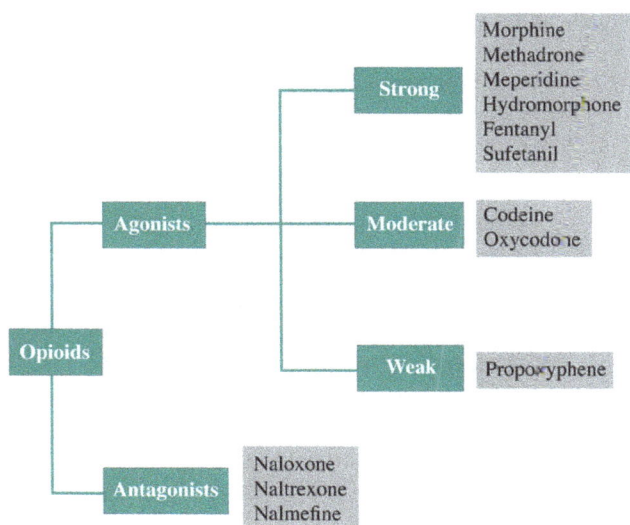

FIGURE 28.2 Opioid classification by the mechanism of action.

TABLE 28.1

Opioid Analgesics

Generic Name	Trade Name
Codeine	(Generic only, also in many combination drugs)
Fentanyl	Actiq
Hydrocodone	Zohydro
Meperidine	Demerol
Methadone	Methadose
Morphine	Avinza, MS-Contin
Oxycodone	Oxycontin, Oxecta
Oxymorphone	Opana
Tapentadol	Nucynta
Tramadol	Ultram

YOU SHOULD REMEMBER

Opioid analgesics include strong medications such as morphine, codeine, oxycodone, hydrocodone, fentanyl, methadone, and many others. The legitimate use of these drugs effectively reduces pain, but they can also cause extreme euphoria or an intense "high," leading to dependence and addiction. Overdose effects include severe respiratory depression that can result in respiratory arrest, coma, and death. Opioid dependence and withdrawal are indicated by nausea, confusion, drowsiness, extreme sweating, and constipation.

Check Your Knowledge

1. What are the different classifications of opioids?
2. What are some examples of potent opioids?
3. What are some examples of opioid antagonists?

MEDICAL TERMINOLOGY OF OPIOIDS

Some medical terms must be addressed here to better understand the concept of the adverse effects of opioids. They include

- **Addiction**: Environmental, genetic, and psychosocial factors influence a primary and chronic disease. It is often concurrent with maladaptive behaviors, including continual drug use, even though harm and severe consequences occur.
- **Abuse**: Use of a drug without a prescription, without therapeutic intent, or in a way other than prescribed to achieve certain feelings or effects of the drug.
- **Misuse:** Involves any drug use outside of the prescription goals or self-medication for reasons outside of the intent of the prescription.
- **Diversion:** Transfer of a controlled substance, such as an opioid analgesic, from a lawful to an unlawful distribution method.
- **Dependence** occurs due to chronic drug use, and use must continue to avoid withdrawal symptoms.
- **Opioid withdrawal** causes diarrhea, restlessness, eye tearing, muscle pain, sweating, tachycardia, and insomnia. It occurs after the abrupt cessation of a drug after the body has become physically dependent on it.
- **Tolerance refers to** either the need for significantly increased amounts of a drug to achieve the desired effect or intoxication, or a marked reduction in effect with continual use of the same amount of the drug. Additionally, tolerance is described by the duration or pain intensity that can be tolerated before extreme pain responses are initiated. It is influenced by cultural perceptions, expectations, fatigue, gender, physical and mental health, anger, apprehension, role behaviors, boredom, and sleep deprivation. Various factors, including alcohol consumption, regular use of pain medications, hypnosis, distracting activities, strong beliefs or religious faith, and warmth, can heighten tolerance.

Mechanism of Action

Opioids mimic the actions of endogenous opioid peptides by interacting with mu, delta, or kappa opioid receptors. These receptors are coupled with G1 proteins, and the effects of opioids are primarily inhibitory. Most clinically relevant opioid analgesics bind to mu-opioid receptors in the central and peripheral nervous systems as agonists to achieve analgesia.

Clinical Indications

Opioids are used for acute, chronic, postsurgical, cancer-related, and vascular pain. The FDA has also approved their use as antitussive and antidiarrheal medications.

Loperamide, known as the over-the-counter medication Imodium A-D, is an opioid used to treat diarrhea and irritable bowel syndrome. It achieves this effect by decreasing intestinal motility and increasing absorption time. Opioids such as codeine and dextromethorphan are effective as cough suppressants. Codeine is considered the standard cough suppressant against which new drugs are tested. Additionally, codeine serves as a valuable antitussive, providing the added benefits of analgesia and sedation. The use of opioids for analgesia is controversial due to the risks of addiction and tolerance.

Contraindications

Opioids are contraindicated in cases of true allergies, severe respiratory depression, acute or chronic bronchial asthma, increased intracranial pressure, acute psychiatric fluctuations, uncontrolled suicide risk, or a heightened risk of prescription misuse. Additionally, opioids can cause spasm of the biliary sphincter and should be avoided in patients with cholecystitis.

Adverse Effects

Adverse effects of opioid use include drowsiness, constipation, euphoria, nausea, vomiting, dizziness, and slowed breathing. A person using opioids over time can develop tolerance, physical dependence, and opioid use disorder, with the risk of overdose and death.

YOU SHOULD REMEMBER

The route of administration for opioids can include oral, intravenous, epidural, subdural, rectal, and topical (patches) methods.

YOU SHOULD REMEMBER

Fentanyl is used to treat acute and severe pain resulting from significant trauma or surgery, as well as chronic pain associated with cancer. It is also administered alongside other medications just before or during an operation to enhance the effectiveness of the anesthetic.

YOU SHOULD REMEMBER

Phantom limb pain is the perception of pain or discomfort in a limb that is no longer present. It most commonly occurs after an amputation and should be distinguished from other clinical conditions, such as residual limb pain. Phantom limb pain originates from the actual site of the amputated limb and typically resolves with wound healing.

Check Your Knowledge

1. What are the characteristics of opioid withdrawal and tolerance?
2. What are the contraindications for opioids?
3. What are the adverse effects of opioids?

OPIOID ANTAGONISTS

Opioid antagonists are medications that block the effects of opioids. Common types include naloxone and naltrexone. Other opioid receptor antagonists include nalmefene, methylnaltrexone, nalbuphine, nalorphine, and naltrexone. These agents are used to treat opioid overdose, opioid use disorder, alcohol use disorder, and opioid-induced constipation.

CLINICAL INDICATIONS

Opioid antagonists are indicated for various conditions, including the emergency treatment of narcotic overdose, diagnosing addiction, alcohol use disorders, and narcotic abuse.

Naloxone comes in intravenous, intramuscular, and intranasal formulations and is FDA-approved for use in opioid overdose and the reversal of respiratory depression associated with opioid use.

MECHANISM OF ACTION

The opioid antagonist binds to opioid receptors in the central or peripheral nervous system but does not activate them. Instead, it prevents opioids from acting on the receptors.

ADVERSE EFFECTS

Adverse effects of opioid antagonists include respiratory depression, headache, itching, dizziness, stomach pain, nausea, vomiting, constipation, muscle spasms, tachycardia, fever, chills, sweating, biliary spasm, urinary retention, dysphoria, euphoria, addiction, and dependence.

CONTRAINDICATIONS

Opioid antagonists should be avoided in patients with hypersensitivity to these agents, liver impairment, biliary spasms, and urinary retention. They are not advised for patients with respiratory depression and increased intracranial pressure.

PHARMACOLOGICAL TREATMENT FOR OPIOID USE DISORDER

Pharmacotherapy for opioid-use disorder can help decrease cravings for opioid abuse, reduce symptoms, help ensure the person is safe, and decrease the risk of opioid overdose. Methadone, buprenorphine, and naloxone are the first-line treatments.

METHADONE

Methadone is a potent and addictive opioid, similar to other opioids. However, being on methadone differs from being dependent on illegal opioids like heroin. It can effectively treat moderate to severe pain when around-the-clock relief is necessary for an extended period. Methadone is also prescribed for detoxification or to maintain opioid tolerance. This medication may be suitable for individuals whose pain is uncontrolled with morphine, such as neuropathic pain, those who cannot tolerate morphine, or those experiencing renal failure. It has a half-life of 30 hours, making it a long-acting agent.

The adverse effects include headache, weight gain, flushing, dry mouth, heavy sweating, constipation, sexual dysfunction, hallucinations, and confusion. Potential risks of methadone include arrhythmias and respiratory depression, so proper dosing is essential. It is safer for patients to take methadone under medical supervision than to take heroin of unknown purity. The use of methadone to treat opioid-use disorder in the U.S. is available only through opioid treatment programs (methadone clinics).

NALTREXONE

Naltrexone is an opioid antagonist used to treat both alcohol-use disorder and opioid dependence. It works by blocking the effects of opioids, particularly the euphoric and rewarding sensations. Additionally, naltrexone is employed in the treatment of alcohol use disorder.

The adverse effects of naltrexone include abdominal cramps, nausea, dizziness, headaches, anxiety, myalgia, and insomnia. There is a risk of liver damage and elevated liver function test results. The preferred dosage form is an intramuscular injection administered once a month by a healthcare provider.

BUPRENORPHINE

Buprenorphine is a synthetic opioid used to treat chronic pain when other pain medications have failed or cannot be tolerated. It is also indicated for opioid detoxification and maintenance. *Additionally,* buprenorphine may benefit patients with a brief history of opioid dependence and lower requirements for opioid agonists. As a long-acting medication, it can detoxify and maintain opioid tolerance.

The most common adverse effects of buprenorphine include headaches, nausea, vomiting, constipation, insomnia, blurred vision, anxiety, and depression. Some adverse effects can be severe and may require immediate medical attention. CNS depression, respiratory depression, orthostatic hypotension, and liver impairment may occur.

YOU SHOULD REMEMBER

Naloxone can reverse the effects of a fentanyl overdose if it is administered quickly. It is a fast-acting medication that is available over-the-counter. This medication can be taken once daily or once monthly to treat alcohol-use disorder or opioid use disorder.

Check Your Knowledge

1. What are the mechanisms of action of opioid antagonists?
2. What are the clinical indications for opioid antagonists?
3. What are the adverse effects of opioid antagonists?

CLINICAL CASE STUDIES

Clinical Case Study

A 46-year-old golfer was involved in a car accident that resulted in multiple bone fractures. His medical history indicates that he suffered from chronic lower back pain due to playing golf and had previously undergone a laminectomy on his L4 and L5 spinal regions. Following his back surgery, the patient had been using oxycodone, but this medication was discontinued as he began to recover from the car accident. Instead, fentanyl was administered through continuous infusion, with dose escalation to manage the patient's acute pain and need for sedation.

CRITICAL THINKING QUESTIONS

1. What are the mechanisms by which opioids act?
2. When do the adverse effects of opioid analgesics occur?
3. What are the contraindications for opioids?

FURTHER READING

Almela Rojo, P. (2020). *Opioids: From Analgesic Use to Addiction.* Intechopen.

Aronson, J.K. (2009). *Meyler's Side Effects of Analgesics and Anti-inflammatory Drugs.* Elsevier Science.

Ballantyne, J.C., and Tauben, D.J. (2013). *Expert Decision Making on Opioid Treatment.* Oxford University Press.

Cruiciani, R.A., and Knotkova, H. (2013). *Handbook of Methadone Prescribing and Buprenorphine Therapy.* Springer.

Deer, T.R., Leong, M.S., Buvanendran, A., Gordin, V., Kim, P.S., Panchal, S.J., and Ray, A.L. (2013). *Comprehensive Treatment of Chronic Pain by Medical, Interventional, and Integrative Approaches.* Springer.

Freye, E., and Levy, J.V. (2008). *Opioids in Medicine — A Comprehensive Review on the Mode of Action and the Use of Analgesics in Different Pain States.* Springer.

Javaherian, A., and Latifpour, P. (2012). *Acetaminophen: Properties, Clinical Uses and Adverse Effects.* Nova Biomedical.

Kaplan, S. (2015). *A Comprehensive Guide to Non-Steroidal Anti-Inflammatory Drugs.* Nova Biomedical.

Lussier, D., and Beaulieu, P. (2015). *Adjuvant Analgesics* (American Pain Library). Oxford University Press.

Macintyre, P.E., and Schug, S.A. (2021). *Acute Pain Management — A Practical Guide*, 5th Edition. CRC Press.

Murphy, J.L., and Rafie, S. (2019). *Chronic Pain and Opioid Management: Strategies for Integrated Treatment.* American Psychological Association.

Parnham, M.J. (2012). *Non-Opioid Analgesics in the Treatment of Acute Pain.* Springer/Birkhauser.

Sinatra, R.S., Jahr, J.S., and Watkins-Pitchford, J.M. (2010). *The Essence of Analgesia and Analgesics.* Cambridge University Press.

Stolberg, V.B. (2016). *Painkillers: History, Science, and Issues* (The Story of a Drug). Greenwood.

U.S. Department of Health and Human Services, Substance Abuse and Mental Health Administration. (2020). *Medications for Opioid Use Disorder — For Healthcare and Addiction Professionals, Policymakers, Patients, and Families.* Lulu. com

Wakeman, S.E., and Rich, J.D. (2021). *Treating Opioid Use Disorder in General Medical Settings.* Springer.

29 Antineoplastic Agents

LEARNING OUTCOMES

After studying the chapter, readers should be able to

1. Distinguish between benign and malignant tumors
2. Review the classifications of chemotherapy agents.
3. Explain the biological and monoclonal therapies.
4. Identify the clinical indications of alkylating and antimetabolite agents.
5. Describe the adverse effects of antimicrotubule agents.
6. List the trade names of fluorouracil, teniposide, bleomycin, and doxorubicin.
7. Explain the anticancer antibiotics and their indications.
8. Discuss the contraindications of alkylating agents and plant alkaloids.
9. Explain the adverse effects of anticancer antibiotics.
10. Review fibrile neutropenia and myelosuppression.

OVERVIEW

Cancer is a disease in which some of the body's cells grow uncontrollably and spread to other parts. The mechanisms behind cancer development are not fully understood. Genetic and environmental factors, age, and gender play a role in cancer development. It can begin in almost any cell or tissue in the body. Normally, human cells grow and multiply to form new cells as needed. When cells age or become damaged, they die, and new cells replace them. In 2024, there were 2,001,140 new cancer cases and 611,720 cancer deaths in the United States. Treatments include surgery, chemotherapy, radiotherapy, immunotherapy, and targeted therapy. In recent decades, millions of cancer patients have been cured by chemotherapy alone. Chemotherapy is one of the most effective cancer treatments, alongside surgery and radiotherapy. It inhibits cell proliferation, helping to prevent metastasis. However, it may cause toxicity in normal cells. Combining chemotherapy and radiation can shrink the tumor before surgery or indicate curative intent in cancers such as head and neck, lung, and colorectal cancer.

MALIGNANT TUMORS

A tumor, or neoplasm, is an abnormal mass of cells in the body that arises when cells divide excessively or fail to die at the appropriate time. Tumors can be either benign or malignant. **Benign tumors** remain in their original location without invading other body areas and do not spread to nearby structures or distant sites. Typically, benign tumors grow slowly and have well-defined borders.

Malignant tumors are cancerous and characterized by cells that grow uncontrollably, spreading locally and to distant sites. They can disseminate to remote areas through the bloodstream or the lymphatic system. **Metastasis** can occur anywhere in the body but is most commonly observed in the lungs, brain, liver, and bones.

Cancer treatments may include surgery, chemotherapy, radiotherapy, immunotherapy, and targeted therapy. Chemotherapy, together with surgery and radiotherapy, is one of the most effective ways to treat cancer. It inhibits cell **proliferation**, preventing metastasis. However, it may also cause toxicity to normal cells. Combining chemotherapy and radiation can shrink the tumor before surgery or aim for curative intent in cancers such as head and neck, lung, and colon. This chapter will concentrate on chemotherapy.

CATEGORIES OF CHEMOTHERAPY AGENTS

The three main medication categories are conventional chemotherapy, hormonal therapy, and biological therapy. Conventional chemotherapy, also known as systemic chemotherapy, is a cancer treatment that uses drugs to destroy cancer cells. It remains a primary treatment option for cancer.

Cancer chemotherapy agents are classified by their chemical structure and function. Some well-known classes include alkylating agents, plant alkaloids, antimetabolites, anthracyclines, topoisomerase inhibitors, and corticosteroids. Hormonal therapy employs hormones to slow or halt the growth of hormone-sensitive cancers. Biologic therapies specifically target specific areas, leading to various toxicities based on treatment goals.

CONVENTIONAL (TRADITIONAL) THERAPY

Conventional therapy is a well-established treatment used by most healthcare professionals. It contrasts with alternative or complementary therapies, which are less commonly practiced. Examples of traditional cancer treatments include chemotherapy, radiation therapy, and surgery.

Whenever a new cell is created, it undergoes a typical process to mature fully. This process consists of several stages known as the **cell cycle** (see Figure 29.1). The cell cycle includes four phases: G1, S, G2, and M. The G1 phase involves the cell preparing its DNA for synthesis in the S phase. The G2 phase is when the cell readies itself for **mitosis** by producing RNA and proteins, followed by the M phase, during which cellular division occurs, resulting in

DOI: 10.1201/9781003461915-32

CELL CYCLE

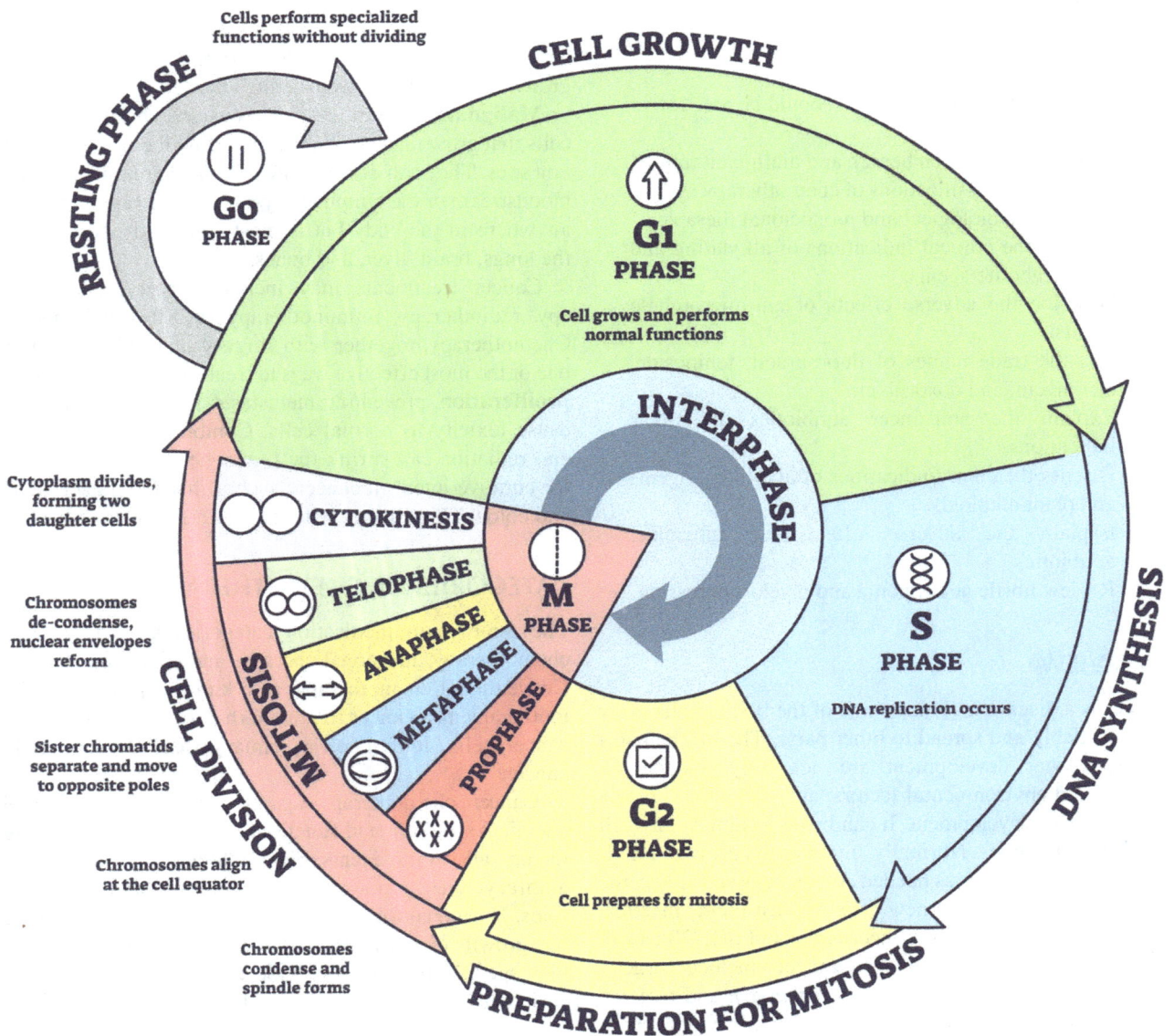

FIGURE 29.1   The cell cycle.

two cells. Depending on their mechanism of action, chemotherapy agents can influence various points in the cell cycle, with some affecting all phases.

YOU SHOULD REMEMBER

Chemotherapy drugs target cells at various phases of the cell cycle. Cancer cells replicate more rapidly than normal cells, making them more susceptible to chemotherapy. However, these drugs also damage normal cells along with cancer cells, resulting in several side effects.

HORMONAL THERAPY

Hormones are substances, either proteins or steroids, produced by the body to help regulate the functions of specific cell types. For example, certain body parts rely on sex hormones such as estrogen, progesterone, and testosterone to function correctly. Other hormones include thyroid hormones, cortisol, adrenaline, and insulin. Different organs or glands generate various types of hormones. Hormonal therapy is a form of cancer treatment that uses hormones to slow or stop the growth of hormone-sensitive cancers. These medications usually cause less severe organ damage compared to other therapies; however, they are not without side effects. Corticosteroids can also be effective against cancer DNA in various types of cancer, including lymphoma and **multiple myeloma**.

Some cancers depend on hormones for growth. Therefore, treatments that block or modify these hormones can sometimes help slow or stop the progression of these cancers. Hormone therapy is primarily used to treat specific types of breast and prostate cancer that rely on sex hormones for growth. However, a few other cancers may also respond to hormone therapy.

Hormone therapy is classified as a systemic treatment because its target hormones circulate throughout the body. The medications used in hormone therapy travel through the body to identify and target these hormones. This distinguishes it from treatments focusing solely on a specific body part, such as surgery and radiation therapy. These are referred to as local treatments, affecting only one body area.

Check Your Knowledge

1. What are the three main categories of medications used for chemotherapy?
2. What are the four stages of the cell cycle?
3. What are the indications for corticosteroids and testosterone in cancer treatment?

BIOLOGIC THERAPY

Biological therapy, often called **immunotherapy**, is a cancer treatment that utilizes the body's immune system to combat cancer cells. It enhances the immune system's ability to recognize and eliminate these cells. Biological therapies include targeted approaches such as monoclonal antibodies, tyrosine kinase inhibitors (TKIs), and immune-mediated therapies, prompting the immune system to focus on cancerous cells. The side effects of these therapies can vary depending on specific targets.

Monoclonal Antibodies

Monoclonal antibodies are specialized antibodies that target specific antigens on cell surfaces. They are produced by cloning a unique cell designated for a particular target. Several therapeutic **monoclonal antibodies** are currently available on the market. Rituximab and trastuzumab received approval in the late 1990s for treating **lymphoma** and breast cancer, marking a significant advancement in cancer treatment.

Tyrosine Kinase Inhibitors

TKIs target and block tyrosine kinases—enzymes involved in cell growth, differentiation, and survival. TKIs are a form of cancer treatment often categorized as chemotherapy, although they operate differently from traditional chemotherapy. Several have been approved in the United States. However, the side effects can vary from mild to life-threatening, affecting various organ systems. In some

cases, severe adverse effects may necessitate the early discontinuation of life-saving cancer therapies. Cardiotoxicity is recognized as a significant consequence of several agents.

Immune-Mediated Therapy

Immune-mediated therapy combats cancer by targeting the immune system to enhance or initiate its response. Many of these agents are monoclonal antibodies focusing on components outside cells. A vital component of the immune system is the T cell**,** which can **recognize and signal the destruction of** cancer cells in the body. These drugs are comparable to pressing the gas pedal in a car. The anticancer agents strengthen the immune system and improve T-cell signaling, thus increasing the quantity and duration of T-cell attacks on foreign cells, including cancer cells.

YOU SHOULD REMEMBER

Immunotherapy aims to enhance cancer's immunogenicity by equipping the body with adequately functioning immune cells and related molecules to stimulate and strengthen the immune response. It also makes cancer more susceptible to the effects of anticancer immunity.

Check Your Knowledge

1. What are three examples of biological therapies?
2. What effect do monoclonal antibodies have on cancer cells?
3. What effect does immune-mediated therapy have on cancer cells?

CHEMOTHERAPEUTIC DRUG CLASSIFICATIONS

Various agents are used to treat cancer and are classified into the following categories. Table 29.1 lists multiple agents used in chemotherapy.

ALKYLATING AGENTS

Alkylating agents were the first drugs developed for cancer treatment and continue to be used in chemotherapy today. Although these agents can effectively target many types of cancer, they are especially beneficial for slow-growing cancers. However, they are generally ineffective against rapidly growing cells. Alkylating agents represent the largest category of anticancer drugs. Table 29.2 lists the generic and trade names of alkylating drugs.

Clinical Indications

Alkylating agents are primarily used to treat cancer. They are effective against leukemia, lymphoma, and multiple

TABLE 29.1
Chemotherapeutic Agents

Alkylating Agents	Plant Alkaloids	Antimetabolites	Antitumor Antibiotic
Chlorambucil	Actinomycin D	Purine antagonists	Doxorubicin
Cyclophosphamide	Docetaxel	Pyrimidine antagonists	Mitoxantrone
Thiotepa	Mitomycin	Folate antagonist	Bleomycin
Busulfan			

TABLE 29.2
Alkylating Agents

Generic Name	Trade Name
Chlorambucil (nitrogen mustard)	Leukeran
Cyclophosphamide	Cytoxan
Ifosfamide	Ifex
Mechlorethamine (nitrogen mustard)	Mustargen
Melphalan	Alkeran
Alkyl Sulfonates	
Busulfan	Busulfex
Oxaliplatin	Eloxatin
Nitrosoureas	
Carmustine	BCNU
Lomustine	CCNU
Fotemustine	Mustophoran
Ethyleneimines	
Altretamine	Hexalen
Thiotepa (injection)	Thiotepa
Triazenes	
Dacarbazine	DTIC-Dome
Diaminazene aceturate	Berenil
Temozolomide	Temodar

myeloma, as well as solid tumors such as brain, lung, breast, melanoma, testicular, ovarian, uterine, and bladder cancer.

Mechanism of Action

Alkylating agents inhibit the transcription of DNA into RNA, halting protein synthesis. This results in abnormal base pairing and inhibits cell division, ultimately leading to cell death. While this action occurs in all cells, alkylating agents primarily target rapidly dividing cells without sufficient DNA repair time.

Adverse Effects

The adverse effects of alkylating agents include loss of appetite, dry mouth, sore throat, nausea, vomiting, diarrhea, baldness, myelosuppression, and hemorrhagic cystitis.

Contraindications

Alkylating agents should be avoided in patients with known allergies to these medications, renal failure, liver impairment, bone marrow suppression, pregnancy, and lactation.

YOU SHOULD REMEMBER

All alkylating agents have significant toxicities, with the predominant effects occurring in the bone marrow and gastrointestinal tract. Most agents have been shown to cause transient elevations in serum aminotransferase levels in some patients.

Check Your Knowledge

1. What are the trade names for busulfan and oxaliplatin?
2. What are the contraindications of alkylating agents?
3. What are the adverse effects of alkylating agents?

PLANT ALKALOIDS

Plant alkaloids, also known as antimicrotubular agents, inhibit cell growth by halting cell division. These drugs disrupt the cellular structures that support chromosome movement during cell division, making them useful for cancer treatment. Microtubules are crucial components of the intracellular **cytoskeleton**, exhibiting unique polymerization dynamics essential for various cellular functions, including cell division and regulation. Antimicrotubule agents include polymerizing agents (such as paclitaxel and docetaxel) and depolymerizing drugs (like vincristine and vinorelbine). Table 29.3 lists the generic and trade names of antimicrotubular agents.

Clinical Indications

Antimicrotubule agents are primarily utilized for treating various cancers, including lung cancer, melanoma,

TABLE 29.3
Plant Alkaloid Agents

Generic Name	Trade Name
Etoposide	VePesid, Etopophos
Teniposide	Vumon
Vinblastine sulfate (injection)	Vinblastine Sulfate
Vincristine	Vincasar PFS
Vinorelbine	Navelbine

esophageal cancer, breast cancer, ovarian cancer, prostate cancer, and other solid tumors.

Mechanism of Action

Antimicrotubule agents inhibit cell division by disrupting microtubules, which are crucial for cell structure, signaling, and transport. These drugs can treat various cancers.

Adverse Effects

The adverse effects of antimicrotubule agents include mouth ulcers, vomiting, diarrhea, chest pain, arrhythmia, and joint pain. A significant limitation of using microtubule agents is the high incidence of neuropathy, which is often the most frequent adverse effect in combination regimens.

Contraindications

Antimicrotubule agents are contraindicated in cases of known hypersensitivity, during pregnancy, and in patients with AIDS-related **Kaposi's sarcoma**.

Check Your Knowledge

1. What are the trade names for teniposide and vinorelbine?
2. What is the mechanism of action for antimicrotubule agents?
3. What are the clinical indications for antimicrotubule agents?

ANTIMETABOLITES

Antimetabolites are a crucial class of cancer drugs that disrupt nucleic acid synthesis in cancer cells. They play a significant role in cancer chemotherapy. There are three categories of antimetabolites, also known as antimetabolite antagonists: purine antagonists, pyrimidine antagonists, and folic acid antagonists. Table 29.4 lists the generic and trade names of these antimetabolites.

Clinical Indications

Antimetabolite agents are used to treat lung cancer, breast cancer, and certain forms of leukemia and lymphoma.

Mechanism of Action

Antimetabolites disrupt the synthesis of DNA components; they are structural analogs of purine and pyrimidine bases and folate **cofactors**, which play a crucial role in several steps

of purine and pyrimidine biosynthesis. Consequently, depleting nucleotides hampers DNA replication. However, some compounds can be incorrectly incorporated into nucleic acids, leading to structural abnormalities that cause cell death through various mechanisms, including DNA breaks.

Adverse Effects

Antimetabolites may cause nausea, vomiting, diarrhea, loss of appetite, myelosuppression, hair loss, fatigue, skin rash, and elevated liver enzymes. These symptoms vary depending on the specific drug and the patient's dosage.

Contraindications

Antimetabolites are contraindicated in patients with known allergies to the drug, peptic ulcers, bone marrow suppression, liver dysfunction, and renal impairment. They should be avoided during pregnancy and breastfeeding due to their teratogenic properties.

Check Your Knowledge

1. What are the trade names for fluorouracil and mercaptopurine?
2. What are the contraindications of antimetabolites?
3. What are the adverse effects of antimetabolites?

ANTICANCER ANTIBIOTICS

Anticancer antibiotics are chemicals that microorganisms produce with anticancer activity. The classification of anticancer antibiotics mainly includes doxorubicin, daunorubicin, bleomycin, mitomycin, and dactinomycin. Table 29.5 presents the anticancer antibiotics.

Clinical Indications

Bleomycin is used to treat cancer in the head and neck regions, Hodgkin's lymphoma, and testicular carcinoma. It is also utilized as a sclerosing agent for malignant pleural effusions. Daunorubicin is indicated in combination with cytarabine for treating **acute myeloid leukemia** in adults and pediatric patients aged 1 year and older. Doxorubicin is typically used for solid tumors and is effective against malignant lymphoma and liver, gastric, breast, lung, and

TABLE 29.4
Antimetabolites

Generic Name	Trade Name
Capecitabine	Xeloda
Fluorouracil	Adrucil
Mercaptopurine	Purinethol
Methotrexate	Rheumatrex, Trexall

TABLE 29.5
Anticancer Antibiotics

Generic Name	Trade Name
Bleomycin	Blenoxane
Daunorubicin	Cerubidine
Doxorubicin	Adriamycin PFS
Epirubicin	Ellence
Idarubicin	Idamycin
Mitomycin	Mutamycin
Plicamycin	Mithracin
Valrubicin	Valstar

soft tissue sarcomas. It can also be beneficial for acute leukemia. Adriamycin is utilized to treat many childhood cancer patients. Mitomycin is effective for various solid tumors. Plicamycin is used in the treatment of testicular embryonal cancer and is also employed for **glioma** and lymphoma.

Mechanism of Action

Anticancer antibiotics hinder cell growth by interfering with DNA and the genetic material within cells, inhibiting uncontrolled proliferation, aggressive growth, and metastasis. Doxorubicin is derived from the bacterium Streptomyces peucetius. It functions by damaging cellular DNA and eliminating cancer cells. Additionally, it inhibits an enzyme required for cell repair and division.

Adverse Effects

The adverse effects of anticancer antibiotics may include soreness, redness, or peeling of the hands and feet. Hand–foot syndrome manifests as pain, tingling, numbness, and skin dryness. It can also result in a rash, itching, or skin darkening.

Contraindications

Anticancer antibiotics are contraindicated in cases of severe liver impairment, recent myocardial infarction, congestive heart failure, acute infections, and hypersensitivity to these agents.

YOU SHOULD REMEMBER

Cardiac toxicity from doxorubicin can develop within days of administration, affecting approximately 11% of patients. This may present as reversible myopericarditis, left ventricular dysfunction, or arrhythmias.

Check Your Knowledge

1. What are the categories of anticancer antibiotics?
2. What are the clinical indications for bleomycin and daunorubicin?
3. What are the adverse effects of anticancer antibiotics?

SPECIFIC CHEMOTHERAPY ADVERSE EFFECTS

The specific adverse effects and long-term consequences of chemotherapy require substantial attention. Chemotherapy-induced adverse effects are often only partially practical, frequently fail to address potential longer-term consequences, or may even lead to additional adverse effects. This chapter discusses neutropenic fever and myelosuppression only.

FEBRILE NEUTROPENIC

Neutropenic fever occurs when an oral temperature exceeds or equals 101°F (38.3°C) for at least an hour, accompanied by an absolute neutrophil count of less than 1500 cells/μL. The risk of bacteremia rises with profound neutropenia. In most instances, the infectious etiology cannot be identified, and it may be referred to as a fever of unknown origin. The definition of fever of unknown origin encompasses neutropenic cases with a fever exceeding 38.3°C, without any signs, symptoms, or a microbiologically defined infection.

Bacteria primarily cause infections, though viruses and fungi can also play a role. Common bacterial pathogens include gram-positive organisms from the *Staphylococcus*, *Streptococcus*, and *Enterococcus groups. Additionally,* infections have been associated with drug-resistant pathogens such as *Pseudomonas aeruginosa*, *Acinetobacter* species, *Stenotrophomonas maltophilia*, *Escherichia coli*, and *Klebsiella* species.

Neutropenic fever is the most common and serious complication associated with hematopoietic cancers or in patients undergoing chemotherapy for cancer. For high-risk patients who present with neutropenic fever, intravenous antibiotic therapy should be administered within 1 hour after triage and monitored for more than 4 hours before discharge.

MYELOSUPPRESSION

The bone marrow fails to produce enough blood cells or platelets during myelosuppression. This condition heightens the risk of blood disorders such as anemia, infections, and bleeding complications. Most patients experience myelosuppression due to chemotherapy, but certain viruses and blood cancers can also trigger it.

Myelosuppression is the most common dose-limiting side effect of chemotherapy, although not all agents induce it. Neutrophils are particularly affected by myelosuppression due to their short lifespan of approximately 8 hours and their rapid turnover into new cells. Following chemotherapy, a low point in the neutrophil count is typically observed around 10–14 days later. The neutrophil count generally recovers within 3–4 weeks after chemotherapy. Neutrophils play a critical role in the body's defense against infections. When the absolute neutrophil count falls below 500, the risk of infection increases; this condition is known as neutropenia.

YOU SHOULD REMEMBER

Tumor lysis syndrome is a serious condition caused by the rapid breakdown of tumor cells, resulting in hypocalcemia, hyperkalemia, hyperphosphatemia, and hyperuricemia. These conditions can lead to severe acute kidney damage, arrhythmias, and even death. Tumor lysis syndrome is commonly observed in patients undergoing treatment for hematologic malignancies, such as non-Hodgkin lymphomas and acute leukemias, particularly after the initiation of chemotherapy. However, it can also occur spontaneously.

CLINICAL CASE STUDIES

Clinical Case Study 1

A 39-year-old woman diagnosed with an aggressive form of right breast cancer underwent surgery and subsequently received chemotherapy, radiation therapy, and hormone therapy. Unfortunately, the cancer metastasized to her vertebral bones. She was an ideal candidate for biologic therapy and was treated with atezolizumab monotherapy, followed by paclitaxel.

CRITICAL THINKING QUESTIONS

1. What categories of chemotherapy agents exist?
2. What are the adverse effects of antimetabolites?
3. What are the indications for hormone therapy in the treatment of various cancers?

Clinical Case Study 2

A 69-year-old man was diagnosed with colorectal cancer that had metastasized to his lungs, bones, and liver. He received three chemotherapy treatments that combined fluorouracil and oxaliplatin. Follow-up CT scans showed that these treatments had reduced multiple liver tumors. The patient passed away 3 months after the treatments.

CRITICAL THINKING QUESTIONS

1. What are the trade names for fluorouracil and oxaliplatin?
2. What are antimetabolites, and how are they classified?
3. What are the mechanisms of action for antimetabolites?

FURTHER READING

Chabner, B.A., and Longo, D.L. (2018). *Cancer Chemotherapy, Immunotherapy and Biotherapy*, 6th Edition. Wolters Kluwer.

Chu, E., and DeVita, Jr., V.T. (2024). *Physicians' Cancer Chemotherapy Drug Manual*, 24th Edition. Jones & Bartlett Learning.

Goldberg, G.S., and Airley, R. (2020). *Cancer Chemotherapy — Basic Science to the Clinic*, 2nd Edition. Wiley Blackwell.

Javle, M., and Jivraj Borad, M. (2023). *Handbook of Targeted Cancer Therapy and Immunotherapy — Gastrointestinal Cancer*. Lippincott, Williams, and Wilkins.

Jones, R.J., and McCormick, T.M. (2024). *Rogue Cells — A Conversation on the Myths and Mysteries of Cancer*. Johns Hopkins University Press.

Marik, P. E. (2023). *Cancer Care — The Role of Repurposed Drugs and Metabolic Interventions in Treating Cancer*. Front Line Covid-19 Critical Care Alliance.

National Cancer Institute. (2019) *Chemotherapy and You*. NCI.

Olsen, M., LeFebvre, K., Walker, S., and Prechtel Dunphy, E. (2023). *Chemotherapy and Immunotherapy Guidelines and Recommendations for Practice*, 2nd Edition. Oncology Nursing Society.

Priestman, T.J. (2012). *Cancer Chemotherapy — An Introduction*, 3rd Edition. Springer.

Rodgers, G., and Young, N.S. (2024). *The Bethesda Handbook of Clinical Hematology*, 5th Edition. Wolters Kluwer.

Skeel, R.T., and Khleif, S.N. (2011). *Handbook of Cancer Chemotherapy*. Wolters Kluwer/Lippincott, Williams, and Wilkins.

Stegall, J. (2023). *Cancer Secrets — An Integrative Oncologist Reveals How to Fight Cancer Using the Best of Modern Medicine and Natural Therapies*, 2nd Edition. Cancer Secrets.

30 Vitamins and Minerals

LEARNING OUTCOMES

After studying the chapter, readers should be able to

1. Explain the functions of vitamins in the body.
2. Identify the classifications of fat-soluble vitamins.
3. Review the effects of retinoids on the human body.
4. Describe the contraindications and adverse effects of vitamin A.
5. Discuss the impact of vitamin D deficiency in children and older adults.
6. Identify the mechanism of action of vitamin E.
7. Describe the clinical indications of vitamin K.
8. Review water-soluble vitamins and explain vitamin B12 deficiency.
9. Explain the essential and trace minerals in the human body.
10. List the microminerals and their requirements.

OVERVIEW

Vitamins are essential for bodily functions and must be consumed in small amounts through the diet. Although they do not provide energy, they aid in energy metabolism, tissue growth, development, and maintenance. Fat-soluble vitamins dissolve in organic solvents, including vitamins A, D, E, and K. The water-soluble vitamins consist of the B vitamin complex and vitamin C, which dissolve in the body's water. Vitamin deficiency can occur and significantly affect health. Initially, a deficiency and its associated signs and symptoms can be reversed by increasing the intake of the specific vitamin. Hypervitaminosis, or vitamin toxicity, generally results from consuming excessive amounts of vitamins A or D. Minerals are nutrients needed in small quantities to support the body's health. They include calcium, iron, iodine, and magnesium. Minerals naturally exist as inorganic substances that the human body cannot produce. A mineral is considered a nutrient when inadequate intake leads to a decline in biological function or overall health.

VITAMINS' FUNCTIONS AND DISORDERS

Vitamins are essential for various bodily functions, including strengthening bones, supporting the immune system, facilitating wound healing, and regulating hormone levels. They also ensure the proper functioning of cells and organs. Multivitamins can assist those who struggle to meet their daily nutritional needs through diet alone, but they cannot replace a balanced diet. A healthy and varied diet that includes all five food groups is sufficient for optimal health.

It is advisable to obtain vitamins from a diverse range of wholesome, unrefined foods; however, some can cause toxicity if consumed in excess. Vitamins and minerals are classified as *micronutrients* because they are essential in only small amounts. Micronutrients do not provide energy but are crucial for metabolic processes that enable the body to extract energy from *macronutrients* (carbohydrates, proteins, and fats). There are 13 vitamins, 8 of which belong to the B vitamin group.

Vitamin deficiency can lead to nutritional disorders. However, vitamin D deficiency is a notable exception, as it is commonly observed in children and older adults. Vegans who avoid animal products may become deficient in vitamin B12 because it is primarily found in animal sources. Deficiencies of B vitamins, such as biotin or pantothenic acid, are rare. Individuals at higher risk for vitamin deficiencies, including those who have undergone **bariatric surgery**, are on hemodialysis, or consume large amounts of alcohol, may benefit from taking a daily multivitamin.

VITAMIN REGULATION IN THE BODY

A deficiency in a vitamin can hinder specific metabolic processes in cells, ultimately disrupting metabolic balance both within the body and externally. Except for vitamin C, all water-soluble vitamins play a catalytic role, acting as **coenzymes** for enzymes involved in energy transfer or the metabolism of macronutrients. Fat-soluble vitamins are crucial for specialized functions in differentiated tissues. Some fat-soluble vitamins help form biological membranes or support their integrity. They may also operate at the genetic level by regulating the synthesis of enzymes.

Many people are unaware that consuming excessive vitamins or minerals can lead to serious adverse effects and toxicity. In the United States, vitamin toxicities occur more frequently than vitamin deficiencies. Individuals should always consult their healthcare providers before taking vitamin or mineral supplements.

YOU SHOULD REMEMBER

Manufacturing or marketing adulterated or misbranded dietary supplement products is illegal, and the FDA can remove these products from the market.

Check Your Knowledge

1. What are the roles of vitamins in the human body?
2. How does the body manage vitamins?

DOI: 10.1201/9781003461913-33

FAT-SOLUBLE VITAMINS

Fat-soluble vitamins consist of vitamins A, D, E, and K. They dissolve in **organic solvents** and are absorbed and transported similarly to fats. These vitamins are vital for various physiological processes, including immune function, vision, bone health, and blood coagulation. The body stores fat-soluble vitamins, which can lead to conditions such as **keratomalacia** or result in potential toxicity.

VITAMIN A

Vitamin A (retinoic acid) is essential for normal vision, development, growth, immune function, and reproduction. Over 90% of vitamin A is stored in the liver, while smaller quantities are found in the eyes, adipose tissue, bone marrow, kidneys, and testes. **Retinoids** and provitamin A carotenoids are converted into active vitamin A. Retinoids represent the biologically active forms of vitamin A, including **retinol** and **retinal**. These consist of preformed retinoids and provitamin A **carotenoids** that can be transformed into active vitamin A.

Retinoids support growth, development, cell differentiation, vision, and immune function, playing a vital role in embryonic development. Vitamin A is essential for forming the eyes, limbs, cardiovascular, and nervous systems. A deficiency in vitamin A during early gestation can lead to congenital disabilities and fetal death. Retinoic acid is necessary for the production, structure, and function of epithelial cells in the lungs, trachea, skin, and gastrointestinal tract. Vitamin A is crucial for maintaining the normal differentiation of cells in essential eye structures, such as the cornea and retina.

Vitamin A deficiency is mainly caused by insufficient dietary intake; however, it can also originate from malabsorption associated with intestinal diseases such as short bowel syndrome, celiac disease, cystic fibrosis, chronic diarrhea, bile duct obstruction, pancreatic disorders, and certain surgeries like gastric bypass. These factors hinder the body's ability to absorb vitamin A effectively. Moreover, heavy alcohol consumption can additionally contribute to this deficiency.

Clinical Indications

Vitamin A is indicated for patients with deficiencies such as nutritional night blindness, **xerophthalmia**, and keratomalacia. Xerophthalmia refers to insufficient tear production, while keratomalacia is characterized by night blindness, dry eyes, **Bitot spots**, corneal softening, and corneal cloudiness (see Figure 30.1).

Mechanism of Action

Vitamin A influences cell proliferation, differentiation, and function, likely due to its effects on gene expression, the regulation of specific gene transcription, or changes in the structural and functional organization of membranes.

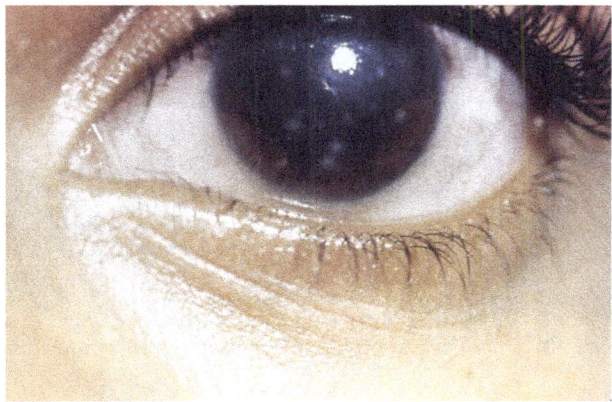

FIGURE 30.1   Keratomalacia.

Adverse Effects

Consuming more than 3,000 mcg of oral vitamin A supplements daily over a long period can lead to headaches, nausea, diarrhea, bone thinning, joint and bone pain, congenital disabilities, liver damage, and skin irritation.

Contraindications

Taking high doses of vitamin A may lead to severe liver damage. Combining high doses of vitamin A supplements with other drugs that can harm the liver might increase the risk of liver failure.

YOU SHOULD REMEMBER

In the United States, vitamin A deficiency is common among low-income individuals, alcoholics, older adults, and patients with chronic diarrhea, Crohn's disease, cystic fibrosis, celiac disease, and acquired immunodeficiency syndrome.

YOU SHOULD REMEMBER

Age-related macular degeneration is the leading cause of blindness in the developed world. While its exact cause remains unclear, it is believed to result from cellular damage to the retina due to oxidative stress. Vitamin A reduces the risk of developing advanced macular degeneration by 25%.

Check Your Knowledge

1. What role does vitamin A play in the human body?
2. What are the indications for vitamin A?
3. What are the adverse effects of vitamin A?

Vitamin D

Vitamin D is a fat-soluble vitamin that belongs to a group of compounds, including vitamins D1, D2, and D3. The body naturally produces vitamin D when exposed to sunlight. It can also be obtained from certain foods and supplements to maintain adequate blood levels. Vitamin D has several vital functions, the most important of which is regulating the absorption of calcium and phosphorus.

Clinical Indications

Vitamin D is essential for maintaining healthy bones and teeth, supporting muscle strength, enhancing immune system function, aiding brain function, and regulating inflammation. It also treats hypoparathyroidism, **rickets**, osteomalacia in adults, and **familial hypophosphatemia**. Furthermore, vitamin D plays a role in preventing osteoporosis (see Figure 30.2).

Mechanism of Action

Vitamin D metabolites bind to hormone receptors in the cell nucleus. This action helps maintain calcium balance, regulates parathyroid hormone, promotes renal reabsorption, increases the intestinal absorption of calcium and phosphorus, and reduces inflammation.

Adverse Effects

Long-term use of high doses of vitamin D supplements may lead to adverse effects, such as nausea, vomiting, dehydration, cardiac arrhythmia, and confusion.

Contraindications

The use of pharmacological vitamin D or excessive dietary intake of vitamin D is contraindicated in patients with hypercalcemia, malabsorption syndrome, abnormal sensitivity to the toxic effects of vitamin D, and hypervitaminosis. It should be avoided in patients with cystic fibrosis, celiac disease, Crohn's disease, ulcerative colitis, and liver disease.

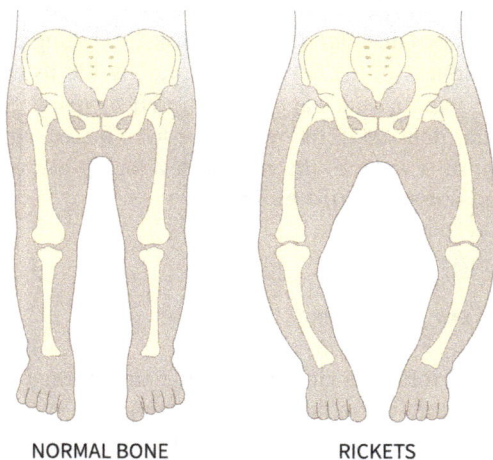

NORMAL BONE   RICKETS

FIGURE 30.2   Rickets in children.

YOU SHOULD REMEMBER

Some individuals may benefit from vitamin D supplements and adequate calcium intake to support their bone health. However, they do not need large amounts of vitamin D to achieve these benefits. Excess can be harmful. For instance, a 2010 study published in *JAMA* indicated that very high doses of vitamin D in older women were linked to an increased risk of falls and fractures. Moreover, supplements with excessive vitamin D can be toxic in rare cases. They may cause hypercalcemia, where too much calcium accumulates in the blood, potentially leading to deposits in the arteries or soft tissues. This condition may also make individuals more susceptible to painful kidney stones.

Check Your Knowledge

1. What are the types of vitamin D, and how does the body obtain it?
2. What are the mechanisms of action for vitamin D?
3. What are the adverse effects of vitamin D?

Vitamin E

Vitamin E is a fat-soluble vitamin with several forms, but only alpha-tocopherol is utilized by the human body. Its primary role is to act as an **antioxidant**. It is essential for reproduction, blood health, brain function, vision, and skin integrity. Vitamin E deficiency is extremely rare in humans; it typically results from irregular dietary fat absorption or metabolism. The two types of vitamin E are **tocopherols** and **tocotrienols**. Good sources of vitamin E include almonds, peanuts, avocados, wheat germ, canola oil, cottonseed oil, safflower oil, sunflower oil, and sunflower seeds.

Clinical Indications

Supplementation with vitamin E is crucial for patients with Alzheimer's disease, cancer, cystic fibrosis, **cholestasis**, severe liver disease, gastrointestinal surgeries, and inadequate diets. Vitamin E may help prevent or reduce atherosclerosis and lower the incidence of ischemic heart disease. A deficiency in vitamin E can cause the early breakdown of red blood cells and lead to hemolytic anemia, which is particularly serious in preterm infants, making supplemental vitamin E essential for these individuals. Additionally, a lack of vitamin E can weaken immune function and lead to neurological changes in the spinal cord and peripheral nervous system. These changes have been observed in cases where the deficiency stemmed from a genetic abnormality affecting lipoprotein synthesis, which hinders the transport and distribution of vitamin E. Routine vitamin E supplementation is not recommended for children or adults who are not deficient.

Mechanism of Action

Vitamin E is an antioxidant that prevents the formation of reactive oxygen species when fats undergo oxidation and during the propagation of free radical reactions.

Adverse Effects

Vitamin E can lead to adverse effects when consumed in high doses (400 units or more daily) or for an extended period. Symptoms may include nausea, diarrhea, intestinal cramps, fatigue, blurred vision, dizziness, gonadal dysfunction, and headaches. The supplement may also increase the risk of bleeding. If someone plans to undergo surgery, they should stop taking vitamin E 2 weeks prior.

Contraindications

Patients who take oral anticoagulants, pregnant women, individuals with iron deficiency anemia, and those with hyperthyroidism should avoid vitamin E supplements.

YOU SHOULD REMEMBER

Age-related macular degeneration and cataracts are among the most common causes of significant vision loss in older adults. Their causes are often unknown, but the cumulative effects of oxidative stress are suggested as contributing factors. If so, antioxidant nutrients like vitamin E may help prevent or treat these conditions.

Check Your Knowledge

1. What is the primary function of vitamin E?
2. What are the clinical indications for vitamin E?
3. What are the contraindications for vitamin E?

VITAMIN K

Vitamin K is essential for blood clotting and bone health. **Prothrombin** is a vitamin-dependent protein that plays a role in blood clotting. **Osteocalcin** is another protein that requires vitamin K to produce healthy bone tissue. There are two primary forms of vitamin K, known as *quinones*. These include vitamin K1 (phylloquinone) in plant foods like leafy greens and **menaquinone** in animal and fermented foods. The most biologically active form is phylloquinone, the primary dietary form.

Clinical Indications

Phylloquinone is administered intravenously, intramuscularly, or subcutaneously to treat coagulation disorders resulting from the improper formation of coagulation factors II, VII, IX, and X due to vitamin K deficiency or interference with vitamin K activity. Additionally, vitamin K is used to address hypoprothrombinemia secondary to conditions

such as sprue, ulcerative colitis, celiac disease, intestinal resection, pancreatic cystic fibrosis, regional enteritis, or hypoprothrombinemia stemming from a disruption in vitamin K metabolism.

Mechanism of Action

Vitamin K functions by carboxylating glutamate residues in proteins, aiding in the regulation of blood clotting, bone metabolism, and vascular biology.

Adverse Effects

Vitamin K's common adverse effects include allergic reactions, dysphagia, flushing, chest pain, and sweating. When administered intravenously, vitamin K1 may cause bronchospasm and cardiac arrest. However, oral vitamin K does not lead to severe reactions.

Contraindications

Vitamin K should be used cautiously in patients with hereditary hypoprothrombinemia, renal impairment, gallbladder stones, cystic fibrosis, celiac disease, and Crohn's disease. It should also be avoided in patients taking blood thinners, those undergoing long-term hemodialysis, and patients with a hypersensitivity reaction to vitamin K, which may manifest as an eczema-like skin rash. Caution is necessary when using vitamin K in neonates, patients with hereditary hypoprothrombinemia, renal impairment, over-anticoagulation due to heparin, and patients with hypersensitivity to vitamin K.

YOU SHOULD REMEMBER

High doses of vitamin K analogs during the first few days of life can result in hyperbilirubinemia in premature infants. This condition may lead to severe hemolytic anemia, hemoglobinuria, and **kernicterus**, potentially causing brain damage or even death.

Check Your Knowledge

1. What is the importance of vitamin K?
2. What are the indications for vitamin K?
3. What are the adverse effects of vitamin K?

WATER-SOLUBLE VITAMINS

Water-soluble vitamins include the vitamin B complex and vitamin C. The vitamin B complex consists of thiamine, riboflavin, niacin, pantothenic acid, pyridoxine, biotin, folate, and cobalamin. These vitamins function as coenzymes that support various bodily processes and play essential roles in the body. Table 30.1 summarizes the effects of deficiencies in water-soluble vitamins.

TABLE 30.1

Water-Soluble Vitamins and Deficiency Results

Vitamin	Deficiency Results
B1 (Thiamine)	Beriberi (peripheral neuropathy, weakness, anorexia, weight loss), Wernicke–Korsakoff syndrome
B2 (Riboflavin)	Ariboflavinosis (inflammation of the mouth and tongue, cracks at the corner of the mouth)
B3 (Niacin)	Pellagra (dementia, dermatitis, diarrhea, and eventually, death)
B5 (Pantothenic acid)	Fatigue, GI tract disturbances, impaired muscle function, weakness
B6 (Pyridoxine)	Peripheral neuropathy, seizure, depression, confusion, dermatitis, cracked lips
B7 (Biotin)	Brittle nails, hair loss, scaly skin rashes around eyes, nose, and mouth
B9 (Folic acid)	Birth defects, dyspnea, megaloblastic anemia, diarrhea, nausea, abdominal pain, and loss of appetite
B12 (Cobalamin)	Numbness of extremities, pernicious anemia, pale skin, confusion, memory fog, fatigue, dyspnea, trouble walking
C (Ascorbic acid)	Bleeding gums, poor wound healing, bruising easily, and anemia

YOU SHOULD REMEMBER

Pernicious anemia can occur due to a vitamin B12 deficiency caused by an autoimmune disorder that prevents the body from absorbing this vitamin. If left untreated, it may lead to serious medical complications, including irreversible damage to the nervous system. Pernicious anemia necessitates lifelong B12 injections.

Check Your Knowledge

1. What are the types of water-soluble vitamins?
2. What are the signs and symptoms of vitamin B3 and folic acid deficiency?
3. What are the signs and symptoms of vitamin C deficiency?

MINERALS

Minerals are nutrients required in small amounts to sustain the body's health. A mineral is classified as a nutrient when inadequate intake reduces biological function or overall health. A primary macromineral must be consumed in an amount of 100 mg or more daily. Essential macrominerals include calcium and phosphorus. In contrast, a micromineral is needed in amounts less than 100 mg daily. Examples of trace minerals are iron, zinc, and copper. Minerals are crucial for numerous bodily functions and structures; for instance, calcium and phosphorus provide rigidity and strength to teeth and bones. Skeletal components also serve as storage for the minerals that the body requires. Minerals facilitate proper muscle contraction and nerve function. Some minerals act as enzyme cofactors, helping to maintain the correct acid-base balance in bodily fluids. They are also essential for blood clotting, tissue repair, and growth.

MACROMINERALS

Macrominerals are minerals needed in relatively large quantities in the diet, particularly calcium, phosphorus, sodium, potassium, chloride, magnesium, and sulfur. They are essential for various homeostatic functions throughout human life. Macrominerals are crucial to the biological structure of the human body, play a key role in metabolic processes, and are involved in nearly all bodily functions. However, excessive intake of macrominerals over a prolonged period can be toxic. Various health issues can arise from deficiencies in these minerals.

A macromineral deficiency refers to a lack of one or more essential minerals that the body needs in significant amounts. Deficiencies in macrominerals can lead to various health problems, including osteoporosis, muscle weakness, and electrolyte imbalances. Insufficient macrominerals may also contribute to several chronic diseases. Hormonal changes resulting from a macromineral deficiency can lead to issues such as weight gain, obesity, and an increased risk of cancer in the future. Maintaining a balanced diet is essential for ensuring adequate intake of these macrominerals.

MICROMINERALS

Microminerals, also known as trace minerals, are vital **inorganic** substances present in small amounts, typically less than 5 mg, in the human body. These include iron, iodine, copper, manganese, fluoride, zinc, selenium, chromium, and molybdenum, all of which are essential for proper development, function, and overall health. Iron plays a crucial role in the formation of red blood cells, while copper, manganese, selenium, and zinc help protect against damage from **free radicals**. Additionally, iodine is vital for the production of thyroid hormones and the regulation of metabolism.

CLINICAL CASE STUDIES

Clinical Case Study 1

A 35-year-old woman with a 2-year history of short bowel syndrome (SBS) was brought to the emergency room, showing signs of vitamin A deficiency. Four weeks before her admission, she developed generalized, erythematous scaly patches primarily on both arms and her upper back but did not exhibit typical ophthalmic signs of vitamin A deficiency. She had been suffering from SBS due to extensive small bowel resection, which resulted in chronic diarrhea and significant weight loss over the past 14 months.

CRITICAL THINKING QUESTIONS

1. What functions do retinoids have in the body?
2. What causes vitamin A deficiency?
3. What are the clinical indications for vitamin A?

Clinical Case Study 2

A 61-year-old man presents with fatigue, numbness in his limbs, and a pale complexion. Blood tests reveal low levels of vitamin B12 and an increased mean corpuscular volume, leading to a diagnosis of pernicious anemia.

CRITICAL THINKING QUESTIONS

1. What are the clinical symptoms associated with vitamin B12 deficiency?
2. How can vitamin B12 deficiency be addressed?

FURTHER READING

Arpilor, L. (2023). *Vitamin Deficiency Symptoms – Insight into the Signs and Symptoms of Various Vitamin Deficiencies and their Health Impacts.* Arpilor.

Davis, M. (2020). *Eat Your Vitamins — Your Guide to Using Natural Foods to Get the Vitamins, Minerals, and Nutrients Your Body Needs.* Adams Media.

DiNicolantonio, J., and Land, S. (2021). *The Mineral Fix: How to Optimize Your Mineral Intake for Energy, Longevity, Immunity, Sleep, and More.* DiNicolantonio.

Gaines, S.R. (2022). *The Mineral Deficiency Antidote — How to Boost Up Your Energy-Consuming Ability.* Gaines.

Gomes, C., and Rautureau, M. (2021). *Minerals latu sensu and Human Health — Benefits, Toxicity and Pathologies.* Springer.

Gualtieri, A.F. (2017). *Mineral Fibres: Crystal Chemistry, Chemical-Physical Properties, Biological Interaction, and Toxicity* (EMU Notes in Mineralogy 18). European Mineralogical Society.

Liebermann, S., and Bruning, N. (2007). *The Real Vitamin and Mineral Book*: The Definitive Guide to Designing Your Personal Supplement Program, 4th Edition. Avery.

Martin, R.J.S. (2015). *Magnesium Deficiency — Weight Loss, Heart Disease and Depression — 13 Ways Curing Your Magnesium Deficiency Can Rejuvenate Your Body.* Bright Ideas Editorial.

Newman, M., and Mason, F. (2016). *Vitamin B12 Deficiency and Chronic Illness.* Newman.

Ochsenham, P., and Vormann, J. (2015). *The Magnesium Deficiency Crisis*: Is This the World's Number 1 Mineral Deficiency? Madhouse MEDIA.

Persky, W. (2014). *Vitamin D and Autoimmune Disease — How Vitamin D Prevents Autoimmune Disease.* Persky Farms.

Purser, D., and Larkin, J. (2019). *Vitamin Deficiency Symptoms and Cures — Modern Deficiency Illness — Using Intracellular Micronutrient Results.* Purser.

Rydon, R. (2017). *Profiles of the Nutrients: 2. Minerals and Trace Elements.* Rydon.

Shatzel, J.J. (2021). *Iron Deficiency — A Patient's Guide to the Most Common Nutrient Deficiency in the World.* Shatzel.

Walters, C. (2013). *Minerals for the Genetic Code — An Exposition & Analysis of the Dr. Olree Standard Genetic Periodic Chart & The Physical, Chemical & Biological Connection.* Acres U.S.A.

Wartian Smith, P. (2019). *What You Must Know About Vitamins, Minerals, Herbs, and So Much More — Choosing the Nutrients that Are Right for You*, 2nd Edition. Square One.

31 Antibacterial Agents

LEARNING OUTCOMES

After studying the chapter, readers should be able to

1. Explain the clinical indications of sulfa drugs.
2. Review the classifications of penicillins.
3. Identify the broad-spectrum and extended-spectrum penicillins (ESPs).
4. Review the characteristics of aminopenicillins.
5. Summarize the categories of cephalosporins.
6. Describe the contraindications of tetracyclines.
7. Describe the adverse effects of macrolides.
8. Identify the adverse effects of fluoroquinolones.
9. Review how aminoglycosides work.
10. Discuss the adverse effects of vancomycin.

OVERVIEW

Antibiotics are vital medications that effectively treat infections caused by bacteria, prevent the spread of infections, and reduce the complications of serious diseases. They are effective for most bacterial infections. However, they are ineffective against viral infections and other microorganisms. Each antibiotic targets specific types of bacteria. It is crucial to identify the cause of the illness when selecting an antibiotic for treatment. The overuse and misuse of antibiotics are significant contributors to the development of antibiotic resistance. The general public, healthcare providers, and hospitals can collaborate to ensure the proper use of these drugs, helping to slow the development of antibiotic resistance. For example, several decades ago, *Staphylococcus aureus* was susceptible to penicillin. However, over time, strains of this bacterium developed an enzyme that breaks down penicillin, making the drug ineffective. Other bacteria have also developed resistance to various antibiotics. Antibiotics are powerful medicines used to treat certain illnesses; however, they do not cure everything, and the use of unnecessary antibiotics can be detrimental.

SULFONAMIDES

Sulfonamides are among the oldest recognized classes of antimicrobial agents. These synthetic, **bacteriostatic** antimicrobials have a broad spectrum of activity against gram-positive and gram-negative organisms. However, many strains of particular species may exhibit resistance. Bacterial sensitivity remains consistent across various sulfonamides, and resistance to one sulfonamide implies resistance to all. Sulfonamides include mafenide, sulfacetamide, sulfadiazine, sulfamethizole, sulfamethoxazole, sulfanilamide, and sulfasalazine. Table 31.1 lists the various sulfonamides.

TABLE 31.1
Various Sulfonamides

Generic Names	Trade Names
Mafenide	Sulfamylon
Sulfacetamide	Klaron, Ovace
Sulfadiazine	Lantrisul, Neotrizine
Sulfamethizole, sulfamethoxazole	Bactrim, Septra
Sulfanilamide	Azulfidine, Diamox, Gantrisin
Sulfasalazine	Azulfidine, Delzicol, Dipentum
Sulfisoxazole	Gantrisin Pediatric, Truxazole

Clinical Indications

Sulfonamides, commonly known as sulfa drugs, remain widely used to treat bacterial infections. They effectively target urinary tract infections, otitis media, acute and chronic bronchitis, and **traveler's diarrhea**. Topical sulfonamides are also effective in treating vaginitis, burns, and eye infections. Trimethoprim/sulfamethoxazole successfully manages community-acquired soft tissue infections and urinary tract infections.

Mechanism of Action

Sulfa drugs serve as competitive inhibitors of the bacterial enzyme dihydropteroate synthetase, which typically utilizes para-aminobenzoic acid to produce essential folic acid. Inhibiting this reaction is critical for these organisms to synthesize folic acid. Without this mechanism, bacteria cannot replicate and survive.

Adverse Effects

The adverse effects of sulfa drugs can arise from both oral and topical use. These medications may cause headaches, nausea, diarrhea, fatigue, dizziness, increased sensitivity to sunlight, and skin blisters. Additionally, hyperkalemia and hypoglycemia may also occur with their use.

Contraindications

Sulfa drugs are contraindicated in patients with a history of allergic reactions or those with porphyria. The use of sulfonamides during the first trimester of pregnancy may lead to congenital malformations. They should not be administered to infants under 2 months of age.

Check Your Knowledge

1. What are the indications of sulfonamides?
2. What are the mechanisms by which sulfonamides act?
3. What are the contraindications for sulfonamides?

DOI: 10.1201/9781003461913-34

PENICILLINS

Penicillin is a class of antibiotics effective against many types of bacteria and is the most commonly prescribed antibiotic. However, particular bacterial species, including enterococci, have developed resistance to penicillin. Therefore, due to this emerging resistance, penicillin should be reserved for susceptible organisms. Enterococcal infections are now treated with penicillin, streptomycin, or gentamicin. Additionally, specific gram-negative rods are resistant to penicillin. Two significant differences between types of penicillin are the methods of production and the types of bacteria against which they are effective. The kinds of penicillin include

- Natural penicillin isolates are used to make medications.
- Semisynthetic penicillins can alter penicillin's natural form to make more effective antibiotics. They include penicillinase-resistant penicillins, aminopenicillins, and ESPs.
- Combination penicillins are often combined with other antibiotics to enhance the effectiveness of a single drug. Table 31.2 lists various classes of penicillins.

Some bacteria produce enzymes that can inactivate **beta-lactam** antibiotics. Therefore, beta-lactamase inhibitors, such as clavulanate or sulbactam, are administered in conjunction with specific penicillins to counteract these enzymes. Typical combinations of penicillin for these purposes include the following:

- Amoxicillin/clavulanate
- Piperacillin/tazobactam
- Ampicillin/sulbactam

Clinical Indications

Penicillin has a broad spectrum of clinical applications. It effectively combats infections caused by gram-positive cocci, rods, most **anaerobes**, and gram-negative cocci. Second-generation penicillins, such as ampicillin and amoxicillin, are effective against *Proteus mirabilis*, *Haemophilus influenzae*, *Salmonella*, *Shigella*, and *E. coli*. Third-generation penicillin, like carbenicillin, are prescribed for infections caused by gram-negative bacteria. Fourth-generation penicillins, including piperacillin, target the same bacterial strains as third-generation penicillins, such as *Klebsiella*, enterococci, *Pseudomonas aeruginosa*, and *Bacteroides fragilis*. Penicillin cannot cross the blood–brain barrier and only treats specific bacterial meningitis infections.

Mechanism of Action

Penicillins inhibit bacterial transpeptidase enzymes, which are involved in synthesizing the bacterial cell wall, known as penicillin-binding proteins. They prevent these enzymes from forming cross-links between peptide chains in the cell wall. Consequently, penicillins activate the bacteria's natural autolytic system, leading to cell lysis and death. Beta-lactamase inhibitors safeguard the beta-lactam ring in penicillin from degradation by beta-lactamase, an enzyme some bacterial species may produce.

Adverse Effects

The adverse effects of penicillin G and V include rash, urticaria, nausea, vomiting, diarrhea, and abdominal pain. Additionally, penicillin G can cause other adverse reactions such as fever, chills, headache, tachypnea, tachycardia, hypotension, flushing, and muscle spasms. Hypersensitivity reactions to penicillin often occur. The allergic response manifests within 20 minutes after the injection is administered. It can present as laryngospasm, urticaria, bronchospasm, edema, pruritus, hypotension, vascular collapse, and even death. Reactions to penicillin injections are treated with corticosteroids, epinephrine injection (EpiPen), antihistamines, bronchodilators, and immunotherapy (desensitization).

Contraindications

Contraindications of penicillin include a previous history of severe allergic reactions to it. It should be avoided in patients with **Stevens–Johnson syndrome**. Penicillins are considered safe for use during pregnancy and lactation. Penicillin has an antagonistic effect with tetracycline, and there is a greater risk of mortality when treating pneumococcal meningitis than when using penicillin alone.

TABLE 31.2
Generic and Brand Names of Penicillins

Generic Names	Brand Names
Penicillin G	Pen G
Penicillin V	Veetids
Amoxicillin	Amoxil
Ampicillin	Omnipen
Nafcillin	Unipen, Nallpen
Oxacillin	Bactocill
Dicloxacillin	Dycill, Dynapen
Piperacillin/tazobactam	Zosyn

YOU SHOULD REMEMBER

Penicillin G can be administered either intramuscularly or intravenously. It is available in vials containing 1, 5, and 20 million units. Due to its short half-life, penicillin G is typically given in divided doses every 4–6 hours.

YOU SHOULD REMEMBER

Rheumatic fever is an inflammatory disease that can develop after a previous infection with Streptococcus bacteria, such as strep throat or scarlet fever. It results from an autoimmune reaction in which the body's immune system mistakenly attacks its tissues.

Check Your Knowledge

1. What are the characteristics and types of penicillins?
2. What are the common combinations of penicillin?
3. What are the contraindications for penicillin?

AMINOPENICILLINS

Aminopenicillins are semisynthetic modifications of natural penicillin that offer a broader spectrum of activity against gram-positive and gram-negative bacteria. While natural penicillin primarily targets gram-positive organisms, aminopenicillins are also effective against certain **rickettsiae**. These antibiotics include ampicillin, amoxicillin, amoxicillin/clavulanic acid, and ampicillin/sulbactam. Adding a β-lactamase inhibitor significantly enhances their efficacy against gram-negative and some β-lactamase-producing gram-positive and anaerobic organisms.

EXTENDED-SPECTRUM PENICILLINS

ESPs are a class of semi-synthetic antibiotics with a broader spectrum of activity than natural penicillins. They exhibit reduced activity against gram-positive and anaerobic organisms relative to penicillin G. Some examples of ESPs include piperacillin/tazobactam and ticarcillin/clavulanate.

PENICILLINASE-RESISTANT PENICILLINS

Penicillinase-resistant penicillins are a class of antibiotics that withstand the enzyme penicillinase produced by certain bacteria. They effectively combat bacteria that produce penicillinase, such as Staphylococcus aureus. These resistant penicillins target strains of staphylococci and treat various infections. Specific conditions effectively managed with penicillinase-resistant penicillins include endocarditis, bacteremia, meningitis, pneumonia, osteomyelitis, cystitis, and pharyngitis. Examples of penicillinase-resistant penicillins include nafcillin, cloxacillin, oxacillin, and dicloxacillin.

CARBAPENEMS

Carbapenems are a class of beta-lactam antibiotics effective against various aerobic and anaerobic gram-positive and gram-negative organisms. Thienamycin was the first carbapenem discovered in 1976. Carbapenems possess the

TABLE 31.3
Various Carbapenems

Generic Names	Trade Names
Biapenem	Omegacin
Doripenem	Doribax
Ertapenem	Invanz
Imipenem/cilastatin	Primaxin
Meropenem	Merrem
Panipenem/betamipron	Carbenin

broadest spectrum of activity and the highest potency against bacteria among all beta-lactam antibiotics. Due to this unique characteristic, they are often reserved for more severe infections or used as "last-line" agents. Table 31.3 lists various carbapenems.

Clinical Indications

Carbapenems are primarily indicated for **empiric monotherapy** in severe conditions, including meningitis, endometritis, **osteomyelitis**, bacteremia, **peritonitis**, bronchitis, pneumonia, **febrile neutropenia**, and nephritis. They are also effective against mixed bacterial infections and aerobic gram-negative bacterial infections resistant to other beta-lactam antibiotics.

Mechanism of Action

Carbapenems can inhibit cell wall synthesis by binding to penicillin-binding proteins. As a result, they cause bacterial cell wall defects and swelling, ultimately leading to bacterial death.

Adverse Effects

Seizures, although uncommon, can be a potential side effect of carbapenem antibiotics, particularly imipenem. They are more likely to occur at higher doses or in patients with certain risk factors, such as renal dysfunction or a history of seizures. The incidence of hypersensitivity reactions to carbapenems is low. However, the adverse effects can include nausea, diarrhea, abdominal pain, and skin reactions.

Contraindications

Carbapenems should be avoided in patients who are hypersensitive to these drugs. Patients with kidney disease must be closely monitored while receiving treatment with a carbapenem.

CEPHALOSPORINS

Cephalosporins are a class of antibiotics used to treat bacterial infections. These beta-lactam antimicrobials are effective against both gram-positive and gram-negative bacteria. The five generations of cephalosporins address meningitis, resistant bacteria, and skin infections. First-generation cephalosporins target gram-positive cocci and certain gram-negative bacteria, including *Escherichia* coli (*E. coli*), *Klebsiella pneumoniae,* and *Proteus mira-bilis.* Second-generation cephalosporins effectively combat *Haemophilus influenzae, Moraxella catarrhalis,* and *Bacteroides* species. Third-generation cephalosporins show reduced activity against most gram-positive organisms but offer improved coverage against *Neisseria species* and *H. influenzae.* Fourth-generation cephalosporins maintain coverage similar to that of the third generation but provide additional protection against Gram-negative bacteria exhibiting antimicrobial resistance, such as those producing beta-lactamases. Fifth-generation cephalosporins are effective against methicillin-resistant Staphylococci and penicillin-resistant pneumococci.

First-generation cephalosporins are used to treat bone infections, biliary complications, respiratory issues, genitourinary infections, otitis media, septicemia, and surgical prophylaxis. Cefazolin is the preferred cephalosporin for surgical prophylaxis.

Second-generation cephalosporins target both gram-positive and gram-negative bacteria. However, they are generally less effective against gram-positive bacteria and more effective against gram-negative bacteria compared to first-generation cephalosporins. These antibiotics inhibit the synthesis of bacterial cell walls. Examples of second-generation cephalosporins include cefuroxime, cefprozil, cephamycin, cefotetan, and cefoxitin. Cefuroxime is also indicated for **Lyme disease** in pregnant women and children. Cephamycin is effective against anaerobes such as *Bacteroides* species. These medications are commonly prescribed to treat bronchiolitis or pneumonia. Other indications for second-generation cephalosporins overlap with those of first-generation cephalosporins, including infections of the bone, respiratory tract, genitourinary tract, biliary tract, bloodstream, otitis media, and surgical prophylaxis. In addition to addressing the gram-negative bacteria targeted by first-generation cephalosporins, second-generation cephalosporins also provide coverage against *H. influenzae* and *Moraxella catarrhalis.*

Third-generation cephalosporins include cefotaxime, cefpodoxime, cefoperazone, ceftazidime, cefdinir, ceftriaxone, and cefixime. These antibiotics effectively target gram-negative infections and resist first- and second-generation beta-lactam antimicrobials. Their activity spectrum against gram-negative bacteria includes *Enterobacter* spp. and *Klebsiella aerogenes.* Ceftriaxone treats meningitis caused by *H. influenzae, Neisseria meningitidis,* or *Streptococcus pneumoniae. Additionally, it* is used to treat gonorrhea and Lyme disease. Notably, ceftazidime provides coverage against *Pseudomonas aeruginosa.*

The fourth-generation cephalosporin includes cefepime, a broad-spectrum antimicrobial capable of penetrating the cerebrospinal fluid. Cefepime possesses an additional quaternary ammonium group, which enhances its ability to penetrate the outer membrane of Gram-negative bacteria. Like cefotaxime and ceftriaxone, cefepime effectively targets *Streptococcus pneumoniae* and methicillin-sensitive *Staphylococcus aureus.* Additionally, similar to ceftazidime,

cefepime is notably effective against *Pseudomonas aeruginosa.* Beyond the gram-negative bacteria addressed by third-generation cephalosporins (*Neisseria* spp., *H. influenzae,* and Enterobacteriaceae), cefepime also provides coverage against beta-lactamase-producing gram-negative bacilli. While cefepime is effective against both Gram-positive and Gram-negative bacteria, it is reserved for treating severe systemic infections.

The fifth-generation cephalosporin includes ceftaroline, a broad-spectrum antibiotic that targets both susceptible gram-positive and gram-negative organisms. Ceftaroline's efficacy against MRSA distinguishes it from other cephalosporins. It is also effective against *Listeria monocytogenes* and *Enterococcus faecalis.*

CLINICAL INDICATIONS

Cephalosporins are used to treat various bacterial infections, particularly in patients with a penicillin allergy. They treat meningitis, sinusitis, strep throat, otitis media, pneumonia, urinary tract infections, and gonorrhea.

Mechanism of Action

Cephalosporins' mechanism of action is similar to that of penicillin. Like penicillins, cephalosporins bind to penicillin-binding proteins, disrupting cell wall enzymes and leading to bacterial death.

Adverse Effects

Cephalosporins are generally considered safe and exhibit low toxicity. Possible adverse effects may include loss of appetite, nausea, vomiting, diarrhea, and abdominal pain. Less common adverse reactions may include hypersensitivity, **pseudomembranous colitis**, vitamin K deficiency, and hemolytic anemia.

Contraindications

Cephalosporins are contraindicated for patients with allergies. Caution should be taken for those with a history of anaphylactic reactions to penicillin or other beta-lactam antimicrobials. The cross-reactivity is thought to arise from similar side chains found in some cephalosporins and certain penicillin products; however, it does not involve the beta-lactam ring.

Check Your Knowledge

1. What are the categories of cephalosporins?
2. What are the indications for cephalosporins?
3. What are the adverse effects of cephalosporins?

TETRACYCLINES

Tetracyclines are classified as antibiotics that inhibit protein synthesis and are considered broad-spectrum. This activity reviews the indications, mechanisms of action, and contraindications of tetracyclines, highlighting their value in

TABLE 31.4
Various Tetracyclines

Generic Name	Trade Name
Chlortetracycline	Aureomycin
Oxytetracycline	Terramycin
Tetracycline	Achromycin
Demeclocycline	Declomycin
Rolitetracycline	Reverin
Lymecycline	Tetralysal
Clomocyline	Megaclor
Methacycline	Rodomycin
Doxycycline	Vibramycin
Minocycline	Minocin
Tigecycline	Tygacil
Eravacycline	Xerava
Omadacycline	Nuzyra

treating bacterial infections. Discovered in the 1940s, they effectively target many gram-positive and gram-negative bacteria and atypical organisms such as **chlamydiae, mycoplasmas**, rickettsiae, and protozoan parasites. Tetracyclines have the advantage of causing minimal adverse effects, contributing to their widespread use in treating human infections. While tetracyclines play crucial roles in human medicine, the rise of microbial resistance has diminished their effectiveness. In clinical practice, these antibiotics have contributed to the selection of resistant organisms. Various tetracyclines are detailed in Table 31.4.

CLINICAL INDICATIONS

Tetracyclines are used to treat infections caused by bacteria, including respiratory tract infections and specific infections of the skin, eyes, intestines, lymphatic system, genitals, and urinary tract. They are also recommended for other diseases transmitted by ticks, lice, mites, and infected animals. Additionally, **tetracyclines** are used prophylactically to prevent malaria. This medication should be taken 2–3 hours before or after using calcium, magnesium, or aluminum products.

Mechanism of Action

Tetracyclines inhibit protein synthesis by blocking aminoacyl-tRNA attachment to the ribosomal acceptor A site, specifically targeting the 30S ribosomal subunit. When the binding process halts, the cell cannot maintain proper function and cannot grow or replicate further.

Adverse Effects

The most common adverse effects of tetracyclines include gastrointestinal issues such as abdominal pain, **anorexia**, nausea, vomiting, diarrhea, and loss of appetite. Some patients may experience **photosensitivity** while taking the medication. Additionally, discoloration of teeth and inhibition of bone growth in children may occur during

tetracycline use. Rare adverse effects of tetracyclines include hepatotoxicity, intracranial hypertension, worsening renal failure, and esophageal strictures.

Contraindications

Tetracyclines should be avoided in patients with known allergies and in children under 8 years old. These medications are contraindicated during pregnancy due to risks such as hepatotoxicity for the mother, permanent tooth discoloration in the fetus, and potential impairment of fetal long bone growth. Moreover, tetracycline is associated with tooth discoloration in children under eight years of age. Therefore, these drugs should generally be avoided in pediatric patients below this age. If necessary, doxycycline is the preferred option.

YOU SHOULD REMEMBER

Antacids containing aluminum, magnesium, calcium, iron, zinc, or sodium bicarbonate may reduce the effectiveness of tetracyclines. Consequently, certain foods rich in these cations and some dairy products can affect absorption.

Check Your Knowledge

1. What is the mechanism of action for tetracyclines?
2. What are the indications for tetracycline?
3. What are the contraindications of tetracyclines?

MACROLIDES

Macrolides are a class of antibiotics used to treat various bacterial infections, including whooping cough, **walking pneumonia**, otitis media, strep throat, and uncomplicated skin infections. Table 31.5 lists different macrolides.

CLINICAL INDICATIONS

Macrolides effectively treat atypical community-acquired pneumonia, *Helicobacter pylori* (as part of triple therapy), chlamydia, and acute **urethritis**. They are also utilized to treat *Legionella* species and *Corynebacterium diphtheriae*. When penicillin is contraindicated, macrolides are preferred for group A streptococcal and pneumococcal infections.

TABLE 31.5
Various Macrolides

Generic Names	Trade Names
Azithromycin	Zithromax
Clarithromycin	Biaxin
Erythromycin	Emicin, ATS
Fidaxomicin	Dificid

Mechanism of Action

Macrolides bind to the 50S subunit of the bacterial ribosome, blocking the initiation of protein synthesis. All macrolides possess a large lactone ring, which inhibits bacterial growth.

Adverse Effects

The most common side effects of macrolides include headaches, seizures, confusion, hallucinations, vertigo, altered taste, **dyspepsia**, nausea, diarrhea, and abdominal pain. Pseudomembranous colitis may develop from macrolide use and can range in severity from mild to life-threatening.

Contraindications

Macrolides are contraindicated for patients who have had an allergic reaction to them. They should be avoided during pregnancy and breastfeeding. Additionally, macrolides are contraindicated for patients with myasthenia gravis, elderly individuals, and those with cardiovascular disease.

Check Your Knowledge

1. What are the brand names of azithromycin, clarithromycin, and fidaxomicin?
2. What are the indications for macrolides?
3. What are the contraindications for macrolides?

FLUOROQUINOLONES

Fluoroquinolones are a group of broad-spectrum systemic antibacterial drugs commonly used to treat respiratory and urinary tract infections. Currently available in the United States, they include ciprofloxacin, delafloxacin, gemifloxacin, levofloxacin, moxifloxacin, norfloxacin, and ofloxacin. These medications are well absorbed when taken orally and are generally well tolerated, with a low incidence of adverse effects. Fluoroquinolones are classified into three generations: first, second, and third.

The first generation includes cinoxacin and nalidixic acid. Today, these medications are used less frequently due to low serum levels and an increased risk of bacterial resistance. They are only recommended for uncomplicated urinary tract infections and are not advised for patients with poor renal function.

Second-generation drugs have enhanced their activity against gram-negative bacteria and offer broad clinical applications for treating complicated urinary tract infections, sexually transmitted diseases, pneumonia, and skin infections. These drugs include ciprofloxacin, enoxacin, lomefloxacin, norfloxacin, and ofloxacin. Ciprofloxacin is particularly effective against *P. aeruginosa* and can treat osteomyelitis caused by susceptible organisms. Ofloxacin exhibits excellent activity against *Chlamydia trachomatis*.

The third generation, which includes levofloxacin, gatifloxacin, moxifloxacin, and sparfloxacin, has expanded activity against gram-positive organisms, particularly both penicillin-sensitive and penicillin-resistant *S. pneumoniae*,

---

TABLE 31.6

Various Fluoroquinolones

Generic Name	Trade Name
Ciprofloxacin	Cipro
Delafloxacin	Baxdela
Gatifloxacin	Zymaxid
Gemifloxacin	Factive
Levofloxacin	Levaquin
Moxifloxacin	Avelox
Norfloxacin	Noroxin
Ofloxacin	Floxin

---

as well as atypical pathogens such as *Mycoplasma pneumoniae* and *Chlamydia pneumoniae*. These antibiotics help treat community-acquired pneumonia, acute sinusitis, acute exacerbations of chronic bronchitis, urinary tract infections, and gonorrhea. Table 31.6 lists various fluoroquinolones.

CLINICAL INDICATIONS

Fluoroquinolones should be prescribed only to patients who need broad-spectrum antibiotics, specifically those with severe β-lactam allergies or those requiring treatment for organisms resistant to first-line agents. They are suitable for various bacterial infections, including sinusitis, bronchitis, pneumonia, urinary tract infections, septicemia, osteomyelitis, prostatitis, typhoid fever, and anthrax.

Mechanism of Action

Fluoroquinolones block DNA replication and inhibit bacterial cell synthesis and division, ultimately leading to the death of the bacteria. They function by inhibiting two enzymes involved in bacterial DNA synthesis. These DNA topoisomerases are absent in human cells but are crucial for bacterial DNA replication, allowing these agents to be both specific and **bactericidal**.

Adverse Effects

The adverse effects of fluoroquinolones are generally manageable. The most common adverse effects include nausea, dyspepsia, and vomiting. However, severe side effects can involve tendonitis, **Achilles tendon** rupture, peripheral neuropathy, myalgia, fatigue, arthralgia, exacerbation of myasthenia gravis, difficulty walking, burning pain, depression, memory loss, insomnia, and changes in vision, hearing, taste, and smell.

Contraindications

Fluoroquinolones are contraindicated for patients with hypersensitivity. They should be avoided in children under 18 due to the risk of cartilage damage. Caution is advised when using fluoroquinolones in older adults with renal impairment and those taking systemic corticosteroids. Fluoroquinolones should not be administered alongside calcium supplements, which can decrease their bioavailability, or with other medications that prolong the QT interval.

Check Your Knowledge

1. What classifications are there for fluoroquinolones?
2. What are the uses of fluoroquinolones?
3. What are the adverse effects of fluoroquinolones?

AMINOGLYCOSIDES

Aminoglycosides are natural or semi-synthetic antibiotics derived from **actinomycetes**. They gained widespread use as first-line agents during the early days of antimicrobial drugs but were ultimately replaced in the 1980s by cephalosporins, carbapenems, and fluoroquinolones. Aminoglycosides are broad-spectrum, bactericidal antibiotics frequently prescribed for children, primarily for infections caused by gram-negative bacteria. Table 31.7 lists various aminoglycosides.

CLINICAL INDICATIONS

Aminoglycosides are used in empiric therapy alongside other antibiotics for patients with severe conditions such as **endocarditis**, sepsis, and complicated genitourinary infections. Generally, aminoglycosides should not be administered for more than 48 hours due to the risk of toxicity associated with prolonged use. Aminoglycoside monotherapy is recommended for treating **tularemia** in children and for multidrug-resistant gram-negative pathogens.

Mechanism of Action

Aminoglycosides exhibit bactericidal activity. They bind to the bacterial ribosomal 30S subunit, specifically the 16S rRNA, a component of this subunit. This binding causes misreading of the genetic code and disrupts translation, thus hindering the bacteria's ability to perform protein synthesis.

Adverse Effects

The primary adverse effects of aminoglycosides include ototoxicity, neuromuscular blockade, and nephrotoxicity. Gentamicin, tobramycin, and streptomycin are more likely to cause **vestibular** damage, while kanamycin and amikacin result in more significant **cochlear** damage. Kanamycin has been discontinued in the United States.

TABLE 31.7

Various Aminoglycosides

Generic Names	Trade Names
Gentamicin	Cidomycin, Garamycin
Amikacin	Amikin
Tobramycin	Nebcin
Neomycin	(Generic only)
Plazomicin	Zemdri
Paromomycin	Humatin
Streptomycin	Rimosidin

Contraindications

Aminoglycosides should be avoided in patients with hypersensitivity to them and in those with myasthenia gravis due to the risk of prolonged neuromuscular blockade.

YOU SHOULD REMEMBER

Neomycin warns of several potential adverse effects, including ototoxicity, nephrotoxicity, neurotoxicity, and respiratory paralysis.

Check Your Knowledge

1. What are the brand names for gentamicin, amikacin, and streptomycin?
2. What are the indications for aminoglycosides?
3. What is the mechanism through which aminoglycosides act?

VANCOMYCIN

Vancomycin (Vancocin®) is a tricyclic glycopeptide antibiotic that targets bacteria in the intestines. It is the preferred treatment for MRSA infections and is typically administered as empirical therapy for hospitalized patients suspected of having these infections.

CLINICAL INDICATIONS

Vancomycin treats and prevents various bacterial infections caused by gram-positive bacteria, including MRSA. It is also effective against infections from streptococci, staphylococci, and enterococci. Additionally, it is used for patients with rheumatic fever or heart valve disease who are allergic to penicillin. Furthermore, vancomycin is prescribed for *Clostridioides difficile*-associated diarrhea, pseudomembranous colitis, *Staphylococcus enterocolitis*, osteomyelitis, bacteremia, and endocarditis.

Mechanism of Action

Vancomycin is bactericidal because it inhibits the polymerization of peptidoglycans in the bacterial cell wall. As a result, it prevents the synthesis and polymerization of the peptidoglycan layer. This inhibition weakens the bacterial cell walls, ultimately leading to the leakage of intracellular components and, consequently, bacterial cell death.

Adverse Effects

The adverse effects of intravenous vancomycin injection include hypersensitivity reactions, hypotension, and nephrotoxicity. Symptoms may involve pruritus, flushing, skin rashes, chills, drug fever, eosinophilia, and reversible neutropenia.

Contraindications

Vancomycin is contraindicated for patients with known hypersensitivity. Rapid bolus administration can lead to hypotension, shock, and cardiac arrest. Caution is advised when administering it to patients with inflammatory bowel diseases such as ulcerative colitis, Crohn's disease, kidney disease, or hearing loss.

YOU SHOULD REMEMBER

The signs and symptoms of bacteremia include high fever, chills, tachycardia, confusion, hypotension, nausea, vomiting, and hyperventilation. The most common bacterial causes of bloodstream infections are *E. coli*, *Staphylococcus aureus*, and *Streptococcal* species.

CLINICAL CASE STUDIES

Clinical Case Study 1

A 9-year-old girl visits the walk-in clinic for an injection of penicillin G. A medical assistant administers the intramuscular injection, and the patient is instructed to remain in the office for 20 minutes. After 10 minutes, she experiences severe allergic reactions.

CRITICAL THINKING QUESTIONS

1. What are the indications for penicillin G?
2. What are the signs and symptoms of allergic reactions?
3. What treatments are available for allergic reactions to penicillin G?

Clinical Case Study 2

A 78-year-old man was admitted to a local hospital due to fever, chills, tachycardia, confusion, hypotension, nausea, vomiting, and hyperventilation. The patient began treatment with vancomycin, ceftriaxone, and metronidazole as empirical therapy for bacteremia. His condition declined, prompting a switch from ceftriaxone to cefepime. Blood cultures confirmed the presence of *Staphylococcus aureus,* identifying the methicillin-resistant strain of the bacterium. The patient showed improvement following his treatment.

CRITICAL THINKING QUESTIONS

1. What are the indications for vancomycin?
2. What are the mechanisms of action for vancomycin?
3. What are the adverse effects of vancomycin?

FURTHER READING

Alton, J., and Alton, A. (2018). *Alton's Antibiotics and Infectious Disease — The Layman's Guide to Available Antibacterials in Austere Settings*. Alton First Aid LLC.

Alves, R. (2016). *Aminoglycosides — Pharmacology, Clinical Uses and Health Effects*. Nova Science Publishers Inc.

Arsic, B., Novak, P., Kragol, G., Barber, J., Grazia Rimoli, M., and Sodano, F. (2018). *Macrolides — Properties, Synthesis, and Applications*. De Gruyter.

Bagchi, D., Das, A., and Downs, B.W. (2022). *Viral, Parasitic, Bacterial, and Fungal Infections — Antimicrobial, Host Defense, and Therapeutic Strategies*. Academic Press.

Gafar Hossion, A. (2014). *Vancomycin — Biosynthesis, Clinical Uses and Adverse Effects* (Pharmacology Research Safety Testing and Regulation). Nova Science Publishers Inc.

Gallagher, J.C., and MacDougall, C. (2022). *Antibiotics Simplified*, 5th Edition. Jones & Bartlett Learning.

Gauthier, T.P. (2022). *Learn Antibiotics — A Collection of Resources for Antimicrobial Drugs*. Gauthier.

Hauser, A.R. (2022). *Antibiotic Basics for Clinicians*, 3rd Edition. Wolters Kluwer.

Levine, B. (2022). *EMRA Antibiotic Guide*, 20th Edition. Emergency Medicine Residents' Association.

Lopez, J. (2023). *Tetracycline Handbook — A Clinician's Guide to Antifungal and Bacterial Infection Mastery*. Lopez.

Mahmud, R., Ai Lian Lim, Y., and Amir, A. (2017). *Medical Parasitology — A Textbook*. Springer.

Rusu, A., Uivarosi, V., and Tanase, C. (2023). *Focus on Antibiotics — New Challenges and Steps Forward in Discovery and Development*. MDPI.

Sarkar, D. (2020). *Sulfonamides — An Overview*. Nova Science Publishers Inc.

Walsh, C., and Wencewicz, T. (2016). *Antibiotics — Challenges, Mechanisms, Opportunities*. ASM Press.

32 Antiviral Agents

LEARNING OUTCOMES

After studying the chapter, readers should be able to

1. Review the general characteristics of viruses.
2. Explain the types of influenza viruses.
3. Describe the indications and adverse effects of acyclovir.
4. Review causes of coronavirus disease 2019 (COVID-19) therapies and the vaccines used in the United States.
5. Discuss the human papillomavirus (HPV) immunizations.
6. List the drugs that are available to treat hepatitis B.
7. Outline four stages for addressing acquired immunodeficiency syndrome (AIDS).
8. Describe the adverse effects of abacavir.
9. Explain the adverse effects of lopinavir/ritonavir.
10. Review the contraindications of tipranavir.

OVERVIEW

Viruses are tiny infectious agents that rely on living cells to multiply. They may use an animal, plant, or bacterial host to survive and reproduce. As such, there is some debate about whether viruses should be considered living organisms. A virus outside a host cell is known as a virion. They consist of small genetic material, either DNA or RNA, enclosed in a protective shell known as a capsid. Some viruses also possess an envelope. Viruses cannot reproduce without a host. Common diseases caused by viruses include influenza, human herpes viruses, enteroviruses, hepatitis viruses, HPVs, and COVID-19. They are obligate intracellular parasites that must reside within a living host cell to survive and replicate. Viruses invade healthy cells, multiply within them, and utilize their DNA and RNA to produce more viruses. The growth cycle of viruses relies on host cell enzymes and cellular substrates for viral replication. Except for human immunodeficiency virus (HIV) and certain types of viral hepatitis, viruses typically cause self-limiting illnesses that usually do not require treatment with specific antivirals. Current antiviral medications target influenza, herpes, and hepatitis. Essentially, viruses are particles that infect and replicate within cells.

GENERAL CHARACTERISTICS OF VIRUSES

Viruses are infectious agents that exhibit traits of both living and nonliving entities. They can infect humans, animals, plants, and even other microorganisms. Viruses that specifically target bacteria are called **bacteriophages**,

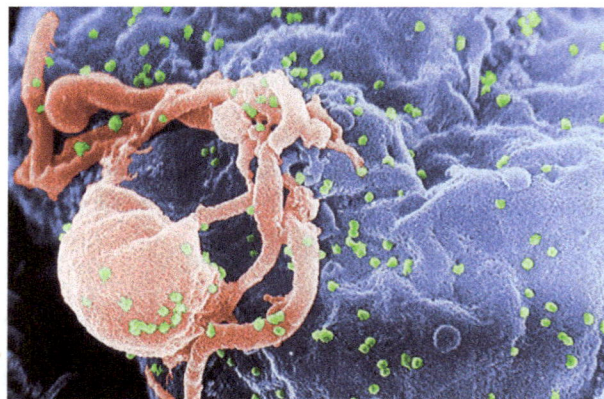

FIGURE 32.1 The HIV virus.

while those that attack fungi are known as **mycophages**. Viruses can only replicate inside the cells of their hosts. They have a simple structure that includes a **genome**, a **capsid**, and an **envelope**. The genomes, which form the inner core of the **virion**, consist of either single-stranded or double-stranded deoxyribonucleic acid (DNA) or ribonucleic acid (RNA) molecules, but not both. The viral capsid is a protein coat that surrounds and protects the genome while facilitating the fusion process between the virion and the host cell. Fusion occurs when virions attach to host cells, preparing them for infection. The envelope is the outermost layer of the virion and is present in some, but not all, viruses. A depiction of HIV is shown in Figure 32.1. The HIV-1 virion consists of a core comprising identical strands of viral RNA molecules and viral enzymes, all enclosed within a core capsid structure primarily composed of the structural viral protein p24.

Viruses are classified based on their size, shape, and type of genetic material. Examples include herpesviruses, influenza viruses, enteroviruses, hepatitis viruses, orthopoxviruses, HPVs, and coronaviruses. Additionally, some viruses exhibit unique characteristics, such as retroviruses and oncoviruses.

Effective drug therapies are available for only a limited range of viral infections. The development of vaccines has significantly reduced the incidence of several viral infections. Vaccines for diseases such as polio, measles, chickenpox, hepatitis B, smallpox, and COVID-19 serve as notable examples.

INFLUENZA VIRUSES

Influenza viruses belong to the Orthomyxoviridae family, which includes influenza A and B as the primary causes of the flu. Strains of influenza A can also result in avian flu

DOI: 10.1201/9781003461913-35

TABLE 32.1

Drugs for Influenza Infections

Generic Names	Trade Names
Amantadine	Symmetrel
Baloxavir	Xofluza
Rimantadine	Flumadine
Oseltamivir	Tamiflu
Peramivir	Rapivab
Zanamivir	Relenza

(bird flu or H5N1) and swine flu (H1N1). Seasonal influenza is an acute respiratory infection that may occasionally require hospitalization. Most patients recover from symptoms within a week without needing medical treatment.

However, it can lead to severe illness or death, particularly among high-risk groups such as infants, older adults, pregnant women, and individuals with severe medical conditions. Seasonal epidemics mainly occur during the winter. Table 32.1 provides examples of medications for influenza. Amantadine and rimantadine are no longer recommended for preventing or treating influenza A due to the rising resistance of circulating viral strains. This discussion of oseltamivir (Tamiflu) illustrates a drug used to treat influenza.

OSELTAMIVIR

Oseltamivir is an antiviral agent primarily used to treat and prevent influenza types A and B. The recommended dosage for treatment is 75 mg, taken orally twice daily for 5 days in adults and adolescents aged 13 and older. For prevention, the recommended dosage is 75 mg, taken orally once daily for at least 10 days. Oseltamivir should be started within 48 hours of symptom onset for treatment and within 48 hours of exposure for prophylaxis.

Clinical Indications

Oseltamivir treats infections caused by influenza A or B. It may alleviate flu symptoms such as fever, headache, cough, weakness, runny or stuffy nose, and sore throat by 1 day. Additionally, oseltamivir is used to prevent influenza in individuals in close contact with someone infected with the flu.

Mechanism of Action

Oseltamivir inhibits the neuraminidase enzyme, which is essential for the release of new virus particles from infected cells. By blocking this enzyme, oseltamivir reduces the spread of the influenza virus in the respiratory tract.

Adverse Effects

Oseltamivir is typically well tolerated. Common adverse effects of this drug include nausea, vomiting, abdominal pain, diarrhea, headache, insomnia, and dizziness. Severe adverse effects can include delirium, hallucinations, and unusual behavior.

Contraindications

Oseltamivir is contraindicated for patients who have a known hypersensitivity to the drug or any of its components. Caution is advised for patients with renal impairment.

Check Your Knowledge

1. What are the clinical indications for oseltamivir?
2. What is the mechanism of action of oseltamivir?
3. What are the adverse effects of oseltamivir?

HUMAN HERPESVIRUS

Human herpesvirus (HHV) is a large family of viruses that includes several types, such as herpes simplex virus type 1 (HSV-1), HSV-2, varicella–zoster virus, cytomegalovirus, and Epstein–Barr virus. Although there is some structural overlap between the two types of HSV, HSV-1, commonly called oral herpes, primarily affects the mouth and face.

In contrast, HSV-2 infection is most commonly associated with blisters on both male and female genitalia, known as **genital herpes**. It is also linked to perioral blisters. HSV is highly contagious and can be severe or potentially life-threatening for immunocompromised patients or newborn infants. The most effective way to prevent transmission to newborns is typically through **cesarean section** delivery for any mother with active genital herpes lesions.

HHV-3, also known as **varicella–zoster virus**, primarily causes chickenpox (varicella) during childhood, remains **dormant** for many years, and can later reactivate in adulthood as shingles. These lesions typically appear in a similar pattern on the body (see Figure 32.2). Chickenpox is generally a self-limiting illness in children. It is highly contagious and spreads quickly through direct contact with open lesions. The most common clinical manifestation of shingles is skin lesions that follow nerve pathways, known as dermatomes, along the skin surface. The lesions are extremely painful.

Table 32.2 lists medications for herpesvirus infections. The varicella virus vaccine is now routinely recommended for healthy children over 1 year old who have not previously had chickenpox. Two vaccines are available to prevent herpes zoster in individuals aged 50 and older. The discussion of acyclovir covers a drug used to treat HSV infections.

ACYCLOVIR

Acyclovir is an agent used to treat HSV infections, genital herpes, and HSV encephalitis.

Clinical Indications

Acyclovir is used to treat the symptoms of chickenpox, **shingles**, and herpes virus infections of the genitals, lips, mouth, skin, brain, and newborns. It is also used to prevent recurrent genital herpes infections.

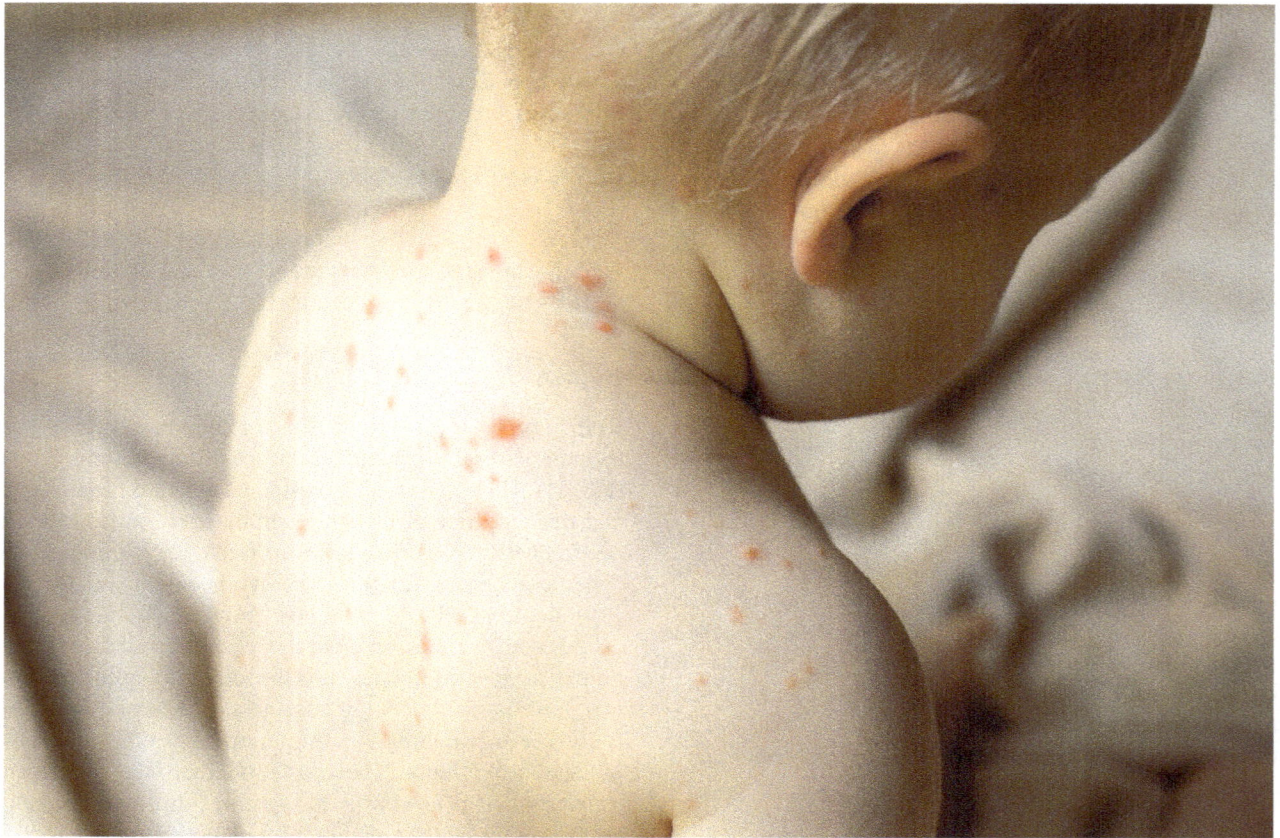

FIGURE 32.2 Chickenpox lesions.

TABLE 32.2

Medications for Herpesvirus Infections

Generic Name	Trade Name
Acyclovir	Zovirax
Brincidofovir	Tembexa
Cidofovir	Vistide
Famciclovir	Famvir
Foscarnet	Foscavir
Ganciclovir	Cytovene
Valacyclovir	Valtrex
Valganciclovir	Valcyte
Docosanol	Abreva
Penciclovir	Denavir
Trifluridine	Viroptic

Mechanism of Action

Acyclovir is an antiviral drug that explicitly inhibits herpesvirus DNA polymerase by serving as an analog to deoxyguanosine triphosphate.

Adverse Effects

Acyclovir's adverse effects may include malaise, headaches, confusion, dizziness, agitation, alopecia, seizures, sensitivity to sunlight, itchy, dry, flaky skin, abdominal pain, nausea, vomiting, and diarrhea.

Contraindications

Acyclovir is contraindicated in patients with hypersensitivity to the medication. It should be used cautiously when administered intravenously to patients with renal disease. Although it can cross the placenta, it may still be used cautiously in pregnant individuals.

Check Your Knowledge

1. What are the types of HHVs?
2. What are the differences between chickenpox and shingles?
3. What are the indications for acyclovir?
4. What are the adverse effects of acyclovir?

CORONAVIRUSES

Coronaviruses are a large family of viruses that usually cause mild to moderate illnesses of the upper respiratory tract in humans. However, three coronaviruses have resulted in more severe and sometimes fatal diseases in humans: the severe acute respiratory syndrome coronavirus (SARS-CoV), which surfaced in November 2002 and causes SARS; and SARS-CoV-2, which appeared in 2019 and causes COVID-19.

As of today, COVID-19 has led to global cases and 6,979,269 deaths. In the United States alone, there have been over 1 million fatalities. While rare, transmission of COVID-19 from animals to humans has occurred; however, it is more commonly spread through close person-to-person contact, primarily via respiratory droplets. Additionally, the virus can spread through contact with contaminated surfaces. High transmission risks exist in long-term care facilities, nursing homes, ships, jails, prisons, churches, gymnasiums, and restaurants.

Most patients with COVID-19 may experience few symptoms or none at all. However, severe illnesses and deaths have primarily occurred among older adults, individuals with obesity, smokers, and patients with concurrent conditions such as diabetes mellitus, cancer, immunocompromised status, and sickle cell disease. Severe complications from COVID-19 include acute respiratory distress syndrome, arrhythmias, cardiomyopathy, myocarditis, thromboembolism, pulmonary embolism, disseminated intravascular coagulation, and sepsis.

The treatment for COVID-19 primarily consists of supportive care, which includes mechanical ventilation, corticosteroids, immunomodulators, and vasopressors. Remdesivir and nirmatrelvir/ritonavir are two medications approved for treating COVID-19 in the United States and will be discussed in this chapter.

REMDESIVIR

Remdesivir is an antiviral medication for COVID-19. It is administered intravenously to hospitalized adults and children who weigh at least 3 pounds.

Clinical Indications

Remdesivir treats COVID-19 in hospitalized patients who need supplemental oxygen, mechanical ventilation, or extracorporeal membrane oxygenation. It shortens recovery time for these individuals and is not recommended for patients with mild illness who do not require hospitalization. Remdesivir may also be given outside the hospital to individuals at high risk of progressing to severe COVID-19.

Mechanism of Action

Remdesivir inhibits viral RNA polymerases, preventing the replication of RNA viruses such as SARS-CoV-2. Integrating into viral RNA chains, remdesivir disrupts viral RNA synthesis, ultimately hindering viral replication and reducing the viral load in infected individuals.

Adverse Effects

The adverse effects of remdesivir include nausea, vomiting, diarrhea, and increased liver enzymes. Additionally, remdesivir may lead to an infusion-related reaction characterized by low blood pressure, nausea, vomiting, sweating, and shivering.

Contraindications

Remdesivir should be avoided in patients with known hypersensitivity to this agent or its components. It is contraindicated in patients with severe liver or kidney damage.

Check Your Knowledge

1. What are the indications for remdesivir?
2. What are the adverse effects of remdesivir?
3. What are the contraindications of remdesivir?

NIRMATRELVIR/RITONAVIR

The Nirmatrelvir/ritonavir combination (Paxlovid) is an antiviral agent used to treat COVID-19 in adults and children (12 years of age and older) at high risk of developing severe illness.

Clinical Indications

Paxlovid is indicated for treating mild to moderate COVID-19 in patients aged 12 and older who are at high risk for progressing to severe disease. This includes individuals with specific underlying medical conditions or those who are immunocompromised.

Mechanism of Action

Paxlovid inhibits the activity of the SARS-CoV-2 main protease (3CL protease), which is essential for viral replication. Inhibiting this enzyme prevents the virus from replicating and spreading within the body, thereby reducing the viral load and slowing disease progression.

Adverse Effects

The side effects of Paxlovid include headaches, fatigue, nausea, vomiting, diarrhea, and elevated liver enzymes.

Contraindications

Paxlovid should be avoided in patients with known hypersensitivity to this agent. It is contraindicated in patients with severe hepatic impairment.

HUMAN PAPILLOMAVIRUSES

HPVs are viral infections that commonly affect the skin or mucous membranes. There are over 100 types of HPV. Some types cause genital **warts**, while others can lead to various cancers (see Figure 32.3). HPV is a double-stranded DNA virus and is the most prevalent sexually transmitted infection in the United States. More than 42 million Americans are infected with different types of HPV, primarily affecting individuals in their late teens and early 20s. Vaccinating all 11- to 12-year-olds can provide protection well before exposure. The CDC recommends two doses of the HPV vaccine for all adolescents at ages 11 or 12. A two-dose schedule is advised for those receiving their first dose before age 15, with the second dose given 6–12 months after the first.

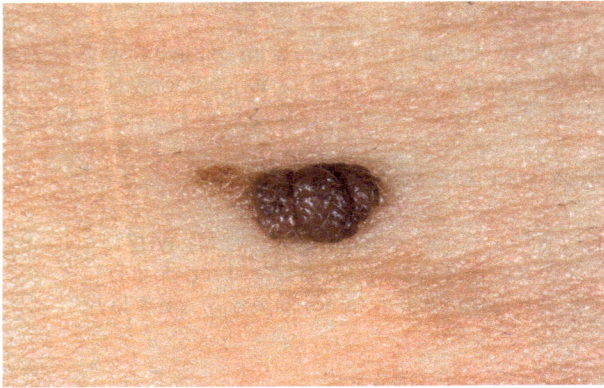

FIGURE 32.3 Human papillomavirus.

In women, HPV complications can range from common problems such as genital warts to, in rare instances, cancers of the cervix, vulva, vagina, and even the head and neck, with cervical cancer being the most prevalent HPV-related cancer. Because early cervical cancer doesn't cause symptoms, women must have regular screening tests to detect any precancerous changes in the cervix that might lead to cancer. Current guidelines recommend that women ages 21–29 have a Pap test every 3 years.

YOU SHOULD REMEMBER

A three-dose schedule is recommended for individuals receiving their first dose after turning 15. In a three-dose series, the second dose should be administered 1–2 months after the first, while the third dose should be given 6 months after the first. The minimum interval between the first and second doses is 4 weeks. If the vaccination schedule is interrupted, there is no need to repeat vaccine doses, as there is no maximum interval.

ENTEROVIRUSES

Enterovirus, a genus within the family *Picornaviridae*, comprises enteroviruses, coxsackieviruses, rhinoviruses, polioviruses, and echoviruses. These viruses are responsible for various illnesses, from the common cold to **poliomyelitis** and aseptic meningitis. In humans, they are among the most common infectious agents worldwide.

Rhinovirus species within the *Enterovirus* genus infect the upper and lower respiratory tracts, and they rarely cause disseminated disease. In infants, rhinovirus infections lead to cell damage in the respiratory tract and alter the immune response. These infections are an independent risk factor for developing recurrent wheezing and asthma. Rhinoviruses are the most prevalent respiratory viruses, responsible for about two-thirds of viral upper respiratory tract infections associated with asthma exacerbations.

There is no specific cure, so treatment focuses on alleviating symptoms, including rest, hydration, and the use of over-the-counter medications.

RETROVIRUSES

Retroviruses represent a large and diverse family of enveloped RNA viruses. They are highly contagious and can lead to severe diarrhea. Before a vaccine was developed, most children had been infected with the virus at least once by age 5. Since the introduction of the vaccine, the incidence of rotavirus has decreased by 80% in the United States. Rotavirus gastroenteritis is common in children under 5 years old. Early signs and symptoms include fever, abdominal pain, vomiting, and 3–7 days of watery diarrhea.

HEPATITIS VIRUSES

Hepatitis viruses are a group of viruses that cause liver inflammation. They can be either acute or chronic. Acute hepatitis is usually treatable, whereas chronic hepatitis can lead to liver failure. Viral hepatitis includes types A, B, C, D, and E, with A, B, and C being the most common types discussed in this chapter.

- Hepatitis A (HepA) replicates in the liver and spreads through the fecal–oral route, either from contact with an infected person or from ingesting contaminated food or water. The incubation period ranges from 2 weeks to 6 months after exposure. Symptoms may include loss of appetite, nausea, vomiting, abdominal pain, fatigue, fever, jaundice, dark urine, and light-colored stools. The Hep A vaccine is an inactivated whole-virus vaccine recommended as part of routine immunizations for all children starting at 1 year of age. It consists of two doses, with the second administered at least 6 months after the first. Individuals with moderate to severe illness should wait until they recover before vaccination.

- Hepatitis B (Hep B) can be either acute or chronic. Chronic hepatitis B can lead to **cirrhosis**, liver failure, and liver cancer. The signs and symptoms of Hep B may include fatigue, abdominal pain, loss of appetite, nausea, vomiting, jaundice, dark urine, and joint pain. Blood tests, liver ultrasounds, and biopsy evaluations are used to confirm chronic hepatitis B. The virus can be found in blood, bodily fluids, semen, and serous fluids. Transmission can occur through sexual contact, **perinatal methods**, and exposure to contaminated needles. The Hep B vaccine is the only vaccine administered to babies at birth. Approximately 66% of individuals infected with the condition are unaware of their status. In 2024, the incidence of new cases was about 12 times higher among Asian Pacific

TABLE 32.3
Drugs for Viral Hepatitis

Generic Name	Trade Name	Indication
Adefovir dipivoxil	Hepsera	HBV
Elbasvir–grazoprevir	Zepatier	HCV
Entecavir	Baraclude	HBV
Glecaprevir/pibrentasvir	Mavyret	HCV
Lamivudine	Epivir	HBV
Sofosbuvir	Sovaldi	HCV
Sofosbuvir/velpatasvir	Epclusa	HCV
Telbivudine	Tyzeka	HBV
Tenofovir alafenamide	Vemlidy	HBV
Peg IFN alfa-2a	Pegasys	HCV, HBV

Islanders than among non-Hispanic Caucasians. Hepatitis B is one of the leading causes of liver cancer.

- Hep C infection has led to an epidemic, with approximately 170 million people infected worldwide and 3–4 million new infections each year. An increasing number of patients with cirrhosis will go on to develop **hepatocellular carcinoma**. Unfortunately, there is currently no vaccine for Hep C. New antiviral medications are the preferred treatment for most individuals with chronic hepatitis C infection, as these drugs can often cure chronic hepatitis C. The current standard therapy for Hep C includes pegylated interferon-α, administered once a week, and daily oral ribavirin for 24–48 weeks. Table 32.3 summarizes the medications used for viral hepatitis.

TENOFOVIR

Tenofovir alafenamide (TAF) is an antiviral drug approved by the FDA for the treatment of chronic HBV in adults and children aged 12 and older. An older version, tenofovir disoproxil, has been reformulated as TAF, which absorbs more rapidly. This means lower drug levels are present in the bloodstream, leading to fewer side effects related to the kidneys and bones.

Clinical Indications

TAF is prescribed to treat chronic HBV infection in patients with compensated liver disease. Additionally, it is effective against HIV and HSV-2.

Mechanism of Action

TAF inhibits HBV polymerase, thereby hindering DNA synthesis and the viral replication of HBV.

Adverse Effects

Tenofovir can lead to adverse effects, including decreased bone mineral density, fever, chills, headaches, dizziness, tachycardia, dyspnea, cough, chest pain, muscle stiffness, back pain, skin redness, loss of appetite, and diarrhea.

Contraindications

TAF should be used cautiously in patients with liver and kidney disease. It is contraindicated in severe kidney disease.

Check Your Knowledge

1. What are the indications for tenofovir?
2. What are the adverse effects of tenofovir?
3. What are the contraindications of tenofovir?

HIV INFECTION AND AIDS

HIV is a devastating viral infection. Over 40 million people are living with HIV worldwide, and over 1.3 million people in the United States. In 2023, there were 34,800 new people diagnosed with HIV in the United States.

HIV is a member of the retrovirus family and belongs to the RNA virus family. It is distinguished by its use of the enzyme reverse transcriptase during replication. This enzyme facilitates the synthesis of complementary DNA molecules from the viral RNA genome. A second enzyme, integrase, facilitates the integration of this viral DNA into the host cell's DNA.

Untreated or treatment-resistant HIV infection ultimately leads to severe immune system failure, resulting in death from **opportunistic infections**. AIDS typically progresses over several years. The World Health Organization (WHO) outlines four stages for addressing AIDS as follows:

- **Stage 1:** asymptomatic infection
- **Stage 2:** mild symptoms: early, general symptoms of the disease
- **Stage 3:** advanced symptoms
- **Stage 4:** severe symptoms, often leading to death

Patients who undergo effective drug therapy usually do not progress through all these stages, or their progression is significantly slowed. Advances in antiretroviral drug therapy have led to an increasing number of long-term survivors of HIV infection. Highly active antiretroviral therapy (ART) refers to the combination of antiretroviral drugs that are now standard for treating patients infected with HIV, starting right after diagnosis. Opportunistic malignancies, such as **Kaposi's sarcoma** and lymphomas, are treated with specific antineoplastic medications, radiation, and surgery when necessary. Table 32.4 lists the antiretroviral drugs used to treat HIV.

ABACAVIR

Abacavir is a nucleoside reverse-transcriptase inhibitor (NRTI) that is used to treat HIV infection in conjunction

TABLE 32.4
Antiretroviral Drugs Used to Treat HIV

Generic Name	Trade Name
Nucleoside Reverse Transcriptase Inhibitors	
Abacavir	Ziagen®
Abacavir/lamivudine	Epzicom®
Abacavir/zidovudine/lamivudine	Trizivir®
Didanosine (enteric coated)	Videx EC®
Didanosine (dideoxyinosine)	Videx®
Emtricitabine	Emtriva®
Lamivudine	Epivir®
Stavudine	Zerit®
Tenofovir disoproxil	Viread®
Tenofovir alafenamide	Descovy®
Tenofovir/emtricitabine	Truvada®
Zidovudine	Retrovir®
Nonnucleoside Reverse Transcriptase Inhibitors	
Delaviridine	Rescriptor®
Efavirenz	Sustiva®
Etravirine	Intelence®
Nevirapine	Viramune®
Protease Inhibitors	
Atazanavir	Reyataz®
Darunavir	Prezista®
Fosamprenavir	Lexiva®
Indinavir	Crixivan®
Lopinavir/ritonavir	Kaletra®
Nelfinavir	Viracept®
Ritonavir	Norvir®
Saquinavir mesylate	Invirase®
Tipranavir	Aptivus®
Fusion Inhibitor	
Enfuvirtide	Fuzeon®
Entry Inhibitor	
Maraviroc	Selzentry®
Capsid Inhibitor	
Lenacapavir	Sunlenca®
Integrase Inhibitors	
Cabotegravir	Apretude®
Dolutegravir	Tivicay®
Elvitegravir	Vitekta®
Raltegravir	Isentress®
Combination Therapies	
Tenofovir Alafenamide/emtricitabine/elvitegravir/ cobicistat	Genvoya®
Tenofovir alafenamide/emtricitabine/bictegravir	Biktarvy®

with other antiretroviral drugs. Like other NRTIs, abacavir use is typically combined with other HIV medications.

Clinical Indications

Abacavir is used to treat HIV-1 infection in combination with other antiretroviral drugs. It is commonly used with abacavir/

lamivudine/zidovudine, abacavir/dolutegravir/lamivudine, and abacavir/lamivudine.

Mechanism of Action

The effect of abacavir is due to its intracellular metabolite, carbovir triphosphate, which interferes with HIV viral RNA-dependent DNA polymerase, thereby inhibiting viral replication.

Adverse Effects

Abacavir's adverse effects include fever, cough, headache, rash, tachypnea, abdominal pain, nausea, vomiting, diarrhea, hematuria, and arthralgia.

Contraindications

Abacavir is contraindicated in patients with hypersensitivity to it, in people carrying the HLA-B*5701 allele, and in those with severe hepatic damage. It should be used with caution in patients with coronary artery disease and patients at increased risk of myocardial infarction due to hyperlipidemia and cardiovascular events.

Check Your Knowledge

1. What are the indications for abacavir?
2. What are the adverse effects of abacavir?
3. What are the contraindications of abacavir?

DIDANOSINE

Didanosine is in the NRTIs class and is used in combination with other medicines for the treatment of HIV infection. It cannot cure or prevent HIV infection or AIDS. But it slows down the destruction of the immune system from HIV. This may help delay the development of problems usually related to AIDS or HIV disease. Didanosine inhibits HIV replication by competing with naturally occurring adenosine for incorporation into the growing viral DNA chain, causing inhibition of the viral polymerase and chain termination. Didanosine is an effective and generally well-tolerated drug in previously untreated and ART-experienced patients with HIV infection. Given once or twice daily, it has a vital role as a component of triple combination regimens for the treatment of patients with symptomatic or asymptomatic HIV infection. Common adverse effects of didanosine include rash, fever, headache, diarrhea, nausea, vomiting, abdominal pain, and **asthenia**. Other adverse effects include severe hypersensitivity reactions, hepatotoxicity, severe metabolic acidosis, peripheral neuropathy, **lipodystrophy,** and pancreatitis.

EMTRICITABINE

Emtricitabine is in the NRTIs class used for the treatment of HIV infection in adults, children, and infants. It is always used in combination with other HIV medicines. Emtricitabine is also used in combination therapy for pre-exposure prophylaxis to prevent HIV infection. The drug works by inhibiting HIV reverse transcriptase, preventing transcription of HIV RNA to DNA. The adverse effects of emtricitabine are similar to those of other NRTIs and include headache, muscle weakness, arthralgia, fatigue, fever, abdominal pain, nausea, vomiting, diarrhea, depression, anxiety, insomnia, rhinitis, cough, and pharyngitis. One of the specific adverse effects of emtricitabine is hyperpigmentation of the palms and soles, especially in African American patients.

STAVUDINE

Stavudine is in the NRTI class with activity against type 1 HIV. It is used in combination with other agents in the treatment of HIV infection. Stavudine slows the progression of HIV in those who have advanced symptoms, early symptoms, or no symptoms at all. It helps to decrease the amount of HIV in the body, inhibits the HIV reverse transcriptase enzyme competitively, and acts as a chain terminator of DNA synthesis. Common adverse effects include diarrhea, headache, nausea, vomiting, and rash. Less common but potentially severe adverse reactions include an allergic reaction, pancreatitis, hepatotoxicity, tingling in the hands or feet, and peripheral neuropathy.

TENOFOVIR

Tenofovir is an NRTI used in combination with other drugs to treat HIV. It works for both HIV infection and chronic hepatitis B virus infections. Tenofovir blocks the effectiveness of reverse transcriptase, an enzyme required for each virus to make copies of itself. Blocking reverse transcriptase can reduce the amount of the virus in the blood. The more common adverse effects of tenofovir include headache, nausea, vomiting, rash, back pain, and depression.

ZIDOVUDINE

Zidovudine is used for treating HIV-1. It is in the NRTI class of drugs. It was the first medication to be approved by the FDA for the treatment of HIV. The adverse effects of zidovudine include headaches, myalgias, insomnia, nausea, vomiting, diarrhea, lactic acidosis, peripheral myopathy, hepatotoxicity, and bone marrow suppression. Zidovudine is contraindicated for patients with severe life-threatening hypersensitivity reactions to the drug, including anaphylaxis and Stevens–Johnson syndrome. Anaphylaxis may present with dyspnea, diffuse rash, and hypotension. Stevens–Johnson syndrome is an excruciating blistering skin condition. Prompt discontinuation of zidovudine therapy is critical if any of these symptoms arise.

NEVIRAPINE

Nevirapine is a drug used in the management and treatment of HIV. It is in the NNRTI class of medications and is utilized in conjunction with other ART agents. This medication is often used as part of newborn antiretroviral regimens to prevent perinatal transmission of HIV. Nevirapine can cause severe, life-threatening liver impairment and allergic reactions. The adverse effects include fever, sore throat, chills, rash, fatigue, nausea, loss of appetite, vomiting, hematuria, pale stools, jaundice, stomach pain, arthralgia, edema of the face, dysphasia, hoarseness, and dyspnea.

DARUNAVIR

Darunavir is in the class of protease inhibitors. It is used in combination with ritonavir and other drugs for the treatment of HIV. As a second-generation protease inhibitor, darunavir is designed to combat resistance to standard HIV therapy. The adverse effects of darunavir include loss of appetite, nausea, myalgia, arthralgia, and dark urine. Body fat redistribution may occur. Hypersensitivity reactions may also occur with use; darunavir must be discontinued in these cases. This medication should be used with caution in patients with severe liver disease, hypokalemia, and hypomagnesemia.

FOSAMPRENAVIR

Fosamprenavir is in the class of protease inhibitors. It is an antiretroviral agent used for the treatment and postexposure prophylaxis of HIV-1 infection. It may slow down the destruction of the immune system caused by HIV or delay the complications of AIDS. However, this medicine cannot cure or prevent HIV infection. The adverse effects of fosamprenavir include chills, fever, white spots in the mouth, sore throat, skin rash, bruising, itching, sweating, fatigue, chest pain, cough, nausea, vomiting, diarrhea, back pain, and hematuria.

INDINAVIR

Indinavir is an antiretroviral protease inhibitor used for the treatment and prevention of HIV infection and AIDS. The FDA approved indinavir in 1996 as it was among the first HIV protease inhibitors authorized in the United States. Indinavir has decreased the likelihood of death due to AIDS progression. It is no longer available in the United States or Europe as it has many adverse effects, and there are better drugs.

LOPINAVIR/RITONAVIR

Lopinavir/ritonavir is a protease inhibitor structurally related to ritonavir. The FDA approved it as a coformulation with ritonavir under the brand name Kaletra.

Clinical Indications

Lopinavir/ritonavir is specifically used for the treatment of HIV-1 infection in adults and children 2 weeks and older. It is no longer recommended as a component of initial therapy for the treatment of HIV due to the release of integrase inhibitor combination products.

Mechanism of Action

Lopinavir/ritonavir works with the lopinavir component, binding to the catalytic site of the HIV protease, thereby preventing the cleavage of viral polyprotein precursors into mature, functional proteins necessary for viral replication. The result is the formation of immature, noninfectious viral particles. The ritonavir component inhibits the CYP3A metabolism of lopinavir, allowing increased plasma levels of lopinavir.

Adverse Effects

It can cause severe, life-threatening adverse effects, which include allergic reactions such as toxic epidermal necrolysis, Stevens–Johnson syndrome, liver impairment, pancreatitis, hyperglycemia, hypertriglyceridemia, uremia, hyperbilirubinemia, dysrhythmia, and skin rash. Other possible adverse effects include headache, fatigue, abdominal pain, nausea, and diarrhea. Coadministration with drugs that are CYP3A inducers may decrease lopinavir plasma concentrations, potentially leading to a loss of virologic response and the development of resistance.

Contraindications

Lopinavir/ritonavir is contraindicated in patients with hypersensitivity, dysrhythmia, porphyria, pancreatitis, diabetes mellitus, heart disease, and hemophilia. Other adverse effects include diarrhea, dyslipidemia, and hepatic disorders.

RITONAVIR

Ritonavir is a protease inhibitor used together with other medicines for the treatment of HIV infection and AIDS in adults and children. It may delay the complications of AIDS or HIV infection from occurring. Ritonavir cannot prevent HIV from spreading to other people. Ritonavir binds to the protease active site and inhibits the enzyme's activity. This inhibition prevents cleavage of the viral polyproteins. The adverse effects of ritonavir include nausea, vomiting, abdominal pain, diarrhea, rhabdomyolysis, malaise, dizziness, hyperlipidemia, hypertriglyceridemia, and insomnia. Ritonavir contraindications include patients with a known allergic reaction to the medication. Hypersensitivity reactions may result in toxic epidermal necrolysis and Stevens–Johnson syndrome.

Check Your Knowledge

1. What are the indications for lopinavir/ritonavir?
2. What are the adverse effects of lopinavir/ritonavir?
3. What are the contraindications of lopinavir/ritonavir?

TIPRANAVIR

Tipranavir is a protease inhibitor that is used to treat HIV-1 when it is resistant to more than one protease inhibitor. It is a sulfonamide-containing dihydropyran and a nonpeptidic protease inhibitor that targets the HIV protease. It is administered along with ritonavir to treat HIV.

Clinical Indications

Tipranavir is indicated for combination antiretroviral treatment of HIV-1-infected adult patients with evidence of viral replication who are highly treatment-experienced or have HIV-1 strains resistant to multiple protease inhibitors.

Mechanism of Action

Tipranavir inhibits the processing of the viral Gag and Gag-Pol polyproteins in HIV-1-infected cells, preventing the formation of mature virions. It binds to the active site of the protease enzyme with fewer hydrogen bonds than peptidic protease inhibitors, resulting in increased flexibility and better adjustment to amino acid substitutions at the active site. The drug's strong hydrogen-bonding interaction with the amide backbone of the protease active site Asp30 may lead to its activity against resistant viruses.

Adverse Effects

The most common adverse effects in adults are abdominal pain, diarrhea, nausea, vomiting, **pyrexia**, fatigue, and headache. It may also lead to high triglyceride or cholesterol levels or increased liver transaminases. The most frequent adverse effects in children are similar to those observed in adults; however, the rash is more common in children than in adults.

Contraindications

Tipranavir is contraindicated in patients with moderate or severe hepatic impairment. Extreme caution should be utilized in patients with chronic hepatitis B or hepatitis C coinfection with HIV. It is also contraindicated for use with drugs that are highly dependent on CYP3A for clearance or are potent CYP3A inducers.

ENFUVIRTIDE

Enfuvirtide is a medication administered subcutaneously for the treatment of HIV infection in adults and children 6 years of age or older, weighing at least 24 lb, whose infection is not well-controlled by ongoing treatment with other HIV medicines.

Clinical Indications

Enfuvirtide injection is indicated for use to treat HIV-1 infection in combination with other antiviral agents in patients previously treated with other HIV medications but who have evidence of HIV-1 replication despite being on ART.

Mechanism of Action

Enfuvirtide is a fusion inhibitor that interferes with HIV-1's penetration into body cells. It has potent, selective inhibition of the membranes of viruses and cells, specifically inhibiting the function of the gp41 transmembrane glycoprotein of HIV-1.

Adverse Effects

Enfuvirtide may cause itching, swelling, pain, tingling, discomfort, tenderness, redness, bruising, or pain at the injection site. It may also cause tiredness, myalgia, nausea, loss of appetite, weight loss, stomach pain, diarrhea, arm or leg pain, flu-like symptoms, nasal congestion, cold sores, dry mouth, eye pain, itching, redness, or tearing.

Contraindications

Enfuvirtide is contraindicated in patients with known hypersensitivity reactions. It should be used with caution in patients with coagulation disorders or those who have rash, hypotension, and elevated serum liver transaminases, as those symptoms may represent a hypersensitivity reaction to the medication.

ENTRY INHIBITORS AND INTEGRASE INHIBITORS

Entry inhibitors are a relatively new class of antiretroviral medications for treating HIV-1 infections. They are also known as fusion inhibitors and are used in combination therapy. Patients with HIV who have become resistant to NRTIs, NNRTIs, and PIs will likely benefit from entry inhibitors because they are a different class of drugs. Currently, there are several entry inhibitors that the FDA has approved for HIV treatment. An example is maraviroc, where adverse effects include fever, chills, cough, rash, fatigue, dizziness, nausea, and diarrhea.

INTEGRASE INHIBITORS

Integrase inhibitors are therapeutic medications active against both HIV-1 and HIV-2. They are now a preferred drug class for HIV treatment. They were discovered in 2007. The HIV-1 integrase is the primary enzyme in the replication mechanism of retroviruses. It transfers virally encoded DNA into the host chromosome, a necessary event in retroviral replication. These drugs destroy the virus over time, especially newer integrase inhibitors such as dolutegravir, bictegravir, and cabotegravir. Older integrase inhibitors, including raltegravir and elvitegravir, have shown some resistance to HIV. The adverse effects of integrase inhibitors include mouth lesions, fatigue, fever, nausea, loss of appetite, and hematuria.

Check Your Knowledge

1. What are the indications for enfuvirtide?
2. What are the adverse effects of enfuvirtide?
3. What are the contraindications of enfuvirtide?

DOLUTEGRAVIR

Dolutegravir is a medication approved for the treatment of HIV infection in adults and children weighing at least 30 kg who are at least 12 years old, either as a standalone therapy or in combination with other antiretroviral agents.

Clinical Indications

Dolutegravir is indicated for the treatment of type 1 HIV infection in treatment-naive individuals and those who have previously received ART. It is also used for pre-exposure prophylaxis to reduce the risk of sexually acquired HIV-1 infection in high-risk adults.

Mechanism of Action

Dolutegravir is an integrase inhibitor that interferes with the integration of viral DNA into the host genome, thereby inhibiting HIV replication. It selectively inhibits the catalytic activity of the HIV-1 integrase enzyme, preventing the incorporation of viral genetic material into human chromosomal DNA.

Adverse Effects

Common adverse effects associated with dolutegravir include headache, insomnia, fatigue, nausea, diarrhea, and abnormal dreams. Additionally, it may cause hypersensitivity reactions, hepatotoxicity, and immune reconstitution syndrome. Rare but serious side effects include severe skin reactions, liver problems, and psychiatric symptoms.

Contraindications

Dolutegravir is contraindicated in individuals with known hypersensitivity reactions to the drug. It should be used with caution in patients with severe hepatic impairment.

CLINICAL CASE STUDIES

Clinical Case Study 1

A 32-year-old woman visited her gynecologist for her annual checkup. In her medical history, she has had two partners in the past 2 years. She has not received HPV vaccinations. She was asymptomatic until 3 months ago, but has since experienced watery, bloody vaginal discharge that can be heavy and has a foul odor, along with pelvic pain and pain during intercourse. Pap smears and HPV tests showed abnormalities, and a colposcopy and biopsy confirmed cervical cancer.

CRITICAL THINKING QUESTIONS

1. How common are HPVs?
2. What are the complications of HPV?
3. What are the recommendations for immunization for HPV?

Clinical Case Study 2

In 2001, a 35-year-old man was admitted to the local hospital. Upon admission, he reported experiencing diarrhea, headaches, and fatigue for at least 30 days. He was diagnosed with HIV and began ART with stavudine, abacavir, and lopinavir/ritonavir. During this period, his 25-year-old wife and 6-month-old daughter were also diagnosed with HIV infection.

CRITICAL THINKING QUESTIONS

1. What is the mechanism of action for stavudine?
2. What are the adverse effects of abacavir?
3. What are the contraindications of lopinavir/ritonavir?

FURTHER READING

Adams, J., and Merluzzi, V.J. (2013). *The Search for Antiviral Drugs — Case Histories from Concept to Clinic*. Birkhauser.

American Academy of HIV Medicine, and Hardy, W.D. (2023). *Fundamentals of HIV Medicine 2023*. Oxford University Press.

De Clercq, E., and Walker, R.T. (2012). *Antiviral Drug Development — A Multidisciplinary Approach* (NATO Science Series A: 143). Springer.

Gholamrezanezhad, A., and Dube, M.P. (2023). *Coronavirus Disease 2019 (Covid-19) — A Clinical Guide*. Wiley-Blackwell.

Jefferson, L. (2023). *Human Papillomavirus (HPV) 1 – Unlocking the Secrets of HPV – Your Roadmap to Health and Empowerment*. Jefferson.

Kazmierski, W.M. (2012). *Antiviral Drugs — From Basic Discovery through Clinical Trials*. Wiley.

Lee, K. (2023). *The History of HIV/AIDS — Discovery and Early Cases — From Patient Zero to Millions — HIV/AIDS Awareness*. Lee.

Li, J.J. (2022). *Conquest of Invisible Enemies — A Human History of Antiviral Drugs*. Oxford University Press.

Liu, X., Zhan, P., Menendez-Arias, L., and Poongavanam, V. (2021). *Antiviral Drug Discovery and Development* (Advances in Experimental Medicine and Biology, Book 1322). Springer.

Rouse, B.T., and Dence, C.W. (2011). *Herpes Simplex Virus — Pathogenesis, Immunobiology and Control* (Current Topics in Microbiology and Immunology, 179). Springer.

Sandler, E. (2023). *Antiviral Drugs — From Basic Discovery through Clinical Trials*. Hayle Medical.

Skalka, A.M. (2018). *Discovering Retroviruses — Beacons in the Biosphere*. Harvard University Press.

Thomas, H.C., Lok, A.S., Locarnini, S.A., and Zuckerman, A.J. (2013). *Viral Hepatitis*, 4th Edition. Pearson/Wiley-Blackwell.

Wilson, V.G. (2022). Viruses — Intimate Invaders. Springer.

33 Antifungal Agents

LEARNING OBJECTIVES

After studying this chapter, readers should be able to

1. Review the characteristics of fungi.
2. Describe the systemic fungal infections and their treatments.
3. Explain the clinical indications of amphotericin B.
4. Describe the mechanism of action of griseofulvin.
5. Identify the adverse effects of griseofulvin.
6. Discuss the clinical indications of nystatin.
7. Summarize the adverse effects of fluconazole.
8. Explain the clinical indications of itraconazole.
9. Describe the mechanism of action of ketoconazole.
10. Identify the contraindications of ketoconazole.

OVERVIEW

Fungi are living organisms that release spores into the air. Only a few fungi pathogenic to humans possess the virulence required to infect a healthy host. Most are relatively harmless unless they come into contact with an immunocompromised patient, whose weakened immune system allows them to invade the body. Primary pathogens can establish infections in healthy hosts, while opportunistic pathogens cause disease in individuals with compromised immune defense mechanisms. There are two main categories of antifungal agents: systemic and superficial. Fungal infections, known as mycoses, are significantly more common in their superficial form than in their systemic form. Infecting fungi can be categorized as either exogenous or endogenous. Exogenous fungi enter through airborne, cutaneous, or percutaneous exposure. Endogenous infections occur through colonization by members of the

normal flora or reactivation of a previous infection. Specific drugs are designated for either superficial or systemic mycoses, and a few are effective for both.

FUNGAL INFECTIONS

Fungi, including **molds** and mushrooms, can be single-celled or multicellular organisms. Particular species are part of the normal flora found in the mouth, skin, and urogenital tract. Patients with immunosuppression may experience frequent fungal infections, necessitating aggressive pharmacotherapy. For these individuals, systemic fungal infections can quickly become life-threatening.

SYSTEMIC MYCOSES

Systemic fungal infections are classified into two categories: *opportunistic* and *nonopportunistic* infections. Opportunistic mycoses primarily affect immunocompromised or debilitated patients, including **candidiasis**, **aspergillosis**, **cryptococcosis**, and **mucormycosis**. Although relatively uncommon, nonopportunistic mycoses can occur in any patient, including *sporotrichosis, blastomycosis, coccidioidomycosis, cryptococcus, pneumocystis jirovecii*, and *histoplasmosis*.

SYSTEMIC ANTIFUNGAL AGENTS

Treating systemic mycoses can be challenging because these infections often resist treatment. As a result, they may require prolonged therapy with medications that can have significant toxicities. Table 33.1 summarizes the causes of fungal infections and the recommended drugs for each.

TABLE 33.1

Causes of Fungal Infection and Drug of Choice

Causative Organism	Drug of Choice
Aspergillosis (*Aspergillus* species)	Voriconazole
Blastomycosis (*Blastomyces dermatitidis*)	Amphotericin B or itraconazole
Candidiasis (*Candida* species)	Amphotericin B or fluconazole (either one with or without flucytosine)
Coccidioidomycosis (*Coccidioides immitis*)	Amphotericin B or fluconazole
Cryptococcosis (*Cryptococcus neoformans*)	Amphotericin B with or without flucytosine (for chronic suppression, fluconazole is the drug of choice)
Histoplasmosis (*Histoplasma capsulatum*)	Amphotericin B or itraconazole (for chronic suppression, itraconazole is the drug of choice)
Mucormycosis (*Mucor*)	Amphotericin B
Paracoccidioidomycosis (*Paracoccidioides brasiliensis*)	Amphotericin B or itraconazole
Sporotrichosis (*Sporothrix schenckii*)	Amphotericin B or itraconazole

---

DOI: 10.1201/9781003461913-36

Amphotericin B

Amphotericin B is the oldest type of antifungal drug. The discovery of amphotericin B and its therapeutic applications are considered one of the most critical scientific milestones of the 20th century. Despite its potential toxicity, it remains valuable in treating invasive fungal diseases because of its broad spectrum of activity, low resistance rate, and excellent clinical and pharmacological effectiveness.

Clinical Indications

Amphotericin B is a vital medication for treating severe systemic fungal infections. Despite the emergence of several new antifungal agents, especially second-generation triazoles and echinocandins, it remains the most commonly used antifungal in intensive care settings. Amphotericin B treats severe infections caused by Blastomyces, Histoplasma, Cryptococcus, Aspergillus, Coccidioides, and Candida species, including blastomycosis, histoplasmosis, cryptococcosis, aspergillosis, coccidioidomycosis, and systemic candidiasis.

Mechanism of Action

Amphotericin B primarily binds to ergosterol in the fungal cell membrane, increasing its permeability. It also generates free radicals that cause oxidative damage to the cells, further enhancing membrane permeability. Additionally, amphotericin B stimulates phagocytic cells, helping to clear fungal infections.

Adverse Effects

Due to the risk of potentially fatal adverse effects, the patient must be admitted to a hospital. Common adverse effects include fever, chills, wheezing, nausea, headache, hypoxia, hypotension, nephrotoxicity, hypokalemia, anemia, bone marrow suppression, and delirium.

Contraindications

Amphotericin B is contraindicated in patients with renal and hepatic impairment, drug hypersensitivity, and during lactation. Amphotericin B should be used during pregnancy only if indicated.

Check Your Knowledge

1. What are the indications for amphotericin B?
2. What are the adverse effects of amphotericin B?
3. What are the contraindications for amphotericin B?

SUPERFICIAL MYCOSES

Superficial mycoses are fungal infections that affect the outer layers of the skin, hair, and nails, as well as mucous membranes such as the mouth, throat, and vagina. They are caused by various types of fungi, including **dermatophytes**, yeasts, and molds.

Superficial mycoses encompass tinea capitis (scalp), tinea corporis, tinea barbae (face and facial hair), tinea manuum (hands), tinea cruris (groin and inner thighs), and tinea pedis (athlete's foot). Candidal infections affect the skin and mucous membranes, leading to conditions such as oral **thrush**, diaper rash, and vaginal yeast infections. The four central dermatophyte infections are named based on their locations on the body:

- Tinea pedis (**athlete's foot**) usually responds to topical therapy. The patient should dry the feet after bathing, wear absorbent cotton socks, and change shoes often.
- Tinea corporis (**ringworm** of the body) usually responds to a topical azole or allylamine. After symptoms have cleared, treatment should continue for at least 1 week. If the infection is severe, a systemic antifungal agent may be required.
- **Tinea cruris (jock itch)** responds well to topical treatments, which should continue for at least 1 week after symptoms have cleared. A systemic antifungal drug and systemic or topical glucocorticoids may be needed if severely inflamed.
- Tinea capitis, or scalp ringworm, is the most challenging form to treat, as topical medications are usually ineffective. The standard treatment is oral *griseofulvin* for 6–8 weeks, while oral *terbinafine* may be more effective for 2–4 weeks.

Vulvovaginal candidiasis is a common fungal infection primarily caused by Candida albicans, Candida glabrata, and diabetes. Factors that increase the risk include pregnancy, obesity, HIV infection, and the use of oral contraceptives, anticancer medications, systemic glucocorticoids, immunosuppressants, and systemic antibiotics.

SUPERFICIAL ANTIFUNGAL AGENTS

Treatment for superficial fungal infections includes oral and topical agents (see Table 33.2). Oral therapies consist of griseofulvin, terbinafine, fluconazole, ketoconazole, and itraconazole. Topical therapies include nystatin, selenium sulfide, tolnaftate, haloprogin, miconazole, clotrimazole, and sodium thiosulfate. Compliance with what is sometimes a long course of treatment is essential for successful outcomes and good personal hygiene.

Griseofulvin

Griseofulvin is an antifungal medication available in capsule or tablet form. It treats various yeast and fungal infections affecting the skin, hair, or nails. This medication should be taken with a glass of water.

Clinical Indications

Griseofulvin treats fungal and yeast infections in the skin, nails, or hair. It is the drug of choice in treating tinea capitis and onychomycosis. It can be used for superficial fungal infections resistant to treatment with topical antifungal

TABLE 33.2
Superficial Antifungal Agents

Generic Name	Trade Name
Butenafine	Mentax
Ciclopirox	Loprox, Penlac
Griseofulvin	Gris-PEG
Naftifine	Naftin
Nystatin	Nystop, Mycostatin
Tavaborole	Kerydin
Terbinafine	Lamisil
Tolnaftate	Aftate, Tinactin
Undecylenic acid	Fungi-Nail, Gordochom
Butoconazole	Femstat, Bynazole
Clotrimazole	Lotrimin AF, Mycelex
Econazole	Spectazole
Efinaconazole	Jublia
Fluconazole	Diflucan
Itraconazole	Sporanox
Ketoconazole	Nizoral
Luliconazole	Luzu
Miconazole	Micatin, Oravig
Oxiconazole	Oxistat
Sulconazole	Exelderm
Terconazole	Terazol
Tioconazole	Monistat-1, Vagistat-1

medications. It is also indicated in **onychomycosis**, although newer antifungals such as terbinafine, itraconazole, and fluconazole have largely replaced it. The majority of onychomycosis cases are caused by **Tinea rubrum** and **Tinea interdigital**. Treatment is partly dependent on the rate of nail growth. Toenails grow more slowly than fingernails, sometimes taking 12–18 months to fully develop, which can lead to a decreased treatment success rate.

Mechanism of Action

Griseofulvin interacts with microtubules, affecting the formation of the mitotic spindle. This interference ultimately inhibits mitosis in dermatophytes. Therefore, griseofulvin is a **fungistatic** drug against *Trichophyton, Microsporum, and Epidermophyton* species.

Adverse Effects

Overall, griseofulvin has few adverse effects. This medication can cause nausea, vomiting, diarrhea, headaches, and allergic reactions. Other adverse effects include photosensitivity, **petechiae**, hepatotoxicity, insomnia, fatigue, confusion, and **urticaria**. It may worsen lupus or porphyria.

Contraindications

Patients with hypersensitivity, porphyria, and liver failure should avoid griseofulvin. Pregnant women should refrain from using griseofulvin during their first trimester. Patients should wait at least 1 month after completing treatment with griseofulvin before trying to conceive (pregnancy category C).

Check Your Knowledge

1. What are the indications for griseofulvin?
2. What is the mechanism of action of griseofulvin?
3. What are the adverse effects of griseofulvin?

NYSTATIN

Nystatin is an antifungal medication that combats various yeast and Candida species. It is primarily used to treat skin and oropharyngeal candidiasis.

Clinical Indications

Nystatin treats cutaneous, mucocutaneous, and gastrointestinal mycotic infections, particularly those caused by Candida species. It is available in coated pills (500,000 units), oral suspension (100,000 units/mL), cream, or ointment. It can also be combined with corticosteroids or antibiotics.

Mechanism of Action

Nystatin binds to ergosterol, leading to the leakage of intracellular components and, ultimately, cell death.

Adverse Effects

The oral form may irritate the mouth and cause nausea, vomiting, or diarrhea, while the topical forms may produce local irritation.

Contraindications

Nystatin oral administration is contraindicated in patients with a history of hypersensitivity to this agent. It is generally not used for the treatment of systemic mycoses. Nystatin should be given to patients with diabetes, liver and kidney impairment, pregnant women, and those who are breastfeeding.

FLUCONAZOLE

Fluconazole belongs to the triazole family and is a commonly used antifungal agent. It treats severe fungal or yeast infections.

Clinical Indications

Fluconazole is used to treat vaginal, oropharyngeal, and esophageal candidiasis, peritonitis, and systemic *Candida* infections, including candidemia, disseminated candidiasis, pneumonia, and **cryptococcal meningitis**. Prophylaxis is also known to decrease the incidence of candidiasis in patients receiving bone marrow transplantation, radiation therapy, or cytotoxic chemotherapy.

Mechanism of Action

Fluconazole is an azole agent that inhibits ergosterol synthesis to increase cellular permeability. It is primarily fungistatic.

Adverse Effects

While most patients tolerate fluconazole, typical side effects include nausea, vomiting, abdominal pain, loss of appetite, fever, headache, dizziness, changes in taste, and skin rash. Other potential impacts of fluconazole may include anaphylaxis, hepatotoxicity, fatigue, myalgia, seizures, insomnia, tremors, neutropenia, agranulocytosis, thrombocytopenia, hypercholesterolemia, hypertriglyceridemia, hypokalemia, dyspepsia, excessive sweating, alopecia, and Stevens–Johnson syndrome.

Contraindications

Fluconazole is contraindicated in patients with hypersensitivity to the drug. It should also be avoided in individuals with liver and kidney disease, hypokalemia, and those who are breastfeeding. The pregnancy category is C for low doses and D for high doses.

ITRACONAZOLE

Itraconazole is a broad-spectrum antifungal agent effective against a wide range of fungal infections. It belongs to a class of antifungals known as triazoles—itraconazole functions by slowing the growth of the fungi responsible for the infection.

Clinical Indications

Itraconazole is a medication used to manage and treat fungal infections. It is used as an alternative to amphotericin B. The oral solution is used explicitly for treating oropharyngeal or esophageal candidiasis, while capsules are used for histoplasmosis, aspergillosis, blastomycosis, and onychomycosis.

Mechanism of Action

Itraconazole is a broad-spectrum antifungal agent that inhibits ergosterol synthesis, helping maintain the fungi's cell membrane. Its half-life ranges from 34 to 42 hours, and it is excreted through urine and feces.

Adverse Effects

Itraconazole is a relatively safe medication. However, it may cause fever, nausea, vomiting, diarrhea, rash, headache, dizziness, swelling, muscle or joint pain, hair loss, abdominal pain, and edema. It may also cause cardiac suppression and liver injury.

Contraindications

There are specific contraindications for the use of itraconazole. The primary one is heart failure or a history of heart failure due to itraconazole's potential cardiotoxic effects. Another contraindication is liver failure or disease, as itraconazole can lead to hepatotoxicity. This drug is also contraindicated in pregnant patients, as it has shown teratogenic and embryotoxic effects in animal studies. Researchers have found that itraconazole can cause eye defects in infants whose mothers used the drug during pregnancy. The pregnancy category is C.

KETOCONAZOLE

Ketoconazole is an imidazole antifungal medication used to manage and treat fungal infections. This activity outlines ketoconazole's indications, actions, and contraindications as an essential tool in this treatment.

Clinical Indications

Ketoconazole treats severe fungal or yeast infections, such as candidiasis, blastomycosis, paracoccidioidomycosis, coccidioidomycosis, and histoplasmosis. The cream and shampoo of ketoconazole are applied to the skin. Over-the-counter ketoconazole comes as a shampoo for the scalp. Ketoconazole cream is typically applied once daily for 2–6 weeks.

Mechanism of Action

Ketoconazole inhibits the synthesis of ergosterol and lanosterol, a key precursor in ergosterol biosynthesis. Ergosterol is essential for maintaining the integrity of fungal membranes. It is effective against most fungi responsible for systemic mycoses and superficial infections. At high doses, ketoconazole can competitively bind to androgen receptors, including those for testosterone and dihydrotestosterone, potentially reducing the activity of testosterone and dihydrotestosterone in prostate cancer.

Adverse Effects

Nausea, vomiting (which can be reduced by taking the medication with food), hepatotoxicity, inhibition of testosterone synthesis, reversible sterility, reduction of estradiol synthesis, rash, itching, dizziness, fever, chills, constipation, diarrhea, photophobia, and headache.

Contraindications

Ketoconazole should be avoided by patients who are hypersensitive to the drug. It is also contraindicated for individuals with liver disease due to its association with hepatotoxicity, which can be fatal. Furthermore, it is contraindicated in cases of adrenal insufficiency, as high doses of ketoconazole hinder adrenocortical function. The medication is contraindicated for patients with known hypersensitivity reactions. Ketoconazole should never be co-administered with HMG-CoA reductase inhibitors, including atorvastatin, rosuvastatin, simvastatin, pravastatin, fluvastatin, lovastatin, and pitavastatin, as it can elevate the risk of myopathy in children under 2 years old. The pregnancy category is C.

Check Your Knowledge

1. What are the indications for fluconazole?
2. What are the adverse effects of fluconazole?
3. What is the mechanism of action of itraconazole?
4. What are the adverse effects of ketoconazole?

CLINICAL CASE STUDIES

Clinical Case Study 1

A 72-year-old man was admitted to the local hospital with dyspnea and respiratory failure, necessitating admission to the intensive care unit and intubation. Despite receiving maximal therapy and appropriate anti-infective treatment for all identified pathogens, the patient's condition continued to decline, ultimately leading to his death. An autopsy revealed the presence of Aspergillus species and mucormycosis in multiple organs, including the lung, heart, and pancreas.

CRITICAL THINKING QUESTIONS

1. What are the different classifications of systemic fungal infections?
2. What is the preferred medication for treating aspergillosis and mucormycosis?
3. What are the adverse effects of amphotericin B?

Clinical Case Study 2

A 73-year-old woman visited a dermatologist due to discoloration, thickening, brittleness, and crumbling nails. Her nails appeared white, black, yellow, and green. Based on these clinical symptoms, along with microscopic examinations and laboratory tests, she was treated with griseofulvin for onychomycosis.

CRITICAL THINKING QUESTIONS

1. What causes onychomycosis, and how is it treated?
2. What is the mechanism of action of griseofulvin?
3. What are the adverse effects of griseofulvin?

FURTHER READING

Basak, A., Chakraborty, R., and Mandal, S.M. (2016). *Recent Trends in Antifungal Agents and Antifungal Therapy*. Springer.

Fernandes, P.B. (2013). *New Approaches for Antifungal Drugs*. Birkhauser.

Ghannoum, M.A., and Perfect, J.R. (2019). *Antifungal Therapy*, 2nd Edition. CRC Press.

Grayson, M.L. (2017). *Kucers' The Use of Antibiotics — A Clinical Review of Antibacterial, Antifungal, Antiparasitic, and Antiviral Drugs*, Three Volume Set, 7th Edition. CRC Press.

Hospenthal, D.R., and Rinaldi, M.G. (2015). *Diagnosis and Treatment of Fungal Infections* (Infectious Disease), 2nd Edition. Springer.

Imran, M., Asim Farid, M., and Adnan Asghar, M. (2015). *Antibacterial and Antifungal Drug Discovery — Benzimidazole Derivatives*. Lap Lambert Academic Publishing.

Iqbal, Z, Ali Saleh Jaber, A., and Mirza, M.A. (2017). *Vaginal Drug Delivery Approach for Antifungal Therapy*. Lap Lambert Academic Publishing.

Kim, J.H., Cheng, L.W., and Land, K. (2022). *Advances in Antifungal Development — Discovery of New Drugs and Drug Repurposing*. Mdpi Ag.

Krysan, D.J., and Moye-Rowley, W.S. (2023). *Antifungal Drug Resistance — Methods and Protocols* (Methods in Molecular Biology, Book 2658). Humana Press.

Kumar, A. (2022). *Phytoconstituents and Antifungals* (Developments in Applied Microbiology and Biotechnology). Academic Press.

Manzoor, N. (2024). *Advances in Antifungal Drug Development — Natural Products with Antifungal Potential*. Springer.

Rasheed, A. (2015). *Synthesis of Anti-fungal Drugs*. CreateSpace Independent Publishing Platform.

Rasheed, A. (2017). *Analysis of Antifungal Drugs*. CreateSpace Independent Publishing Platform.

Sokovic, M., and Liaras, K. (2020). *Antifungal Compounds Discovery — Natural and Synthetic Approaches*. Elsevier.

Torrado Duran, J.J., Serrano, D.R., and Capilla, J. (2020). *Antifungal and Antiparasitic Drug Delivery*. Mdpi Ag.

34 Antiprotozoal Agents

LEARNING OUTCOMES

After studying the chapter, readers should be able to

1. List some examples of protozoal infections.
2. Review the malaria lifecycle and the most common forms.
3. Identify drugs that are used to treat malaria.
4. Describe the adverse effects of chloroquine.
5. Summarize the clinical indications of quinine.
6. Compare the characteristics of giardiasis with amebiasis.
7. Discuss the characteristics of Chagas' disease.
8. Review the adverse effects of metronidazole.
9. Explain the contraindications of pyrimethamine.
10. Describe the clinical indications of nifurtimox.

OVERVIEW

Parasitic infectious diseases significantly impact health, society, and the economy in tropical regions worldwide. Diseases caused by protozoa, such as malaria and schistosomiasis, account for the majority of parasite-related morbidity, mortality, and annual deaths. These pathogens are prevalent in tropical and subtropical countries and contribute to some of the most common intestinal infections in the United States. Many protozoal infections are diagnosed through stool sample examinations, which are essential for surgical pathologists. Common protozoan diseases include malaria, giardiasis, and toxoplasmosis, while less common ones encompass African trypanosomiasis and amoebic dysentery. Each condition affects the body in unique ways. Protozoal infections represent serious illnesses that significantly contribute to global mortality and morbidity.

PROTOZOAL INFECTIONS

The incidence of protozoal infections in the United States is rising. This trend may be linked to increased American international travel and heightened immigration from countries where infectious protozoa are endemic. *Entamoeba histolytica, Trichomonas vaginalis, Giardia lamblia*, and various Plasmodium species are the most commonly encountered infections. Other protozoal infections in the United States encompass amebiasis, **cryptosporidiosis**, leishmaniasis, toxoplasmosis, **trichomoniasis**, and American trypanosomiasis.

MALARIA

Malaria is a potentially life-threatening disease caused by parasites transmitted through the bites of infected female **Anopheles mosquitoes**, primarily affecting tropical and subtropical regions. Although it is preventable and treatable, early diagnosis and effective treatment are crucial. The malaria lifecycle occurs within the female Anopheles mosquito. As the mosquito feeds on a human host's blood, sporozoites enter the bloodstream and invade liver cells. The **merozoites** then infect the red blood cells of the human host. Symptoms include chills, fever, and sweating, which peak, decline, and peak again every 48 hours. Since hypnozoites remain in the liver, relapse is expected following the end of an acute attack; however, after two or more years, this process ceases completely. Medications that eliminate the hypnozoites can prevent relapse.

The most common forms of malaria are those caused by **Plasmodium vivax and Plasmodium falciparum.** *P. vivax* is more severe, causing approximately 10% of deaths among infected patients. Many of its strains are now resistant to various medications. When treatment is delayed, the disease may progress quickly to irreversible shock and death. Malaria is primarily treated with *chloroquine, iodoquinol, primaquine, quinine, quinidine gluconate, mefloquine,* and antibacterial drugs such as tetracyclines and *clindamycin*.

ANTIMALARIAL DRUGS

The antimalarial agents used to treat malaria are selected based on treatment goals and the drug resistance of the Plasmodium strain causing the infection. For prophylaxis against malaria, chloroquine is the preferred treatment for any parasite sensitive to the drug. However, parasites in many parts of the world have developed resistance to chloroquine, rendering the drug ineffective. If impractical, atovaquone is combined with proguanil, doxycycline, or mefloquine. The primary medications used to treat malaria are listed in Table 34.1.

CHLOROQUINE

Chloroquine is used to manage and treat uncomplicated malaria and for prophylaxis in regions with chloroquine-sensitive

---

TABLE 34.1

Drugs to Treat Malaria

Generic Name	Trade Name
Artemether–lumefantrine	Coartem
Atovaquone–proguanil	Malarone
Chloroquine	Aralen
Hydroxychloroquine	Plaquenil
Mefloquine	(generic only)
Primaquine	(generic only)
Quinine	Qualaquin

DOI: 10.1201/9781003461913-37

malaria, specifically certain strains of *P. falciparum*, *P. ovale*, *P. vivax*, and *P. malariae*. It can be found in Mexico, the Panama Canal, the Caribbean, East Asia, and some Middle Eastern countries. Chloroquine is classified as a medication in the sulfonamide group.

Clinical Indications

Chloroquine prevents and treats malaria and liver infections caused by amebiasis. This medication should be administered parenterally. Furthermore, chloroquine may benefit as an antitumor agent in conjunction with chemotherapy and radiation for cancer treatment.

Mechanism of Action

Chloroquine can transform heme into nontoxic metabolites and accumulate in parasitized erythrocytes, possibly explaining its selective effects against erythrocytic forms of *Plasmodium*.

Adverse Effects

Although chloroquine has relatively few side effects when taken in low doses, higher doses may cause severe adverse effects, including blurred vision, reduced visual acuity, and diplopia. Additionally, high doses may lead to bleeding gums, anxiety, paranoia, seizures, hallucinations, and suicidal ideation.

Contraindications

Chloroquine is contraindicated for patients with retinopathy and visual field changes, hypersensitivity, or severe liver disease. Although it has not been formally assigned a pregnancy category, chloroquine is safe during pregnancy and in children.

YOU SHOULD REMEMBER

Chloroquine is not used to treat severe or complicated malaria or to prevent malaria in areas where resistance to chloroquine is recognized.

Check Your Knowledge

1. What are the indications for chloroquine?
2. What are the adverse effects of chloroquine?
3. What are the contraindications for chloroquine?

MEFLOQUINE

Mefloquine is an antimalarial medication used to prevent and treat malaria caused by infections with *Plasmodium vivax* and *Plasmodium falciparum*.

Clinical Indications

Mefloquine is used to treat malaria and to prevent malaria infection in regions where other medicines are known to be ineffective.

Mechanism of Action

Membrane-bound mefloquine may inhibit merozoite invasion and interact with proteins involved in parasite membrane lipid trafficking and nutrient uptake. Mefloquine binds to heme, forming a complex that may also be toxic to the parasite.

Adverse Effects

Dose-related effects of mefloquine include nausea, dizziness, syncope, gastrointestinal disturbances, nightmares, altered vision, headaches, cardiac dysrhythmias, CNS toxicity, vertigo, confusion, psychosis, convulsions, hallucinations, depression, and suicidal ideation.

Contraindications

Mefloquine use is contraindicated in individuals with a known hypersensitivity to this medication. It is also contraindicated for those with current depression, generalized anxiety disorder, psychosis, schizophrenia, other primary psychiatric disorders, or a history of seizures. Its pregnancy category is C.

QUININE

Quinine remains an important antimalarial drug and is one of the oldest drugs for treating malaria by eliminating the parasite or preventing its growth. However, its continued use faces challenges due to poor tolerability, low compliance with complex dosing regimens, and the availability of more effective antimalarial drugs.

Clinical Indications

Quinine is used to treat malaria caused by Plasmodium falciparum. It kills the parasite or prevents it from growing. A combination of quinine with doxycycline, tetracycline, or clindamycin is recommended as a second-line treatment for uncomplicated malaria.

Mechanism of Action

It concentrates in parasitized red blood cells and may selectively target erythrocytic parasites; it kills plasmodia by causing heme to accumulate within them. It is active against the erythrocytic forms of *Plasmodium* but has little effect on hepatic forms and sporozoites. It is most effective against chloroquine-resistant *P. falciparum*.

Adverse Effects

The adverse effects of quinine include **tinnitus**, headaches, anxiety, blurred vision, hypoglycemia, abdominal pain, nausea, diarrhea, hemolysis, and hemolytic anemia.

Contraindications

Quinine should be avoided in patients with optic neuritis, preexisting tinnitus, **myasthenia gravis**, glucose-6-phosphate dehydrogenase deficiency, and atrial fibrillation. It is also contraindicated for patients with known hypersensitivity to quinine. The pregnancy category for this drug is C.

Check Your Knowledge

1. What are the indications for quinine?
2. What is quinine's mechanism of action?
3. What are the contraindications of quinine?

PRIMAQUINE

Primaquine is the only widely available treatment for preventing relapsing *Plasmodium vivax* malaria and is also used to treat **Pneumocystis pneumonia**. It is classified as an aminoquinoline, N-substituted diamine, and aromatic ether.

Clinical Indications

In the United States, primaquine is used to treat vivax malaria, in combination with other antimalarial agents. It is also used as a prophylactic against vivax malaria. Primaquine has been proposed as a major agent to prevent relapse after the therapy of acute attacks of vivax malaria with more effectiveness than chloroquine or artemisinin.

Mechanism of Action

Primaquine binds to protozoal DNA, preventing the production of DNA, RNA, and subsequent protein synthesis. It is effective against several stages of Plasmodium development, including liver schizonts, hypnozoites, and gametocytes, with the most significant activity observed against P. ovale and P. vivax.

Adverse Effects

Common adverse effects of primaquine include nausea, diarrhea, abdominal pain, headaches, anorexia, black or tarry stools, bleeding gums, skin rashes, and itching.

Contraindications

Contraindications for primaquine treatment include pregnancy and acute illness with a predisposition to granulocytopenia. The medication is also contraindicated in patients with known hypersensitivity to pyrimethamine or any component of the formulation. Additionally, the drug is contraindicated in patients with documented megaloblastic anemia due to folate deficiency.

Check Your Knowledge

1. What are the indications for primaquine?
2. What are the adverse effects of primaquine?
3. What are the contraindications for primaquine?

GIARDIASIS

Giardiasis is an infection caused by *Giardia lamblia* that primarily affects the small intestine. In the United States, more people become ill from *Giardia* than any other parasite impacting this organ. Each year, over 1 million individuals fall sick due to this parasite. Nearly half of those infected remain asymptomatic. Giardia can spread through contaminated water and food or contact with an infected individual. The signs and symptoms may include fatigue, nausea, vomiting, diarrhea, abdominal pain, foul-smelling belching, and flatulence.

Metronidazole is the first-line treatment for giardiasis, typically taken for 5–7 days. Tinidazole serves as an alternative treatment to metronidazole and may be effective with a single dose.

AMEBIASIS

Amebiasis is an infection caused by **Entamoeba histolytica**. Although primarily affecting the intestine, the disease can also impact the liver. *Entamoeba histolytica* is a significant source of morbidity and mortality in the developing world. This condition *accounts* for an estimated 35 to 50 million cases of symptomatic disease and approximately 100,000 deaths annually. The parasite's potent cytotoxic activity appears to underlie the pathogenesis of the disease, although the exact mechanism remains unknown. The condition is typically asymptomatic but may include symptoms such as bloody diarrhea, abdominal pain, and weight loss. The primary therapy for symptomatic amebiasis involves hydration, metronidazole, or tinidazole.

LEISHMANIASIS

Leishmaniasis is a parasitic disease caused by infection with *Leishmania* parasites, transmitted through an infected sandfly's bite. The parasites reside on or within another organism. Although different forms of leishmaniasis affect patients in various ways, the most common type results in skin sores. The disease can be classified as cutaneous, mucocutaneous, or visceral. The most severe form is visceral leishmaniasis, which can be fatal.

Anyone who lives in or travels to an area where *Leishmania* parasites are present and is bitten by an infected sand fly is at risk for leishmaniasis. It is typically more common in rural regions but can also occur in certain cities. Signs and symptoms may include nodules or ulcers, fever, hepatosplenomegaly, liver dysfunction, hypoalbuminemia, lymphadenopathy, pancytopenia, and internal bleeding.

TRYPANOSOMIASIS

Human African trypanosomiasis (HAT), commonly known as sleeping sickness, is a vector-borne parasitic disease caused by protozoans from the genus *Trypanosoma*. *It is* transmitted to humans through the bites of tsetse flies, which acquire the parasites from infected humans or animals. Tsetse flies inhabit sub-Saharan Africa, with only particular species capable of transmitting the disease. American trypanosomiasis, or Chagas disease, primarily occurs in Latin America. It is caused by a different subgenus of Trypanosoma and is transmitted by a different vector. The characteristics of this disease are distinctly different from those of HAT.

Initial symptoms include swollen bumps around the bite, fever, skin rash, and pain in muscles and joints. Advanced symptoms include headaches, chills, confusion, difficulty walking, and trouble staying awake. The parasites then invade the heart and the brain, which may result in cardiomyopathy, megacolon, and death.

The choice of treatment depends on the form and stage of the disease. The earlier the disease is treated, the better the chances of a cure. Drugs such as fexinidazole, pentamidine, eflornithine, nifurtimox, and melarsoprol treat trypanosomiasis. Fexinidazole serves as a first-line oral treatment for acute sleeping sickness.

TOXOPLASMOSIS

Toxoplasmosis is caused by *Toxoplasma gondii*, typically transmitted through the consumption of undercooked meat. When the disease is congenital, it can impact the brain, eyes, liver, and other organs, and it is usually fatal. Signs and symptoms, when they occur, include retinal damage and encephalitis.

Check Your Knowledge

1. What are the signs and symptoms of Giardiasis?
2. What are the main therapies for amebiasis?
3. What are the signs and symptoms of leishmaniasis?
4. What is the cause of Chagas disease

PHARMACOTHERAPY FOR OTHER PROTOZOAL INFECTIONS

The drugs of choice for treating other protozoal infections, such as giardiasis, amebiasis, leishmaniasis, trichomoniasis, and toxoplasmosis, include eflornithine, iodoquinol, melarsoprol, metronidazole, miltefosine, nifurtimox, nitazoxanide, paromomycin, pentamidine, pyrimethamine, sodium stibogluconate, suramin, and tinidazole (see Table 34.2).

TABLE 34.2
Antiprotozoal Drugs

Generic Name	Trade Name
Eflornithine	Ornidyl
Iodoquinol	Yodoxin
Melarsoprol	Arsobal
Metronidazole	Flagyl
Miltefosine	Impavido
Nifurtimox	Lampit
Nitazoxanide	Alinia
Paromomycin	Humatin
Pentamidine	NebuPent, Pentam
Pyrimethamine	Daraprim
Sodium stibogluconate	Pentostam
Suramin	Germanin
Tinidazole	Tindamax

METRONIDAZOLE

Metronidazole is an antibiotic used to treat specific infections of the vagina, stomach, liver, skin, joints, brain and spinal cord, lungs, heart, or bloodstream.

Clinical Indications

Metronidazole treats *Trichomonas vaginalis*, *Entamoeba histolytica*, *Giardia lamblia*, Blastocystis, and Balantidium coli. It also treats intestinal amebiasis and liver amebiasis. The extended-release tablets are used to treat vaginal infections.

Mechanism of Action

Metronidazole inhibits protein synthesis by interacting with DNA, leading to a loss of helical DNA structure and strand breakage. Consequently, it induces cell death in susceptible organisms.

Adverse Effects

The adverse effects of metronidazole include headache, confusion, dizziness, a metallic taste, nausea, vomiting, diarrhea, vaginitis, genital pruritus, abdominal pain, xerostomia, dysmenorrhea, urinary tract infections, bacterial infections, flu-like symptoms, pharyngitis, and sinusitis.

Contraindications

Metronidazole should be avoided in patients with hypersensitivity to the drug. It is contraindicated during the first trimester of pregnancy. Avoid consuming alcohol while taking metronidazole tablets, liquid, suppositories, or vaginal gel, as well as for 2 days after completing treatment. The pregnancy category for metronidazole is B.

YOU SHOULD REMEMBER

Alcohol should be avoided during treatment and for up to 48 hours to 14 days after treatment completion, depending on the source.

Check Your Knowledge

1. What are the clinical indications for metronidazole?
2. What are the adverse effects of metronidazole?
3. What are the contraindications of metronidazole?

NIFURTIMOX

Nifurtimox is a nitrofuran antimicrobial agent. Therapy with nifurtimox is generally prolonged, and another drug is often substituted due to toxicity. The toxic effects of nifurtimox are typically reversible.

Clinical Indications

Nifurtimox is used to treat Chagas disease (American trypanosomiasis), a chronic protozoal infection caused by Trypanosoma cruzi that can lead to severe disability and death from gastrointestinal and cardiac complications. Nifurtimox is rarely associated with serum aminotransferase elevations during therapy and has not been linked to clinically apparent liver injury in humans.

Mechanism of Action

Nifurtimox has two distinct actions. It can form a nitro-anion radical metabolite that interacts with parasitic nucleic acids, leading to the breakdown of DNA, or it may produce superoxide anions and hydrogen peroxide, which are toxic to the parasite and inhibit trypanothione reductase (a parasite-specific antioxidant defense enzyme).

Adverse Effects

The adverse effects of nifurtimox include anorexia, nausea, vomiting, abdominal pain, peripheral neuropathy, weight loss, rash, memory loss, insomnia, **vertigo**, and headache. Hemolysis may occur in patients with a deficiency of glucose-6-phosphate dehydrogenase.

Contraindications

Active or a history of peripheral neuropathy, active or a history of seizures and cerebral impairments (such as behavioral disorders, epilepsy, or psychoses), and hepatic or renal impairment. The pregnancy category of nifurtimox is not formally assigned due to a lack of human studies.

Pyrimethamine

Pyrimethamine is an antiparasitic medication approved by the FDA for treating toxoplasmosis. It is typically used in conjunction with a sulfonamide for treatment.

Clinical Indications

Pyrimethamine is an antiparasitic used to treat severe parasitic infections such as toxoplasmosis affecting the body, brain, or eyes, and to prevent toxoplasmosis in individuals with HIV. It should also be taken with folinic acid to minimize its adverse effects.

Mechanism of Action

Pyrimethamine inhibits the dihydrofolate reductase of plasmodia, thereby blocking the biosynthesis of purines and pyrimidines, which are essential for DNA synthesis and cell multiplication. This results in the failure of nuclear division during schizont formation in erythrocytes and the liver.

Adverse Effects

The adverse effects of pyrimethamine include hypersensitivity reactions, **hyperphenylalaninemia**, anorexia, vomiting, megaloblastic anemia, leukopenia, thrombocytopenia, pancytopenia, atrophic glossitis, hematuria, cardiac rhythm disorders, and pulmonary eosinophilia.

Contraindications

Pyrimethamine is contraindicated in patients with known hypersensitivity to this medication or any component of its

formulation. Additionally, the use of the drug is contraindicated in patients with documented megaloblastic anemia due to folate deficiency. The pregnancy category of pyrimethamine is C.

Sodium Stibogluconate

Sodium stibogluconate is an agent used to treat leishmaniasis. It is available only for injection and belongs to the medicines known as pentavalent antimonials. Widespread resistance has limited its effectiveness.

Clinical Indications

Sodium stibogluconate has been the mainstay in treating visceral, cutaneous, and mucosal leishmaniasis for approximately half a century. However, the FDA has not approved it, and it is available only from the CDC in the U.S.

Mechanism of Action

The mechanism of action of sodium stibogluconate is still being determined. However, it may directly inhibit DNA topoisomerase I, hindering DNA replication and transcription.

Adverse Effects

The adverse effects of sodium stibogluconate include **myalgia**, joint stiffness, bradycardia, changes in electrocardiogram results that may precede severe dysrhythmias, liver and kidney impairment, shock, and, rarely, sudden death.

Contraindications

Sodium stibogluconate is classified under pregnancy category B, and its contraindications include hepatic or renal disease, pneumonia, heart disease, complicating secondary infections, and specific prior treatments for leishmaniasis.

Check Your Knowledge

1. What are the clinical indications for nifurtimox?
2. What are the side effects of nifurtimox?
3. What are the contraindications for pyrimethamine?

CLINICAL CASE STUDIES

Clinical Case Study 1

A 47-year-old man returned from a visit to the Panama Canal and developed a cyclic fever that lasted for 7 days. His wife observed changes in his mood and behavior. The patient experienced fatigue and dyspnea. The attending physician in the emergency department ordered blood tests and imaging studies. A malaria smear returned positive, and he was diagnosed with severe malaria alongside multi-organ failure. Chloroquine and primaquine were administered intravenously. By the third day, the patient began to show improvement.

CRITICAL THINKING QUESTIONS

1. What are the characteristics of a malaria infection?
2. What is the mechanism by which chloroquine acts?
3. What are the contraindications of primaquine?

Clinical Case Study 2

A 68-year-old man experiencing nausea, watery diarrhea, bloating, and abdominal cramps was brought to the emergency department. He had returned from the port co 10 days earlier. A physical examination revealed abdominal tenderness; laboratory tests indicated low total protein and albumin levels. A *Giardia duodenalis* trophozoite was identified in a fecal smear, confirming a diagnosis of giardiasis. Metronidazole was administered for 10 days, resulting in the complete resolution of the patient's symptoms.

CRITICAL THINKING QUESTIONS

1. What are the clinical indications for metronidazole?
2. What are the adverse effects of metronidazole?
3. What are the contraindications of metronidazole?

FURTHER READING

Ali, E. (2013). *Anti-Malaria Drugs and Plasmodium falciparum Gametocytes Prevalence — and Drug Resistance Following Treatment of Uncomplicated Plasmodium Falciparum Malaria*. Lap Lambert Academic Publishing.

Al-Mamun, A., Ahmed, T., and Kumar Kundu, S. (2013). *Metronidazole Extended Release Solid Dosage Form — Development of Extended Release Formulation and In Vitro Bioequivalence Study of Metronidazole Solid Dosage Form*. Lap Lambert Academic Publishing.

Ariey, F., Gay, F., and Menard, R. (2019). *Malaria Control and Elimination* (Methods in Molecular Biology). Human Press.

Becker, R.B. (2022). *The Role of Antimalarial Medications in Combating Malaria*. Becker.

Blokdijk, G.J. (2018). *Chloroquine Phosphate*, 2nd Edition. CreateSpace Independent Publishing Platform.

Eugenia, S. (2024). *Comprehensive Guide to Giardiasis — Understanding, Management, and Holistic Health*. Eugenia.

Farid, A. (2021). *Cryptosporidiosis — A Problem to be Solved*. Lap Lambert Academic Publishing.

Frank, K. (2022). *Metronidazole — Guide to Treat Bacterial, Pneumonia and Other Respiratory Tract Infections; Certain Infections of Skin, Eye, Lymphatic, Intestinal, Genital and Urinary Systems*. Frank.

Hurst, K. (2023). *Amebiasis — Diagnosis, Prevention and Treatment*. American Medical Publishers.

Kendrekar, P. (2022). *Drug Development for Malaria — Novel Approaches for Prevention and Treatment*. Wiley-VCH.

Kumar, S., and Pal, S. (2019). *Structure-Based Drug Design for Malaria*. Lap Lambert Academic Publishing.

Staines, H.M., and Krishna, S. (2012). *Treatment and Prevention of Malaria — Antimalarial Drug Chemistry, Action and Use* (Milestones in Drug Therapy). Springer.

35 Antituberculosis Agents

LEARNING OUTCOMES

After studying this chapter, readers should be able to

1. Review mycobacterial infections and nontuberculous mycobacteria (NTM).
2. Explain primary and secondary tuberculosis (TB).
3. List the first-line drugs for treating TB.
4. Identify the adverse effects and contraindications of isoniazid.
5. Discuss the mechanism of action of pyrazinamide.
6. Explain the adverse effects of rifampin.
7. Review the clinical indications of ethambutol.
8. Describe leprosy and the drug of choice for treatment.
9. Identify the adverse effects of dapsone.
10. Explain the *Mycobacterium avium* complex (MAC) and its treatment.

OVERVIEW

Mycobacterial infections encompass various diseases caused by obligate aerobic bacteria, most notably *Mycobacterium tuberculosis*. TB is a chronic pulmonary and systemic disease that spreads through the air when patients with lung TB cough, spit, or sneeze. Although it is more prevalent in developing countries, its incidence is rising in developed nations, likely due to the increase in AIDS cases and the growing number of drug-resistant *Mycobacterium tuberculosis* strains. In addition to immunocompromised individuals, other groups at increased risk include minorities, those of lower socioeconomic status, alcohol users, prisoners, and older adults. Gastrointestinal TB can be contracted from drinking milk contaminated with *Mycobacterium bovis, which is* uncommon in countries where milk is regularly pasteurized. However, it still occurs in nations with infected dairy cows and unpasteurized milk. In the United States, TB primarily affects immigrants from high-burden countries, older adults, racial and ethnic minorities, and individuals with AIDS. Additional risk factors for TB include malnutrition, alcoholism, diabetes mellitus, Hodgkin lymphoma, kidney failure, and silicosis.

MYCOBACTERIAL INFECTIONS

Mycobacteria may be found in the environment, particularly in soil, dust, and water. Additionally, they can be found in human-made environments like tap water and even in medical settings. TB is a systemic infectious disease caused by *Mycobacterium tuberculosis*, affecting the lungs, lymphatic organs, and genitourinary systems. Miliary TB is an extrapulmonary condition. HIV-related miliary TB is rapidly progressive and invariably fatal without treatment.

NTM are widely recognized as opportunistic pathogens that affect patients with underlying illnesses or weakened immune systems. *Mycobacterium leprae* is the causative bacterium of **leprosy**, which impacts the skin, mucous membranes, and peripheral nerves. Leprosy remains a significant health concern in India, Indonesia, and Brazil.

The MAC is the most common type of NTM in humans and is the leading cause of mycobacterial disease. **Atypical mycobacteria** primarily infect immunocompromised patients and children. Active TB infection may present as a cough, chills, fever, night sweats, **hemoptysis**, and weight loss.

PRIMARY AND SECONDARY TB

Primary TB infection is the initial phase in patients without specific immunity, particularly in children and young adults not previously exposed to *Mycobacterium tuberculosis*. Primary TB develops within 5 years of the initial infection, leading to a positive skin response to the purified protein derivative of tuberculin. Although symptoms of primary TB may be minimal, early detection and treatment are essential to prevent immediate complications that carry a high risk of morbidity and mortality and to avert the spread of infection.

Secondary or active TB is a multi-organ disease resulting from a primary infection or the reactivation of latent TB. Accordingly, active TB can be classified into primary TB and reactivation TB. Reactivation TB is the most common form of active TB, accounting for 90% of cases. The lungs are the most frequently affected organs, while other significantly impacted systems include the gastrointestinal, musculoskeletal, lymphoreticular, skin, liver, and reproductive systems.

DIAGNOSIS OF TB

The diagnosis of pulmonary TB includes the patient's history of exposure to the disease, a skin TB test, a chest X-ray, blood tests, and a sputum culture. In contrast to secondary TB, diagnosing progressive primary TB in adults can be challenging. Acid-fast smears and sputum cultures should be performed on patients suspected of having TB, with culturing growth expected within 3–6 weeks.

DOI: 10.1201/9781003461913-38

PHARMACOTHERAPY OF TB

The treatment for TB involves various first- and second-line medications (see Table 35.1). The most common agents for active TB include isoniazid and other first-line drugs. Patients may feel better within a few weeks of starting these medications. However, treating TB requires more time than treating different bacterial infections.

ISONIAZID

Isoniazid (INH) is an antibiotic used in the first-line treatment of active TB infections. For decades, INH has been a crucial drug in TB treatment regimens and prevention. If a patient is receiving treatment for active TB, the medication regimen must continue throughout the entire course, even if symptoms begin to improve. It is essential to take all doses.

Clinical Indications

INH is an antibiotic indicated for the first-line treatment of active *Mycobacterium tuberculosis* infection.

TABLE 35.1

Drugs for Treating Tuberculosis

Generic Name	Trade Name
First-Line Drugs	
Ethambutol (EMB)	Myambutol
Isoniazid (INH)	Hydra, Hyzyd, Isovit
Pyrazinamide (PZA)	Rifater, Tebrazid
Rifabutin	Mycobutin
Rifampin (RIF)	Rifadin, Rimactane
Rifapentine	Priftin
Second-Line Drugs	
Amikacin	Amikin
Bedaquiline	Sirturo
Capreomycin	Capastat Sulfate
Cycloserine	Seromycin
Delamanid	Deltyba
Ethionamide	Trecator
Kanamycin	Kantrex
Levofloxacin	Levaquin
Moxifloxacin	Avelox
Ofloxacin	Floxin
Para-aminosalicylic acid (pas)	Paser
Pretomanid	Dovprela

Active TB infection may present with chills, fever, night sweats, cough, hemoptysis, and weight loss.

Mechanism of Action

INH is a prodrug that inhibits the formation of the mycobacterial cell wall. A bacterial in *Mycobacterium tuberculosis* must activate it. The antimicrobial activity of INH is selective for mycobacteria, likely due to its inhibition of mycolic acid synthesis. The drug inhibits an **enoyl reductase** called InhA by forming a **covalent adduct** with the nicotinamide adenine dinucleotide cofactor.

Adverse Effects

Isoniazid can cause severe adverse effects, such as liver damage, particularly in patients over 50 years old. The side effects of isoniazid may also include pruritus, nausea, diarrhea, jaundice, and hepatitis. Additionally, isoniazid is associated with peripheral neuropathy, paresthesia, and sensory impairment, which may require discontinuation. Rare side effects can include blurred vision or vision loss, with or without eye pain, convulsions, fever, sore throat, joint pain, mental depression, mood changes or other psychological alterations, skin rash, and unusual bleeding or bruising.

Contraindications

INH is contraindicated in cases of severe hypersensitivity reactions, including drug-induced hepatitis, previous isoniazid-related liver damage, significant adverse effects such as drug fever, chills, or arthritis, and any acute liver disease.

PYRAZINAMIDE

Pyrazinamide (PZA) is an antimicrobial drug used to treat active TB, typically during the first 2 months of therapy. It is administered alongside another antimicrobial medication and has a narrow spectrum of action.

Mechanism of Action

The mechanism of action of PZA remains unclear. It converts to pyrazinoic acid in susceptible strains of *Mycobacterium,* resulting in a lowered environmental pH. Unlike most antibiotics, it exhibits little to no activity against growing tubercle bacilli. PZA is primarily effective against nongrowing microorganisms, and its activity increases as metabolic activity diminishes.

Adverse Effects

Arthralgia and gout are common adverse effects of PZA. The most dangerous adverse effect is dose-related hepatotoxicity. Less common adverse effects include nausea, vomiting, anorexia, **sideroblastic anemia**, skin rash, pruritus, urticaria, dysuria, **interstitial nephritis**, malaise, fever, and porphyria.

Contraindications

PZA is contraindicated in cases of known hypersensitivity, severe liver damage, and acute gout. Liver transaminases and uric acid levels must be monitored before and during treatment.

RIFAMPIN

Rifampin is used alongside other medications to treat TB. It also manages and treats various mycobacterial and Gram-positive bacterial infections, including TB, as well as asymptomatic carriers of *Neisseria meningitidis*. Rifampin is administered concurrently with INH, ethambutol, and PZA for the first 2 months, followed by 4 months of rifampin and INH therapy alone. This treatment regimen lasts 6 months and is 83% effective in eradicating TB.

Clinical Indications

Rifampin is used alongside other medications to treat TB in different areas of the body. It is also administered to patients in close contact with someone with meningococcal disease but who do not show infectious symptoms, helping prevent the bacteria from spreading to others. Additionally, it may function as an adjunctive agent for various infections, including endocarditis and bone and joint diseases.

Mechanism of Action

Rifampin inhibits DNA-dependent RNA polymerase by binding to the polymerase subunit within the DNA/RNA channel, effectively blocking RNA transcription. This effect is believed to be dose-related.

Adverse Effects

Although rifampin is generally well tolerated by patients, it can still cause adverse effects that may be either dose-dependent or dose-independent. Common adverse effects include loss of appetite, nausea, diarrhea, abdominal distention, heartburn, jaundice, chest pain, dizziness, fever, headache, hives, petechiae, agitation, difficulty concentrating, fainting, tachycardia, blurred vision, muscle weakness, and paresthesias. Additionally, saliva, tears, teeth, urine, feces, and sweat may become discolored.

Contraindications

Rifampin should be avoided in patients with a history of allergy to the drug, porphyria, preexisting hepatitis, or cirrhosis of the liver.

ETHAMBUTOL

Ethambutol is a bacteriostatic antimycobacterial medication effective against *Mycobacterium tuberculosis* and other mycobacteria. It is combined with other anti-TB drugs to treat both pulmonary and extrapulmonary TB.

Clinical Indications

Ethambutol is prescribed alongside other anti-TB medications for treating pulmonary TB. It is one of the first-line treatments for TB, along with INH, rifampicin, and pyrazinamide. Ethambutol offers a prolonged duration of action and a moderate therapeutic window. Furthermore, this agent is effective against the MAC.

Mechanism of Action

Ethambutol disrupts the biosynthesis of arabinogalactan in the cell wall, stopping bacilli multiplication. However, the molecular mechanisms underlying this effect remain unclear.

Adverse Effects

The adverse effects of ethambutol include decreased visual acuity, blurred vision, visual field defects, **scotomas**, peripheral neuropathy, hepatotoxicity, mental confusion, hallucinations, disorientation, and psychosis. Patients with impaired renal function due to renal TB may be at higher risk of developing ethambutol-induced optic neuropathy.

Contraindications

Ethambutol should be avoided in patients who are hypersensitive to the medication, those with known **optic neuritis**, preexisting ophthalmological diseases caused by ocular toxicity, or unconscious patients.

LEPROSY

Leprosy, also known as **Hansen's disease**, is a chronic infectious illness caused by a type of bacteria called *Mycobacterium leprae*. It primarily affects various organs and systems in the body. If left untreated, it can lead to progressive and permanent disabilities. The bacteria are transmitted through droplets from the nose and mouth during close and frequent contact with untreated individuals. Leprosy is curable through multidrug therapy, and most new cases detected each year originate from the Southeast Asia Region.

Leprosy affects the skin, peripheral nerves, anterior portion of the eye, nose, hands, testes, and feet. Macular, papular, or nodular lesions may develop on the face, ears, elbows, wrists, and knees. As the disease progresses, the leonine facies resembles a lion's face.

PHARMACOTHERAPY OF LEPROSY

The recommended treatment regimen for leprosy consists of three medications: dapsone, rifampicin, and clofazimine, collectively referred to as multi-drug therapy. The regimen should last from 6 to 12 months.

DAPSONE

Dapsone is a bacteriostatic antibiotic used to treat various systemic and dermatologic conditions, including leprosy. Dermatitis herpetiformis is an example of its use for skin conditions. It exhibits excellent bioavailability when absorbed through the gastrointestinal tract and shows equal efficacy when applied topically.

Clinical Indications

Dapsone treats leprosy and helps control dermatitis herpetiformis. It is combined with one or more other medications to treat leprosy. Additionally, it may be used to treat *pneumocystis* pneumonia, *Toxoplasma gondii infections,* or encephalitis in patients with HIV.

Mechanism of Action

Dapsone inhibits the synthesis of dihydrofolic acid by competing with para-aminobenzoic acid for the active site of dihydropteroate synthetase.

Adverse Effects

The adverse effects of dapsone include headaches, insomnia, blurred vision, anorexia, nausea, vomiting, abdominal pain, jaundice, pancreatitis, decreased hemoglobin, neutropenia, tachycardia, and albuminuria.

Contraindications

The contraindications of dapsone include hypersensitivity to dapsone. It should be used with caution in patients with liver or renal damage, cardiopulmonary disease, peripheral neuropathy, and during pregnancy.

YOU SHOULD REMEMBER

Leprosy is a tropical disease found in over 100 countries, with approximately 200,000 new cases yearly. Its prevalence is less than 1 in 10,000 people and is more common in Brazil, India, and Indonesia.

MYCOBACTERIUM AVIUM COMPLEX

The MAC, found in soil, water, dust, and domestic animals, can lead to lung disease in older adults and those with compromised immune systems. Most individuals require several antibiotics for over a year to eradicate the infection. MAC lung disease is a form of nontuberculous mycobacterial infection. Unlike TB, it is not contagious.

MAC leads to widespread infections in patients with severe T-cell immunodeficiency, with organisms proliferating abundantly in various organs, including the lungs, bones, and intestines. Patients experience fever, night sweats, and weight loss. In rare MAC cases occurring in non-HIV-infected individuals, the organisms primarily target the lungs, leading to a productive cough, sometimes accompanied by fever and weight loss.

MAC is treated with three antimicrobials for at least 12 months after sputum culture conversion. First-line medications usually include azithromycin or clarithromycin, ethambutol, and either rifampin or rifabutin. Additional agents, especially for patients with cavitary or advanced bronchiectatic lung diseases, may include amikacin or fluoroquinolones.

CLINICAL CASE STUDIES

Clinical Case Study 1

A 26-year-old male soldier who served in the Iraq War returned to the United States with symptoms such as fever, weight loss, night sweats, anorexia, cough, dyspnea, and chest pain. Chest X-ray, blood tests, and a TB test confirmed an active TB infection. Treatment begins with a combination of isoniazid, rifampin, and pyrazinamide.

CRITICAL THINKING QUESTIONS

1. What are the adverse effects of isoniazid?
2. What are the contraindications for pyrazinamide?
3. What are the adverse effects of rifampin?

Clinical Case Study 2

A 54-year-old man presented to a dermatologist with a painful, progressive erythematous rash and loss of sensation. The lesions had begun a few months earlier on his face, hands, and trunk. He reported traveling to India 6 years prior, where he came into contact with someone known to have leprosy. His physician confirmed the diagnosis through microscopic examination of a skin smear and biopsy. The patient was subsequently referred to an infectious disease specialist who, advised by the National Hansen's Disease Program, prescribed a triple therapy consisting of dapsone, rifampin, and clofazimine.

CRITICAL THINKING QUESTIONS

1. What are the mechanisms of action of dapsone?
2. What are the possible adverse effects of dapsone?
3. What are the causative pathogens of dapsone?

FURTHER READING

Andrade, I.O. (2024a). *Tuberculosis Volume I*. Andrade.
Andrade, I.O. (2024b). *Tuberculosis Volume II*. Andrade.
Anuradha, C.M. (2023). *Anti-Tuberculosis Drug Design*. Lap Lambert Academic Publishing.
Battista Migliori, G. (2021). *Essential Tuberculosis*. Springer.
Brodston, J. (2015). *Multiple Drug Resistant Tuberculosis — An Analysis of the Standard Protocol Used in the Detection and Treatment of Multiple Drug Resistant Tuberculosis*.

De Souza, M.V.N. (2018). *Tuberculosis Treatment — The Search for New Drugs*. Bentham Books.

Friedman, L.N., Dedicoat, M., and Davies, P.D.O. (2020). *Clinical Tuberculosis*, 6th Edition. CRC Press.

Griffith, D.E. (2019). *Nontuberculous Mycobacterial Disease — A Comprehensive Approach to Diagnosis and Management* (Respiratory Medicine Series). Humana Press.

Heemskerk, D., Caws, M., Marais, B., and Farrar, J. (2015). *Tuberculosis in Adults and Children* (Briefs in Public Health Book 2). Springer.

Laniado Laborin, R. (2015). *Drug Resistant Tuberculosis – Practical Guide for Clinical Management*. Bentham Books.

Lu, P.X., Lu, H.Z., and Yi, Y.X. (2024). *Diagnostic Imaging of Drug Resistance Pulmonary Tuberculosis*. Springer.

Riccardi, G., and Sala, C. (2019). *Tuberculosis Drug Discovery and Development*. MDPI.

Sardana, K., and Khurana, A. (2022). *Jopling's Handbook of Leprosy*, 6th Edition. CPS Publishers and Distributors Pvt Ltd.

Schlossberg, D.L. (2017). *Tuberculosis and Nontuberculous Mycobacterial Infections* (Book 36), 7th Edition. ASM Books.

Teixeira Soares, C. (2021). *Histopathological Diagnosis of Leprosy*. Bentham Books.

36 Immunizations

LEARNING OUTCOMES

After studying the chapter, readers should be able to

1. Review the role of B and T lymphocytes in the immune system.
2. Compare the activated and inactivated vaccines.
3. Identify some examples of live-attenuated vaccines.
4. Explain the polysaccharide vaccines.
5. Describe the influenza infection and its complications.
6. Summarize diphtheria, tetanus, and pertussis vaccine schedules.
7. Review the complications of *Haemophilus influenzae* type B.
8. Explain the human papillomavirus and its prevention.
9. Describe yellow fever and its symptoms.
10. Review Ebola and Japanese encephalitis.

OVERVIEW

The discovery of vaccines is one of the most significant events in medicine. The foundations of modern immunization trace back to the 1790s when Edward Jenner developed the smallpox vaccine using the cowpox virus. In the late 1800s, Louis Pasteur created the first live attenuated bacterial vaccine and developed the first rabies vaccine. Since then, vaccines have been developed for at least 26 diseases. Vaccines have saved countless lives and prevented pain, suffering, and complications from diseases that were once untreatable. The WHO estimates that immunization prevents 2–3 million deaths annually. With expanded global vaccination, an additional 1.5 million deaths could be averted. Failure to immunize can lead to serious illness from vaccine-preventable diseases. High levels of vaccination are essential for community immunity. Although an estimated 91% of children in the United States were vaccinated against measles in 2019, measles outbreaks have been reported in multiple states, partly due to underimmunization.

THE IMMUNE SYSTEM

The immune system is a network of cells, organs, and proteins that can quickly recognize and destroy microorganisms and other agents if they re-enter the body. Abnormalities in the immune system can lead to various conditions, including infections, allergic diseases, immunodeficiencies, and autoimmune disorders. An autoimmune disorder may develop when the immune system malfunctions. This system consists of neutrophils, basophils, eosinophils, monocytes, and lymphocytes, with T lymphocytes (T cells) and B lymphocytes (B cells) playing crucial roles in defending against **antigens**.

Activated T cells play a crucial role in coordinating the immune response. **Macrophages** play a vital role in phagocytizing foreign substances and assisting in T-cell activation. They receive support from **dendritic cells**, which capture antigens and transport them to the **lymph nodes**. Like fibroblasts, **reticular cells** form the reticular fiber stroma, a network supporting other cells within lymphoid organs and tissues.

T lymphocytes regulate immune responses and eliminate cancer cells, while B lymphocytes produce **antibodies**. They are present in blood circulation, lymphoid tissues, and various organs. Most lymphocytes have a short lifespan, averaging from a week to a few months, although some can survive for years. T cells migrate to the **thymus** to mature, while other lymphocytes remain in the bone marrow to develop into B cells.

For a robust immune response, T and B cells collaborate to enhance the production of specialized B cells that can develop into **plasma cells**, which generate antibodies against the vaccine antigen. These antibodies bind to the antigen and help the body eliminate the threat. Once the danger subsides, some developed B cells remain in the body as memory B cells.

Most antigens are large, complex molecules that can be natural or synthetic and are foreign to the body. Therefore, they are considered intruders and are referred to as non-self. Antigens may be complete or incomplete. Whole antigens possess both immunogenicity and reactivity. Immunogenicity is the ability to stimulate B cells to multiply and release antibodies.

VACCINES CLASSIFICATION

Vaccines can be live, attenuated, or inactivated. Live attenuated vaccines are derived from viruses or bacteria. An attenuated vaccine consists of a weakened form of a live pathogen. This pathogen is weakened, rendering it unable to cause disease while stimulating the immune system to produce antibodies, thereby protecting against the full-strength pathogen.

The virus or bacteria are modified to replicate and stimulate immunity without causing the disease. Live-attenuated vaccines include measles, mumps, rubella, varicella, rotavirus, yellow fever, oral typhoid, and intranasal live-attenuated influenza vaccines.

Some vaccines, including polio, hepatitis A, and rabies, are made from inactivated whole viruses. Others consist of subunits from bacteria or viruses, such as acellular pertussis, influenza, hepatitis B, and human papillomavirus. Additionally, some vaccines are derived from bacterial toxoids, including tetanus and diphtheria.

DOI: 10.1201/9781003461913-39

ADMINISTRATION OF VACCINES

Proper vaccine administration is critical to ensuring that vaccination is safe and effective. Comprehensive, skills-based training should be integrated into existing staff education programs, such as new staff orientation and annual education requirements. Each patient's immunization history must be reviewed to determine necessary vaccines.

INTRANASAL

The only vaccine administered intranasally is the live attenuated influenza vaccine. The syringe has a clip that separates the dose into two 0.1 mL sprays. One spray (0.1 mL) is delivered in each nostril while the patient sits and breathes normally. Do not repeat the dose if the patient sneezes or expels it.

Oral

Oral vaccines are easy to administer and convenient for individuals. Their delivery is the most appealing method of administration compared to other routes, as it is non-invasive, safe, and straightforward. This approach demonstrates good patient compliance and clinical practicality. The only oral vaccines available in the United States are the rotavirus, adenovirus, cholera, and oral typhoid vaccines. When multiple vaccines are needed, administer oral vaccines first. Administer the vaccine slowly between the infant's cheek and gum. If the baby spits out the dose, it is not advisable to repeat it. The oral polio vaccine is given in other countries but not in the United States.

Subcutaneous

Subcutaneous vaccines are administered by injecting the vaccine into the fatty tissue beneath the skin, allowing for slower absorption than intramuscular (IM) injections. The preferred subcutaneous site for infants under 12 months is the thigh, while the upper outer triceps is recommended for children 1 year and older. A 5/8-inch, 23- to 25-gauge needle is recommended, and it should be inserted at a 45-degree angle into the subcutaneous tissue.

Intramuscular

Most vaccines should be administered intramuscularly into the **vastus lateralis muscle** in infants. The **deltoid muscle** is the preferred injection site for adolescents and adults. A 22–25 gauge needle that is 5/8 to 1 inch long should be utilized. Examples of IM vaccines include measles, mumps, rubella (MMR), and varicella.

Multiple Vaccine Administration

Multiple vaccines must be administered at separate anatomical sites. The injection sites should be one inch apart within the same muscle group. It is important to remember that oral vaccines should be given before other types. More painful vaccines, such as MMR, pneumococcal conjugate (PCV15 or PCV20), and HPV, should be administered last. Examples of combination vaccines include DTaP (diphtheria–tetanus–pertussis), trivalent IPV (three strains of the inactivated polio vaccine), MMR (measles–mumps–rubella), MMR–varicella, DTaP–Haemophilus influenzae type B (Hib), and Hib–Hep B.

Check Your Knowledge

1. What are the classifications of vaccines?
2. What are the various routes of vaccine administration?
3. How should we administer multiple vaccines?

CHILDREN FROM BIRTH THROUGH 6 YEARS

Hepatitis B is the first vaccine most newborns receive, administered within 24 hours of birth. Infants receive a second dose of the hepatitis B vaccine between 1 and 2 months old, followed by a third dose at 6 months. The initial doses of the vaccines, given at 2 months of age, protect infants from six diseases: DTaP, Hib, polio, pneumococcal, rotavirus, hepatitis A, and the flu. The CDC vaccination schedules for children 6 weeks through 6 years are shown in Figure 36.1.

CHILDREN 7 YEARS THROUGH 17 YEARS

By age 7, children need several booster shots. They will receive the fifth dose of the DTaP (diphtheria, tetanus, and acellular pertussis) vaccine, the fourth dose of the polio vaccine, a second dose of the MMR (measles, mumps, and rubella) vaccine, and the second dose of the varicella (chickenpox) vaccine. The CDC vaccination schedules for children aged 7 through 18 are illustrated in Figure 36.2.

ADULT IMMUNIZATIONS

The process of adult immunization is complex, involving various vaccines and a diverse target population. Routine assessments of adult patients' vaccination needs and recommendations and offerings for necessary vaccines should be integrated into regular clinical care for adults. Despite the availability of vaccines, many adults remain unvaccinated due to a lack of awareness about the importance of adult vaccines or misinformation regarding the vaccines and the diseases they are designed to prevent. Therefore, there is an urgent need to address the issue of adult immunization.

INFECTIOUS DISEASES AND VACCINES

Vaccines are vital for preventing infectious diseases and enhancing the immune system's ability to identify and combat microorganisms. They significantly reduce the incidence of many life-threatening illnesses. Consequently, vaccines have lowered morbidity and mortality rates globally from severe infectious diseases. This chapter explores several contagious diseases in more detail.

Your child needs vaccines as they grow!
2025 Recommended Immunizations for Birth Through 6 Years Old

Want to learn more?
Scan this QR code to find out which vaccines your child might need. Or visit www2.cdc.gov/vaccines/childquiz/

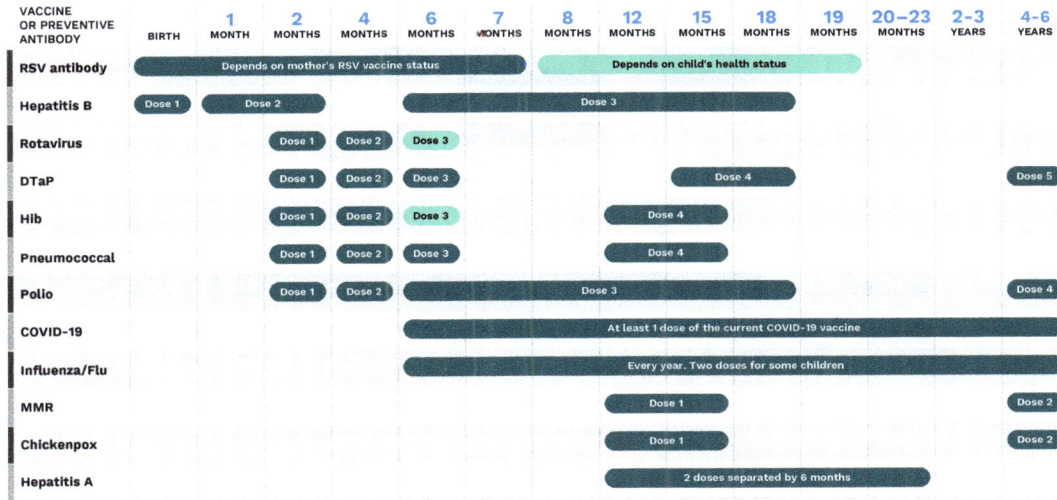

VACCINE OR PREVENTIVE ANTIBODY	BIRTH	1 MONTH	2 MONTHS	4 MONTHS	6 MONTHS	7 MONTHS	8 MONTHS	12 MONTHS	15 MONTHS	18 MONTHS	19 MONTHS	20–23 MONTHS	2–3 YEARS	4–6 YEARS
RSV antibody	Depends on mother's RSV vaccine status						Depends on child's health status						
Hepatitis B	Dose 1	Dose 2			Dose 3									
Rotavirus			Dose 1	Dose 2	Dose 3									
DTaP			Dose 1	Dose 2	Dose 3				Dose 4					Dose 5
Hib			Dose 1	Dose 2	Dose 3			Dose 4						
Pneumococcal			Dose 1	Dose 2	Dose 3			Dose 4						
Polio			Dose 1	Dose 2	Dose 3									Dose 4
COVID-19					At least 1 dose of the current COVID-19 vaccine									
Influenza/Flu					Every year. Two doses for some children									
MMR								Dose 1						Dose 2
Chickenpox								Dose 1						Dose 2
Hepatitis A								2 doses separated by 6 months						

KEY

● ALL children should be immunized at this age

● SOME children should get this dose of vaccine or preventive antibody at this age

Talk to your child's health care provider for more guidance if:

1. Your child has any medical condition that puts them at higher risk for infection.
2. Your child is traveling outside the United States. Visit wwwnc.cdc.gov/travel for more information.
3. Your child misses a vaccine recommended for their age.

CDC U.S. CENTERS FOR DISEASE CONTROL AND PREVENTION

FOR MORE INFORMATION
Call toll-free: 1-800-CDC-INFO (1-800-232-4636)
Or visit: www2.cdc.gov/vaccines/childquiz/

AAFP AMERICAN ACADEMY OF FAMILY PHYSICIANS

American Academy of Pediatrics
DEDICATED TO THE HEALTH OF ALL CHILDREN®

What diseases do these vaccines protect against?

BIRTH–6 YEARS OLD

VACCINE-PREVENTABLE DISEASE	DISEASE COMPLICATIONS
RSV (Respiratory syncytial virus) Contagious viral infection of the nose, throat, and sometimes lungs; spread through air and direct contact	Infection of the lungs (pneumonia) and small airways of the lungs; especially dangerous for infants and young children
Hepatitis B Contagious viral infection of the liver; spread though contact with infected body fluids such as blood or semen	Chronic liver infection, liver failure, liver cancer, death
Rotavirus Contagious viral infection of the gut; spread through the mouth from hands and food contaminated with stool	Severe diarrhea, dehydration, death
Diphtheria* Illness caused by a toxin produced by bacteria that infects the nose, throat, and sometimes skin	Swelling of the heart muscle, heart failure, coma, paralysis, death
Pertussis (Whooping Cough)* Contagious bacterial infection of the lungs and airway; spread through air and direct contact	Infection of the lungs (pneumonia), death; especially dangerous for babies
Tetanus (Lockjaw)* Bacterial infection of brain and nerves caused by spores found in soil and dust everywhere; spores enter the body through wounds or broken skin	Seizures, broken bones, difficulty breathing, death
Hib (Haemosphilus influenzae type b) Contagious bacterial infection of the lungs, brain and spinal cord, or bloodstream; spread through air and direct contact	Depends on the part of the body infected, but can include brain damage, hearing loss, loss of arm or leg, death
Pneumococcal Bacterial infections of ears, sinuses, lungs, or bloodstream; spread through direct contact with respiratory droplets like saliva or mucus	Depends on the part of the body infected, but can include infection of the lungs (pneumonia), blood poisoning, infection of the lining of the brain and spinal cord, death
Polio Contagious viral infection of nerves and brain; spread through the mouth from stool on contaminated hands, food or liquid, and by air and direct contact	Paralysis, death
COVID-19 Contagious viral infection of the nose, thort, or lungs; may feel like a cold or flu. Spread through air and direct contact	Infection of the lungs (pneumonia); blood clots; liver, heart or kidney damage; long COVID; death
Influenza (Flu) Contagious viral infection of the nose, throat, and sometimes lungs; spread through air and direct contact	Infection of the lungs (pneumonia), sinus and ear infections, worsening of underlying heart or lung conditions, death
Measles (Rubeola)† Contagious viral infection that causes high fever, cough, red eyes, runny nose, and rash; spread through air and direct contact	Brain swelling, infection of the lungs (pneumonia), death
Mumps† Contagious viral infection that causes fever, tiredness, swollen cheeks, and tender swollen jaw; spread through air and direct contact	Brain swelling, painful and swollen testicles or ovaries, deafness, death
Rubella (German Measles)† Contagious viral infection that causes low-grade fever, sore throat, and rash; spread through air and direct contact	Very dangerous for pregnant women; can cause miscarriage or stillbirth, premature delivery, severe birth defects
Chickenpox (Varicella) Contagious viral infection that causes fever, headache, and an itchy, blistering rash; spread through air and direct contact	Infected sores, brain swelling, infection of the lungs (pneumonia), death
Hepatitis A Contagious viral infection of the liver; spread by contaminated food or drink or close contact with an infected person	Liver failure, death

FIGURE 36.1 CDC vaccination schedules for children up to 6 years of age.

Older children and teens need vaccines too!
2025 Recommended Immunizations for Children 7–18 Years Old

Want to learn more?
Scan this QR code to find out which vaccines your child might need. Or visit www2.cdc.gov/vaccines/childquiz/

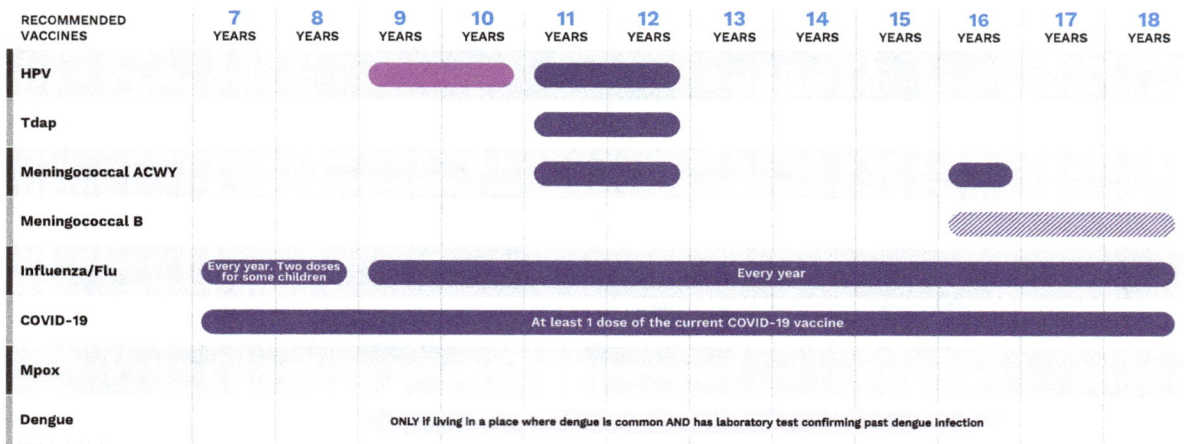

RECOMMENDED VACCINES	7 YEARS	8 YEARS	9 YEARS	10 YEARS	11 YEARS	12 YEARS	13 YEARS	14 YEARS	15 YEARS	16 YEARS	17 YEARS	18 YEARS
HPV			ALL can	ALL can	ALL should	ALL should						
Tdap					ALL should	ALL should						
Meningococcal ACWY					ALL should	ALL should				ALL should		
Meningococcal B										Talk to provider	Talk to provider	Talk to provider
Influenza/Flu	Every year. Two doses for some children		Every year →									
COVID-19	At least 1 dose of the current COVID-19 vaccine →											
Mpox												
Dengue					ONLY if living in a place where dengue is common AND has laboratory test confirming past dengue infection							

KEY

● ALL children in age group should get the vaccine

○ SOME children in age group should get the vaccine

● ALL children in age group can get the vaccine

▨ Parents/caregivers should talk to their health care provider to decide if this vaccine is right for their child

Talk to your child's health care provider for more guidance if:

1. Your child has any medical condition that puts them at higher risk for infection or is pregnant.

2. Your child is traveling outside the United States. Visit wwwnc.cdc.gov/travel for more information.

3. Your child misses any vaccine recommended for their age or for babies and young children.

CDC U.S. CENTERS FOR DISEASE CONTROL AND PREVENTION

FOR MORE INFORMATION
Call toll-free: 1-800-CDC-INFO (1-800-232-4636)
Or visit: www2.cdc.gov/vaccines/childquiz/

AAFP AMERICAN ACADEMY OF FAMILY PHYSICIANS

American Academy of Pediatrics DEDICATED TO THE HEALTH OF ALL CHILDREN®

What diseases do these vaccines protect against?

7-18 YEARS OLD

VACCINE-PREVENTABLE DISEASE	DISEASE COMPLICATIONS	NUMBER OF VACCINE DOSES
HPV (Human papillomavirus) Contagious viral infection spread by close skin-to-skin touching, including during sex	Genital warts and many types of cancers later in life, including cancers of the cervix, vagina, penis, anus, and throat	2 or 3 doses
Tetanus (Lockjaw)* Infection caused by bacterial spores found in soil and dust everywhere; spores enter the body through wounds or broken skin	Seizures, broken bones, difficulty breathing, death	1 dose at age 11-12 years Additional doses if missed childhood doses 1 dose for dirty wounds
Diphtheria* Illness caused by a toxin produced by bacteria that infects the nose, throat, and sometimes skin	Swelling of the heart muscle, heart failure, coma, paralysis, death	1 dose at age 11-12 years Additional doses if missed childhood doses
Pertussis (Whooping Cough)* Contagious bacterial infection of the lungs and airway; spread through air and direct contact	Infection of the lungs (pneumonia), death; especially dangerous for babies	1 dose at age 11-12 years Additional doses if missed childhood doses 1 dose every pregnancy
Meningococcal** Contagious bacterial infection of the lining of the brain and spinal cord or the bloodstream; spread through air and direct contact	Loss of arm or leg, deafness, seizures, death	2 doses. Additional doses may be needed depending on medical condition or vaccine used.
Influenza (Flu) Contagious viral infection of the nose, throat, and sometimes lungs; spread through air and direct contact	Infection of the lungs (pneumonia), sinus and ear infections, worsening of underlying heart or lung conditions, death	1 dose each year 2 doses in some children aged 6 months through 8 years
COVID-19 or flu. Spread through air and direct contact Contagious viral infection of the nose, throat, or lungs; may feel like a cold	Infection of the lungs (pneumonia); blood clots; liver, heart or kidney damage; long COVID; death	1 or more doses of the current COVID-19 vaccine depending on health status. For more information: www.cdc.gov/covidschedule
Mpox Contagious viral infection spread through close, often skin-to-skin contact, including sex; causes a painful rash, fever, headache, tiredness, cough, runny nose, sore throat, swollen lymph nodes	Infected sores, brain swelling, infection of the lungs (pneumonia), eye infection, blindness, death	2 doses
Dengue Viral infection spread by bite from infected mosquito; causes fever, headache, pain behind the eyes, rash, joint pain, body ache, nausea, loss of appetite, feeling tired, abdominal pain	Severe bleeding, seizures, shock, damage to the liver, heart, and lungs, death	3 doses

*Tdap protects against tetanus, diphtheria, and pertussis   **Healthy adolescents: Meningococcal ACWY vaccine (2 doses); Meningococcal B vaccine (2 doses if needed).

FIGURE 36.2 CDC vaccination schedules for children 7 years through 18 years.

INFLUENZA

Influenza is a highly contagious viral illness that affects individuals of all ages. The most common types are influenza A and B. Typical symptoms of influenza include chills, fever, sore throat, nasal congestion, ear infections, cough, headache, and muscle aches. Complications from influenza can include pneumonia, bronchitis, sinusitis, acute respiratory distress syndrome, encephalitis, myocarditis, and sepsis.

The influenza vaccine should be administered annually in the fall and winter. There are two types of influenza vaccines: live attenuated (LAIV) and inactivated (IIV). The inactivated vaccine is given as an IM injection. For children aged 6 months to 8 years, a priming series of two doses of the flu vaccine is recommended, with doses spaced 4 weeks apart. Children under eight who have not received two doses in previous years will need two during the current season. After completing the initial priming doses in subsequent years, only one dose will be required. LAIV is administered intranasally, delivering 0.1 mL in each nostril for children at least 2 years old.

POLIOMYELITIS

Poliomyelitis, commonly known as polio, is a viral infectious disease that can lead to **paralysis**. It is highly contagious and caused by a virus that invades the nervous system, potentially resulting in total paralysis within hours. The virus is transmitted from person to person and primarily spreads through the fecal–oral route, although it can also be transmitted, less frequently, through contaminated food or water. Symptoms include fever, fatigue, headache, vomiting, stiff neck, and limb pain. One in 200 infections results in irreversible paralysis, typically affecting the legs. Among those who become paralyzed, 5%–10% may die if the muscles necessary for breathing are compromised.

Polio is considered eradicated in the United States, but it may still exist in other countries. The inactivated poliovirus vaccine is available as a standalone vaccine (IPOL) and in several combination products known as DTaP–IPV combinations.

IPV combination vaccines are administered as a 0.5 mL dose via intramuscular injection and are part of the routine immunization schedule for infants. The primary series starts at 2, 4, 6, and 18 months, followed by a final dose between 4 and 6 years of age. The individual IPV vaccine, IPOL, can also be given by subcutaneous injection. Oral polio vaccines are utilized in other countries.

DIPHTHERIA, TETANUS, AND PERTUSSIS

Diphtheria is a severe infection caused by strains of *Corynebacterium diphtheriae* that produce toxins. It can lead to a sore throat, a thick gray membrane covering the tonsils and throat, chills, fever, malaise, dysphagia, decreased appetite, dyspnea, myocarditis, arrhythmia, and potentially death. Severe cases may also result in lymph node enlargement. Treatment for diphtheria involves antibiotics, such as penicillin or erythromycin. In more severe instances, hospitalization and intubation may be necessary to manage breathing difficulties. Additionally, diphtheria antitoxin must be administered to neutralize the toxin and prevent further bodily harm.

Tetanus is a severe and acute disease that affects the nervous system, caused by **spores** of *Clostridium tetani*. These spores are typically found in soil and enter the body through cuts or wounds. Once inside, they grow and produce toxins that lead to muscle spasms and convulsions. The disease often begins with muscle spasms in the jaw, resulting in lockjaw, and then spreads to other muscles. The neonatal form of tetanus is usually caused by an infection of the infant's umbilical stump, which rarely occurs in the U.S. Anyone can contract tetanus; however, the disease is most prevalent and severe in newborns and pregnant women who have not received adequate immunization with tetanus–toxoid-containing vaccines. Tetanus during pregnancy or within 6 weeks after delivery is termed maternal tetanus, while tetanus occurring within the first 28 days of life is referred to as neonatal tetanus.

Pertussis is a highly contagious bacterial infection known for causing severe, uncontrollable coughing fits. Also referred to as whooping cough, this condition is caused by the bacterium *Bordetella pertussis*. The bacterium produces toxins that lead to a runny nose, sneezing, fever, and rapid coughing, often followed by a whooping sound when the patient inhales. Complications may include pneumonia, bronchitis, otitis media, seizures, respiratory failure, and encephalopathy due to reduced oxygen supply to the brain; in severe cases, this can even lead to death.

ROTAVIRUS

Rotavirus is a gastrointestinal disease caused by a virus from the Reoviridae family. It can lead to severe diarrhea, vomiting, fever, and dehydration. The virus spreads through the fecal–oral route and primarily affects children aged 3–35 months. Vaccination against rotavirus is recommended for all children starting at 2 months old. The rotavirus vaccine is contraindicated for individuals with a history of severe allergic reactions and intussusception. Infants with **severe combined immunodeficiency** should not receive the vaccine.

HAEMOPHILUS INFLUENZAE TYPE B

Haemophilus influenzae type B (Hib) is a bacterium that can cause serious infections, particularly in infants. This bacterium has six serotypes, with type b (Hib) being the most prevalent, accounting for 90% of infections. Hib primarily causes pneumonia, septicemia, meningitis, septic arthritis, **epiglottitis**, and **cellulitis**, contributing to approximately 3 million cases of severe disease each year. The increasing antibiotic resistance of Hib makes vaccination the only effective strategy for reducing the global incidence of Hib disease. In adults, Hib is less common than in children and can lead to bronchitis, sinusitis, and, in rare cases, meningitis or septicemia.

Hib vaccination is recommended at 2 months of age. Three doses are required at 2, 4, and 6 months, followed by a booster between 12 and 18 months. All Hib-containing products are administered via intramuscular injection, with a dosage of 0.5 mL per product.

YOU SHOULD REMEMBER

Due to the Hib vaccine, invasive *Haemophilus influenzae* type b is rare in the United States. The incidence has dropped dramatically since the pre-vaccine era, primarily affecting infants and young children.

Check Your Knowledge

1. What are the types of influenza vaccines?
2. What do you know about polio vaccines, and how are they administered in the US?
3. What are the schools of diphtheria, tetanus, and pertussis?
4. What are the contraindications of rotavirus vaccines?
5. What are the complications of *Haemophilus influenzae* type B?

MEASLES/MUMPS/RUBELLA

Measles, also known as **rubeola**, is a highly contagious viral illness. The virus spreads through direct contact with droplets from an infected person's coughs or sneezes. Symptoms typically appear 7–14 days after exposure to the virus, including a distinct rash and fever. Other symptoms may include **Koplik spots**, a dry cough, a runny nose, conjunctivitis, and a rash that develops 3–5 days after the initial symptoms. Complications can include diarrhea, ear infections, pneumonia, encephalitis, seizures, and even death.

Mumps is a contagious disease caused by a virus that spreads through airborne transmission and direct contact with droplets or saliva. It commonly occurs during childhood but is largely preventable with a vaccine. Mumps can affect the salivary glands and remains a prevalent illness in many countries, making vaccination crucial for personal protection. Common signs and symptoms include swollen salivary glands under one or both ears, difficulty chewing, fever, headache, fatigue, loss of appetite, muscle aches, and malaise. Complications of mumps include **orchitis**, **oophoritis**, **mastitis**, and encephalitis, which are more common in adults than in children.

Rubella is a contagious viral infection best known for its distinctive red rash. It is also referred to as German measles or 3-day measles. Rubella spreads through airborne transmission and respiratory droplets. Although most patients may experience mild symptoms or none at all, it can cause severe complications for unborn babies whose mothers become infected during pregnancy. Consequently, congenital disabilities, including eye, heart, and neurological abnormalities, may occur in newborns.

The measles/mumps/rubella vaccine (MMR) is a live attenuated virus vaccine. Another live attenuated virus vaccine includes measles/mumps/rubella and varicella antigens (MMRV). Children 12 months and older should receive the MMR vaccine, followed by a booster at ages 4–6. The MMR vaccine is contraindicated in children who have a severe allergic reaction, a history of anaphylaxis to neomycin, congenital immunodeficiency, HIV infection, leukemia, lymphoma, or malignancy. The contraindications and precautions for MMRV and MMR are identical, except for one detail: MMRV contraindications include a personal or family history of seizures. In these cases, separate MMR and varicella vaccines should be administered.

H2 VARICELLA

Varicella (chickenpox) is a contagious disease caused by the varicella–zoster virus (VZV), a member of the herpesvirus family. Following the primary infection, VZV persists in the nerve ganglia as a latent infection. The primary infection results in chickenpox, whereas a recurrent infection causes herpes zoster (also known as shingles). Chickenpox is marked by a rash that develops small, itchy blisters that eventually scab over. The rashes initially appear on the face and then spread to the trunk and extremities.

Chickenpox is a common illness among children, typically occurring during the winter and spring months. In the United States, it results in over 9,000 hospitalizations annually. Its highest prevalence is found in children aged 4–10 years. A complication of chickenpox can be a secondary bacterial infection that may present as cellulitis, **impetigo**, or **erysipelas**. Disseminated primary varicella is frequently seen in immunocompromised individuals and has a high mortality rate. CNS complications are rare but can occur as encephalitis. A primary varicella infection during pregnancy can also affect the fetus, who might later develop chickenpox. Additionally, the virus has the potential to cause **congenital varicella syndrome**.

The varicella vaccine is a live, attenuated vaccine that protects against a common and contagious viral illness in children. Children under 13 should receive two doses: the first between 12 and 15 months and the second between 4 and 6 years. Individuals aged 13 and older who have never had chickenpox or received the vaccine should get two doses at least 28 days apart. Those with cancers, HIV/AIDS, or other diseases affecting the immune system should avoid the varicella vaccine.

HEPATITIS A

Viral hepatitis is classified into five types: A, B, C, D, and E. Vaccines are available only for types A and B, so the focus here will be on these two. Hepatitis A is a highly contagious liver disease caused by a virus that enters the mouth and replicates in the liver, leading to symptoms such as abdominal pain, nausea, malaise, fever, and jaundice. The incubation period lasts about 4 weeks. Hepatitis A spreads through the fecal–oral route by contacting an infected person or consuming contaminated food or water. The Hep A vaccine can prevent it.

The hepatitis A vaccine is an inactivated whole-virus vaccine recommended for children aged 12 months to 18 years and others at an increased risk of infection. The vaccination series consists of two doses, with the second administered at least 6 months after the first. The HepA vaccine should be delivered via intramuscular injection

HEPATITIS B

Hepatitis B is a severe liver infection that affects people worldwide. The acute form often resolves without treatment. However, chronic hepatitis B is an incurable disease that can lead to liver damage. Approximately 2 billion people are infected, and each year, 1 million patients die from it, despite being preventable. Hepatitis B is transmitted through contact with infected blood or certain bodily fluids. The virus is most commonly transmitted from an infected pregnant person to their baby during childbirth due to blood exchange between mother and baby. It can also be transmitted through unsterile medical or dental equipment, unprotected sex, or unsterile needles, or by sharing personal items such as razors, toothbrushes, nail clippers, and body jewelry. Chronic hepatitis B can lead to cirrhosis, liver failure, and liver cancer.

The Hep B vaccine is the only vaccine administered to newborns at birth. It is given via intramuscular injection. Newborns whose mothers have had hepatitis B during pregnancy should receive the vaccine. Three doses are recommended: the first dose is typically administered within 24 hours after birth, the second dose is given 1–2 months after the first, and the third dose is administered between 6 and 18 months. Four doses are permitted if a combination product is used for the subsequent doses. Contraindications and precautions include severe allergic reactions to components such as yeast protein and latex found in the packaging of prefilled syringes and moderate to severe illness.

The routine vaccination series includes three doses of the hepatitis B vaccine, typically administered at 0, 1–2, and 6–18 months of age. Four doses are allowed if a combination product is used for follow-up vaccinations. Catch-up schedules for certain age groups recommend two doses of specific products. Contraindications and precautions include severe allergic reactions to components (like yeast protein and latex in prefilled syringe packaging) and moderate to severe illness.

YOU SHOULD REMEMBER

The lower incidence of hepatitis B in the United States than in Asia and Africa is due to improved access to healthcare and more effective use of vaccinations and other preventive measures.

Check Your Knowledge

1. What are the signs and symptoms of measles?
2. What are the complications of mumps?
3. When should the MMR vaccine be administered in children?
4. What is the description of congenital varicella syndrome?
5. What vaccine can be given at birth?

HUMAN PAPILLOMAVIRUS

Human papillomavirus (HPV) is a common sexually transmitted infection caused by viruses that can infect the skin and mucous membranes. More than 30 strains of HPV can affect the genitals, including benign types that cause genital warts and cervical cancer. Types 16 and 18 are categorized as high-risk for cervical, penile, and anal cancers.

The HPV vaccine can protect against exposure to the virus. If HPV vaccination begins between 9 and 14, a two-dose series is recommended, with an interval of 6–12 months between doses. If HPV vaccination starts at age 15 or older, a three-dose series is necessary, with doses administered at 0, 1–2 months, and 6 months after the first dose. Individuals with immunocompromising conditions also require a three-dose series. Children with a history of sexual abuse or assault may receive the vaccine starting at 9 years of age. Each dose consists of a 0.5-mL intramuscular injection.

MENINGOCOCCAL DISEASE

Meningococcal disease is a severe bacterial infection caused by the bacterium *Neisseria meningitidis*. The two most common types of meningococcal infections are meningitis and septicemia. Transmission occurs through aerosolized respiratory droplets or direct contact. Signs and symptoms include nausea, vomiting, fever, headache, stiff neck, sensitivity to light, and changes in mental status. Infection can progress rapidly and may be fatal in up to 40% of individuals with **meningococcemia**. Close contacts of patients with meningococcal disease should receive antibiotics to prevent illness. The most serious complications of meningococcal infection include brain damage, hearing loss, and death.

Routine vaccination begins with a dose of MenACWY at ages 11–12, followed by a booster at ages 16–18. For patients with specific medical conditions or high-risk situations, such as HIV infection, MenACWY is given at an earlier age. Contraindications for meningococcal vaccines include severe allergic reactions to vaccine components and serious illness.

PNEUMOCOCCAL DISEASE

Pneumococcal disease is an infection commonly caused by the bacterium *Streptococcus pneumoniae*. It can lead to contagious and potentially severe illnesses, such as pneumonia, meningitis, and sepsis. The vaccine provides protection against infections caused by *Streptococcus pneumoniae*, which can lead to pneumonia, meningitis, and other severe conditions.

The United States utilizes two types of pneumococcal conjugate vaccines (PCVs) and polysaccharide vaccines (Pneumovax 23, Merck, and PCV15). For healthy infants and children, the PCV schedule includes a four-dose series: at 2, 4, 6, and 12–15 months of age. This schedule protects against pneumococcal disease, which can lead to serious illnesses such as pneumonia and meningitis. For adults aged 50 and older, a single Pneumovax 23 dose is typically sufficient, but a second dose may be recommended for certain high-risk groups. Adults aged 19–49 with specific conditions or risk factors may also need the vaccine, and the schedule depends on their prior vaccination history.

These vaccines are contraindicated for individuals with severe allergic reactions to them or any vaccine containing diphtheria toxoid. Patients who are without a spleen, those with **sickle cell disease**, and individuals living with HIV are at an increased risk of pneumococcal disease.

RESPIRATORY SYNCYTIAL VIRUS VACCINES

Respiratory syncytial virus (RSV) causes lung infections that are common in children by age 2. RSV can also infect older adults. The virus spreads through direct contact, droplets, coughing, or sneezing. Symptoms of RSV are usually mild and often resemble those of the common cold. However, it can be severe in infants, particularly in premature babies, older adults, patients with heart or lung disease, or individuals who are immunocompromised. Symptoms include a runny nose, cough, sneezing, fever, dyspnea, and wheezing. While most people recover within 1–2 weeks, the condition can worsen and lead to shortness of breath and low oxygen levels.

RSV vaccines are recommended for all adults aged 75 and older and for adults aged 60–74 who are at increased risk for severe RSV. They should receive a single dose of the RSV vaccine, which may be administered simultaneously with other vaccines. Pregnant women should receive one dose of RSVpreF (Abrysvo) during weeks 32–36 of pregnancy, specifically from September to January, to help prevent severe RSV disease in infants.

COVID-19

The COVID-19 pandemic, caused by the severe acute respiratory syndrome coronavirus 2 (SARS-CoV-2), began with an outbreak in Wuhan, China, in December 2019 and quickly spread to countries worldwide. The virus that causes COVID-19 primarily spreads through tiny droplets of fluid in the air among individuals nearby. While many people with COVID-19 have no symptoms or experience mild illness, older adults and those with specific medical conditions may face hospitalization or even death due to the virus. Keeping up with the COVID-19 vaccine helps prevent serious illness, reduces the need for hospital care related to COVID-19, and lowers the risk of death from the virus.

Treatment can lessen the severity of the infection. Most patients recover without long-term effects, although some may experience symptoms that linger for months.

Typical COVID-19 symptoms usually appear 2–14 days after exposure to the virus. Symptoms may include a dry cough, difficulty breathing, fever, chills, headaches, loss of taste or smell, fatigue, vomiting, and diarrhea. Some patients face a higher risk of severe COVID-19 illness than others, particularly those aged 65 and older, as well as children under 6 months. These age groups have the highest likelihood of needing hospital care for COVID-19. Furthermore, individuals with sickle cell disease, heart disease, kidney disease, or liver disease are also at risk, along with those who are obese and have diabetes mellitus.

The four approved vaccines in the United States are the Pfizer–BioNTech, Moderna mRNA COVID-19, Novavax subunit COVID-19, and Johnson & Johnson's vector COVID-19 vaccines. All of these vaccines are administered intramuscularly in the deltoid muscle. The advantages of receiving COVID-19 vaccines include a reduced risk of severe illness, hospitalization, and death from COVID-19 compared to unvaccinated individuals. Updated COVID-19 vaccines can help restore the diminished protection from earlier vaccinations. These updated vaccine formulations provide increased protection against the recent Omicron subvariants, which are more contagious than earlier strains.

The vaccines' adverse effects include severe allergic reactions, **myocarditis**, and **pericarditis**, primarily among males aged 12–39. These rare risks may be minimized with an extended interval between the first and second doses. Instances of myocarditis and pericarditis have also been reported in individuals who received the Novavax COVID-19 vaccine. A potential cause-and-effect relationship exists between the J&J/Janssen COVID-19 vaccine and a rare but severe adverse event: blood clots with low platelets (thrombocytopenia syndrome).

YOU SHOULD REMEMBER

Vaccines for the SARS-CoV-2 virus have led to notable adverse effects across various organ systems, including fever, fatigue, migraines, and cardiovascular issues such as myocarditis, arrhythmias, pericarditis, acute coronary disease, hypertension, thrombotic events, cardiac arrest, and anemia. These rare adverse effects occur in a small percentage of vaccine recipients.

Check Your Knowledge

1. How many HPV vaccines should be administered between 9 and 14?
2. What is the common cause of pneumococcal disease?

3. How many RSV vaccines are recommended for older adults?
4. What do you know about the COVID-19 pandemic?

EBOLA

Ebola virus disease, also known as **Ebola hemorrhagic fever**, is a severe and often fatal illness caused by the Ebola virus. It spreads through contact with the blood, bodily fluids, or contaminated objects of infected patients or animals. The virus can also be transmitted through contact with infected animals, such as bats and monkeys. Symptoms include fever, headache, muscle pain, fatigue, vomiting, diarrhea, and bleeding. Ebola often results in death. The Ebola outbreak in West Africa began in March 2014 and marked the largest hemorrhagic viral epidemic in history. Nearly 40% of patients who contracted the Ebola virus died. While there is no specific cure for Ebola, supportive care, such as hydration, electrolyte management, and pain relief, can improve survival rates. The U.S. Food and Drug Administration has approved a safe and effective vaccine called ERVEBO to prevent Ebola. This vaccine is a live virus administered intramuscularly and should only be given to individuals who meet specific criteria.

TUBERCULOSIS

Tuberculosis (TB), caused by the bacterium *Mycobacterium tuberculosis*, is a contagious disease that primarily affects the lungs but can also spread to other body parts. While many infections remain inactive or latent, a small percentage can progress to active disease. Approximately 10% of individuals infected with TB may develop symptoms. TB is typically treated with antibiotics and can be fatal without treatment. Over 80% of cases and deaths occur in low- and middle-income countries, and TB exists in every region of the world. TB spreads rapidly in crowded environments or among groups of people. Individuals living with HIV/AIDS and those with weakened immune systems are at a higher risk of contracting TB than those with healthy immune systems. Tuberculosis can be treated with antibiotics; however, some strains of the bacteria have developed resistance to treatment.

Primary infection symptoms include a low fever, cough, and fatigue. Symptoms of active TB disease in the lungs typically start gradually and worsen over several weeks. They may consist of cough, chest pain, fever, chills, night sweats, weight loss, and loss of appetite. Common sites of active TB disease outside the lungs are the kidneys, liver, myocardium, lymph nodes, bones, and joints. The TB BCG vaccine, known as Bacille Calmette-Guérin (BCG), is a live, weakened form of the bacterium that causes TB. It is primarily used to prevent TB disease in high-risk populations.

MPOX

Mpox is a viral illness transmitted through close contact with an infected person, contaminated objects, or infected animals. It is an orthopoxvirus, previously known as monkeypox, and is characterized by symptoms similar to **smallpox**, though less severe. While smallpox was eradicated in 1980, mpox continues to occur in Central and West African countries. Since May 2022, cases have been reported in countries without documented mpox transmission outside the African region. The mpox virus has been detected in squirrels, Gambian rodents, rats, and monkeys. The disease can also spread from human to human. It can be transmitted through contact with bodily fluids, skin lesions, mucosal surfaces, and contaminated objects. Patients may develop rashes that can appear on the face, mouth, hands, feet, chest, penis, testicles, labia, vagina, and anus. The incubation period ranges from 3 to 17 days.

The vaccine is safe and effective at preventing mpox infection and severe illness. It is designed for individuals aged 18 and older who are at high risk for mpox infection. JYNNEOS is approved for intradermal or subcutaneous administration as a two-dose series, with doses given 4 weeks apart.

Check Your Knowledge

1. What is the primary use of the BCG vaccine?
2. How is the mpox vaccine administered?

TRAVEL-RELATED DISEASES

Vaccination against one or more diseases is often recommended before travel. Some countries require proof of immunization for specific diseases, such as yellow fever or hepatitis A. Traveling in developing nations and rural areas may necessitate additional vaccines. It is crucial to be up-to-date on routine vaccinations. Immunization for the following diseases is advised before the trip.

YELLOW FEVER

Mosquitoes transmit **yellow fever**, which can lead to hemorrhagic fever. This disease is endemic to South America and tropical and subtropical regions of Africa. Symptoms include chills, fever, severe headache, nausea, vomiting, jaundice, fatigue, and liver disease, and it can result in death. No treatment is available, but a vaccine exists to prevent the disease.

The vaccine is a live virus recommended for individuals aged 9 months and older traveling to or living in high-risk areas. Most travelers require only one dose. However, patients who have received a bone marrow transplant after the original vaccination, women who were pregnant during immunization, and individuals infected with HIV may need additional doses due to a reduced immune response. The vaccine is administered in a dose of 0.5 mL via subcutaneous or intramuscular injection. The yellow fever vaccine should be avoided by individuals with an allergy to any component of the vaccine, those younger than 6 months of age, as well as HIV-positive individuals and those with immunocompromising conditions.

JAPANESE ENCEPHALITIS

Japanese encephalitis (JE) is a viral infection primarily affecting the brain, caused by the Japanese encephalitis virus (JEV). The virus is transmitted to humans through mosquito bites and is most prevalent in rural areas of Southeast Asia, the Pacific islands, and the Far East. It is also present in pigs and birds, but does not spread from person to person. Most patients infected with JEV are asymptomatic, but fewer than 1% may develop clinical disease. Acute encephalitis is the most commonly recognized clinical manifestation, followed by aseptic meningitis and Parkinsonian syndrome. The JE vaccine is over 90% effective, administered either intramuscularly (IM) or subcutaneously (SC). The inactivated vaccine is the only JE vaccine licensed and available in the United States.

There is no specific treatment for JE; clinical management is supportive. Patients with encephalitis require close monitoring for elevated intracranial pressure, seizures, and the ability to protect their airways. JE can be prevented through vaccination. One JE vaccine, IXIARO, is available in the United States.

RABIES

Rabies is a deadly viral infection that affects the brain. It is transmitted through the saliva of infected animals and is almost always fatal once symptoms appear; however, it can be prevented with prompt treatment. The animals most likely to transmit rabies in the United States include bats, coyotes, foxes, raccoons, and skunks. In developing countries, stray dogs are the most likely carriers of rabies to humans. Once a patient exhibits signs and symptoms of rabies, the disease almost always results in death. Therefore, anyone at risk of contracting rabies should receive vaccinations for prevention. These symptoms may persist for days and then progress to brain dysfunction, which includes headache, fever, excessive salivation, nausea, vomiting, anxiety, confusion, agitation, **hydrophobia**, dysphagia, coma, and death. The rabies vaccine is recommended as pre-exposure prophylaxis before traveling to certain countries and after contact with or a bite from specific animals.

Rabies vaccines are essential for preventing the disease, providing both pre-exposure prophylaxis for individuals at high risk and post-exposure treatment following potential contact with a rabid animal. Those at high risk of close contact with rabid animals include veterinarians, animal control workers, laboratory staff, and international travelers. The human diploid cell vaccine (hdcv) and purified chick embryo cell vaccine (PCEC) are available. A two-dose pre-exposure prophylaxis schedule has replaced the previous 3-dose program, allowing for monitoring for up to 3 years. Doses are administered on days 0 and 7. High-risk groups requiring pre-exposure prophylaxis encompass veterinarians and their staff, animal handlers, rabies researchers, certain laboratory workers, and individuals who may frequently come into contact with the rabies

virus or animals that could carry it. Additionally, travelers whose activities may bring them to areas near potentially rabies-infected animals, where access to medical treatment is limited, should also obtain pre-exposure prophylaxis.

The vaccination series for postexposure prophylaxis in individuals previously vaccinated consists of two doses administered on days 0 and 3. For those who have not been immunized, human rabies immunoglobulin (HRIG) and four doses of rabies vaccine are required on days 0, 3, 7, and 14 for individuals who are not immunosuppressed. Immunosuppressed patients require five doses on days 0, 3, 7, 14, and 28. HRIG is infiltrated around the bite site and provides rapid passive immune protection, with a half-life of approximately 21 days. It is administered only once, on the first day of the postexposure prophylaxis regimen.

Contraindications for pre-exposure prophylaxis include a history of hypersensitivity reactions to the vaccine. However, due to the fatal nature of rabies, there are no contraindications for postexposure prophylaxis. Pregnancy and concurrent immunosuppressive therapy are considered precautions.

TYPHOID FEVER

Typhoid fever, also known as **enteric fever**, is caused by *Salmonella Typhi* and is transmitted through contaminated food or water. It is uncommon in areas with low bacteria carriers and rare in regions where water is treated to remove germs and where human waste disposal is managed correctly. In the United States, occurrences are infrequent; however, most outbreaks occur regularly in Africa, Latin America, and South Asia. Typhoid fever presents a significant health risk, especially to children in developing countries. Symptoms include high fever, headache, fatigue, decreased appetite, abdominal pain, and diarrhea or constipation, with vomiting and diarrhea being more prevalent in children.

Typhoid fever can be prevented through vaccination. There are two main types of typhoid vaccines. The oral vaccine is a live-attenuated option for individuals 6 years and older, but it does not guarantee complete protection. The recommended dosage is one capsule orally every other day for four doses. Capsules should be taken with cool water 1 hour before and 2 hours after meals. All doses must be completed at least 1 week before traveling. With continued exposure, a booster is needed every 5 years. For individuals aged 2 years and older, one dose of the injectable inactivated typhoid vaccine should be administered at least 2 weeks before traveling. If necessary, a booster is required every 2 years.

The live attenuated vaccine is contraindicated during pregnancy and in individuals with a history of allergy to any vaccine component. It should be avoided by immunocompromised individuals, during acute febrile illness, and in the context of therapies, malignant neoplasms, and transplants.

CHOLERA

Cholera is an acute diarrheal infection caused by the bacterium Vibrio cholerae. This highly contagious disease can spread rapidly through contaminated water and food. Cholera can lead to severe acute watery diarrhea, which may be fatal within hours if untreated. Most infected patients do not show symptoms but can spread the bacteria through their feces for 1–10 days. Symptoms typically appear 12 hours to 5 days after infection. Cholera is associated with limited access to safe water, basic sanitation facilities, and poor hygiene practices. Travelers to areas where cholera is endemic or experiencing active epidemics are at risk of cholera infection.

The signs and symptoms include dry mucous membranes, **turgor**, hypotension, tachycardia, and thirst. Additional symptoms, such as muscle cramps, arise from resultant electrolyte imbalances. Untreated cholera can lead to rapid fluid loss, potentially resulting in severe dehydration, hypovolemic shock, and death within hours. The case–fatality ratio for untreated cholera exceeds 50%; however, with adequate and timely rehydration, it can drop to less than 1%. Rehydration is the cornerstone of cholera treatment. Administer oral rehydration solution and, when necessary, intravenous fluids and electrolytes; timely administration in adequate volumes will reduce fatality rates. Antibiotics will decrease fluid requirements and shorten the duration of illness. They are indicated alongside aggressive hydration for severe cases and patients with moderate dehydration and ongoing fluid losses.

A live, attenuated, single-dose oral cholera vaccine (Vaxchora, PaxVax) is licensed in the United States and should be taken at least 10 days before travel. This vaccine is not approved for children. The CDC recommends vaccination for individuals traveling to or residing in areas with active cholera transmission. The cholera vaccine is contraindicated in cases of any acute illness or a history of severe allergic reactions. Pregnant women are especially vulnerable to complications from cholera.

YOU SHOULD REMEMBER

Vaccines must be administered with care and precision. Using expired vaccines, incorrect products, or subtherapeutic doses can put vulnerable individuals at risk of disease. Understanding and learning from reported mistakes is crucial to preventing vaccine medication errors.

Check Your Knowledge

1. What are the characteristics of the yellow fever vaccine?
2. How is the Mpox vaccine administered?
3. How many rabies vaccines are recommended for postexposure treatment?
4. How can you differentiate typhoid fever from cholera?

VACCINE STORAGE

Vaccines must be stored at the appropriate temperature, avoiding extremes that are either too hot or too cold, to protect their quality and potency. While most vaccines require refrigeration, some require freezing. Frozen vaccines should be kept in a freezer at a temperature of 5°F or lower. When placing vaccines in the freezer, it is advisable to leave 2–3 inches of space between the containers and the freezer walls. Another method for maintaining consistent temperatures is to store water bottles in the freezer alongside the vaccines. Generally, it is recommended not to use dormitory-style freezers or combination refrigerator/freezer units. A pharmacy-grade refrigerator is appropriate for vaccine storage, and food should not be stored in the unit. Vaccines should not be placed on the shelves of the freezer doors.

Certain vaccines, such as chickenpox, shingles, and a combination vaccine of Varicella, MMR, or MMRV, must be stored at or below −15°C (5°F) to maintain potency. Some COVID-19 vaccines, like Pfizer–BioNTech, require ultra-cold storage and are shipped with dry ice to remain frozen. Always follow the manufacturer's storage instructions.

CLINICAL CASE STUDIES

Case Study 1

A 35-year-old woman was diagnosed with cervical cancer, which resulted in chronic pelvic pain and bleeding after intercourse. She had never received the HPV vaccine. Her medical history indicates that she had four partners and became sexually active at the age of 16. Moreover, she had not undergone Pap smears in the past.

CRITICAL THINKING QUESTIONS

1. What is the description of human papillomavirus infection?
2. What types of HPV can cause cervical or penile cancers?
3. How many doses are advised for HPV vaccination?

Case Study 2

An infant was brought to the emergency room. His mother stated that he had started experiencing a runny nose, coughing, sneezing, fever, dyspnea, and wheezing. His medical history indicated that he was born prematurely. The patient appeared lethargic and pale, exhibiting moderate respiratory distress. Immunization records are unknown. Following blood tests and a chest X-ray, he was diagnosed with a respiratory syncytial virus infection.

CRITICAL THINKING QUESTIONS

1. What are the risk factors for RSV infections?
2. How many doses of the vaccine are required for RSV?
3. When should pregnant women get the RSV vaccine?

FURTHER READING

Abbas, A.K., Lichtman, A.H., and Pillai, S. (2021). *Cellular and Molecular Immunology*, 10th Edition. Elsevier.

Alter, J. (2010). *Vaccines* (21st Century Skills Innovation Library). Cherry Lake Publishing.

Bansal, R. (2021). *Antibodies and Their Role in Therapeutics* (Biotechnology). Bansal.

Blaxill, M., and Olmsted, D. (2015). *Vaccines 2.0: The Careful Parent's Guide to Making Safe Vaccination Choices for Your Family*. Skyhorse.

Cook, W., and Klaas, K. (2020). *Mayo Clinic Guide to Your Baby's First Years*, 2nd Edition. Mayo Clinic Press.

Cowan, T., and Fallon Morell, S. (2018). *Vaccines, Autoimmunity, and the Changing Nature of Childhood Illness*. Chelsea Green Publishing.

Doan, T., Lievano, F., Swanson-Mungerson, M., and Viselli, S.M. (2021). *Lippincott Illustrated Reviews: Immunology* (Illustrated Reviews Series), 3rd Edition. Wolters Kluwer Health.

Ertl, H.C.J. (2020). *Rabies and Rabies Vaccines*. Springer.

Feder, L., and Hoang, L. (2017). *The Parents' Concise Guide to Childhood Vaccinations — Revised and Updated* (2nd) Edition — *Practical Medical and Natural Ways to Protect Your Child*. Hatherleigh Press.

Gullberg, R.M. (2021). *Health Advice & Immunizations for Travelers — Intended for International Travelers & Travel Clinics*. Gullberg.

Humphries, S., and Bystrianyk, R. (2013). *Dissolving Illusions — Disease, Vaccines, and the Forgotten History*. CreateSpace Independent Publishing Platform.

Ikidde, U. (2017). *Essential Family and Travellers' Health — Health Advice Everyone Should Know*. WriterMotive.

Lenox, D.A. (2022). *Pneumococcal Vaccine: How to Protect Yourself from Pneumococcal Disease*. Leddnox.

Rappuoli, R., and Vozza, L. (2022). *Vaccines in the Global Era: How to Deal Safely and Effectively with the Pandemics of Our Time*. WSPC.

Rees, A.R. (2022). *A New History of Vaccines for Infectious Diseases: Immunization — Chance and Necessity*. Academic Press.

Schama, S. (2023). *Foreign Bodie: Pandemics, Vaccines, and the Health of Nations*. Ecco.

Thomas, P., and Margulis, J. (2016). *The Vaccine-Friendly Plan: Dr. Paul's Safe and Effective Approach to Immunity and Health-from Pregnancy through Your Child's Teen Years*. Ballantine Books.

USAF Counterproliferation Center, and Penny Hill Press Inc. (2016). *The Anthrax Vaccine Debate: A Medical Review for Commanders*. CreateSpace Independent Publishing Platform.

37 Poisons and Antidotes

LEARNING OUTCOMES

After studying the chapter, readers should be able to

1. Describe symptoms of acute acetaminophen poisoning.
2. Explain the signs and symptoms of an aspirin overdose.
3. Discuss opiate toxicity and the treatment.
4. List atypical agents of sedative-hypnotics.
5. Describe the common symptoms of atypical antipsychotic poisoning.
6. Explain the signs and symptoms of digitalis poisoning.
7. Describe the complications of lithium poisoning.
8. Explain the symptoms of lead poisoning.
9. Describe the treatments for cyanide poisoning.
10. List the antidotes for carbon monoxide, cyanide, and opiates.

OVERVIEW

Poisoning refers to injury or death caused by swallowing, inhaling, touching, or injecting various drugs, detergents, venoms, or gases. Many substances, such as drugs and carbon monoxide, become toxic only at higher concentrations or dosages. Others, like chemical detergents, are primarily dangerous when ingested. Children are particularly vulnerable to even small amounts of certain drugs and chemicals. Cyanide, arsenic, and pesticides are hazardous even in small quantities. However, certain medications can become toxic when taken in excessive doses. Accidental poisonings are more common at home, while industrial intoxication often results from chronic exposure. Accidental poisoning usually stems from the unintentional ingestion of toxic substances, with most cases involving children. In many instances, these incidents are preventable, especially the poisoning of young children from drugs and chemicals found at home. Pediatric exposures occur because naturally curious children explore their environment through touch and taste. Such unintentional exposures typically involve small amounts ingested from household substances, primarily analgesics, due to their widespread availability.

SEDATIVE-HYPNOTICS

Sedatives and hypnotics constitute a broad class of medications that reduce the brain's activity. Common types include benzodiazepines and barbiturates. They serve various purposes, from treating anxiety and insomnia to providing anesthesia. Because these medications have the potential for addiction, they must be used with caution.

BENZODIAZEPINES

Benzodiazepines are therapeutically used to treat anxiety, promote sleep, or manage seizures. They function by enhancing the effect of an inhibitory neurotransmitter known as gamma-aminobutyric acid, leading to generalized depression of the central nervous system (CNS). Symptoms of benzodiazepine poisoning include drowsiness, slurred speech, lack of coordination, and coma. While respiratory arrest is possible, overdoses of benzodiazepines are typically not life-threatening unless combined with another CNS depressant, such as alcohol or an opioid. For further details, see **Chapter 11**.

To treat benzodiazepine overdose, flumazenil was once a potential antidote. It is a competitive antagonist at the benzodiazepine receptor. Flumazenil can unmask seizures in patients habituated to benzodiazepines or patients who have taken a co-ingestion of seizure-causing medications. Therefore, the risks of flumazenil usually outweigh its benefits in overdose, and it is rarely indicated.

BARBITURATES

Barbiturates also depress brain function. First synthesized in the early 1900s, they were widely used as sedatives, anesthetics, and anticonvulsants. Overdosage refers to a potentially lethal condition that occurs when an excessive amount of barbiturates is consumed. It is characterized by symptoms such as respiratory arrest, heart failure, coma, and death. For further details, see **Chapter 11**.

ANALGESICS

Analgesics are substances that relieve pain. They function by interrupting brain or spinal cord pain signals and reducing inflammation. They account for more total exposures and deaths reported to poison centers than any other substance. Various classes of medications fall into this category, each with different mechanisms of toxic action, expected symptoms, and treatments.

ASPIRIN

Aspirin, also known as salicylate, inhibits the production of certain natural substances that lead to fever, pain, inflammation, and blood clots. It is also available with other medications, such as antacids, pain relievers, and cough and cold remedies. For more information, see **Chapter 27**.

Aspirin toxicity can occur due to acute or chronic ingestion. The signs and symptoms of aspirin overdose include nausea, vomiting, sweating, tachypnea, tinnitus, metabolic

DOI: 10.1201/9781003461913-40

acidosis, confusion, seizures, pulmonary and brain edema, coma, and death. There is no antidote for aspirin poisoning. Patients with toxicity are often treated with intravenous sodium bicarbonate and dialysis of the kidneys.

Nonsteroidal Anti-Inflammatory Drugs

Nonsteroidal anti-inflammatory drugs (NSAIDs) are a class of medications that reduce fever, pain, and inflammation by inhibiting the production of prostaglandins, which play a role in these processes. Common examples include aspirin, ibuprofen, and naproxen. Although they can cause severe adverse effects, the safety of these drugs varies based on patient factors and medical conditions. Symptoms of NSAID poisoning include nausea, vomiting, and drowsiness. Large overdoses may lead to bleeding, hypotension, seizures, kidney failure, **metabolic acidosis**, and coma. Treatments for NSAID poisoning are provided through supportive care. For more information, see **Chapter 27.**

Acetaminophen

In many countries, acetaminophen is known as N-acetyl-para-aminophenol or paracetamol. It is a nonopioid analgesic and antipyretic medication used to treat pain and fever.

Chapter 27 provides a more detailed discussion of acetaminophen. Symptoms of toxicity include nausea, vomiting, sweating, abdominal pain, jaundice, confusion, kidney damage, coma, and even death. Liver impairment is the most significant manifestation of acetaminophen toxicity, appearing 1–2 days after ingestion.

Treatment for an acetaminophen overdose should start with the induction of emesis or gastric lavage, followed by the administration of activated charcoal. N-acetylcysteine (NAC) is an antidote for both acute and chronic acetaminophen poisoning. NAC can be administered orally or intravenously. It is most effective when administered soon after an acute overdose before liver damage begins. Nevertheless, NAC remains beneficial even after liver injury has started.

YOU SHOULD REMEMBER

In most cases, taking a regular dose of acetaminophen during or after a night of drinking should not cause liver damage. A "normal" dose is up to 1,000 mg over 4–6 hours, totaling no more than 4,000 mg daily.

Opioids and Opiates

Opioids are pain-relieving agents derived from the opium **poppy plant**, including morphine (Duramorph) and heroin. While opiates are not direct derivatives of the poppy plant, they are derived from opioids. Opiates include fentanyl, oxycodone, hydrocodone, and codeine. For more information, refer to **Chapter 28**. The use of opioids, whether

alone or in combination with other drugs, significantly contributes to the drug overdose crisis in the United States. Currently, the vast majority of overdose deaths are linked to illegally manufactured fentanyl and other potent synthetic opioids that can be mixed with various drugs and consumed unknowingly.

Toxicity can cause drowsiness, analgesia, euphoria, apnea, and constipation.

Some synthetic opioids can lead to further overdose symptoms. For instance, meperidine may trigger seizures, while methadone could result in cardiac arrhythmias. However, the primary toxicity for all opioids is considered to be respiratory depression.

Naloxone is an antidote for respiratory depression caused by opioids. It functions as a competitive antagonist at opioid receptors, effectively blocking them so that opioids cannot bind. Treatment for overdose may require intubating the patient and providing mechanical ventilation until the opioid has been metabolized.

YOU SHOULD REMEMBER

Naloxone is available in intramuscular, intravenous, and nasal spray forms. Both intravenous and nasal routes offer rapid effects, reversing opioid toxicity within seconds to minutes.

OTHER TOXIC SUBSTANCES

Medication errors can lead to poisoning. Approximately 530,000 injury incidents occur each year in outpatient clinics due to these errors. A double dose and taking the wrong medication can result in medication poisoning and medication errors in education.

Antihistamines

Antihistamines are common medications that alleviate allergy symptoms. An overdose of antihistamines may result in drowsiness, hallucinations, irritability, tachycardia, **stupor**, blurred vision, seizures, and coma. Treatment is supportive and focuses on removing the unabsorbed drug and maintaining vital functions.

Digitalis

Digoxin is a well-known cardiac glycoside and one of the oldest medications still used to treat heart failure and atrial fibrillation. Although it is considered safe, digoxin has a narrow therapeutic window, and its proper dosing requires attention to various patient characteristics, including age, gender, kidney function, and concurrent use of other medications, to avoid potentially life-threatening toxicity.

Signs and symptoms of digitalis poisoning include headache, malaise, insomnia, altered mental status, abdomin pain, nausea, vomiting, cardiac arrhythmias, seizures, visual abnormalities (such as photophobia and **photopsia**), and dizziness. Treatment may involve activated charcoal, atropine, antiarrhythmics, electrical cardioversion, and advanced life support. Digitalis poisoning can lead to multiple organ failure and recurrence of toxicity after treatment has concluded.

LITHIUM

Lithium is a medication used to treat bipolar disorder. It helps balance substances in the brain that regulate mood, behavior, and thoughts. Signs and symptoms of lithium toxicity include nausea, vomiting, diarrhea, tachycardia, memory loss, delirium, tremors, seizures, and coma. Treatment may involve advanced life support, fluid therapy, benzodiazepines, whole-bowel irrigation, hemodialysis, amiloride, and other interventions. Complications of lithium poisoning can include permanent encephalopathy, nystagmus, ataxia, and choreoathetosis (the occurrence of involuntary movements).

ETHYLENE GLYCOL

Ethylene glycol is a colorless, odorless, sweet-tasting liquid commonly found in antifreeze, coolants, and hydraulic brake fluids. Although its toxicity is rare, when it occurs, it can be life-threatening and may lead to metabolic acidosis. Ethylene glycol is highly toxic and is a major cause of poisoning worldwide. Toxicity can result in central nervous system (CNS) dysfunction, cardiovascular compromise, and acute kidney injury. Symptoms of ethylene glycol toxicity include confusion, ataxia, hallucinations, slurred speech, and coma. These symptoms are most severe 6–12 hours after ingestion. Supportive care and resuscitation are the primary treatments for ethylene glycol toxicity, although dialysis might be necessary.

METHANOL

Methanol is a toxic alcohol that can be found in various household and industrial products. Exposure to this organic compound can be hazardous, potentially leading to significant morbidity and mortality if left untreated. Methanol poisoning most often occurs due to accidental or intentional ingestion. Products containing methanol include windshield wiper fluid, industrial solvents, certain types of antifreeze, carburetor cleaner, copy machine fluid, perfumes, and various fuels used for warming food. Exposures can cause varying degrees of toxicity, potentially requiring a range of treatments from close laboratory monitoring to antidotal therapy and dialysis. Fomepizole or ethanol may be administered as the primary treatment, while dialysis is often recommended, unlike in cases of ethylene glycol toxicity.

Ingesting just 15 mL of methanol can lead to permanent blindness due to severe toxicity affecting the optic nerve. Other symptoms may include nausea, vomiting, headache,

dizziness, and respiratory failure. Treatment involves emesis and gastric lavage, which should only be performed within the first 2 hours after ingestion. Intravenous sodium bicarbonate is administered to treat metabolic acidosis. Additionally, intravenous ethanol therapy should be provided as an antidote.

Check Your Knowledge

1. What are the signs and symptoms of antihistamine toxicity?
2. What are the signs and symptoms of digitalis poisoning?
3. What are the signs and symptoms of lithium poisoning?
4. What are the signs and symptoms of ethylene glycol toxicity?
5. What are the signs and symptoms of methanol poisoning?

CYANIDE

Cyanide is an exceedingly potent and rapid-acting poison, as well as a deadly chemical. It affects the body's ability to utilize oxygen. Cyanide poisoning can result from inhalation or ingestion. It occurs in natural substances found in some foods and certain plants, including the pits and seeds of various common fruits.

Cyanide poisoning is characterized by chest pain, apnea, headache, dizziness, excitement, tachycardia, seizures, coma, and cardiac arrest. It may cause death within 1–15 minutes.

This is a true medical emergency; however, treatment is highly effective if administered quickly. The **antidotes** should be given without delay. Supportive measures should be implemented immediately, particularly artificial respiration with 100% oxygen.

CARBON MONOXIDE

Carbon monoxide (CO) is a colorless, odorless gas produced from burning fuels such as gasoline, wood, propane, or charcoal. Appliances and engines that lack proper ventilation can lead to dangerous CO buildup levels, which are exacerbated in tightly enclosed spaces.

Carbon monoxide poisoning occurs when CO accumulates in the blood. Too much CO in the air displaces oxygen in red blood cells, resulting in severe tissue damage or even death. Each year, nearly 2,500 accidental deaths and suicides occur in the United States. Severe CO poisoning symptoms include cherry-red skin color and mucous membranes. Treatment requires adequate ventilation with high concentrations of oxygen and the removal of CO. If necessary, ventilation should be supported artificially. Pure oxygen must be administered, and cerebral edema should be treated with diuretics and steroids.

ORGANOPHOSPHATES

Organophosphates are chemical compounds utilized as pesticides and insecticides. They inhibit the enzyme acetylcholinesterase, resulting in an accumulation of the neurotransmitter acetylcholine. Toxicity may be achieved between 30 minutes and 2 hours after absorption.

Organophosphates accumulate in body tissues and are gradually eliminated by the liver.

Organophosphate pesticide self-poisoning is a significant clinical and public health issue throughout rural Asia. Of the estimated 500,000 deaths from self-harm in the region each year, approximately 60% result from pesticide poisoning. Symptoms of poisoning include headaches, blurred vision, nausea, vomiting, abdominal pain, and excessive salivation. Treatments include removing contaminated clothing, washing the skin with soap and water, and administering activated charcoal. Oxygen therapy and seizure management are also essential.

Check Your Knowledge

1. What is the characteristic of cyanide toxicity?
2. What is the antidote for carbon monoxide poisoning?
3. What are the signs and symptoms of organophosphates?

LEAD

Lead is present in the environment – air, soil, water, and homes. Much of the exposure results from human activities, such as burning fossil fuels, the historical use of leaded gasoline, certain industrial facilities, and the former application of lead-based paint in homes. Lead and its compounds have been incorporated into various products in and around homes, including paint, ceramics, plumbing pipes and materials, solders, gasoline, batteries, ammunition, and cosmetics.

Children are particularly vulnerable to lead poisoning and can be exposed to lead in various ways, such as lead paint, contaminated soil, water from plumbing made with lead materials, and imported items like pottery and jewelry. Lead is absorbed from the gastrointestinal tract and deposited in tissues throughout the body, including the bones and the brain.

Signs and symptoms include nausea, vomiting, a metallic taste, loss of appetite, headaches, fatigue, weight loss, anemia, and muscle aches. Symptoms affecting the CNS include difficulty concentrating, lack of coordination, irritability, and tremors. Severe lead poisoning can lead to encephalopathy and kidney damage. The first step in treating lead poisoning is to eliminate the source of contamination.

Chelating therapy agents bind to lead and are excreted in urine. There is no antidote for lead poisoning.

ARSENIC

Arsenic is a naturally occurring element found in water and soil. The roots of rice and tea plants quickly absorb it.

TABLE 37.1
Poisons and Antidotes

Poison	Antidotes
Acetaminophen	N-Acetylcysteine
Benzodiazepines	Flumazenil (Romazicon®, others)
Carbon monoxide	Oxygen
Cyanide	Amyl nitrate
Iron	Deferoxamine
Methanol	Ethanol
Opiates	Naloxone (Narcan®, others)
Organophosphates	Atropine (Atropen®) or pralidoxime

Organic arsenic is considered harmless because it is bound to sugars in fruits and vegetables. In contrast, inorganic arsenic is regarded as hazardous, potentially leading to lung, bladder, and skin cancers in adults.

Signs and symptoms of arsenic poisoning include a metallic taste and a garlic-like odor. Other symptoms are nausea, vomiting, gastric bleeding, cardiogenic shock, arrhythmias, pulmonary edema, pancytopenia, hemolytic anemia, convulsions, coma, polyneuropathy, and renal impairment. Treatments include advanced life support, endotracheal intubation, fluid therapy, gastric lavage, chelation therapy, hemodialysis, and blood transfusions. Arsenic poisoning may lead to neurotoxic effects and multisystem organ failure.

Check Your Knowledge

1. What are the signs and symptoms of lead poisoning?
2. What are the first steps in treating lead poisoning?
3. What are the signs and symptoms of arsenic poisoning?

ANTIDOTES

Antidotes are chemical substances that counteract the effects of poisons. An antidote, such as activated charcoal, can absorb toxins in the gastrointestinal tract and prevent their absorption. It may also neutralize the toxin physiologically by counteracting the poison. Specific antidotal therapy is available for only a limited number of poisons.

Specific antidotes, along with their related poisons, are listed in Table 37.1.

FURTHER READING

Anderson, I.B., Benowitz, N.L., Blanc, P.D., Kim-Katz, S.Y., Lewis, J.C., Wu, A.H.B., Olson, K.R., and Smollin, C. (2022). *Poisoning and Drug Overdose*, 8th Edition. McGraw Hill/Medical.
Armstrong, J., and Pascu, O. (2022). *The Toxicology Handbook*, 4th Edition. Elsevier.
Gilbert, S., Mohapatra, A., Bobst, S., Hayes, A., and Humes, S.T. (2020). *Information Resources in Toxicology, Volume 2 — The Global Arena*, 5th Edition. Academic Press.

Hoffman, R.S., Gosselin, S., Nelson, L.S., Howland, M.A., Lewin, N.A., Smith, S.W., and Goldfrank, L.R. (2023). *Goldfrank's Clinical Manual of Toxicologic Emergencies*, 2nd Edition. McGraw Hill/Medical.

Katz, K.D. (2022). *EMRA and AMCT Medical Toxicology Guide*, 2nd Edition. Emergency Medicine Residents' Association.

Klaassen, C.D. (2018). *Casarett & Doull's Toxicology — The Basic Science of Poisons*, 9th Edition. McGraw Hill/Medical.

Levine, B.S., and Kerrigan, S. (2020). *Principles of Forensic Toxicology*, 5th Edition. Springer.

McQueen, C. (2018). *Comprehensive Toxicology*, 3rd Edition. Elsevier Science.

Narayan Reddy, K.S. (2022). *The Essentials of Forensic Medicine & Toxicology*, 35th Edition. Jaypee Brothers Medical Publishers.

Negrusz, A., and Cooper, G. (2013). *Clarke's Analytical Forensic Toxicology*, 2nd Edition. Pharmaceutical Press.

Olson, K.R., Anderson, I.B., Benowitz, N.L., Blanc, P.D., Clark, R.F., Kearney, T.E., Kim-Yatz, S.Y., and Wu, A.B.H. (2017). *Poisoning and Drug Overdose*, 7th Edition. McGraw Hill/Medical.

Roberts, S.M., James, R.C., and Williams, P.L. (2022). *Principles of Toxicology*, 4th Edition. Wiley.

Timbrell, J., and Barile, F.A. (2023). *Introduction to Toxicology*, 4th Edition. CRC Press.

Appendix A
High-Alert Medications List by ISMP

Type of Medication	Examples
Adrenergic agonists, IV	Epinephrine, norepinephrine, phenylephrine
Adrenergic antagonists, IV	Labetalol, metoprolol, propranolol
Anesthetic agents, general, inhaled, and IV	Ketamine, propofol
Antiarrhythmics, IV	Amiodarone, lidocaine
Antithrombotic agents	
Anticoagulants	Low-molecular weight heparin, unfractionated heparin, warfarin
Direct oral anticoagulants and factor Xa Inhibitors	Apixaban, betrixaban, dabigatran, edoxaban, fondaparinux, rivaroxaban
Direct thrombin inhibitors	Argatroban, bivalirudin
Glycoprotein IIb/IIIa inhibitors	Eptifibatide
Thrombolytics	Alteplase, reteplase, tenecteplase
Cardioplegic solutions	Calcium chloride, magnesium chloride, potassium chloride, sodium chloride
Chemotherapeutic agents, parenteral and oral	Dactinomycin, etoposide, ifosfamide, mitomycin, toposar
Dextrose, hypertonic, 20% or greater	
Dialysis solutions, peritoneal and hemodialysis	Dextrose monohydrate, eravacycline, sodium chloride
Epidural and intrathecal medications	Morphine sulfate, sufentanil citrate
Inotropic medications, IV	Digoxin, milrinone
Insulin, subcutaneous and IV	
Liposomal forms of drugs and conventional counterparts	Liposomal amphotericin B; counterpart agent amphotericin B desoxycholate
Moderate sedation agents, IV	Dexmedetomidine, lorazepam, midazolam
Moderate and minimal sedation agents, oral, for children	Chloral hydrate, midazolam
Opioids, including IV, oral, transdermal, neuromuscular blocking agents	Oral: liquid concentrates, immediate- and sustained-release formulations
	Neuromuscular: rocuronium, succinylcholine, vecuronium
Parenteral nutrition preparations	SMOFlipid, Intralipid, Nutrilipid

Appendix B
Drug Dosage Calculations

USING RATIOS AND PROPORTIONS TO CALCULATE DOSAGES

1. Ratios may be written as 3:4, which means three parts of the drug to four parts of the solution or solvent. *Ratios* are used to express the relationship between two or more quantities. Ratios are usually expressed as fractions in drug calculations as follows:

$$\frac{3 \text{ parts of drug}}{4 \text{ parts of a solution}} = \frac{3}{4}$$

Proportions show the relationship between two ratios as follows:

$$\frac{\text{Dose on hand}}{\text{Quantity on hand}} = \frac{\text{Desired dose}}{\text{Desired quantity } (X)}$$

The same formula can be written as follows by using cross-multiplication:

$$\text{Desired quantity } (X) = \frac{\text{Desired dose}}{\text{Dose on hand} \times \text{Quantity on hand}}$$

If the dose on hand is 200 mg, the desired dose is 400 mg, and the quantity on hand is 10 mL, what is the desired amount (X)?

$$\frac{\text{Dose on hand } (300 \text{ mg})}{\text{Quantity on hand } (5 \text{ mL})} = \frac{\text{Desired dose } (600 \text{ mg})}{\text{Desired quantity } (X)}$$

When cross-multiplying, we find
$300 \times X = 5 \text{ mL} \times 600$
$300X = 3{,}000 \text{ mL}$
$X = 10 \text{ mL}$
The dose to be administered is 10 mL.

2. The same proportion method can solve solid dosage calculations as follows:
    If the dose on hand is available as 5 mg tablets and the desired dose is 25 mg/day, how many tablets should be administered each day?

$$\frac{\text{Dose on hand } (50 \text{ mg})}{1 \text{ tablet}} = \frac{\text{Desired dose } (250 \text{ mg})}{\text{Desired quantity } (X)}$$

By cross-multiplying, it is found that
$50 \text{ mg} \times X = 250 \text{ mg} \times 1 \text{ tablet}$
$50X = 250 \text{ mg}$
$X = 5 \text{ tablets/day}$
Therefore, five tablets should be given daily.

CALCULATING DOSAGE BY BODY SURFACE AREA

Using body surface area to calculate pediatric dosages is a very accurate method. Nomograms are charts that use patient weight and height to determine body surface area in square meters (m²). This body surface area (BSA) is then placed into a ratio with the average adult's body surface area (1.73 m²). The following formula is then used:

$$\text{Child's dose} = \frac{\text{Child's BSA in m}^2}{1.73 \text{ m}^2} \times \text{Adult dose}$$

Nomogram scales contain metric and avoirdupois values for height and weight, enabling body surface area to be determined in pounds and inches or kilograms and centimeters without making conversions.

A ruler or straightedge is recommended to determine BSA. After selecting the patient's height and weight, place the ruler or straightedge on the nomogram and connect the two points on the height and weight scales representing the patient's values. Where the ruler or straightedge crosses the center column (BSA), the corresponding reading is the value of the BSA in square meters. Substitute the BSA value in the formula to calculate the dosage for this patient. If this child's BSA is 0.52 m² and the adult dosage of the required drug is 500 mg, use the following formula to determine the child's dose:

$$\text{Child's dose} = \frac{0.52 \text{ m}^2}{1.73 \text{ m}^2} \times \text{Adult dose} (500 \text{ mg})$$

$$= 0.3 \times 500 \text{ mg}$$

$$= 150 \text{ mg (child's dose)}$$

CALCULATING IV INFUSION RATES

To calculate the flow rate using the ratio and proportion method, follow these steps:

1. Determine the number of milliliters the patient will receive per hour.
2. Determine the number of milliliters the patient will receive per minute.
3. Determine the number of drops per minute equal to the milliliters calculated above. The IV set's drop rate must be considered. This is expressed as a ratio of drops per milliliter (gtt/mL).

Example: The prescriber orders 3,000 mL of dextrose 5% in water (D_5W) IV over 24 hours. If the IV set delivers 15 drops per milliliter, how many drops must be administered per minute?

First, calculate mL/hr.

$$\frac{3{,}000 \text{ mL}}{24 \text{ hours}} = \frac{X \text{ mL}}{1 \text{ hour}}$$

$X = 125$ mL/hr or
   125 mL/60 minutes
   Next, calculate mL/min.

$$\frac{125 \text{ mL}}{60 \text{ minutes}} = \frac{X \text{ mL}}{1 \text{ minutes}}$$

$X = 2$ mL/min

Finally, calculate gtt/min using the drop rate per minute of the IV set (IV set drop rate = 15 drops/mL).

$$\frac{15 \text{ gtt}}{1 \text{ mL}} = \frac{X \text{ gtt}}{2 \text{ mL (amount needed/min)}}$$

$X = 30$ gtt/min

Appendix C
Answer Key to Clinical Case Studies

CHAPTER 1: DRUG CLASSIFICATIONS, DRUG NAMES, AND DRUG SOURCES

No case study

CHAPTER 2: DRUG DEVELOPMENT, DRUG APPROVAL, AND DRUG REGULATION

No case study

CHAPTER 3: PHARMACOKINETICS
Clinical Case Study 1

1. Drug absorption may be influenced by stomach pH. When the difference between the plasma pH and the pH at the administration site causes drug molecules to have a greater tendency for ionization in the plasma, absorption is enhanced. Some medications require strong acids to be absorbed. For example, an antibiotic like cefuroxime needs the stomach pH to be low.
2. The efficiency of drug absorption is greatly determined by the surface area available. Absorption is quicker over a larger surface area. Therefore, orally administered drugs are usually absorbed not from the stomach but from the small intestine. The lining of the microvilli of the small intestine causes it to have a vast surface area, making it an ideal site for drug absorption.
3. Large muscles have a high blood flow, so intramuscular injections may be used since they maximize drug absorption. Conversely, the skin has slow absorption due to the poor blood supply in the epidermis, reducing the time for topical drugs to take effect.

Clinical Case Study 2

1. The principal site of drug metabolism is the smooth endoplasmic reticulum of liver cells. Chemicals absorbed in the GI tract perfuse the liver, the first organ. Drugs then enter hepatic circulation through the portal vein, becoming well-metabolized. This effect means some medications can be metabolized to an inactive form on their first pass through the liver. This occurs before they reach the systemic blood circulation.

2. The hepatic microsomal enzyme system, also known as the P450 system, serves as the liver's primary mechanism for drug metabolism. Cytochrome P450 is a vital component of this system. It comprises 12 related enzyme families, three of which (CYP1, CYP2, and CYP3) are responsible for metabolizing drugs, while the remaining nine metabolize endogenous compounds such as steroids and fatty acids. Each cytochrome family has specific subforms that metabolize only particular drugs, identified by abbreviations like CYP1A2, CYP2D6, and CYP3A4. Hepatic microsomal enzymes catalyze a variety of reactions, utilizing drugs as substrates. In some cases, drug metabolism does not result in the breakdown of drugs into smaller molecules; instead, it can lead to the formation of a molecule larger than the parent drug.
3. Diminished hepatic metabolic activity may lead to altered drug action. For instance, infants develop a mature microsomal enzyme system only after reaching 1 year of age. Consequently, those younger than 6 months have immature livers, and drugs must be administered cautiously. In older adults, hepatic activity is often reduced, meaning that lower doses of many medications are adequate for treatment. Liver diseases also necessitate decreasing doses. Certain genetic disorders can result in patients lacking specific metabolic enzymes, which slows their metabolic rates. The consumption of alcohol and tobacco also affects liver metabolism. Herbal supplements, such as St. John's wort, an enzyme inducer, can enhance the response to certain medications. Even grapefruit juice may inhibit enzyme activity.

Clinical Case Study 3

1. Drug excretion removes intact drugs primarily through the kidneys. Kidney disease or reduced cardiac output can influence the clearance of drugs dependent on renal function. The typical process of drug elimination by the kidneys includes glomerular filtration and distal tubular reabsorption. Glomerular filtration occurs when drugs enter the kidneys through the renal arteries. Distal tubular reabsorption occurs when a drug moves to the distal convoluted tubule, where its concentrations exceed the perivascular space.

2. The three processes involved in a drug's urinary excretion are glomerular filtration, passive tubular reabsorption, and active tubular secretion.

3. As individuals age, renal drug excretion declines. By age 80, clearance is typically reduced to approximately half of what it is at age 30.

CHAPTER 4: PHARMACODYNAMICS

CLINICAL CASE STUDY 1

1. Overdosage of diphenhydramine may result in the development of anticholinergic symptoms, seizures, and coma. A fatal outcome following a diphenhydramine overdose does not commonly occur. The patient initially developed seizures followed by cardiac conduction and hemodynamic compromise, resulting in death despite life support measures.

2. Diphenhydramine is an antihistamine that blocks the action of histamine, which causes allergic symptoms.

CLINICAL CASE STUDY 2

1. Synergism is when drugs interact to enhance or magnify one or more adverse effects. The negative impact of synergism is a form of contraindication.

2. Synergism can have pharmacologic benefits. When treating certain infections, for example, two antibiotics may be combined to increase antimicrobial actions against various pathogens.

CHAPTER 5: PHARMACOGENOMICS AND DRUG TOXICITY

CLINICAL CASE STUDY 1

1. Medications with the potential for neurotoxicity include antiseizure, antianxiety, antipsychotic, and antidepressant agents. Some examples of neurotoxic agents are amitriptylinoxide, dibenzepin, opipramol, and oxaprotiline. Amphetamines are the most commonly used illicit drugs after cannabis.

2. Aminoglycosides, tetracyclines, clindamycin, erythromycin, polymyxins, ethambutol, isoniazid, and chloramphenicol may cause severe neurotoxicity.

3. The signs and symptoms include hallucinations, depression, sedation, mania, behavioral changes, seizures, imbalance, hearing loss, resting tremors, and visual alterations. Other symptoms include suicidal ideation, delirium, cognitive decline, sexual dysfunction, and seizures.

CLINICAL CASE STUDY 2

1. Various mechanisms can cause nephrotoxicity, including renal tubular toxicity, glomerulonephritis,

and crystal nephropathy. In detecting early renal damage, blood urea and serum creatinine can evaluate nephrotoxicity and renal dysfunction.

2. Aminoglycosides, NSAIDs, diuretics, proton pump inhibitors (PPIs) (e.g., Nexium and Prilosec), and angiotensin-converting enzyme (ACE) inhibitors can cause kidney damage. Certain medications administered in hospital settings, such as aminoglycosides and vancomycin, can also damage the kidneys.

3. The signs and symptoms of acute kidney injury may include nausea, loss of appetite, fatigue, confusion, **edema**, itching, irregular heartbeat, chest pain, and seizures.

CHAPTER 6: DRUG INTERACTIONS AND ADVERSE DRUG EFFECTS

CLINICAL CASE STUDY 1

1. Drug–drug interaction is the clinical response to administering a combination of drugs that differs from the anticipated effects of the two drugs when given alone. For example, medications used for numbing (local anesthetics) often contain epinephrine and lidocaine to prolong the effect of lidocaine.

2. Drug–disease interactions may develop when a drug affects a pre-existing condition or disease. Some disorders can interact with drugs to increase the risk of adverse effects. For example, aspirin increases bleeding in patients with peptic ulcer disease (PUD); people with hypertension may be at greater risk for tachycardia with oral decongestants found in over-the-counter cough, cold, and allergy products. Sometimes, helpful drugs for one disease are harmful for another disorder. For example, some beta-blockers taken for heart disease or hypertension can worsen asthma or make it difficult for patients with diabetes to know when their blood sugar is too low. Aging can also increase drug interaction risks. For example, sedatives increase the risk of falls in older adults, and lower doses of narcotics are often most effective for pain relief. The anticoagulant warfarin causes more bleeding in elderly patients and usually requires a lower dose too.

3. Food–drug interactions may occur with both prescription and over-the-counter medicines. These include antacids, vitamins, iron pills, herbs, supplements, and beverages. Some nutrients can affect how they metabolize certain drugs by binding with the medicine's ingredients. This reduces their absorption or speeds their elimination. For example, the acidity of fruit juice may decrease the effectiveness of antibiotics such as penicillin. Dairy products may affect the infection-fighting

effects of tetracycline. Grapefruit juice can interfere with some blood pressure and organ transplant medicines by increasing their metabolic breakdown.

CLINICAL CASE STUDY 2

1. Anaphylactic shock is a life-threatening emergency that develops from anaphylaxis.
2. The signs and symptoms include hypotension, narrowing of the airways, dyspnea, wheezing, hives, flushed skin, tachycardia, swollen tongue or lips, nausea, vomiting, dysphagia, and abdominal pain.
3. Treatment of anaphylactic shock includes epinephrine, antihistamines, corticosteroid injections, and oxygen therapy.

CHAPTER 7: DIETARY NUTRITION AND HERBAL REMEDIES

CLINICAL CASE STUDY 1

1. Garlic's benefits are most commonly promoted as a dietary supplement for conditions related to hypertension and hypercholesterolemia. A garlic-rich diet seems to confer a lower risk of developing colon, prostate, stomach, and breast cancer. Garlic's antibacterial properties and antioxidants can clear up the skin by killing acne-causing bacteria.
2. Adverse effects of garlic include garlicky nausea, vomiting, flatulence, tachycardia, flushing, breath and body odor, and insomnia.
3. Large quantities of garlic should be avoided during breastfeeding because they may cause colic in infants or be deadly to children for unknown reasons. Garlic is also contraindicated in people with known hypersensitivity or gastritis. Due to the risk of bleeding, garlic should not be used in patients before or after surgery.

CLINICAL CASE STUDY 2

1. St. John's wort is a native plant with yellow, star-shaped flowers. It is commonly found in Europe, the United States, Africa, and Asia. The leaves are dried, made into tea, or prepared as a dried extract in pills or capsules.
2. St. John's wort plant is indicated to treat depression, anxiety, **obsessive–compulsive disorder,** attention-deficit hyperactivity disorder, **premenstrual syndrome, eczema,** and pain.
3. The adverse effects of St. John's wort include insomnia, restlessness, dizziness, nausea, diarrhea, allergic reactions, xerostomia, erectile dysfunction, and orgasm dysfunction. It is unsafe to use during pregnancy.

CHAPTER 8: MEDICATION ERRORS

CLINICAL CASE STUDY 1

1. The three phases in which errors often occur are prescribing, dispensing, and administering medication.
2. The five medications most commonly associated with severe adverse events are insulin, IV narcotics, IV heparin, IV potassium concentrates, and IV hypertonic sodium chloride solutions. Most of these drugs are kept as floor stock in nursing areas, necessitating double-checking of their use.
3. The seven "rights" of drug administration include the right patient, the right route, the right drug, the correct dose, the right time, the proper technique, and the correct information in the patient's chart.

CHAPTER 9: DRUGS AFFECTING THE CENTRAL NERVOUS SYSTEM

CLINICAL CASE STUDY 1

1. Absence seizures begin simultaneously on both sides of the brain, causing lapses in awareness. The older term for absence seizures was petit mal seizures. These seizures start and end quickly, lasting only a few seconds. Generalized seizures are more dramatic forms, formerly known as grand mal seizures. Signs and symptoms include unconsciousness, groaning, crying out, shaking, sudden collapse, spasms, stiffening, loss of bowel or bladder control, cessation of breathing, and cyanosis.
2. The preferred medications for generalized tonic–clonic seizures are phenytoin, lamotrigine, carbamazepine, and valproic acid.
3. The adverse effects of antiseizure drugs include nausea, vomiting, diarrhea, loss of appetite, weight loss, glossitis, gingival hyperplasia, skin rash, headache, dizziness, nervousness, insomnia, tremors, double or blurred vision, nystagmus, cognitive impairment, and difficulty speaking.

CLINICAL CASE STUDY 2

1. Migraines are classified into several categories. The most common types of migraines are migraines with an aura, migraines without an aura, silent migraines, and chronic migraines.
2. Migraine with aura is less common than migraine without aura. Symptoms begin 30 minutes before the headache starts. The visual symptoms include flashing lights, wavy lines, or brief loss of vision: the patient's inability to speak, numbness, and tingling. People aged 50 or older may experience the aura without other symptoms. A migraine with aura usually lasts for an hour. Migraine without aura is the most common type of migraine,

affecting 75% of patients. It involves a severe headache on one side of the head. Bright lights, loud sounds, and physical activity aggravate this condition. Anxiety, depression, obesity, and snoring are risk factors.

3. For severe migraines, serotonin agonists are usually administered. Narcotic analgesics can stop migraine pain that has not responded to any other medication.

CLINICAL CASE STUDY 3

1. The primary signs and symptoms of PD include muscle rigidity, bradykinesia, tremor, and postural instability. Muscle rigidity leads individuals to have difficulty moving their limbs. The rigidity of the facial muscles can result in a lack of expression or altered expressions. When the pharyngeal muscles are affected, chewing and swallowing may become challenging. Arm swinging may be diminished during walking, with the arms held close to the sides of the body. Bradykinesia is an uncontrollable slowness of voluntary movement and speech, which is highly noticeable and causes difficulty in speaking, chewing, and swallowing. Individuals may struggle with initiating activities and controlling fine muscle movements. They may shuffle their feet while walking, no longer taking normal strides, making walking more difficult. Tremors of the head and hands may develop, resulting in continuous shaking or motions while at rest. These tremors can become severe enough that holding objects becomes challenging. As PD progresses, individuals may exhibit pill-rolling motions, with their thumbs and forefingers rubbing together in circular motions.

2. The adverse effects of carbidopa–levodopa and carbidopa on their own include dyskinesias, nausea, vomiting, constipation, headaches, and asthenia.

3. The contraindications of dopamine agonists include known hypersensitivity, narrow-angle glaucoma, suspicious undiagnosed skin lesions, or a history of melanoma.

CHAPTER 10: DRUGS AFFECTING THE AUTONOMIC NERVOUS SYSTEM

CLINICAL CASE STUDY 1

1. Alpha-2 adrenergic agonists are indicated for treating hypertension, either alone or in combination with diuretic medications. They are also used epidurally to manage severe pain and to assist with menopausal symptoms, attention-deficit hyperactivity disorder, and withdrawal from opioids, alcohol, and nicotine. Additionally, alpha-2 adrenergic

agonists are employed during colonoscopy procedures, for epidural anesthesia, and to mitigate adverse cardiovascular effects associated with acute amphetamine or cocaine intoxication or overdose.

2. The adverse effects of alpha-2 adrenergic agonists include hypotension, peripheral edema, rash, pruritus, dry mouth or eyes, drowsiness, dizziness, constipation, impotence, nausea, vomiting, hepatitis, hallucinations, and depression.

3. Alpha-2 adrenergic agonists should be avoided in children, as well as in pregnant and lactating patients, those with hepatitis, cirrhosis, polyarteritis nodosa, pheochromocytoma, scleroderma, and blood dyscrasia. They are also contraindicated in individuals with depression, myocardial infarction, and stroke.

CLINICAL CASE STUDY 2

1. Muscarinic antagonists are used to treat COPD, asthma exacerbations, peptic ulcers, irritable bowel syndrome, pupil dilation during ophthalmic examinations, bradycardia, and to reduce salivary and respiratory secretions during anesthesia. They are also employed in reversing poisoning from muscarinic mushrooms or organophosphate insecticides.

2. The adverse effects of muscarinic antagonists include dry eyes, xerostomia, urinary retention, hyperthermia, photophobia, and increased intra-ocular pressure. Other adverse effects encompass blurred vision, mydriasis, headaches, coughing, constipation, insomnia, pharyngitis, sinusitis, and allergic conjunctivitis. Severe adverse effects may include worsening of glaucoma, dysrhythmias, and paralytic ileus. *Atropine* can cause ventricular fibrillation, delirium, and coma.

3. Muscarinic antagonists should be avoided in patients with acute angle-closure glaucoma. Atropine must be used carefully in cardiovascular conditions because it increases the heart rate. It should also be used with caution in cases of hyperthyroidism, ulcerative colitis, and paralytic ileus. Muscarinic antagonists must be administered cautiously in patients with Down syndrome.

CHAPTER 11: PSYCHOTHERAPEUTIC AGENTS

CLINICAL CASE STUDY 1

1. The exact cause is unknown. However, it may involve heredity, a decrease in certain neurotransmitters, psychosocial factors, or changes in neuroendocrine function. Patients with anxiety are more likely to develop depression, which can also occur in individuals with other mental disorders. Females are at a higher risk than males for

depressive disorders. Changes in neurotransmitter levels are likely involved in depression, with examples including abnormal regulation of catecholaminergic, cholinergic, glutamatergic, and serotonergic neurotransmission.

2. The main indication for SSRIs is major depressive disorder. However, they are frequently prescribed for generalized anxiety disorder, bulimia nervosa, and posttraumatic stress disorder.

3. The common side effects of SSRIs include insomnia, dry mouth, nausea, constipation, dizziness, loss of appetite, headache, and sexual dysfunction, particularly with paroxetine. These effects are typically mild and tend to subside after the first few weeks of treatment.

CLINICAL CASE STUDY 2

1. First-generation antipsychotics are prescribed for schizoaffective disorders and psychotic disorders. These medications are also utilized for teenagers experiencing early-onset schizophrenia.

2. First-generation drugs may cause photosensitivity, orthostatic hypotension, seizures, sedation, drowsiness, and constipation. Serious adverse effects include agranulocytosis, pancytopenia, anaphylaxis, tardive akathisia, dyskinesia, hypothermia, and adynamic ileus. Non-phenothiazines often lead to tremors, drowsiness, sedation, orthostatic hypotension, and weight gain. Severe adverse effects of non-phenothiazines include laryngospasm, hepatotoxicity, kidney failure, respiratory depression, and sudden death.

3. Second-generation antipsychotics are contraindicated in patients with known hypersensitivity, diabetes mellitus, suicidal thoughts, seizures, heart failure, Alzheimer's disease, metabolic syndrome X, and severe hepatic impairment. *Clozapine* may cause rash, swelling, and angioedema. It also carries black box warnings due to risks, including neutropenia, agranulocytosis, seizures, myocarditis, cardiomyopathy, orthostatic hypotension, bradycardia, and syncope.

CHAPTER 12: LOCAL AND GENERAL ANESTHETICS

CLINICAL CASE STUDY 1

1. Amide local anesthetics are more therapeutic than ester local anesthetics but can produce more toxic reactions than esters. Amides are highly stable in solution, while esters are unstable. Amino-esters are hydrolyzed in plasma by the enzyme pseudocholinesterase, whereas amide compounds undergo enzymatic degradation in the liver and excretion in the urine.

2. Procaine, like cocaine, constricts blood vessels, which reduces bleeding. However, hypersensitivity can develop in patients, leading to anaphylactic symptoms such as urticaria, pruritus, erythema, laryngeal edema, tachycardia, nausea, vomiting, dizziness, syncope, perspiration, and hyperthermia. Treatment involves discontinuing procaine and a personalized regimen of intramuscular epinephrine, supplemental oxygen, intravenous corticosteroids, resuscitative fluids, beta-agonists, and supportive care.

3. Bupivacaine, chloroprocaine, etidocaine, lidocaine, and procaine are often known by their trade names: Marcaine, Nesacaine, Duranest, Xylocaine, and Novocaine.

CLINICAL CASE STUDY 2

1. Remifentanil is administered during the initiation and maintenance of general anesthesia to provide pain relief after surgery and can be used alone or in conjunction with midazolam. Sufentanil is utilized as part of balanced anesthesia or as a primary anesthetic, accompanied by 100% oxygen given simultaneously. It is occasionally indicated, along with bupivacaine, as an epidural anesthetic during labor and delivery.

2. Alfentanil, etomidate, ketamine, propofol, succinylcholine, and rocuronium are used for tracheal intubation.

3. Methohexital may lead to hypotension, laryngospasm, bronchospasm, apnea, respiratory depression, visual hallucinations, vivid dreams, and arrhythmias. Likewise, propofol can result in hypotension and respiratory depression. It may also trigger anaphylaxis, rhabdomyolysis, metabolic acidosis, kidney failure, and heart failure.

CHAPTER 13: SUBSTANCE ABUSE

1. Cocaine directly affects the brain's reward pathway, a system that reinforces pleasurable experiences and encourages users to repeat them. It inhibits this transporter from removing dopamine, accumulating dopamine in the synapse, and generating an amplified signal to the receiving receptors. Furthermore, it establishes a strong connection between cocaine use and feelings of euphoria.

2. Cocaine induces intense euphoria, analgesia, decreased appetite, heightened sensory perception, and illusions of physical strength. Larger doses amplify these effects and can lead to tachycardia, mydriasis, hyperthermia, and sweating. Additional symptoms may include rhinorrhea, redness around the nostrils, and deterioration of the nasal cartilage.

3. Chronic cocaine use can cause significant damage to the cardiovascular system, leading to heart disease, heart attacks, strokes, and even sudden death. Snorting cocaine may result in persistent nasal congestion, nosebleeds, and the potential perforation of the nasal septum. Smoking crack cocaine can cause respiratory issues such as chronic cough, lung damage, and an increased risk of respiratory infections. Prolonged cocaine use can result in neurological damage, leading to memory loss, cognitive impairment, and a heightened risk of strokes and seizures.

CHAPTER 14: DRUGS AFFECTING THE CORONARY AND HEART DISORDERS

CLINICAL CASE STUDY 1

1. The four types of angina pectoris are stable, unstable, Prinzmetal, and microvascular. Stable angina pain is predictable in frequency and duration, and it is relieved by rest and nitroglycerin. Unstable angina increases in frequency and duration and occurs more quickly than stable angina. This indicates worsening coronary artery disease that may progress to myocardial infarction. Prinzmetal's angina, also referred to as variant angina, is caused by spasms of the coronary arteries leading to pain. It may occur spontaneously and is unrelated to physical exercise or emotional stress. Microvascular angina is caused by impairment of vasodilator reserve, resulting in angina-like chest pain in a person with normal coronary arteries.

2. Angina pectoris may occur when the myocardium does not receive sufficient blood and oxygen. The narrowing or hardening of the arteries restricts blood flow to the heart.

3. Treatment for angina involves medications and procedures aimed at reducing ischemia and restoring or improving coronary blood flow. Commonly used nitrates, such as nitroglycerin, can be administered sublingually, orally, transdermally, or as a topical ointment. Isosorbide may lower myocardial oxygen consumption. Beta-adrenergic blockers decrease the heart's workload and oxygen demands through reduced heart rate and peripheral resistance to blood flow. Calcium channel blockers may help prevent spasms in coronary arteries. Antiplatelet drugs reduce platelet aggregation and the risk of coronary occlusion. Antilipemic agents can also lower serum cholesterol or triglycerides.

CLINICAL CASE STUDY 2

1. The primary cause of MI is a blockage in a coronary artery, usually due to atherosclerosis, which results in a blood clot that restricts blood flow to the heart muscle. The risk factors include hypertension, smoking, high cholesterol, diabetes, obesity, an unhealthy diet, age, family history, and gender.

2. The signs and symptoms of MI include pain or pressure beneath the sternum that often radiates outward. This pain may extend to the back, jaw, and left arm. The discomfort is generally more intense and prolonged than angina pectoris and may be accompanied by dyspnea, fatigue, diaphoresis, nausea, vomiting, and syncope.

3. The main objectives of emergency department medical therapy are rapid intravenous thrombolysis and rapid referral for percutaneous coronary intervention, optimization of oxygenation, reduction of cardiac workload, and pain control with morphine. Pharmacotherapy for MI includes aspirin, thrombolytics, fibrinolytics, nitroglycerin, morphine, beta-blockers, and statins.

CHAPTER 15: DRUGS AFFECTING HYPERLIPIDEMIA

CLINICAL CASE STUDY 1

1. Statins serve as the primary treatment for hyperlipidemia and mixed dyslipidemia. They are also utilized for hypertriglyceridemia, atherosclerosis, and the primary prevention of atherosclerotic cardiovascular disease. These medications are recommended for adults aged 40–75 years who have one or more cardiovascular risk factors, including dyslipidemia, diabetes, hypertension, or smoking.

2. While statins are generally effective and safe for most patients, they have been associated with muscle pain, digestive issues, and mental fog in some individuals. Rarely, they can lead to liver damage. Rhabdomyolysis represents the most severe adverse effect linked to statin use. Other possible side effects may include hypothyroidism, headaches, fatigue, dizziness, and insomnia.

3. Ezetimibe is the only medication in the cholesterol-absorption inhibitor class. It works by blocking a protein that decreases cholesterol absorption in the intestine. This action explicitly targets cholesterol without impacting the absorption of other medications or fat-soluble vitamins.

CLINICAL CASE STUDY 2

1. Niacin influences lipoprotein metabolism by decreasing triglyceride synthesis via multiple pathways. In adipocytes, it inhibits the lipolysis of triglycerides and retards the mobilization of free fatty acids to the plasma.

2. Niacin oral tablets are indicated for use as monotherapy or in combination with simvastatin or levastatin to treat primary hyperlipidemia and mixed dyslipidemia. They can also lower the risk of nonfatal myocardial infarctions in patients with a history of hyperlipidemia.

3. Significant adverse effects of niacin include hot flashes, itching, vomiting, diarrhea, fatigue, and headaches, which often lead to discontinuation of the medication. Most patients develop tolerance with prolonged use. Gradually increasing the niacin dosage over several weeks is crucial to minimize the frequency and severity of skin reactions. Other adverse effects include liver toxicity, hyperglycemia, and hyperuricemia. Notably, severe liver toxicity occurs more frequently with high doses of the short-acting formulation of niacin.

CHAPTER 16: DRUGS AFFECTING HYPERTENSION

CLINICAL CASE STUDY 1

1. Hypertension can be classified as primary, secondary, prehypertension, resistant, or hypertensive crisis. Primary hypertension is also known as essential hypertension. When blood pressure is slightly higher than usual, it is considered prehypertension. A regular reading is 120/80 mm Hg. With prehypertension, systolic values can be as high as 139 mm Hg, and diastolic pressure can reach 89 mm Hg. Resistant hypertension is a condition where blood pressure remains elevated despite the use of three or more antihypertensive medications at the maximum tolerated doses. A hypertensive crisis occurs when blood pressure exceeds 180/120 mm Hg, which is an emergency condition that requires immediate hospital access.

2. Risk factors include age, family history, race or ethnicity, and unhealthy lifestyle choices such as a diet high in sodium and low in potassium, a lack of exercise, smoking, and excessive alcohol consumption. Certain diseases, including diabetes and kidney disease, can also increase these risks.

3. Several common medications for hypertension include ACE inhibitors, angiotensin receptor blockers, **calcium channel blockers,** potassium-sparing diuretics, and traditional diuretics.

CLINICAL CASE STUDY 2

1. Risk factors associated with resistant hypertension include chronic kidney disease, advanced age, left ventricular hypertrophy, African-American ethnicity, female gender, and residing in the southeastern United States. Among these, chronic kidney disease may be the most significant contributor to resistant hypertension. Other factors are modifiable, such as obesity, excessive dietary salt, and heavy alcohol consumption. High salt or alcohol intake increases the risk of hypertension and undermines treatment effectiveness. Excessive sodium intake has a particularly detrimental effect on patients who are African-American, elderly, or have chronic kidney disease.

2. ACE inhibitors are well tolerated, but adverse effects, such as a dry cough and allergic reactions, can occur. More severe adverse effects include kidney damage and angioedema, which may require emergency medical care. Other adverse effects include hypotension, hyperkalemia, and hypermagnesemia. ACE inhibitors decrease the production of aldosterone, which causes an increase in potassium levels. One potentially profound side effect that can occur in less than 1% of patients is angioedema. Cholestatic jaundice or hepatitis is another rare but severe adverse effect that can progress to hepatic necrosis and, in some cases, death.

3. ARBs are contraindicated in bilateral renal artery stenosis, aortic stenosis, coarctation of the aorta, and pregnancy. No evidence exists regarding the safe use of ARBs during breastfeeding, and the effects of potential exposure to a nursing infant are unknown.

CHAPTER 17: DRUGS AFFECTING ARRHYTHMIA

CLINICAL CASE STUDY 1

1. Atrial fibrillation is an irregular and rapid heartbeat that can be intermittent, long-lasting, or permanent. It is the most common type of treated heart arrhythmia. The symptoms include angina, dyspnea, palpitations, and fatigue. It is the leading cardiac cause of stroke and death.

2. The risk factors for atrial fibrillation include advanced age, hypertension, underlying lung disease, congenital heart disease, rheumatic heart disease, mitral stenosis, and increased alcohol consumption.

3. Atrial fibrillation can be treated with beta-blockers such as esmolol and atenolol or calcium channel blockers like verapamil or diltiazem. Digoxin may also be prescribed if other medications are unsuitable. Additionally, cardioversion can sometimes be employed.

CLINICAL CASE STUDY 2

1. Drugs like amiodarone, lidocaine, and epinephrine are used after defibrillation attempts to treat

ventricular fibrillation. Amiodarone is a primary antiarrhythmic agent for ventricular fibrillation.

2. Amiodarone's mechanism of action includes blocking potassium's ability to repolarize the heart during phase III of the cardiac action potential. This potassium channel-blocking effect increases action potential duration and a prolonged effective refractory period in cardiac muscle, reducing excitability.

3. Lidocaine is a local anesthetic agent commonly used for local and topical anesthesia. It also has antiarrhythmic and analgesic properties, making it an effective adjunct for tracheal intubation. Lidocaine is most frequently used in the treatment of ventricular arrhythmias, especially those associated with acute myocardial infarction.

CHAPTER 18: DRUGS AFFECTING COAGULATION DISORDERS

CLINICAL CASE STUDY 1

1. Risk factors for PE include a family history of blood clotting disorders, surgery or injury (particularly to the legs), prolonged bed rest, long-distance travel (by air or road), paralysis, older age, obesity, cigarette smoking, cancer treatment, and the use of oral contraceptives.

2. Fondaparinux is used to prevent deep vein thrombosis in high-risk patients after hip fracture surgery, hip or knee replacement surgery, or abdominal surgery. When combined with warfarin, it actively treats deep vein thrombosis with or without pulmonary embolism. It also treats deep vein thrombosis and pulmonary embolism while preventing recurrent deep vein thrombosis and pulmonary embolism.

3. The adverse effects of fondaparinux include mild injection site reactions, anemia, skin rash, edema, nausea, vomiting, insomnia, constipation, and fever.

CLINICAL CASE STUDY 2

1. Atrial fibrillation may result in blood clots forming in the heart, thereby raising the risk of stroke, heart failure, and other heart-related complications.

2. Direct thrombin inhibitors are utilized for treating or preventing acute coronary syndrome, atrial fibrillation, and VTE. They constitute a novel class of medications developed as an effective alternative mode of anticoagulation, particularly for patients experiencing heparin-induced thrombocytopenia, those undergoing dialysis, cardiopulmonary bypass, and the management of thromboembolic disorders.

3. Common adverse effects of DTIs include life-threatening bleeding, chest pain, cardiac arrest, bruising, indigestion, abdominal pain, diarrhea, and hypersensitivity. Other adverse effects may include headache, dizziness, fever, swelling, arm and leg pain, and itching.

CHAPTER 19: ANTI-ANEMIC DRUGS

CLINICAL CASE STUDY 1

1. The causes of iron deficiency anemia include blood loss, insufficient dietary iron, an inability to absorb iron, and pregnancy. Symptoms encompass headaches, palpitations, irritability, dyspnea, fatigue, tachycardia, glossitis, splenomegaly, and *pica*.

2. The adverse effects of ferrous sulfate may include heartburn, stomach pain, loss of appetite, constipation, nausea, vomiting, diarrhea, darkening of the stool, and staining of the teeth.

3. Patients with hereditary hemochromatosis, hemosiderosis, or a history of hemolytic anemia should avoid ferrous sulfate.

CLINICAL CASE STUDY 2

1. Pernicious anemia results from a deficiency of vitamin B12 due to a lack of intrinsic factors. The signs and symptoms include fatigue, pale skin, headaches, depression, dementia, peripheral neuropathy, and weight loss.

2. Cyanocobalamin binds to intrinsic factors, and absorption occurs in the ileum at the end of the small intestine. When administered parenterally, cobalamin bypasses the intestinal barrier, is absorbed rapidly by diffusion, and enters the bloodstream.

3. The adverse effects of cyanocobalamin include anaphylaxis, dyspnea, peripheral vascular thrombosis, optic nerve atrophy, weight gain, pulmonary edema, and congestive heart failure. Other side effects include leg cramps, irregular heartbeats, numbness, hypokalemia, fatigue, and joint pain.

CHAPTER 20: DRUGS AFFECTING THE ENDOCRINE SYSTEM

CLINICAL CASE STUDY 1

1. The signs and symptoms may include thickened skin, an enlarged tongue, a protruding abdomen, an umbilical hernia, stunted growth, intellectual disabilities, and neurological impairments such as hearing and speech defects.

2. The adverse effects of levothyroxine include headaches, rashes, dyspnea, tachycardia, tremors, irregular menstruation, diarrhea, mood changes, and heat intolerance. It can also lead to bone loss and fractures.

3. Levothyroxine is contraindicated in patients with thyrotoxicosis and acute myocardial infarction. It should also be avoided in those with uncorrected adrenal insufficiency, as thyroid hormones can precipitate an acute adrenal crisis by increasing the metabolic clearance of glucocorticoids. Lastly, it is contraindicated for individuals with hypersensitivity to inactive ingredients.

CLINICAL CASE STUDY 2

1. Addison's disease may result from an ACTH deficiency or the atrophy of the adrenal cortex. Other causes include genetics, tuberculosis, and decreased cortisol levels.
2. The adverse effects of corticosteroids depend on the dose, type of steroid, and duration of treatment. They are more common at higher dosages and with prolonged use. Common adverse effects include increased appetite, weight gain, edema, a "puffy' face, acne, fatigue, and mood changes. Other potential adverse effects may consist of agitation, insomnia, diabetes, osteoporosis, hypertension, gastritis, cataracts, and glaucoma.
3. Corticosteroids are contraindicated in patients with hypersensitivity to any component of the formulation, systemic fungal infections, varicella infection, herpes simplex keratitis, concurrent administration of live vaccines, diabetes mellitus, glaucoma, osteoporosis, hypertension, and joint infections. They should also be avoided in patients with PUD and congestive heart failure.

CHAPTER 21: ANTIDIABETIC DRUGS

CLINICAL CASE STUDY 1

1. The signs and symptoms of diabetes mellitus include polyphagia, polydipsia, polyuria, blurred vision, urinary tract infections, skin infections, weight loss, fatigue, and weakness. Additionally, hyperglycemia and electrolyte imbalances can occur.
2. The main types of insulin therapy include **short-, intermediate-, and long-acting insulins**.
3. The adverse effects of Humulin R U-500 include hypoglycemia, chills, sweating, chest pain, allergic reactions, rashes, and injection site reactions. Other adverse effects encompass anxiety, blurred vision, cough, flushing, dizziness, tachycardia, polydipsia, polyphagia, oliguria, weight gain, and seizures.

CLINICAL CASE STUDY 2

1. Meglitinides are primarily used to improve blood glucose control in individuals with type 2 diabetes, alongside diet and exercise. They stimulate the pancreas to secrete more insulin, especially after meals. These medications are not suitable for individuals with type 1 diabetes or diabetic ketoacidosis. Potential side effects of meglitinides may include diarrhea, nausea, vomiting, stomach discomfort, myalgia, arthralgia, weight gain, dizziness, hypoglycemia, and headaches.
2. The adverse effects of sulfonylureas include hypoglycemia, palpitations, shaking, nausea, diaphoresis, dizziness, confusion, fatigue, headache, seizures, and coma.

   Sulfonylureas should be avoided in patients who are hypersensitive to these drugs. They are also contraindicated in individuals with type 1 diabetes or severe renal or hepatic failure. Patients experiencing ketoacidosis should receive insulin rather than an oral antihyperglycemic agent. Additionally, sulfonylureas should be avoided during pregnancy and lactation.
3. Metformin is used to treat type 2 diabetes mellitus and gestational diabetes. It is also prescribed to prevent type 2 diabetes in patients at high risk of developing the condition.

CHAPTER 22: DRUGS AFFECTING THE RESPIRATORY SYSTEM

CLINICAL CASE STUDY 1

1. Pseudoephedrine relieves nasal or sinus congestion caused by the common cold, hay fever, sinusitis, and other respiratory allergies. It is also used to relieve ear congestion caused by otitis media. Some of these preparations are available only with a prescription.
2. Pseudoephedrine is a sympathomimetic with a mixed direct and indirect mechanism of action. It directly stimulates beta-adrenergic receptors and indirectly stimulates alpha-2 adrenergic receptors, releasing endogenous norepinephrine from the granules of neurons.
3. Contraindications for pseudoephedrine include hypersensitivity to the drug and treatment with monoamine oxidase inhibitors currently or within the last 2 weeks. It should be used with caution in patients with hypertension, coronary artery disease, liver or renal impairment, benign prostatic hyperplasia, diabetes mellitus, hyperthyroidism, narrow-angle glaucoma, and mental agitation. Pseudoephedrine should not be used during pregnancy, lactation, or in individuals under 2 years of age. The extended-release form of the drug should not be used until the age of 12 years.

CLINICAL CASE STUDY 2

1. Albuterol targets $\beta2$-adrenergic receptors, promoting bronchial smooth muscle relaxation and

inhibiting the release of immediate hypersensitivity mediators, particularly from mast cells. Although it also influences β1-adrenergic receptors, the effect is minimal and has little impact on heart rate. Anticholinergic agents act as competitive antagonists of the neurotransmitter acetylcholine at receptor sites within the cholinergic system.

2. The adverse effects of albuterol include immediate hypersensitivity reactions, bronchospasm, hypokalemia, and dysrhythmia. These are associated with anticholinergic agents, which encompass upper respiratory tract infection, sinusitis, headache, dry mouth, dizziness, and bronchitis.

3. Albuterol is contraindicated for patients with a history of hypersensitivity to it. Rare cases of hypersensitivity reactions, such as urticaria, rash, and angioedema, have been reported following the use of albuterol sulfate. Caution is required when using anticholinergics, especially with older adults.

CHAPTER 23: DRUGS AFFECTING THE URINARY SYSTEM

CLINICAL CASE STUDY 1

1. Loop diuretics are the first choice for treating hypertension, heart failure, nephrotic syndrome, and pulmonary edema. In patients with heart failure who are resistant to loop diuretics, a thiazide or amiloride may be added to enhance the response. However, high doses of loop diuretics can lead to ototoxicity.

2. The adverse effects of loop diuretics include anaphylaxis, fever, headache, vertigo, dizziness, restlessness, orthostatic hypotension, and syncope. They may also cause severe electrolyte imbalances such as hyponatremia, hypochloremia, hypokalemia, azotemia, metabolic alkalosis, dehydration, and hyperuricemia. Other adverse reactions include skin rash, tinnitus, ototoxicity, photosensitivity, and interstitial nephritis. Additionally, thrombocytopenia, aplastic anemia, hemolytic anemia, leukopenia, agranulocytosis, pneumonitis, pulmonary edema, blurred vision, abdominal pain, anorexia, diarrhea, urticaria, jaundice, pancreatitis, hepatic coma, and impotence may also occur.

3. Loop diuretics should be avoided in patients with severe hypersensitivity, hyponatremia, hypokalemia, azotemia, hypotension, anuria, and hepatic coma. They are also contraindicated in cases of anticipated fluid depletion, such as during surgery.

CLINICAL CASE STUDY 2

1. Osmotic diuretics are used when increased water excretion with minimal sodium excretion is required. They are the first choice for treating cerebral edema. Mannitol's primary indication is the treatment of acute angle-closure glaucoma. Glycerin lowers intraocular pressure and reduces superficial corneal edema. These drugs can also treat congenital glaucoma and keratoplasty after retinal detachment repair.

2. Osmotic diuretics primarily inhibit the reabsorption of water and sodium chloride in the proximal convoluted tubule (PCT), the descending loop of Henle, and the collecting duct, leading to reduced urea and uric acid reabsorption. Mannitol decreases tubular water reabsorption and increases sodium and chloride excretion by enhancing the osmolarity of the glomerular filtrate.

3. Osmotic diuretics should be avoided in patients with kidney failure, anuria, or heart failure. Mannitol and urea are contraindicated for individuals with cranial bleeding. Glycerin must be avoided in patients with known hypersensitivity or diabetes mellitus, and it is also contraindicated for those with hypervolemia and cardiac failure.

CHAPTER 24: DRUGS AFFECTING THE DIGESTIVE SYSTEM

CLINICAL CASE STUDY 1

1. PPIs are the first-line treatment for PUD. They effectively reduce stomach acid to promote ulcer healing. Examples of PPIs include omeprazole, esomeprazole, lansoprazole, pantoprazole, rabeprazole, and dexlansoprazole.

2. Omeprazole reduces stomach acid production and irreversibly blocks an enzyme called H+/K+ ATPase, which regulates acid production. This enzyme, known as the proton pump, is located in the parietal cells of the stomach lining. PPIs inhibit acid production stimulated by meals. Dosing PPIs twice daily allows for the inhibition of more pumps, resulting in more rapid and complete gastric acid suppression.

3. The adverse effects are usually minor. The most commonly reported are headaches, stomach pain, and constipation.

CLINICAL CASE STUDY 2

1. Sulfasalazine is utilized to treat and prevent acute episodes of mild to moderate ulcerative colitis. It alleviates inflammation and other symptoms associated with the disease. Additionally, sulfasalazine is prescribed for Crohn's disease and rheumatoid arthritis.

2. Overall, sulfasalazine is generally well-tolerated. Its adverse effects may include headache, tinnitus, fatigue, insomnia, mouth sores, loss of appetite, gastric upset, nausea, vomiting, and diarrhea.

Severe adverse effects can involve sun sensitivity, hearing loss, mood changes, hematuria, cold sweats, and blurred vision.

3. Sulfasalazine is contraindicated for patients with allergies, porphyria, heart, lung, kidney, or liver disease, and glucose-6-phosphate dehydrogenase deficiency.

CHAPTER 25: DRUGS AFFECTING THE SKELETAL SYSTEM

CLINICAL CASE STUDY 1

1. Adalimumab is indicated for use as either monotherapy or in combination with methotrexate for moderate to severe active rheumatoid arthritis. It also treats juvenile idiopathic arthritis, psoriatic arthritis, and ankylosing spondylitis.

2. Adalimumab's adverse effects typically include injection-site pain, elevated creatine phosphokinase, upper respiratory infections, headaches, and rashes. This drug carries a black box warning because it can suppress the immune system, leading to infections. Latent infections like hepatitis B and tuberculosis may reactivate. TNF blockers have been associated with malignancies such as leukemias and lymphomas. Rarely, fatal cases of hepatosplenic T-cell lymphoma have occurred.

3. Adalimumab contraindications are rare. However, due to its immunosuppressive effects, patients must be monitored for infections and malignancies.

CLINICAL CASE STUDY 2

1. Osteoarthritis (OA) is a degenerative joint disease that leads to inflammation and pain. It impacts the cartilage at the ends of bones. OA is the most prevalent form of arthritis, affecting millions worldwide. This condition occurs when the cushioning cartilage deteriorates over time. While it often involves a limited number of joints, it can also become generalized.

2. The risk factors for OA include obesity, joint deformity, joint injury, hemochromatosis, diabetes mellitus, and ochronosis. The knees and hands are more frequently affected in females, while the hips are more commonly affected in males.

3. Osteoarthritis cannot be reversed, but treatments can reduce pain and help patients move better. OA treatment depends on the severity of the condition, age, general health, and symptoms. Treatment aims to reduce pain and stiffness while improving joint movement. Treatments include medications, lifestyle changes, surgery, and physical therapy. Patients may benefit from medication for mild symptoms. Acetaminophen has been shown to help some patients with OA who experience mild

to moderate pain. However, exceeding the recommended dose can cause liver damage. Systemic NSAIDs are the most common medications for relieving pain and inflammation in OA, and NSAID creams are also utilized as topical treatments. Corticosteroid injections into the joint may alleviate pain for a few weeks.

CHAPTER 26: MUSCLE RELAXANTS

CLINICAL CASE STUDY 1

1. Baclofen is used to help relax specific muscles in the body. It relieves muscle spasms, cramping, and tightness caused by multiple sclerosis or specific spinal injuries.

2. The most common adverse effects associated with baclofen administration include nausea, hypotension, confusion, vertigo, muscle fatigue, tremor, insomnia, urinary retention, and impotence. Abrupt discontinuation of oral baclofen therapy may cause seizures and hallucinations. Gradual dose reduction is recommended to prevent withdrawal symptoms.

3. Baclofen should be avoided in patients who have a hypersensitivity to the drug. It is contraindicated for patients with a history of stroke, Parkinson's disease, or skeletal muscle spasms related to rheumatoid disorders.

CLINICAL CASE STUDY 2

1. Dantrolene is used to relieve muscle spasms and stiffness that may be caused by multiple sclerosis, cerebral palsy, or stroke. Additionally, it is utilized to treat and prevent malignant hyperthermia, which can lead to a life-threatening reaction to certain anesthetic drugs.

2. The adverse effects encompass seizures, allergic reactions, dyspnea, reduced inspiratory capacity, dizziness, muscle fatigue, nausea, and diarrhea. Its oral form includes a black box warning regarding the potential for hepatotoxicity, such as overt hepatitis.

3. Oral dantrolene is contraindicated in patients with liver cirrhosis, nonalcoholic steatohepatitis, and hepatitis B or C infections. However, there are no contraindications for using IV dantrolene to treat malignant hyperthermia.

CHAPTER 27: ACETAMINOPHEN AND NSAIDS

CLINICAL CASE STUDY 1

1. Acetaminophen is one of the most widely used over-the-counter analgesics and antipyretics, primarily utilized to treat pain and fever. Effective

pain management is critical, and acetaminophen can be used for mild to moderate pain or in conjunction with an opioid analgesic for severe pain. Additionally, it can help treat common colds and the flu. Injectable acetaminophen is available for managing mild to moderate pain in patients over 2 years old.

2. Acetaminophen's common adverse effects include hypersensitivity reactions, skin rash, nephrotoxicity, elevated blood urea nitrogen and creatinine levels, anemia, leukopenia, neutropenia, and pancytopenia. It may also lead to hyperammonemia, hyperchloremia, hyperuricemia, increased serum glucose, and elevated bilirubin and alkaline phosphatase levels. At higher doses, this agent can cause impairment and liver failure. Other adverse effects include headache, nausea, vomiting, constipation, pruritus, and abdominal pain.

3. Contraindications for acetaminophen include hypersensitivity to the drug and severe liver impairment. Acetaminophen should be avoided during alcohol consumption because its use can be linked to liver failure, which may occasionally require liver transplants or lead to fatalities.

CLINICAL CASE STUDY 2

1. Ibuprofen primarily inhibits the precursors of prostaglandins. The cyclooxygenase pathway plays a significant role in its intended uses.

2. Ibuprofen increases the risk of gastrointestinal complications, including bleeding, perforation, and ulceration. This risk is further elevated in older patients and those with pre-existing gastrointestinal conditions.

3. Ibuprofen is contraindicated for patients with a known history of hypersensitivity or allergic reactions to the drug itself, other NSAIDs, or aspirin. It raises the risk of serious cardiovascular events such as myocardial infarction and stroke. Due to the potential for Reye syndrome, it should be avoided in children with viral upper respiratory infections.

CHAPTER 28: OPIOID ANALGESICS

CLINICAL CASE STUDY

1. Opioids mimic the actions of endogenous opioid peptides by interacting with mu, delta, or kappa opioid receptors. These receptors are coupled to G1 proteins, and the activities of opioids are primarily inhibitory. Most clinically relevant opioid analgesics act as agonists on mu-opioid receptors within the central and peripheral nervous systems to achieve analgesia.

2. The adverse effects of opioid use include drowsiness, constipation, euphoria, nausea, vomiting, dizziness, and slowed breathing. Over time, a person using opioids may develop tolerance, physical dependence, and opioid use disorder, along with an increased risk of overdose and death.

3. Opioids are contraindicated for true allergies, severe respiratory depression, acute or chronic bronchial asthma, increased intracranial pressure, acute psychiatric fluctuations, uncontrolled suicide risk, or a heightened risk of prescription misuse. Opioids can induce a spasm of the biliary sphincter and should be avoided in patients with cholecystitis.

CHAPTER 29: ANTINEOPLASTIC AGENTS

CLINICAL CASE STUDY 1

1. The three main medication categories are conventional chemotherapy, hormonal therapy, and biological therapy. Traditional chemotherapy, often called systemic chemotherapy, is a cancer treatment that uses drugs to kill cancer cells. It remains a crucial treatment option for cancer.

2. Antimetabolites may lead to symptoms such as nausea, vomiting, diarrhea, loss of appetite, myelosuppression, hair loss, fatigue, skin rashes, and elevated liver enzymes. The severity of these symptoms varies depending on the specific drug and the patient's dosage.

3. Hormone therapy is mainly used to treat specific types of breast and prostate cancer that rely on sex hormones for growth, although a few other cancers might also respond to hormone therapy.

CLINICAL CASE STUDY 2

1. The trade names of fluorouracil and oxaliplatin include Adrucil and Eloxatin.

2. Antimetabolites are a significant class of cancer drugs that disrupt nucleic acid synthesis in cancer cells. They play a crucial role in cancer chemotherapy. There are three categories of antimetabolites, also known as antimetabolite antagonists: **purine antagonists, pyrimidine antagonists, and folic acid antagonists.**

3. Antimetabolites interfere with the synthesis of DNA constituents; they are structural analogs of either purine and pyrimidine bases or folate cofactors, which are involved in several steps of purine and pyrimidine biosynthesis. Therefore, depleting nucleotides inhibits DNA replication. However, some of these compounds can be fraudulently incorporated into nucleic acids, causing structural abnormalities that result in cell death through various mechanisms, including DNA breaks.

CHAPTER 30: VITAMINS AND MINERALS

CLINICAL CASE STUDY 1

1. Retinoids support growth, development, cell differentiation, vision, and immune function, playing a vital role in embryonic development. Vitamin A is essential for forming the eyes, cardiovascular system, limbs, and nervous system.
2. Vitamin A deficiency is mainly caused by inadequate dietary intake; however, it can also result from malabsorption due to intestinal diseases such as short bowel syndrome, celiac disease, cystic fibrosis, chronic diarrhea, bile duct obstruction, pancreatic disorders, and certain surgeries like gastric bypass. These factors hinder the body's ability to absorb vitamin A effectively. Additionally, heavy alcohol consumption can contribute to this deficiency.
3. Vitamin A is recommended for patients with deficiencies, including nutritional night blindness, xerophthalmia, keratomalacia, and corneal cloudiness.

CLINICAL CASE STUDY 2

1. The clinical symptoms linked to vitamin B12 deficiency include numbness in the extremities, pale skin, confusion, brain fog, fatigue, shortness of breath, and difficulty walking.
2. After diagnosing pernicious anemia, the patient must be given lifelong vitamin B12 injections.

CHAPTER 31: ANTIBACTERIAL AGENTS

CLINICAL CASE STUDY 1

1. The most common applications for penicillin G include pneumonia, strep throat, staph infections, diphtheria, meningitis, gonorrhea, and syphilis. Penicillin G also prevents rheumatic fever, chorea, rheumatic heart disease, and acute glomerulonephritis.
2. The allergic reaction occurs within 20 minutes of the injection being administered. It is manifested by laryngospasm, urticaria, bronchospasm, edema, pruritus, hypotension, vascular collapse, and death.
3. Penicillin injection reactions are treated with corticosteroids, epinephrine injection (EpiPen), antihistamines, bronchodilators, and immunotherapy (desensitization).

CLINICAL CASE STUDY 2

1. Vancomycin treats and prevents various bacterial infections caused by gram-positive bacteria, including MRSA. It is also effective against infections from streptococci, staphylococci, and enterococci. Additionally, it is utilized in patients with rheumatic fever or heart valve disease who are allergic to penicillin. Furthermore, vancomycin is prescribed for *Clostridioides difficile*-associated diarrhea, pseudomembranous colitis, S*taphylococcus enterocolitis,* osteomyelitis, bacteremia, and endocarditis.
2. Vancomycin is bactericidal because it inhibits the polymerization of peptidoglycans in the bacterial cell wall. As a result, it prevents the synthesis and polymerization of the peptidoglycan layer. This inhibition weakens the bacterial cell walls and ultimately leads to the leakage of intracellular components, resulting in bacterial cell death.
3. The adverse effects of intravenous vancomycin injection include hypersensitivity reactions, hypotension, and nephrotoxicity. Symptoms may include pruritus, flushing, skin rashes, chills, drug fever, eosinophilia, and reversible neutropenia.

CHAPTER 32: ANTIVIRAL AGENTS

CLINICAL CASE STUDY 1

1. Human papillomaviruses are the most common sexually transmitted infection in the United States. Over 42 million Americans are infected with various types of human papillomavirus (HPV), mainly impacting individuals in their late teens and early 20s.
2. The complications of HPV can vary from common issues like genital warts to, in rare cases, cancers of the cervix, vulva, vagina, and even the head and neck, with cervical cancer being the most common HPV-related cancer.
3. The **CDC recommends that all adolescents receive two doses of the HPV vaccine between the ages of 11 and 12**. A two-dose schedule is advised for individuals **who get their first dose before turning 15**, with the second dose given 6–12 months after the first.

CLINICAL CASE STUDY 2

1. Stavudine slows the progression of HIV in individuals with advanced symptoms, early symptoms, or no symptoms at all. It helps reduce the quantity of HIV in the body, competitively inhibits the HIV reverse transcriptase enzyme, and serves as a chain terminator in DNA synthesis.
2. Abacavir's adverse effects include fever, cough, headache, rash, tachypnea, abdominal pain, nausea, vomiting, diarrhea, hematuria, and arthralgia.

3. Lopinavir/ritonavir is contraindicated in patients with hypersensitivity, dysrhythmia, porphyria, pancreatitis, diabetes mellitus, heart disease, and hemophilia. Other adverse effects include diarrhea, dyslipidemia, and hepatic disorders.

CHAPTER 33: ANTIFUNGAL AGENTS

CLINICAL CASE STUDY 1

1. Systemic fungal infections are classified into two categories: opportunistic and nonopportunistic infections. Opportunistic mycoses primarily affect immunocompromised or debilitated patients, including *candidiasis, aspergillosis, cryptococcosis,* and *mucormycosis.* Although relatively uncommon, nonopportunistic mycoses can occur in any patient, including *sporotrichosis, blastomycosis, coccidioidomycosis, cryptococcus, pneumocystis jirovecii,* and *histoplasmosis.*
2. The drug of choice for the treatment of aspergillosis is voriconazole, and for mucormycosis, it is amphotericin B.
3. Due to the risk of potentially fatal adverse effects, the patient must be admitted to the hospital. Common adverse effects include fever, chills, wheezing, nausea, headache, hypoxia, hypotension, nephrotoxicity, hypokalemia, anemia, bone marrow suppression, and delirium.

CLINICAL CASE STUDY 2

1. The majority of onychomycosis is caused by *Tinea rubrum* and *Tinea interdigital.* Treatment is partly dependent on the rate of nail growth. Toenails grow more slowly than fingernails, sometimes taking 12–18 months to fully develop, resulting in a decreased treatment success rate.
2. Griseofulvin interacts with microtubules, impacting the formation of the mitotic spindle. This interference ultimately inhibits mitosis in dermatophytes. Therefore, griseofulvin is a fungistatic drug against *Trichophyton, Microsporum, and Epidermophyton* species.
3. Overall, griseofulvin has few adverse effects. This medication can cause nausea, vomiting, diarrhea, headaches, and allergic reactions. Other adverse effects include photosensitivity, petechiae, hepatotoxicity, insomnia, fatigue, confusion, and urticaria. It may worsen lupus or porphyria.

CHAPTER 34: ANTIPROTOZOAL AGENTS

CLINICAL CASE STUDY 1

1. Malaria is a potentially life-threatening disease caused by parasites transmitted through the bites of infected female Anopheles mosquitoes. It primarily affects tropical and subtropical regions. While it is preventable and treatable, early diagnosis and effective treatment are crucial. The lifecycle of malaria occurs in the female Anopheles mosquito. Sporozoites enter the bloodstream and invade the liver cells as the mosquito feeds on a human host's blood. Subsequently, the merozoites infect the red blood cells of the human host. Symptoms include chills, fever, and sweating that peaks, declines, and peaks again every 48 hours. Since hypnozoites can remain in the liver, relapse is expected following the end of an acute attack. However, this process completely ceases after two or more years. Medications that eliminate hypnozoites can prevent relapse. The most common forms of malaria are caused by *Plasmodium vivax* and *Plasmodium falciparum, with Plasmodium falciparum* being more severe and resulting in approximately 10% of infected patients' deaths.
2. Chloroquine can convert heme to nontoxic metabolites and concentrate in parasitized erythrocytes, possibly explaining its selective actions against the erythrocytic forms of Plasmodium.
3. Contraindications for primaquine treatment include pregnancy and acute illness with a predisposition to granulocytopenia. The medication is also contraindicated in patients with known hypersensitivity to pyrimethamine or any component of the formulation. Furthermore, the drug is contraindicated in patients with documented megaloblastic anemia due to folate deficiency.

CLINICAL CASE STUDY 2

1. Metronidazole is effective against Trichomonas *vaginalis, Entamoeba histolytica, Giardia lamblia,* blastocysts, and Balantidium coli. It also treats intestinal amoebiasis and liver amoebiasis. The extended-release tablets are utilized for treating vaginal infections.
2. The adverse effects of metronidazole include headache, confusion, dizziness, a metallic taste, nausea, vomiting, diarrhea, vaginitis, genital pruritus, abdominal pain, xerostomia, dysmenorrhea, urinary tract infections, bacterial infections, flu-like symptoms, pharyngitis, and sinusitis.
3. Metronidazole should be avoided in patients with hypersensitivity to the medication. It is contraindicated during the first trimester of pregnancy. Avoid consuming alcohol while taking metronidazole tablets, liquid, suppositories, or vaginal gel, as well as for 2 days after completing treatment. The pregnancy category for metronidazole is B.

CHAPTER 35: ANTITUBERCULOSIS AGENTS

CLINICAL CASE STUDY 1

1. Isoniazid can lead to severe adverse effects, including liver damage, especially in patients over 50 years old. The side effects of isoniazid may also include pruritus, nausea, diarrhea, jaundice, and hepatitis. Furthermore, isoniazid is linked to peripheral neuropathy, paresthesia, and sensory impairment, which may require discontinuation. Rare side effects may involve blurred vision or vision loss, with or without eye pain, convulsions, fever, sore throat, joint pain, mental depression, mood changes, or other psychological alterations, skin rash, and unusual bleeding or bruising.

2. PZA is contraindicated for individuals with known hypersensitivity, severe liver damage, or acute gout. Liver transaminases and uric acid levels should be monitored before and during treatment.

3. Although rifampin is generally well tolerated by patients, it can still cause adverse effects that may be either dose-dependent or dose-independent. Common adverse effects include loss of appetite, nausea, diarrhea, abdominal distention, heartburn, jaundice, chest pain, dizziness, fever, headache, hives, petechiae, agitation, difficulty concentrating, fainting, tachycardia, blurred vision, muscle weakness, and paresthesia. Saliva, tears, teeth, urine, feces, and sweat may also be discolored.

CLINICAL CASE STUDY 2

1. Dapsone inhibits dihydrofolic acid synthesis by competing with para-aminobenzoic acid for the active site of dihydropteroate synthetase.

2. The adverse effects of dapsone include headaches, insomnia, blurred vision, anorexia, nausea, vomiting, abdominal pain, jaundice, pancreatitis, decreased hemoglobin, neutropenia, tachycardia, and albuminuria.

3. The contraindications of dapsone include hypersensitivity to the drug. Caution is warranted in patients with liver or renal damage, cardiopulmonary disease, peripheral neuropathy, and during pregnancy.

CHAPTER 36: IMMUNIZATIONS

CLINICAL CASE STUDY 1

1. HPV is a common sexually transmitted infection caused by viruses that can infect the skin and mucous membranes. More than 30 strains of HPV can affect the genitals, including harmless types that lead to genital warts and cervical cancer. Types 16 and 18 are classified as high-risk for cervical, penile, and anal cancers.

2. Types 16 and 18 are classified as high-risk for cervical, penile, and anal cancers.

3. If HPV vaccination begins between 9 and 14, a recommended two-dose series should be administered within 6 to 12 months between doses. If vaccination starts at age 15 or older, a three-dose series is necessary, with doses given at 0, 1–2, and 6 months after the first dose.

CLINICAL CASE STUDY 2

1. The risk factors for RSV infection include premature babies, older adults, patients with heart or lung disease, and individuals who are immunocompromised.

2. RSV vaccines are recommended for all adults aged 75 and older and for those aged 60 to 74 who are at an increased risk of severe RSV. They should receive a single dose of the RSV vaccine, which can be administered simultaneously with other vaccines.

3. Pregnant women should receive one dose of RSVpreF (Abrysvo) between weeks 32 and 36 of pregnancy, specifically from September to January, to help prevent severe RSV disease in infants.

CHAPTER 37: POISONS AND ANTIDOTES

NO CASE STUDY

Glossary

Abrasion: A minor, superficial scrape or rub of the skin caused by friction against a rough surface, leading to the stripping away of the top layers.

Absence of seizure: A generalized onset seizure that simultaneously begins on both sides of the brain. It is most common in children and typically does not cause any long-term problems. An older term is petit mal seizures.

Achilles tendon: The tendon connecting calf muscles to the heel bone.

Achondroplasia: A genetic disorder characterized by abnormal bone growth, resulting in short stature and disproportionately short limbs.

Acinar cells: Specialized epithelial cells located in various glands, especially the exocrine pancreas, that form clusters known as acini.

Acne: A common skin condition that happens when hair follicles under the skin become clogged.

Actinomycetes: A diverse group of filamentous, Gram-positive bacteria recognized for their distinctive "ray fungus" appearance and their vital role in soil ecosystems.

Action potential: The change in electrical potential associated with the passage of an impulse along the membrane of a muscle cell or nerve cell.

Active transport: The movement of ions or molecules across a cell membrane into a region of higher concentration, assisted by enzymes and requiring energy.

Active tubular secretion: The process in the kidney's renal tubules, where substances like hydrogen ions, potassium, drugs, and other wastes are actively transported from the blood in the peritubular capillaries into the tubular fluid.

Acute myeloid leukemia: A cancer of the myeloid blood cells involving rapid growth of abnormal bone marrow and blood.

Acute pain: A brief, sharp, and localized discomfort that typically occurs from a specific injury, trauma, or medical procedure.

Addison's disease: A condition that impacts the function of the adrenal cortex. It causes a reduction in the production of two hormones by the adrenal cortex: cortisol and aldosterone.

Additive effect: The combined impact of two or more agents' actions equals the sum of their individual effects.

Adenylyl cyclase: An enzyme that produces cyclic adenosine monophosphate (cAMP), a second messenger involved in regulating numerous cellular processes.

Adverse effect: A negative and unintended outcome caused by a medical treatment, drug, or other intervention.

Adynamic ileus: A condition in which foods and drinks do not usually pass through the intestines (also called paralytic ileus).

Affinity: The extent or fraction to which a drug binds to receptors at any given drug concentration or the strength with which the drug binds to the receptor.

Agonist: A substance that acts like another substance and stimulates an action.

Agranulocytosis: A serious condition where the body has an alarmingly low level of white blood cells.

Akathisia: A condition characterized by agitation, distress, and restlessness, which can sometimes occur as a side effect of antipsychotic and antidepressant medications.

Aldosterone: A corticosteroid hormone that stimulates the absorption of sodium by the kidneys and regulates water and salt balance.

Aldosteronism: A condition marked by excessive production of the hormone aldosterone by the adrenal glands.

Allergen: A substance capable of triggering an allergic reaction.

Allergic drug reactions: The immune system overreacts to a medication, causing symptoms like hives, rashes, itching, and breathing difficulties.

Alopecia: The partial or complete loss of hair from areas of the body where it normally grows; baldness.

Alveoli: Tiny air sacs at the end of the bronchioles.

Alzheimer's disease: A brain disorder that gradually destroys memory and thinking skills and, eventually, the ability to perform even simple tasks.

Amebiasis: A common parasitic infection of the intestines caused by amoebae from the *Entamoeba* group.

Amenorrhea: The absence of menstrual periods for more than three months in women who are not pregnant, menopausal, or breastfeeding.

Amide: An organic or inorganic compound that contains a nitrogen atom bonded to a carbonyl carbon (C=O) or a similar group involving a carbonyl.

Amnesia: A partial or total loss of memory.

Amputation: The surgical removal or loss of a body part such as a finger, toe, hand, foot, arm, or leg.

Amyotrophic lateral sclerosis: A progressive nervous system disease that affects nerve cells in the brain and spinal cord, leading to loss of muscle control.

Anaerobes: Any organism that does not require molecular oxygen for growth.

Anaphylactoid reactions: Severe, systemic, immediate responses that resemble anaphylaxis but are not caused by the typical IgE-mediated immune response.

Anaphylaxis: An excessive localized enlargement of an artery caused by weakening of the artery wall.

Androgen: A male sex hormone that is a steroid and includes testosterone and androsterone.

Angioedema: Swelling under the skin.

Angiotensinogen: A protein mainly produced in the liver that plays a vital role in the renin-angiotensin-aldosterone system (RAAS), which controls blood pressure and fluid balance.

Ankylosing spondylitis: A long-lasting inflammatory arthritis that mainly affects the spine and sacro-iliac joints.

Anopheles mosquito: A genus of mosquitoes that includes the main vectors for malaria, a parasitic disease affecting humans and animals.

Anorexia: Loss of appetite.

Antagonism: A drug or chemical that activates a receptor to produce a biological response, while an antagonist binds to a receptor but does not activate it.

Antagonist: A substance that acts against and blocks an action.

Antepartum: The period before giving birth.

Antibodies: Specialized immune proteins produced by white blood cells called B cells (or plasma cells) in response to an antigen.

Antidote: A substance that neutralizes the harmful effects of poison.

Antigens: Any molecule or substance that provokes an immune response in the body, usually by being identified as foreign.

Antioxidant: A substance that stops and repairs cell damage caused by harmful molecules called free radicals.

Anuria: The absence of urine production. It can occur due to shock, severe blood loss, or heart or kidney failure. Medications or toxins can also cause it. Anuria is a medical emergency and can be life-threatening.

Apathy: A state of lacking feeling or emotion.

Aphonia: A voice disorder that causes an individual to lose their voice or be unable to produce sound.

Aplastic anemia: A rare but severe blood disorder that occurs when the bone marrow is unable to produce enough new blood cells.

Apolipoprotein B: A protein that plays a key role in transporting cholesterol and triglycerides in the bloodstream.

Arrhythmia: A irregular heartbeat or a problem with the rate or rhythm of the heart.

Arterioles: A tiny blood vessel that branches from an artery and delivers blood to capillaries.

Arthralgia: Pain in a joint caused by injury, infection, inflammation, and immune disorders, such as rheumatoid arthritis, lupus, Sjogren's syndrome, or systemic lupus erythematosus.

Aspergillosis: A disease caused by Aspergillus molds, which are common fungi found in the environment that release spores into the air.

Asthenia: A feeling of weakness, lack of energy, and exhaustion.

Asthma: A long-term lung condition that causes inflammation and tightening of the muscles around the airways, making it hard to breathe.

Asystole: The state of *total cessation of electrical activity from the heart.*

Ataxia: The loss of complete control of bodily movements.

Athlete's foot: A fungal infection that affects the skin on the feet.

Atresia: The absence or closure of a normal body opening, passage, or tubular structure.

Atrial fibrillation: The most common type of heart arrhythmia, an irregular and often very rapid heart rhythm that can lead to blood clots in the heart.

Atrioventricular node: A small mass of heart tissue connecting the electrical systems of the atria and ventricles.

Atrophy: Decrease in size of a body part, cell, organ, or other tissue.

Atypical mycobacteria: Also known as non-tuberculous mycobacterial infections, are caused by bacteria found in soil, water, and animals.

Automation: The use of technology to perform tasks or processes with minimal or no human intervention, aiming to increase efficiency, productivity, and consistency.

Azotemia: A medical condition marked by increased levels of nitrogen waste products, like urea and creatinine.

Bactericidal: Having the ability to kill bacteria.

Bacteriophages: A type of virus that infects and multiplies inside specific bacteria, effectively destroying them.

Bacteriostatic: The agent prevents the growth of bacteria.

Basal metabolic rate: Refers to the amount of energy the body needs to maintain homeostasis.

Bechet's disease: A rare, chronic inflammatory disorder that affects multiple organs and systems in the body. It is characterized by recurring, painful ulcers in the mouth, genitals, and eyes, along with other systemic symptoms.

Benign tumors: A non-cancerous growth of abnormal cells that develops slowly and stays confined to one area, meaning it does not spread or invade other parts of the body.

Beta-lactam: A class of antibiotics that share a common chemical structure, the beta-lactam ring. They are effective against a wide range of bacteria and are among the most widely used antibiotics worldwide.

Bioavailability: The proportion of a drug or other substance that enters the blood circulation when introduced into the body and can have an active effect.

Bioinformatics: An interdisciplinary field that applies computational tools, statistics, and mathematical modeling to analyze and interpret biological data such as DNA, RNA, and proteins.

Biologic therapy: A type of treatment that uses substances derived from living organisms, such as proteins, antibodies, or cells, to treat diseases.

Biotechnology: The use of living organisms, biological systems, and processes to develop products and technologies that solve human problems, especially in health, agriculture, and industry.

Biotransformation: The process of changing a drug or substance within the body.

Biphasic anaphylaxis: The recurrence of anaphylaxis symptoms within minutes to several hours of the initial reaction's complete resolution, but without re-exposure to the trigger.

Bitot spots: Are small, white or yellowish patches that appear on the conjunctiva. They are a sign of vitamin A deficiency.

Black cohosh: A member of the buttercup family, its flowers and roots were commonly used in traditional Native American medicine. Today, it's a famous women's health supplement claimed to help with menopause symptoms, fertility, and hormonal balance.

Blepharospasm: A neurological disorder characterized by involuntary, repetitive eye movements, typically causing rapid blinking or eye closure.

Blood urea nitrogen: A lab test that measures urea levels in the blood. Urea is a waste product created when the body breaks down protein.

Blood-brain barrier: A network of blood vessels and tissue made up of closely packed cells that helps prevent harmful substances from reaching the brain; it allows certain substances, such as water, oxygen, carbon dioxide, and general anesthetics, to pass through.

Bone marrow: A soft, highly vascularized modified connective tissue that fills the cavities of most bones and exists in two forms.

Bowman's capsule: A cup-shaped sac in the kidney surrounding the glomerulus, the first step in filtering blood to create urine.

Bradycardia: A type of arrhythmia, or abnormal heart rhythm, that happens when the heart beats fewer than 60 times a minute.

Bradykinesia: Slowness of movement.

Brainstem: The central trunk of the brain, consisting of the medulla oblongata, pons, and midbrain, and continuing downward to form the spinal cord.

Brand name: A unique word or phrase, a signature, that identifies and distinguishes a product, service, or company from competitors in its category.

Bronchial tree: The bronchi are the branches of the tree within the lungs.

Bronchitis: An infection of the bronchi.

Bulimia nervosa: A serious eating disorder marked by repeated episodes of binge eating followed by inappropriate behaviors to prevent weight gain. A person's self-esteem is heavily and unduly influenced by their body shape and weight.

Bundle of His: A bundle of specialized muscle fibers in the heart that carries electrical signals from the atrioventricular (AV) node to the ventricles.

Calcitonin: A polypeptide hormone secreted by the parafollicular or C cells of the thyroid gland; it helps regulate plasma calcium levels and reduces the rate of bone resorption.

Candidiasis: A fungal infection caused by an overgrowth of the yeast Candida. It can affect different parts of the body in various ways.

Cannabinoids: The chemical compounds that are the active principles of marijuana.

Capsid: The protein shell that encloses a virus's genetic material, acting as a protective coating around the viral nucleic acid, allowing it to attach to host cells and deliver its genome during infection.

Carboxylesterase: An enzyme that catalyzes the hydrolysis of esterified chemicals, including pesticides, drugs, and lipids.

Cardiac cycles: The series of events that occur between the start of one heartbeat and the start of the next.

Cardiac output: The amount of blood the heart pumps in a minute.

Cardiogenic shock: A life-threatening condition in which the heart suddenly cannot pump enough blood to meet the body's needs.

Cardiomyocytes: The cells that make up the myocardium and are responsible for the heart's contraction and relaxation.

Cardiotoxicity: The harmful effects of certain drugs or treatments on the heart muscle, leading to damage or impaired cardiac function.

Cardioversion: A medical procedure used to restore a normal heartbeat by delivering an electrical shock to the heart.

Carotenoid: A natural pigment that appears yellow, orange, and red in plants, algae, and photosynthetic bacteria, giving foods like carrots and tomatoes their bright colors.

Carrier proteins: Are integral membrane proteins that help transport specific molecules across cell membranes.

Cartilage: A strong, flexible connective tissue that protects the joints and bones. It acts as a shock absorber throughout your body.

Catabolism: The breakdown of proteins into absorbable monomers for further degradation or reassembly.

Cataplexy: A sudden, brief loss of voluntary muscle control, often triggered by intense emotions such as laughter, surprise, or fear, and most commonly linked to narcolepsy.

Catecholamine: Sympathomimetic amines, including dopamine, epinephrine, and norepinephrine, are essential for physiological responses to stress.

Cecum: A pouch that forms the first part of the large intestine.

Cellulitis: A common bacterial skin infection that impacts the deeper layers of skin and underlying tissues.

Cerebral cortex: The outermost layer of the cerebrum, made up of gray matter, controls higher-level cognitive functions, including thought, memory, language, reasoning, and consciousness.

Cerebral edema: Swelling of the brain. It is a relatively common phenomenon with numerous causes.

Cerebral palsy: A group of disorders that affect a person's ability to move and control their muscles, caused by abnormal brain development or brain damage.

Cesarean section: A surgical procedure performed to deliver a baby through incisions made in the mother's abdomen and uterus.

Chagas disease: A potentially life-threatening disease caused by the parasite *Trypanosoma cruzi*.

Chamomile: An aromatic European plant of the daisy family with white and yellow flowers.

Chelating agent: A compound that binds to and helps remove metal ions from the body.

Chelation therapy: A medical procedure that uses a chemical solution to remove heavy metals and other toxins from the body.

Chlamydiae: A common sexually transmitted infection caused by the bacterium Chlamydia trachomatis. It can affect people of all ages, but is most common in young people, especially women.

Cholecystokinin: A polypeptide hormone secreted in the small intestine; it stimulates gallbladder contraction and the secretion of pancreatic enzymes.

Cholera: A bacterial disease that causes severe diarrhea, vomiting, and dehydration. It is caused by the bacterium Vibrio cholerae and is spread through contaminated food or water.

Cholestasis: A condition where the flow of bile from the liver to the small intestine is obstructed or blocked.

Chronic pain: Ongoing or recurring pain that lasts longer than three months. It can differ in severity and location and may be accompanied by other symptoms such as fatigue, sleep issues, and emotional distress.

Circumflex artery: A coronary artery that supplies oxygenated blood to the left side of the heart.

Cirrhosis: A chronic liver condition that develops when the liver slowly becomes damaged and healthy liver cells are replaced by scar tissue. This scarring, also known as fibrosis, hinders the liver's ability to function properly.

Clearance: The rate at which the active drug is eliminated from the body; for most drugs, at steady state, clearance stays constant so that drug input equals drug output.

Clubbing of fingers: A physical sign marked by bulbous swelling of the tips of one or more fingers.

Cochlear: A hollow tube located in the inner ear of higher vertebrates, typically coiled like a snail shell, which contains the sensory organ responsible for hearing.

Coenzymes: Small, organic, non-protein molecules that assist enzymes (protein catalysts) in chemical reactions.

Cofactor: A non-protein chemical compound bound to a protein and required for the protein's biological activity.

Collecting tubule: The final section of a long, winding tube that collects urine from the nephrons (cell structures in the kidney that filter blood and produce urine) and channels it into the renal pelvis and ureters. Also known as a collecting duct.

Competitive antagonists: A substance that binds to the same site on a receptor as an agonist but does not activate the receptor, instead blocking the agonist's ability to bind and produce a biological effect.

Complement fixation: Combining complement with the antigen-antibody complex, rendering the complement inactive or fixed.

Conjunctivitis: An inflammation or infection of the conjunctiva, the transparent layer covering the white of the eye and the inner eyelid.

Conn's syndrome: Primary aldosteronism resulting from hyperplasia or tumors.

Consent form: A legal document signed to provide formal permission for a specific action, such as a medical procedure, research participation, or data use, after being fully informed of its details, risks, and benefits.

Convulsion: A condition in which muscles contract and relax quickly, causing uncontrolled shaking of the body. Head injuries, high fevers, some medical disorders, and certain drugs can cause convulsions. They may also occur during seizures caused by epilepsy.

Coronary arteries: The coronary arteries wrap around the outside of the heart. They send oxygen-rich blood into the heart's muscle tissues.

Coronary sinus: A large vein in the heart that carries deoxygenated blood from the myocardium to the right atrium.

Corona sulcus: A groove on the heart's surface that separates the atria from the ventricles.

Corpus luteum: A mass of cells that forms in an ovary. It is a temporary organ that appears every menstrual cycle and disappears if fertilization does not occur.

Corticotropin: A hormone produced by the pituitary gland that stimulates the adrenal glands to release cortisol.

Covalent adduct: A complex formed when a covalent bond joins two species.

Creatine phosphate: A high-energy phosphate compound located in muscle cells.

Creatinine: A waste product produced by muscle metabolism. It is filtered and excreted by the kidneys.

Cretinism: A birth defect caused by a lack of thyroid hormone during prenatal development, characterized by short stature, intellectual disability, bone dystrophy, and a slow basal metabolic rate.

Crohn's disease: A long-term inflammatory condition affecting the intestines, especially the colon and ileum, associated with ulcers and fistulas.

Cross-tolerance: A pharmacological phenomenon where a person's tolerance to one drug causes a reduced response to another drug with a similar mechanism of action.

Cryptococcal meningitis: A fungal infection of the meninges that cover the brain and spinal cord. It is caused by the fungus Cryptococcus neoformans.

Cryptococcosis: A fungal infection caused by the yeast-like fungus Cryptococcus neoformans. It mainly affects the lungs and central nervous system.

Cryptosporidiosis: A contagious disease that leads to watery diarrhea, caused by parasites called Cryptosporidium.

Crystal nephropathy: A condition where crystals form in the kidneys, damaging the renal tubules.

Cumulative effect: When multiple doses of a drug are given before previous doses are fully cleared from the body. This leads to the gradual buildup of the drug's concentration, and therefore its effect, over time. This is especially important for medications that are administered over an extended period.

Cushing's syndrome: Characterized by a moon-shaped face, fat buildup around the torso, loss of fat from the extremities, and easy bruising caused by high cortisol levels. These elevated cortisol levels result either from excess production of the hormone ACTH due to an adrenal or pituitary tumor or from taking high doses of corticosteroid hormones.

Cyanosis: A bluish color of mucous membranes and skin.

Cystic fibrosis: A long-term, inherited genetic condition that impacts the body's mucus-producing glands. It results in the production of thick, sticky mucus that blocks the airways, pancreas, and other organs.

Cystoscopy: A medical procedure that enables a doctor to examine the bladder and urethra using a thin, flexible tube called a cystoscope.

Cytochrome P450: A group of enzymes that metabolize foreign substances and liver drugs.

Cytoskeleton: A dynamic network of protein filaments—including actin filaments, intermediate filaments, and microtubules—located in the cytoplasm of all cells. It provides structural support, maintains cell shape, and facilitates cell movement and internal transport of organelles.

Cytotoxic T cells: A type of white blood cell that detects and destroys infected, cancerous, or damaged cells in the body.

Defibrillation: A procedure that uses an electric shock to restore a normal heartbeat and treat life-threatening heart arrhythmias.

Degenerative disc disease: A condition that affects the intervertebral discs, the cushions that sit between the vertebrae in the spine.

Degenerative disease: Chronic conditions marked by the gradual decline of tissues and organs over time. This decline results in reduced function and structural integrity.

Delirium tremens: A rapid onset of confusion usually caused by withdrawal from alcohol.

Delirium: An acutely disturbed state of mind that occurs in fever, intoxication, and other disorders and is characterized by restlessness, illusions, and incoherence of thought and speech.

Dendritic cells: A type of immune cells that play a vital role in the body's immune response.

Depolarization: Reversing the charge across a cell membrane of neurons and causing an action potential.

Dermatitis herpetiformis: A chronic, intensely itchy, blistering skin manifestation of gluten-sensitive enteropathy, commonly known as celiac disease.

Dermatophytes: A group of fungi that cause superficial infections of the skin, hair, and nails. They are characterized by their ability to digest keratin, a protein found in these tissues.

Designer drugs: A structural or functional analog of a controlled substance intended to imitate the pharmacological effects of the original drug while avoiding being classified as illegal and undetectable in standard drug tests.

Diabetes insipidus: An uncommon condition where the kidneys cannot conserve water, causing excessive thirst and frequent urination of large amounts of pale, watery urine.

Diabetic ketoacidosis: A dangerous complication of diabetes mellitus in which the chemical balance of the body becomes dangerously acidic.

Diaphoresis: Excessive sweating, often occurring as a symptom of illness or a side effect of medication.

Diphtheria: A serious bacterial infection, predominantly caused by Corynebacterium diphtheriae, that is characterized by a thick, greyish membrane, or pseudomembrane, in the throat.

Diplopia: A visual disturbance, commonly known as double vision, where a person sees two images of a single object..

Dissolution rate: The speed at which a solid solute dissolves into a solvent.

Distal convoluted tubule: A short segment of the nephron in the kidney that plays a critical role in regulating electrolytes and fluid volume.

Diuresis: An increase in the amount of urine made by the kidney and passed from the body.

Diurnal body rhythms: A 24-hour cycle synchronized to an external zeitgeber, such as the light–dark cycle.

Dong Quai: An aromatic herb of the parsley family, native to China and Japan, whose root is used to treat premenstrual syndrome, menstrual cramps, menopausal symptoms, and other gynecological issues.

Dopamine receptors: G-protein coupled receptors that bind the neurotransmitter dopamine and are crucial for motor control, motivation, memory, emotion, and neuroendocrine functions.

Dormant: Inactive but capable of becoming active.

Double-blind study: A scientific trial in which neither the participants nor the researchers know who is receiving the experimental treatment and who is receiving a placebo or control.

Drug interactions: When two or more drugs react with each other, either prescription or over-the-counter.

Drug recalls: An action by a manufacturer or the Food and Drug Administration (FDA) to withdraw a drug from the market because of safety issues.

Drug standard: It refers to established guidelines and reference materials, such as pharmacopoeial reference standards, that specify the required quality, purity, and strength of drugs to ensure public safety. These standards are created by authoritative organizations and are essential for verifying a drug's safety, efficacy, and consistent performance through standardized testing and analysis.

Drug tolerance: A pharmacological condition where a person's body or brain has a diminished response to a drug or medication after repeated use. As a result, a larger dose is required to achieve the same effect that was once produced by a smaller amount.

Dura mater: One of the layers of connective tissue that make up the meninges of the brain.

Dwarfism: Short stature that results from a genetic or medical condition; generally defined as an adult height of 4 feet 10 inches or less.

Dysmenorrhea: Painful menstrual cramps or periods.

Dyspareunia: Painful sexual intercourse due to medical or psychological causes.

Dyspepsia: A discomfort or pain in the upper abdomen that can occur during or after eating.

Dysphasia: A neurological condition that affects the ability to understand, produce, or use language.

Dyspnea: Difficulty breathing, gasping, and breathlessness

Dystonia: A state of abnormal muscle tone resulting in muscular spasms and abnormal posture, typically due to neurological disease or a side effect of drug therapy.

Ebola hemorrhagic fever: A severe and often fatal viral disease caused by the Ebola virus.

Echinacea: A coneflower of North America. It is used in herbal medicine mainly for its antibiotic and wound-healing properties.

Eclampsia: A form of toxemia of pregnancy characterized by albuminuria, hypertension, and convulsions.

Edema: A medical condition where excess fluid accumulates in the body's tissues, causing swelling. It can affect various parts of the body, including the legs, feet, ankles, arms, and face.

Efficacy: The maximum response that can be achieved with a drug.

Eicosanoids: Lipid-based signaling molecules that have a distinctive role in innate immune responses, such as inflammation, blood pressure regulation, and brain function.

Electroencephalogram: A test that measures the electrical activity in the brain. It is performed by placing small electrodes on the scalp that record brain activity, specifically brain waves.

Electrolytes: Minerals that carry an electric charge when dissolved in water, found in bodily fluids like blood, urine, and sweat.

Emesis: The forceful ejection of some or all of the stomach's contents through the mouth.

Emphysema: A progressive, chronic lung condition where the alveoli are damaged or destroyed. This causes the tiny air sacs to rupture, forming a single large air pocket instead of multiple small ones. Air becomes trapped in these damaged areas and prevents proper oxygen transfer throughout the body.

Empiric monotherapy: A treatment that uses a single antibiotic, or β-lactam agent, before the results of blood cultures and antibiotic susceptibility tests are known.

Endocarditis: An inflammation of the heart chambers and the valves' inside lining.

Endocardium: The innermost layer of the heart that lines the chambers and covers the heart's valves, papillary muscles, and chordae tendineae.

Endogenous: Anything that originates from within the body, system, or cell itself, rather than from an external source.

Endometrium: The inner lining, consisting of the functional and basal endometrium.

Endoplasmic Reticulum: A network of interconnected membranes inside a cell that helps produce, modify, and move proteins and lipids, acting like a factory assembly line; it can be either "rough" with attached ribosomes for protein creation, or "smooth" for lipid synthesis, depending on its structure.

Endothelial cells: The cells that line the inside of blood vessels, lymph vessels, and the heart.

Entamoeba histolytica: A protozoan that causes intestinal amebiasis and extraintestinal manifestations.

Enteric fever: A bacterial infection caused by Salmonella enterica serotype Typhi. It is typically spread through contaminated food or water.

Envelope: A viral envelope is the outermost layer of certain viruses, composed of a lipid bilayer derived from the host cell's membrane during the budding process.

Epicardium: A layer of epithelial cells that covers the surface of the heart.

Epidermal necrolysis: A rare and serious skin reaction marked by widespread peeling of the outer skin layer.

Epidermis: The outermost layer of skin that acts as the first barrier against harmful substances entering the body.

Epidural block anesthesia: An epidural nerve block is a procedure used to block pain by injecting anesthetic medication into the epidural space. The epidural space is the area between the inner wall of the vertebral column and the outermost layer of the dura mater that surrounds the spinal cord.

Epiglottitis: A rare but life-threatening infection of the epiglottis

Epilepsy: A neurological disorder characterized by sudden recurring episodes of sensory disturbance, loss of consciousness, or convulsions caused by abnormal electrical activity in the brain.

Epinephrine: A neurotransmitter and a hormone. It plays a vital role in the body's "fight-or-flight" response.

Epiphyses: The rounded end of a long bone that forms a joint with another bone.

Epistaxis: A nosebleed is a loss of blood from the tissue lining the inside of the nose.

Erysipelas: An acute bacterial skin infection.

Erythrocytes: A type of blood cell produced in the bone marrow and found in the blood. They contain a protein called hemoglobin, which transports oxygen from the lungs to all parts of the body.

Ester: A chemical substance made when an acid and an alcohol combine and water is removed.

Ethylene glycol: A synthetic liquid substance that absorbs water. It is odorless but has a sweet taste. It makes antifreeze and de-icing solutions for cars, airplanes, and boats.

Euphoria: A feeling or state of intense excitement and happiness.

Euthyroid goiters: An enlargement of the thyroid gland without alterations in thyroid hormone levels.

Exogenous: Originating outside the body, referring, for example, to drugs (exogenous chemicals) or to phenomena, conditions, or disorders caused by external factors.

Extrasystoles: A prematurely occurring beat of one of the heart's chambers that leads to momentary arrhythmia but leaves the fundamental rhythm unchanged.

Factor Xa inhibitors: A class of blood thinners that prevent blood clots from forming or treat existing clots.

Familial hypophosphatemia: A rare inherited disorder that causes low levels of phosphate in the blood.

Familial Mediterranean fever: A rare, inherited autoinflammatory disorder characterized by repeated episodes of fever, abdominal pain, chest pain, and joint pain.

Fat embolus: Happens when fat globules enter the bloodstream, often after a bone fracture or other trauma.

Fatty acids: Long chains of carbon atoms with a carboxylic acid group at one end, serving as the fundamental components of fats, oils, and other lipids.

Febrile neutropenia: A severe condition that occurs when a patient with neutropenia develops a fever.

Fertilization: The process where a male sperm fuses with a female gamete to form a zygote, marking the initiation of development of a new individual organism; essentially, the union of two reproductive cells to create a new life.

Fetal-placental barrier: The tissues between maternal and fetal blood must be crossed to enable exchange.

Fetus: The developing offspring of a human or other mammal during prenatal stages that occur after the embryo stage (in humans, starting at eight weeks after conception).

Fibroblasts: A type of cell that helps form connective tissue, a fibrous material that supports and connects other tissues or organs in the body. They produce collagen proteins that help maintain the structural framework of tissues.

Fibromyalgia: A chronic disorder characterized by widespread musculoskeletal pain, fatigue, and tenderness in localized areas.

First-order kinetics: A constant percentage of the drug is eliminated per unit time. The elimination rate is proportional to the amount of drug in the body. The higher the concentration, the more drug is eliminated per unit of time.

First-pass effect: A process of drug metabolism where a significant amount of the drug is broken down before it reaches the body's systemic circulation. It is most common with drugs taken orally, as the absorbed drug is transported to the liver and metabolized before it can circulate throughout the body.

Floor stock: A supply of medications kept in patient care areas, like nursing stations, rather than in the central pharmacy.

Follicle-stimulating hormone: A hormone produced by the anterior pituitary in response to gonadotropin-releasing hormone (GnRH) from the hypothalamus.

Formaldehyde: A chemical commonly used to kill germs and to preserve laboratory specimens and tissues. It is also used to make building materials (such as wood), glue, fabric, paint, fertilizers, pesticides, and other substances.

Formic acid: A colorless, corrosive liquid with a strong, pungent odor.

Free radicals: Any molecular species capable of independent existence that contains an unpaired electron in an atomic orbital.

Full agonists: Have high efficacy and produce a complete response while occupying a relatively small proportion of receptors.

Fungi: A group of eukaryotic microorganisms, some of which are capable of causing superficial, cutaneous, subcutaneous, or systemic disease.

Fungistatic: Anti-fungal agents that inhibit the growth of fungus (without killing the fungus).

G cells: Cells in the pyloric antrum of the stomach that produce the hormone gastrin, which stimulates the stomach to release acid to help digest food.

Gallstone: Solid cholesterol, bilirubin, and bile deposits in the gallbladder or bile ducts.

Gastrectomy: A surgical procedure in which part or all of the stomach is removed.

Gastric lavage: A medical procedure that removes the stomach's contents using a tube.

Gastrin: A polypeptide hormone secreted by specific cells of the pyloric glands; it strongly stimulates the secretion of gastric acid and pepsin; it also weakly stimulates the secretion of pancreatic enzymes and gallbladder contraction.

Gastrotomy: The operation of cutting into the stomach.

Gender dysphoria: A condition where a person feels significant distress and discomfort because their gender identity doesn't match the sex they were assigned at birth.

Generalized seizures: Those that originate from multiple brain foci, characterized by general rather than localized neurologic symptoms; they may be tonic-clonic and may progress from a focal seizure.

Generic name: The official nonproprietary name of a drug under which it is licensed and identified by the manufacturer.

Genetic engineering: Involves directly changing an organism's DNA to add, remove, or alter genes to give it desired traits.

Genital herpes: A chronic sexually transmitted infection caused by the herpes simplex virus.

Genome: The complete set of genetic information in an organism. It provides all of the information the organism requires to function.

Genomics: The study of an organism's complete set of genes, known as its genome.

Gestational diabetes: A condition during pregnancy that involves a defect in how the body processes and uses glucose in the diet; the pancreas is not engaged, while the placenta is implicated.

Ghrelin: A hormone secreted by the stomach cells that promotes hunger, decreases after eating, and promotes growth hormone secretion.

Giardiasis: A microscopic parasite that causes giardiasis, a common intestinal illness characterized by diarrhea.

Ginger: A flowering plant is used as a spice for its distinct flavor and for medicinal purposes due to compounds called gingerols.

Gingival hyperplasia: Overgrowth of the gums.

Ginkgo: An extract of ginkgo leaves that is held to enhance mental functioning by increasing blood circulation to the brain.

Ginseng: The root of a plant found in countries such as China, Russia, and America, which some people believe is good for their health.

Glaucoma: A group of eye diseases that cause vision loss by damaging the optic nerve.

Glioma: A type of brain tumor that originates from glial cells, which support and nourish neurons in the brain. Gliomas can be benign (non-cancerous) or malignant (cancerous)

Glomerular filtration rate: The main measure of kidney function, indicating how much blood the glomeruli filter each minute to eliminate waste and excess fluid from the body.

Glomerular filtration rate: A measure of kidney function that estimates how much blood passes through the filters each minute.

Glomerulonephritis: A condition in which the tissues in the kidney become inflamed and have problems filtering waste from the blood.

Glomerulus: A network of vascular tufts encased in the capsule of the kidney.

Glossitis: An inflammation of the tongue that causes swelling, redness, and pain.

Glucagon: A hormone produced by the pancreas that plays a crucial role in regulating blood sugar levels.

Glucocorticoids: Involved in the metabolism of carbohydrates, proteins, and fats; they possess anti-inflammatory properties.

Gluconeogenesis: The process of producing glucose from noncarbohydrate sources, such as amino acids and glycerol. It primarily occurs in the liver and kidneys when the carbohydrate supply is low and cannot meet the body's energy needs.

Glycogen: An extensively branched glucose polymer that humans use as an energy reserve.

Glycogenolysis: The process of breaking down glycogen, a stored energy source in the liver and muscles, to produce glucose.

Glycolaldehyde: A simple organic molecule regarded as the smallest sugar molecule. It is often viewed as a potential precursor to more complex biomolecules and may have played a role in the origin of life due to its presence in interstellar space.

Glycolic acid: A chemical compound used in skincare products and as a metabolite in ethylene glycol poisoning.

Goblet cells: Specialized epithelial cells in the body's lining, such as the respiratory and digestive tracts, that secrete mucins and create a protective mucus layer.

Goldenseal: A North American woodland plant from the buttercup family, known for its bright yellow root used in herbal medicine.

Gonadotropins: Hormones that regulate the reproductive system, including the development, growth, and function of the ovaries and testes.

Gonorrhea: A sexually transmitted infection (STI) caused by the bacterium Neisseria gonorrhoeae.

Gout: A condition where faulty uric acid metabolism causes arthritis, especially in the smaller bones of the feet, deposits of chalkstones, and episodes of sudden pain.

Grand mal seizures: A type of epilepsy marked by tonic-clonic seizures in two stages: the tonic phase, during which the body becomes rigid, and the clonic phase, characterized by uncontrolled jerking. Tonic-clonic seizures may or may not be preceded by an aura and are often followed by headache, confusion, and sleep.

Graves' disease: An autoimmune disorder that causes the thyroid gland to produce too much thyroid hormone (hyperthyroidism).

Gynecomastia: Overdevelopment or enlargement of breast tissue in men or boys.

Half-life: The time required for half of a drug to undergo a process.

Hallucinations: Experiences of perceptions that are not real.

Hansen's disease: is a long-term infectious illness caused by Mycobacterium leprae. It impacts the skin, peripheral nerves, mucous membranes of the upper respiratory tract, and the eyes. Leprosy can develop at any age, from childhood to old age.

Hapten: A small molecule attached to a larger carrier, like a protein, can trigger antibody production and bind specifically to it.

Hay fever: A common allergic reaction to airborne substances such as pollen, dust mites, and pet dander.

Heberden nodes: Bony, knot-like growths that develop on the joints nearest the fingertips, indicating osteoarthritis in the hand.

Hemarthrosis: A condition of articular bleeding that occurs in the joint cavity.

Hematopoiesis: The process of creating blood cells and bone marrow cells in the body.

Hematopoietic growth factors: A group of hormone-like proteins that regulate the production and development of blood cells.

Hemifacial spasm: A neurological condition marked by involuntary, repetitive muscle contractions on one side of the face.

Hemochromatosis: A genetic disorder that causes the body to absorb and store too much iron. This can lead to damage in organs such as the liver, heart, and pancreas.

Hemoglobin A1c: A protein in red blood cells that indicates long-term blood sugar control.

Hemolytic anemia: A condition where red blood cells are destroyed faster than they can be produced.

Hemoptysis: Coughing or spitting up blood from the respiratory tract.

Hemosiderosis: The accumulation of excess iron in the body as a direct result of multiple blood transfusions.

Hepatocellular carcinoma: Primary liver tumor that is the most common type of liver cancer in adults

Hepatotoxicity: Liver damage resulting from exposure to harmful substances, including certain medications, herbs, chemicals, and toxins.

Herbal supplements: Products made from plants or their oils, roots, seeds, berries, or flowers. They have been used for hundreds of years. Herbal supplements are thought to have healing effects.

Herniated disc: Occurs when the soft, jelly-like center of an intervertebral disc (the cushion between vertebrae) pushes out through a tear in the outer layer of the disc.

High alert medications: Drugs that carry a higher risk of causing serious patient harm if used in error. Although errors may or may not be more common with these drugs, the consequences of a mistake are more severe for patients.

Hirsutism: A condition that causes excessive, dark, or coarse body hair to grow in a male-like pattern, such as on the face, chest, and back. It is the most common endocrine disorder among women.

HMG-CoA reductase inhibitors: Lipid-lowering medications used in the primary and secondary prevention of coronary heart disease.

Homogeneous system: A system of linear equations where all constant terms are zero, meaning the right-hand side of every equation is 0.

Human papillomaviruses: A group of viruses that can cause warts and some types of cancer.

Huntington's disease: A rare, inherited, and progressive brain disorder that causes uncontrolled movements, emotional problems, and loss of thinking ability.

Hydramnios: A condition that occurs when too much amniotic fluid builds up during pregnancy.

Hydrophobia: An intense fear of water, especially when linked to painful laryngeal spasms caused by rabies.

Hydrostatic pressure: The force that a fluid exerts in a confined space due to gravity. In physiology, hydrostatic pressure plays a crucial role in maintaining blood pressure and facilitating the movement of fluid in and out of blood vessels.

Hyperacusis: A condition marked by an unusually low tolerance to sounds, leading to discomfort or pain from everyday noises that most people can tolerate.

Hyperammonemia: A dangerous metabolic condition where there are abnormally high levels of ammonia in the blood, potentially leading to brain damage or death.

Hyperglycemia: An abnormality in the circulating blood, especially in patients with diabetes mellitus.

Hyperhidrosis: A condition marked by excessive sweating that disrupts daily activities and social interactions.

Hyperphenylalaninemia: A genetic disorder that impairs the body's ability to break down the amino acid phenylalanine.

Hypertensive crisis: A sudden, severe spike in blood pressure, 180/120 mm Hg or higher. It is a medical emergency and can cause a heart attack, stroke, or other serious health problems.

Hyperthyroidism: A condition where the thyroid gland produces excess thyroid hormone.

Hypertonic: Having a higher level of tone, tension, or tonicity.

Hyperuricemia: A condition with too much uric acid in the blood.

Hypoesthesia: A partial or total loss of sensation in the body. It can involve a tingling sensation and make a person less sensitive to touch, temperature, vibration, or pain.

Hypoglycemia: A condition where blood glucose levels drop below normal. It happens when the body lacks enough glucose to satisfy its energy needs.

Hypogonadism: A condition that occurs when the body's sex glands, or gonads, produce little to no sex hormones.

Hypomanic episodes: A period of elevated mood, increased energy, and activity that doesn't reach the intensity of a full manic episode. It is a key feature of bipolar disorder, a mental health condition marked by alternating episodes of mania and depression.

Hypoxia: A medical condition that occurs when the body's tissues do not have enough oxygen to maintain homeostasis.

Iatrogenic disease: An adverse condition or injury caused unintentionally by medical treatment or procedures.

Idiopathic: A disease with an unknown cause.

Idiosyncratic reaction: Adverse effects that cannot be explained by the known mechanisms of action of the offending agent occur in most patients at any dose and develop unpredictably in susceptible individuals only.

Ileocecal valve: A sphincter muscle at the junction of the ileum and colon.

Immune-mediated therapy: Treatments that utilize or modify the body's immune system to fight disease.

Immunotherapy: A cancer treatment that uses the patient's own immune system to fight cancer. It works by boosting the immune system's ability to recognize and destroy cancer cells.

Impetigo: A common and highly contagious skin infection caused by bacteria, usually Streptococcus pyogenes (group A strep) or *Staphylococcus aureus.*

Impotence: The inability to achieve or maintain an adequate erection for sexual intercourse. Also known as erectile dysfunction.

Infiltration: Local anesthesia is achieved by injecting the anesthetic solution directly into the area of terminal nerve endings.

Influx: The process of a substance entering the cell.

Infundibulum: A funnel-shaped structure in the brain that links the hypothalamus to the pituitary gland.

Ingredients: Any substance intended for incorporation into a finished drug product.

Inorganic: Chemical substances that typically lack carbon-hydrogen (C-H) bonds and often contain metals, nonmetals, and other elements.

Inotropy: Refers to the myocardium contractions and changes the force of those contractions.

Insomnia: A sleep disorder involving trouble falling asleep or staying asleep.

Insulin-like growth factors: A group of hormones that play crucial roles in growth, development, and metabolism.

Intermittent porphyria: A rare inherited disorder that affects the body's ability to produce heme, an essential component of red blood cells.

Interstitial cell-stimulating hormone: A hormone made by the anterior pituitary gland in both males and females. It is also called luteinizing hormone (LH).

Interstitial nephritis: A condition where there is inflammation within the kidneys.

Interstitial nephritis: An inflammation of the renal interstitium, the space between the kidney tubules.

Intrinsic factor: A protein the stomach produces that helps the body absorb vitamin B12.

Investigational drugs: Experimental medications that have not yet been approved by the Food and Drug Administration (FDA) for general use.

Ionization: The process by which neutral atoms or molecules gain or lose electrons, forming electrically charged ions.

Japanese encephalitis: A viral brain infection transmitted by infected mosquitoes.

Jaundice: Yellow staining of the skin and sclerae of the eyes by abnormally high blood levels of the bile pigment bilirubin.

Jock itch: Also known as tinea cruris, is a fungal infection that affects the groin area.

Juvenile idiopathic arthritis: A chronic autoimmune condition that causes joint inflammation, stiffness, and pain in children under 16.

Kaposi's sarcoma: A cancer that causes red or purple patches of abnormal tissue to grow under the skin. It is a rare cancer often linked to HIV infection.

Keratomalacia: An eye disorder characterized by drying and clouding of the cornea in people with undernutrition.

Keratoplasty: A surgical procedure that replaces a damaged or diseased cornea with a healthy donor cornea.

Koplik spots: Small, blue-white spots with red halos that appear on the inner lining of the cheeks (buccal mucosa), usually opposite the upper molars. They are considered a hallmark sign of measles and typically appear two to three days before the characteristic measles rash.

Lactiferous ducts: A branching network of tubes within the female breast that carry milk from the lobules of the mammary gland to the nipple, where it can be expressed to an infant.

Legend drug: A drug, chemical, or substance approved by the U.S. Food and Drug Administration (FDA) that can only be provided to patients with a prescription from a licensed healthcare practitioner, as required by federal or state law.

Leishmaniasis: A parasitic disease caused by the bite of an infected sandfly and the infection with Leishmania parasites.

Leukocytes: Also known as white blood cells (WBCs), are cells in the blood that are part of the immune system and help the body fight infections and diseases.

Leukotrienes: A group of inflammatory mediators produced by leukocytes and involved in allergic and inflammatory responses.

Leydig cells: Specialized cells in the testes responsible for producing testosterone and other androgens.

Libido: A person's overall sexual drive or desire for sexual activity. Biological, psychological, and social factors influence libido.

Ligand: A ligand is a substance that forms a complex with a biomolecule to serve a biological purpose.

Lipid profile: A blood test that can measure the amount of cholesterol and triglycerides in the blood.

Lipogenesis: A metabolic process that converts fatty acids and glycerol into fats, or triglycerides, for storage in adipose tissue.

Lipophilic: Having a strong attraction for fats, oils, or lipids.

Lipotropic extracts: The body promotes, encourages, and supports the removal of fat from the liver. It also helps ensure that excess fat is burned off for fuel and energy instead of being stored away.

Liver transplantation: A surgical procedure that replaces a diseased or damaged liver with a healthy one from a donor.

Loading dose: An initial larger dose of a drug given at the start of treatment, before switching to a lower maintenance dose.

Loop of Henle: Extended U-shaped tubule portion that conducts urine within each nephron.

Luteinizing hormone: A hormone that plays a vital role in the reproductive system and is produced in the pituitary gland.

Lyme disease: An inflammatory disease caused by a spirochete *(Borrelia burgdorferi)* that is transmitted by ticks and usually characterized by an initial rash followed by flu-like symptoms, including fever, joint pain, and headache. If left untreated, the disease can result in chronic arthritis and nerve and heart dysfunction.

Lymph nodes: Small, bean-shaped immune system clusters that filter lymphatic fluid and help the body fight infection and disease.

Lymphoma: A type of blood cancer affecting the lymphatic and immune systems.

Macrocytic anemia: A type of anemia in which red blood cells are larger than usual.

Macrophages: A type of white blood cell that surrounds and kills microorganisms, removes dead cells, and stimulates the action of other immune system cells.

Macula densa: A group of specialized cells in the nephron that monitors sodium levels in the fluid passing through the kidney's convoluted tubule.

Maintenance doses: The maintenance rate [in mg/h] of drug administration equals the rate of elimination at a steady state.

Malaria: A life-threatening disease caused by parasites that infect red blood cells and are transmitted by the bite of an infected mosquito. A single-celled parasite of the genus Plasmodium causes it.

Mammotropin: A hormone that encourages the growth of mammary glands and stimulates milk production.

Mania: A mental health disorder marked by a period of abnormally elevated or irritable mood, increased energy and activity, racing thoughts, and less need for sleep, often leading to risky or out-of-character actions such as extravagant spending, substance use, or reckless behavior.

Mast cells: Connective tissue cells that contain many granules rich in histamine and heparin.

Mastitis: An inflammation of the breast tissue, typically caused by a bacterial infection.

Mastocytosis: A rare disease that happens when your body makes too many mast cells.

Maximal efficacy: The greatest effect a drug can produce, regardless of the dose. It indicates the maximum therapeutic benefit achievable with a specific drug and is a key concept in pharmacology.

Measles: A highly contagious viral disease that causes a rash, fever, cough, runny nose, and other symptoms.

Medication administration record: A written record of medications prescribed for and administered to a patient.

Medication error: Mistakes in prescribing, dispensing, and giving medications.

Medication reconciliation: The process of identifying the most accurate list of all medications that the patient is taking.

Medullary paralysis: A serious, often deadly result of general anesthesia caused by high doses of anesthetic agents that suppress the breathing centers in the brain's medulla oblongata, leading to breathing cessation and potentially death.

MedWatch program: This initiative enables healthcare professionals and consumers to report serious issues they suspect are linked to the medical products they prescribe, dispense, or use.

Melanoma: The most invasive skin cancer with the highest risk of death. At the same time, it is a severe *skin cancer* and highly curable if diagnosed early.

Melatonin: A hormone that aids in controlling the body's sleep-wake cycle and circadian rhythm.

Melena: Black, tarry stools that indicate bleeding in the upper gastrointestinal tract, which includes the esophagus, stomach, and duodenum.

Menaquinone: A fat-soluble vitamin that plays an essential role in bone health, blood clotting, and heart function.

Meningococcemia: A bloodstream infection caused by Neisseria meningitidis.

Menstrual periods: The monthly hormonal changes from the start of one period to the beginning of the next.

Merozoites: The individual forms of a malaria parasite that invade red blood cells after being released from liver cells.

Metabolic syndrome X: A group of conditions that increase the risk of developing chronic diseases such as heart disease, stroke, and type 2 diabetes.

Metabolites: Substances produced by metabolism the chemical processes within an organism, or by the breakdown of substances by the body.

Metallic taste: A loss or alteration of the sense of smell can lead to a metallic taste in the mouth. It may result from oral hygiene issues, nutrient deficiencies, infections, or taking certain supplements.

Methyl alcohol: A colorless, flammable, and highly toxic organic compound. It is used in the production of antifreeze, pesticides, windshield wiper fluid, paint thinner, certain types of fuel, and other substances. It is also known as wood alcohol.

Microvascular angina: A type of chest pain caused by dysfunction in the small blood vessels (microvessels) in the heart.

Microvilli: Tiny, finger-like projections on the surface of many cells, especially epithelial cells in the small intestine, that increase surface area for absorbing nutrients and other substances.

Migraine headaches: Moderate to severe headaches, often on one side of the head.

Military tuberculosis: A serious and potentially deadly form of tuberculosis that happens when the bacteria Mycobacterium tuberculosis spreads through the bloodstream and affects multiple organs.

Milk thistle: A flowering herb related to the daisy and ragweed family. It is native to Mediterranean countries.

Mineralocorticoids: A class of steroid hormones that control the body's water and salt balance.

Minimum effective concentration: The lowest plasma drug level required to reach sufficient concentration at the receptors to produce the intended pharmacologic effect.

Mitosis: A process in which a cell replicates its chromosomes and then divides into two identical daughter cells.

Mitral valve: One of four valves in the heart that keep blood flowing in the right direction. Each valve has flaps called leaflets that open and close once during each heartbeat.

Mixed dyslipidemia: A condition where various blood lipid (fat) levels are elevated or low at the same time.

Modulation: The functional and morphological fluctuation in response to changing environmental conditions.

Molds: A type of fungi, a microscopic organism that grows in long filaments called hyphae.

Monoclonal antibodies: Laboratory-produced proteins that imitate natural antibodies generated by the immune system.

Morning sickness: A common symptom experienced by pregnant women, characterized by nausea and vomiting that often occurs in the morning.

Mpox: Previously known as monkeypox, it is a viral illness caused by the monkeypox virus, a species of the genus Orthopoxvirus.

Multiple myeloma: A type of cancer affecting plasma cells that is usually incurable.

Mumps: A contagious viral infection that causes painful swelling of the salivary glands, usually in the cheeks and jaw. It is spread by contact with saliva or respiratory droplets from an infected person, and it's as contagious as the flu.

Muscular dystrophy: A collection of genetic disorders that lead to gradual muscle weakness and loss of mobility.

Myasthenia gravis: A long-term autoimmune disease where antibodies attack the connection between nerves and muscles, leading to weakness in skeletal muscles.

Mycophages: An organism that feeds on fungi (a type of fungivore), or, more specifically, in some scientific contexts, a virus that infects and kills mycobacteria.

Mycoplasma: A genus of bacteria that lacks a cell wall. They are the smallest known free-living organisms and are found in a wide range of hosts, including humans, animals, and plants.

Mycoses: Fungal infections that impact the skin, hair, and nails. They are caused by various fungi, including dermatophytes, yeasts, and molds.

Mydriasis: Prolonged dilatation of the pupil of the eye.

Myelosuppression: A condition that occurs when the bone marrow's activity is reduced, resulting in a decrease in red blood cells, white blood cells, and platelets.

Myocarditis: An inflammation of the heart muscle (myocardium).

Myoclonic seizure: Characterized by brief, jerking muscle or muscle group spasms. They often occur with atonic seizures, which cause sudden muscle limpness.

Myofascial pain syndromes: A chronic condition that causes muscle pain and affects fascia, the thin connective tissue surrounding muscles (epimysium).

Myoglobin: An iron-rich protein in muscle that stores and transports oxygen to muscle cells.

Myxedema: Severe hypothyroidism that causes decreased mental status, hypothermia, and other symptoms due to slowed function in multiple organs; it is a medical emergency with a high risk of death.

Narrow therapeutic index: Also known as closed-angle or angle-closure glaucoma, the eye pressure increases rapidly compared to the more common open-angle glaucoma.

Narrow-angle glaucoma: Also known as closed-angle or angle-closure glaucoma and is characterized by a quick increase in eye pressure compared to the more common open-angle glaucoma.

Necrosis: The death of most or all of the cells in an organ or tissue due to disease, injury, or failure of the blood supply.

Necrotizing enterocolitis: The death of tissue in the intestine. It occurs most often in premature or sick babies.

Nephrons: The individual kidney units that filter blood and produce urine from waste products.

Nephrotoxicity: Rapid deterioration in kidney function due to the toxic effects of medications and chemicals.

Neurodegenerative: A group of chronic, progressive disorders marked by the gradual loss and death of neurons in the central nervous system, resulting in impaired functions, movement, cognition, or strength.

Neuroleptic malignant syndrome: A life-threatening neurological emergency related to antipsychotic use, characterized by a distinct clinical picture of mental status changes, rigidity, fever, and dysautonomia.

Neurons: The fundamental cells of the nervous system that transmit information using electrical and chemical signals between different parts of the brain, spinal cord, and body.

Neuropathic pain: A chronic condition caused by nerve damage, often described as burning, tingling, or "pins and needles" sensations.

Neurotoxicity: Damage to the brain or peripheral nervous system from exposure to toxic substances.

Neutropenia: A condition where there are abnormally low levels of neutrophils in the blood.

Neutropenic fever: A serious medical emergency marked by a fever and an extremely low level of neutrophils, a type of white blood cell that fights infection.

New Drug Application: The request a pharmaceutical company submits to the U.S. Food and Drug Administration (FDA) for permission to sell a new drug in the United States.

Nociceptive pain: A receptor in the form of a naked dendrite that responds to a pain stimulus.

Nocturnal seizures: A form of epilepsy that can cause abnormal movement or behavior during sleep.

Norepinephrine: A chemical messenger that functions as a hormone and a neurotransmitter in the body.

Nystagmus: Rapid involuntary movements of the eyes.

Obsessive-compulsive disorder: A mental condition that leads individuals to experience unwanted, recurring thoughts and feelings.

Ochronosis: A rare condition where tissues build up ochre or blue-black pigment.

Oliguria: A reduced amount of urine production.

Onychomycosis: A fungal infection of the nail that causes discoloration, thickening, and separation from the nail bed.

Oocyte: The female gamete cell.

Oophoritis: The inflammation of an ovary or ovaries, usually occurring as a complication of pelvic inflammatory disease.

Open-angle glaucoma: A chronic, irreversible, and progressive optic neuropathy that occurs when the drainage canals of the eye become clogged. This leads to a gradual increase in eye pressure and a loss of peripheral vision. It is the most common type of glaucoma.

Opioid analgesics: A powerful medication used to relieve moderate to severe pain. These drugs, sometimes called narcotics, work by influencing the central and peripheral nervous systems to block pain signals.

Opioid crisis: A devastating consequence of the opioid epidemic that includes increases in opioid misuse and related overdoses.

Optic neuritis: An inflammatory condition that demyelinates the optic nerve and degrades vision in one or both eyes.

Oral thrush: A fungal infection in the mouth caused by an overgrowth of the Candida albicans fungus.

Orange Book: A publicly available list maintained by the FDA that details all pharmaceutical drugs proven to be safe and effective.

Orchitis: The inflammation of the testicle unilaterally or bilaterally, usually caused by viruses and bacteria.

Orexin: A neuropeptide that regulates arousal, wakefulness, and appetite.

Organic solvents: Volatile, carbon-based liquids that dissolve other substances.

Orthostatic hypotension: A condition where blood pressure drops suddenly after rising from a sitting or lying position.

Osmoreceptors: Sensory neurons in the brain that detect changes in the body's osmotic pressure and help maintain fluid balance.

Osmosis: The movement of a solvent, usually water, through a selectively permeable membrane into an area of higher solute concentration, driven by the need to equalize concentrations on both sides of the membrane.

Osmotic pressure: The minimum pressure required to prevent the inward flow of pure solvent into a solution across a semipermeable membrane.

Osteoarthritis: A common joint disease that causes pain, stiffness, and swelling in the affected joints.

Osteocalcin: A protein found in bone that plays a role in bone formation, calcium homeostasis, and energy metabolism.

Osteoclasts: Large, multinucleated cells that are essential for bone resorption (breakdown).

Osteogenesis imperfecta: A hereditary disorder marked by fragile bones that easily fracture from minimal trauma.

Osteomalacia: Softening of the bones, typically through vitamin D or calcium deficiency.

Osteomyelitis: An infection of the bone that can cause inflammation, pain, and damage to the bone tissue. It can affect any bone in the body, but it is most common in the long bones (e.g., femur, tibia) and feet. In children, it can also affect the spine.

Osteophytes: Bony outgrowths that develop on the edges of bones, usually in joints. They are also called bone spurs.

Osteoporosis: A condition where bones become weak and more likely to break. It occurs when the amount and density of bone tissue decrease or when the bone's structure changes.

Ototoxicity: Ear poisoning from drugs or chemicals that can harm hearing and balance.

Ovulation: A phase in the menstrual cycle when the ovary releases an egg (ovum).

Oxidation: A process that occurs when atoms or groups of atoms lose electrons.

P-glycoprotein: A protein that acts as an ATP-powered efflux transporter, actively removing a variety of substances, including drugs and toxins, from cells.

Paget's disease: A long-term condition that affects bone renewal, causing bones to break down and rebuild faster than normal, which results in larger, weaker, and misshapen bones that are more prone to fracture.

Palpitations: The sensation of your heart pounding, racing, or fluttering in the chest.

Pancytopenia: A condition characterized by a lower-than-normal count of red blood cells, white blood cells, and platelets in the blood.

Panic disorder: An anxiety condition characterized by unexpected and repeated episodes of intense fear, accompanied by physical symptoms.

Parafollicular cells: A type of cell in the thyroid gland that produces calcitonin. Also called C-cells.

Paraganglioma: A rare tumor that develops from cells called paraganglia, which are located near blood vessels and nerves throughout the body.

Paralysis: The inability to move all or part of the body.

Paranoia: A mental disorder characterized by extreme fear and distrust of others. A paranoid individual may believe that people are trying to cause them harm.

Paresthesia: A sensation of numbness, burning, or tingling, often in the extremities like the hands and feet.

Parkinson's disease: A progressive nervous system disorder characterized by tremors, muscular rigidity, and slow, imprecise movements, mainly affecting middle-aged and older adults. It results from degeneration of the brain's basal ganglia and a deficiency of the neurotransmitter dopamine.

Paroxysmal: Happen suddenly with intense bursts or attacks.

Partial agonists: This agonist causes less-than-maximal activation even when it occupies the entire receptor population and, therefore, cannot produce the maximum response.

Partial prothrombin time: Measures how long it takes for a patient's blood to clot and is used to evaluate the function of clotting factors.

Partial seizure: A partial (focal) seizure happens when abnormal electrical activity impacts a small area of the brain. It is considered a simple partial seizure when the seizure does not affect awareness.

Passive transport: The movement of molecules across a cell membrane from an area of higher concentration to a lower concentration, without the cell using any energy.

Passive tubular reabsorption: The process by which the kidney tubules return useful substances and water from the glomerular filtrate back into the bloodstream without needing the body to use energy.

Pathogens: Microorganisms or biological agents that can cause disease in a host.

Pedal edema: The swelling and accumulation of fluid in the feet and ankles.

Pepsin: An essential stomach enzyme that breaks down proteins into smaller peptides and amino acids during digestion.

Peptic ulcer: A sore in the stomach lining or the duodenum.

Perception: The process by which humans and animals organize, interpret, and make sense of their sensory information and surroundings to understand the world and act accordingly.

Pericarditis: An inflammation of the pericardium, the sac-like membrane that surrounds and safeguards the heart.

Periodic idiopathic paralysis: Sudden, temporary episodes of muscle weakness or paralysis caused by genetic defects in ion channels of muscle cells, often involving potassium levels.

Peritonitis: A serious, life-threatening condition characterized by inflammation of the peritoneum, the tissue lining the abdominal wall and covering most abdominal organs.

Pernicious anemia: A condition in which the body cannot properly absorb vitamin B12, leading to a lower-than-normal number of red blood cells.

Peroxisomes: Specialized for carrying out oxidative reactions using molecular oxygen.

Pertussis: A very contagious bacterial disease that leads to intense coughing fits.

Petechiae: Tiny bleeding spots under the skin or in the mucous membranes.

Pharmacodynamics: The effects of drugs in the body and the mechanism of their action.

Pharmacoepidemiology: A field that studies how medications are used, their effects, and safety in populations. It combines principles from epidemiology, pharmacology, and statistics to explore the following

Pharmacogenomics: The branch of genetics concerned with how an individual's genetic attributes affect the likely response to therapeutic drugs.

Pharyngeal edema: Swelling of the pharynx, the area at the back of the throat.

Pheochromocytoma: A tumor of the chromaffin cells, most often located in the medulla of the adrenal gland.

Phlebitis: Inflammation of a vein, typically in the legs, caused by infection, injury, or irritation.

Phospholipids: A group of lipids that make up the main component of cell membranes and are essential for life.

Photosensitivity: Having a chemical, electrical, or other response to light.

Physiological agonist: A naturally occurring substance in the body, such as endorphins or acetylcholine, that binds to specific receptors to imitate a physiological response.

Pica: A disorder where individuals compulsively eat non-food items that lack nutritional value.

Pituitary gland: The endocrine gland at the base of the brain that secretes hormones to control metabolism, growth, sexual maturation, reproduction, and blood pressure.

Placebo: A harmless pill, medicine, or procedure prescribed more for the psychological benefit of the patient than for any physiological effect.

Placenta previa: A condition in which the placenta implants low in the uterus, partially or fully covering the cervix.

plasma cells: White blood cells (B lymphocytes) that produce antibodies to fight infection

Plasma membrane: A selectively permeable barrier that surrounds all cells, separating the cell's interior from the external environment.

Plasma osmolarity: The number of particles of solute per liter of solution, whereas the term osmolality refers to the number of solute particles per kilogram of solvent.

Plasma proteins: A group of proteins found in blood plasma, the liquid part of blood. They are essential for maintaining blood volume, transporting nutrients and hormones, and supporting the immune system.

Plasmodium falciparum: A life-threatening disease caused by the parasite *P. falciparum*, transmitted by mosquitoes. *P. falciparum* is the most dangerous, often leading to severe complications and death, while the other three usually cause less serious infections but can result in dormant liver stages and recurrent symptoms.

Platelet aggregation: The process where platelets stick together and form a clot to help stop bleeding.

Pneumocystis pneumonia: An opportunistic lung infection caused by the fungus-like organism *Pneumocystis jirovecii*.

Pneumonitis: An inflammation of the lungs.

Polar molecules: Electrons are distributed unevenly, forming partial positive and negative ends, similar to a small magnet.

Poliomyelitis: An uncommon inflammatory disease that causes muscle weakness affecting both sides of the body.

Polycystic kidney disease: A genetic disorder that causes fluid-filled sacs, called cysts, to grow in the kidneys. These cysts can begin to develop at any age, from before birth to adulthood. It is a chronic condition that can result in kidney failure.

Polydipsia: Excessive thirst or excess drinking.

Polymorph: An organism with more than one form or type resulting from discontinuous variation.

Polymorphism: The ability of something to exist or function in different ways or forms.

Polypharmacy: The simultaneous use of multiple drugs by a single patient for one or more conditions.

Polyuria: The excessive production and excretion of urine.

Porphyria: A rare hereditary disorder where the blood pigment hemoglobin is abnormally metabolized. Porphyrins are excreted in the urine, turning it dark; other symptoms include mental disturbances and extreme skin sensitivity to light.

Porphyrin: Ring-shaped organic molecules found in many living organisms, forming essential components such as the iron-containing heme in hemoglobin and the magnesium-containing chlorophyll in plants.

Post-traumatic stress disorder: A mental health condition that can develop after experiencing or witnessing a traumatic event.

Potentiation: When a mixture of two or more drugs produces a response greater than expected, exceeding the sum of their individual effects.

Preeclampsia: A pregnancy disorder marked by high blood pressure and often a significant amount of protein in the urine; it starts after 20 weeks of pregnancy.

Premenstrual syndrome: Any of a complex of symptoms (including emotional tension and fluid retention) experienced by some women in the days immediately before menstruation.

Primary bronchi: The first and most extensive branches of the respiratory tree, and are the two primary tubes that branch off from the trachea and supply air to each lung.

Primary hypertension: Unknown causes of hypertension.

Prinzmetal's angina: Characterized by chest pain at rest with transient ischemic electrocardiographic changes in the ST segment, with a prompt response to nitrates. Pt is also called vasospastic or variant angina.

Prodrome: An early symptom indicating the onset of a disease or illness.

Prodrugs: A compound with little or no pharmacological activity that metabolizes inside the body and converts into a pharmacologically active drug compound.

Prolactin-inhibiting hormone: A substance produced by the hypothalamus that prevents the release of prolactin from the anterior pituitary gland.

Proliferation: A rapid multiplication of parts or the increase in the number of something.

Prostaglandin: A group of lipids made at tissue damage or infection sites involved in dealing with injury and illness. They control inflammation, blood flow, clot formation, and labor induction.

Proteonomics: The study of proteomes, which are proteins produced in an organism, system, or biological context.

Prothrombin: A blood test that measures how long it takes for a blood sample to clot.

Prototype drug: The main example of a drug class used as a reference to understand the similar chemical structure, mechanism of action, and effects of other drugs in that class.

Protozoa: A single-celled eukaryotic organism that feeds heterotrophically, meaning it consumes solid particles like bacteria rather than photosynthesizing like plants.

Proximal convoluted tubule: A renal tubule segment in the kidney that reabsorbs water and solutes from the glomerular filtrate and returns them to the bloodstream.

Pruritus: The medical term for itching.

Pseudomembranous colitis: A severe colon inflammation caused by an overgrowth of the bacterium Clostridioides difficile.

Psoriasis: The medical term for itching.

Psoriatic arthritis: An inflammatory autoimmune condition that impacts both the skin (psoriasis) and the joints.

Pulmonary arteries: Blood vessels that carry deoxygenated blood from the right side of the heart to the lungs.

Pulmonary edema: A serious condition where excess fluid accumulates in the lungs, making it hard to breathe.

Pulmonary embolectomy: A significant surgical procedure to remove a pulmonary embolism.

Pulmonary embolism: A blockage in a lung artery caused by a blood clot that has traveled from elsewhere in the body. It is a life-threatening medical emergency that requires immediate treatment.

Pulmonary hypertension: A serious condition that happens when blood pressure in the lungs is abnormally high.

Pulmonary trunk: A short, thick artery that carries deoxygenated blood from the heart to the lungs.

Pulmonary valve: A semilunar valve in the heart that controls blood flow from the right ventricle to the pulmonary artery.

Pupillary miosis: The abnormal constriction of the eye's pupil, causing it to remain small even when lighting conditions change.

Purkinje fibers: Specialized nerve cells that transmit electrical signals to the heart's ventricles, causing them to contract. This contraction enables blood to flow from the heart to the rest of the body.

Pyrexia: The elevation of core body temperature above normal.

QT syndromes: A heart rhythm disorder characterized by a prolonged QT interval on an electrocardiogram (ECG).

Quadriplegia: A type of paralysis marked by partial or total loss of function in all four limbs and the torso.

Rabies: A lethal viral disease that targets the central nervous system and spreads through the bite or scratch of an infected animal.

Raynaud's phenomenon: A condition marked by spasms of the arteries in the extremities, especially the fingers (Raynaud's phenomenon). It is usually triggered by constant cold or vibration and results in pallor, pain, numbness, and in severe cases, gangrene.

Receptor: A cell or group of cells that detects stimuli.

Red Man Syndrome: An infusion-related reaction to vancomycin. It typically involves pruritus and an erythematous rash affecting the face, neck, and upper torso.

Regional anesthesia: A type of pain relief used during surgery that numbs a large area of the body, such as from the waist down.

Renal artery stenosis: A condition where the arteries that supply blood to the kidneys become narrowed.

Renal corpuscle: A filtration unit in the kidney that filters blood and forms urine.

Repolarization: The process by which a cell's membrane potential returns to its normal resting state after a nerve impulse or action potential.

Resistant hypertension: High blood pressure that stays above target despite the use of three different antihypertensive medications taken at the maximum tolerated doses.

Restless legs syndrome: A chronic neurological disorder that causes an uncontrollable urge to move the legs.

Reticular cells: A type of fibroblast that provides structural support and is involved in the immune response.

Retinal: A vitamin A derivative used in skincare to boost skin cell turnover, increase collagen, and reduce wrinkles and breakouts. It is also the molecule in the eye's retina responsible for detecting light and converting it into electrical signals for vision.

Retinal detachment: A condition in which the retina separates from the underlying tissue.

Retinoids: A class of chemical compounds that are natural derivatives of vitamin A or are chemically related to it.

Retinopathy: A disease of the retina. There are several types of retinopathy, but all affect the small blood vessels in the retina.

Reye's syndrome: A rare but serious condition that affects the brain and liver, usually in children and teenagers. It is often triggered by a viral infection, such as influenza or chickenpox, and is exacerbated by the use of aspirin.

Rhabdomyolysis: A serious condition caused by muscle injury, in which the release of muscle fiber contents into the bloodstream can lead to kidney failure.

Rheumatic fever: An inflammatory disease that develops as a complication of a prior Group A streptococcal infection, such as strep throat or scarlet fever.

Rheumatic heart disease: A condition where the heart valves are permanently damaged by rheumatic fever.

Rheumatoid arthritis: A long-lasting autoimmune disorder that leads to joint pain, swelling, stiffness, and decreased function.

Rhinorrhea: A condition involving excess mucus in the nasal cavity.

Rickets: A bone disease that occurs in children when they do not get enough vitamin D, calcium, or phosphorus.

Rickettsiae: A diverse group of obligately intracellular Gram-negative bacteria found in ticks, lice, fleas, mites, chiggers, and mammals. They are characterized by soft, weak bones that can cause deformities, such as bowed legs, knock-knees, and an enlarged skull. Rickets is most common in children in developing countries where sunlight exposure is limited or dietary deficiencies are present.

Ringworm: A fungal infection that affects the skin, hair, or nails. Also known as tinea.

Root cause analysis: A structured process for identifying the underlying reasons behind a problem, rather than just fixing its symptoms, to develop effective, long-term solutions and prevent it from happening again.

Rubella: Also called German measles; it is a highly contagious viral infection that is typically mild.

Rubeola: A highly contagious viral infection that causes a distinctive rash. Also known as measles.

Sandflies: A common name for any species or genus of flying, biting, blood-sucking dipteran (fly) found in sandy areas.

Sarcoplasmic reticulum: A specialized form of the endoplasmic reticulum found in muscle cells that functions as a storage and release compartment for calcium ions.

Saw palmetto: A small palm with fan-shaped leaves with sharply toothed stalks, native to the southeastern US.

Schizoaffective disorders: A mental health condition that combines symptoms of schizophrenia and a mood disorder, such as depression or bipolar disorder.

Scoliosis: A medical condition characterized by a sideways curvature of the spine.

Scotomas: A visual field abnormality or a blind spot.

Secondary bronchi: The initial division of the main bronchi within the lungs.

Secondary hypertension: High blood pressure that's caused by another medical condition. It can be caused by conditions that affect the kidneys, arteries, heart, or endocrine system.

Serotonin syndrome: A serious drug reaction caused by medications that increase serotonin levels in the body. Serotonin is a chemical naturally produced by the body. It is essential for the proper functioning of nerve cells and the brain.

Serum sickness: An immune complex-mediated hypersensitivity reaction that typically presents with fever, rash, polyarthritis, or polyarthralgia.

Shingles: Also called herpes zoster or zona. Any person who has had chickenpox can develop shingles. This condition erupts along the course of the affected nerve, producing lesions anywhere on the body, and may cause severe nerve pain.

Sickle cell disease: An inherited condition where red blood cells have an abnormal crescent shape, clog small blood vessels, and live shorter lives than normal red blood cells. It is caused by a mutation in one of the genes for hemoglobin. It is most common among people of West and Central African descent. Also called sickle cell anemia.

Side effects: Any effect of a drug, chemical, or other medicine that is in addition to its intended effect, especially an effect that is harmful or unpleasant.

Sideroblastic anemia: A type of anemia that results from abnormal utilization of iron during erythropoiesis.

Sinoatrial node: A small mass of tissue made up of nerve fibers embedded in the musculature of the higher right atrium that originates the impulses stimulating the heartbeat.

Sinus dysrhythmia: A condition where the heart rate varies, typically by more than 10 beats per minute.

Skin abrasions: An abrasion is a superficial rub or wearing off of the skin, usually caused by a scrape or a brush burn.

Sleep apnea: A sleep disorder where people stop breathing or breathe very shallowly during sleep.

Slurred speech: Also known as dysarthria. Speech impairment can be caused by conditions such as brain injury, stroke, multiple sclerosis, Parkinson's disease, and cerebral palsy. It can also result from substances like alcohol, opioids, barbiturates, and benzodiazepines.

Smallpox: A contagious and deadly viral disease, caused by the variola virus, that causes a characteristic rash leading to scabs and scarring.

Smooth endoplasmic reticulum: An organelle in eukaryotic cells that carries out multiple functions, including the production of lipids, steroids, and carbohydrates, detoxification of drugs and metabolic wastes, and regulation of calcium ions.

Social phobia: An anxiety disorder that causes people to experience persistent fear and anxiety in social situations.

Solubility: The ability of a substance to dissolve in a solvent to form a uniform solution.

Solute: A substance that is dissolved in another substance (a solvent), forming a solution.

Somatotropin: A hormone synthesized in and secreted by the anterior lobe of the pituitary gland that promotes the growth of the long bones in the limbs and increases the synthesis of protein essential for growth.

Sphenoid bone: One of the seven bones that articulate to form the orbit. Its shape resembles that of a butterfly or bat with its wings extended.

Sphingolipids: A specialized group of lipids essential to the composition of the plasma membrane of many cell types; however, they are primarily localized within the nervous system.

Spinal anesthesia: A type of regional anesthesia that blocks pain signals in the lower body by injecting local anesthetic medication into the spinal fluid.

Spores: A single-celled reproductive unit produced by certain plants, fungi, and bacteria that can develop into a new organism without fusing with another cell. It serves as a means of asexual reproduction and often resists harsh conditions thanks to its thick protective wall.

Sporozoites: The infective, motile stage of the malaria parasite (genus *Plasmodium*) that is transmitted from an infected mosquito to a vertebrate host via a bite.

St. John's wort: A plant with yellow flowers used in traditional European medicine as far back as the ancient Greeks.

Stable angina: Chest pain of cardiac origin that has not changed in character, frequency, intensity, or duration for 60 days.

Statins: A class of medications that lower cholesterol and reduce the risk of cardiovascular disease.

Status asthmaticus: A severe, life-threatening exacerbation of asthma that does not respond to conventional treatment.

Status epilepticus: A seizure that lasts longer than 5 minutes, or having more than one seizure within 5 minutes, without returning to an average level of consciousness between episodes, is called status epilepticus. This is a medical emergency that may lead to permanent brain damage or death.

Sterols: Unsaturated solid alcohols in the steroid group, including cholesterol and ergosterol. They are found in animals, plants, and fungi, especially as components of cell membranes.

Stevens-Johnson syndrome: A rare, severe skin reaction that occurs when the immune system overreacts to a trigger, such as a medication or infection.

Stillbirth: The death or loss of a baby before or during delivery.

Strabismus: A condition where the eyes do not align properly and point in different directions.

Striatum: The part of the basal ganglia that controls motor and action planning, decision-making, motivation, reinforcement, and reward perception.

Stroke volume: The amount of blood ejected from the ventricle with each contraction; in adults, it is typically about 70 mL.

Stupor: A state of near-unconsciousness or insensibility.

Subarachnoid hemorrhage: A life-threatening medical emergency that occurs when a blood vessel ruptures and bleeds into the space between the brain and the arachnoid membrane.

Subarachnoid: The layer located between the arachnoid and the pia mater.

Subdural hematoma: A blood collection that forms between the dura mater and the brain.

Substance abuse: The harmful or hazardous use of psychoactive substances, including alcohol and illicit drugs. One of the critical impacts of illegal drug use on society is the negative health consequences experienced by its members.

Substance misuse: A serious public health challenge. It includes the use of illegal drugs and the inappropriate use of legal substances, such as alcohol.

Substance P: A small protein-like molecule that functions as a neurotransmitter and neuromodulator.

Substrate: The surface or material where an organism lives and grows, such as soil or rock, or the specific reactant molecule that an enzyme binds to at its active site to catalyze a chemical reaction.

Sudden infant death syndrome: The term used to describe the sudden death of a baby younger than 1 year of age that does not have a known cause, even after a full investigation.

Superagonist: A molecule or compound that binds to a receptor and produces a biological response that is greater than the maximum response that can be achieved by the natural (endogenous) agonist for that receptor.

Sympathetic ganglia: A series of ganglia that form the sympathetic chains along each side of the spinal cord.

Sympathomimetic: A drug or substance that mimics the effects of the sympathetic nervous system, often leading to increased heart rate, higher blood pressure, and bronchodilation.

Synergism: The interaction or cooperation of two or more organizations, substances, or other agents to create a combined effect greater than the sum of their individual effects.

Synovial joints: Freely movable joints that allow for a wide range of motion between bones. Also known as diarthroses.

Systemic lupus erythematosus: A disease where a person's immune system attacks and injures the body's organs and tissues. Almost every system of the body can be affected by SLE.

Tachycardia: A heart rate of more than 100 beats per minute at rest.

Tachyphylaxis: A reduction in response to a drug caused by prior exposure to the agent, which can sometimes be offset by increasing the dose.

Tachypnea: Rapid, shallow breathing, quicker than normal for a person's age and condition.

Tardive dyskinesia: A neurological disorder characterized by involuntary movements of the face and jaw.

Tarsometatarsal joints: A set of five joints located in the midfoot, connecting the distal row of tarsal bones (cuboid, cuneiforms) to the proximal row of metatarsal bones.

Teratogen: Any agent that causes an abnormality following fetal exposure during pregnancy. They are usually discovered after an increased prevalence of a particular birth defect.

Terpenes: The main components of essential oils that give cannabis its aroma qualities.

Tertiary bronchi: The third level of branching in the bronchial tree, located at the edge of the lungs.

Tetanus: A bacterial infection that causes muscle spasms and lockjaw, also known as trismus.

Therapeutic drug monitoring : A clinical practice of measuring the concentration of a specific medication in a patient's bloodstream at scheduled intervals. This process is used to adjust the medication dose and optimize patient care by ensuring that drug levels are effective while minimizing the risk of dangerous toxicity.

Therapeutic index: A ratio that compares the blood concentration at which a drug becomes toxic and the concentration at which the drug is effective. The larger the therapeutic index, the safer the drug is.

Therapeutic window: The dosage range or concentration of a drug in the body that offers safe and effective treatment.

Threshold: The lowest level of a stimulus or condition needed to trigger a specific physiological or psychological effect.

Thrombocytopenia: A condition characterized by an abnormally low number of platelets.

Thrombocytosis: A condition characterized by an abnormally high number of platelets in the blood.

Thrombolytics: Drugs used to break down blood clots.

Thrombopoietin: A protein hormone that plays a crucial role in regulating platelet production.

Thrombosis: The formation of a blood clot in a blood vessel.

Thrombus: A blood clot that forms inside one of the veins or arteries.

Thrush: A fungal infection in the mouth caused by an overgrowth of *Candida albicans* yeast.

Thymus: A gland in the chest that plays a vital role in the immune system by producing and maturing immune cells, particularly T lymphocytes, and secreting thymosin.

Thyroidectomy: A surgical procedure to remove part or all of the thyroid gland.

Thyrotoxicosis: A condition resulting from excessive concentrations of thyroid hormones in the body, such as hyperthyroidism.

Thyrotropin: A hormone secreted by the anterior lobe of the pituitary gland that specifically targets and stimulates the thyroid gland.

Time-response relationships: A coordinate graph showing how the size of a dose (stimulus) relates to the response of a biological system.

Tinea interdigital: A fungal infection that affects the skin between the toes (interdigital spaces).

Tinea rubrum: A common superficial fungal infection that invades skin, hair, and nails, leading to symptoms like scaling, redness, and flaking, often in the feet (athlete's foot), groin (jock itch), and nails.

Tinnitus: Ringing or buzzing in the ears.

Tocopherols: A group of organic compounds that act as antioxidants and are essential for human health.

Tocotrienols: A group of compounds belonging to the vitamin E family.

Tonic-clonic seizure: The classic type of epileptic seizure consists of two phases to a tonic-clonic seizure—the tonic phase and the clonic phase.

Tophus: A deposit of monosodium urate crystals in the blood of people with longstanding high levels of uric acid in the blood.

Toxic concentration: The level of a substance that can cause harmful effects on a living organism. The specific concentration that is considered toxic depends on the substance itself, the route and duration of exposure, and the sensitivity of the organism. A central principle of toxicology is that "the dose makes the poison," meaning that any substance can be harmful in sufficiently high concentrations.

Toxicology: The scientific study of poisons and their effects on living organisms.

Toxoplasmosis: A contagious disease caused by the single-celled protozoan parasite *Toxoplasma gondii*. Although most people do not show symptoms, the disease can be serious and even deadly in people with weakened immune systems.

Trancelike state: A half-conscious state, seemingly between sleeping and waking.

Transduction: The process of transferring genetic material from one microorganism to another through a viral agent.

Traveler's diarrhea: A gastrointestinal infection that affects many travelers to regions with poor sanitation and limited access to safe drinking water. It's caused by eating food or drinking water contaminated with bacteria, viruses, or parasites.

Trichomoniasis: A sexually transmitted infection caused by a parasite called Trichomonas vaginalis.

Triglycerides: A type of fat (lipid) found in the bloodstream and stored in fat cells. They consist of three fatty acids connected to a glycerol molecule.

Truncal obesity: Abdominal central obesity, in which there is excessive visceral fat around the stomach and abdomen.

Tuberculosis: A contagious bacterial infection that spreads through the air and typically affects the lungs. The bacteria Mycobacterium tuberculosis causes it.

Tubular reabsorption: The kidneys remove water and solutes from the filtrate and return them to the bloodstream. It is the second major step in the formation of urine.

Tubular secretion: A process in the kidneys that moves waste products and other substances from the blood into the renal tubules.

Tularemia: A rare but potentially fatal bacterial infection caused by *Francisella tularensis*. It is primarily transmitted through contact with infected animals, their carcasses, or their environment

Turgor: Refers to skin elasticity, which is the skin's ability to return to its normal shape after being pulled or pinched. A decrease in skin turgor, often indicated by the skin remaining tented or returning slowly to its normal state, is a late sign of dehydration.

Tyramine: A monoamine, meaning it has a chemical structure similar to neurotransmitters like dopamine and norepinephrine.

Tyrosine kinase: A crucial role in cell signaling and regulation. They are proteins that catalyze the transfer of a phosphate group from adenosine triphosphate (ATP) to the amino acid tyrosine on specific protein substrates.

Tyrosine: An amino acid that helps the body produce neurotransmitters, hormones, and melanin.

Ulcerative colitis: A chronic inflammatory bowel disease that causes inflammation in the colon and rectum.

Unstable angina: Chest pain of cardiac origin that is variable, usually increasing in frequency and intensity, and with irregular timing.

Urethritis: The inflammation of the urethra, the tube that carries urine out of the body.

Urticaria: A skin condition characterized by raised, itchy, and red welts (wheals).

Valerian: A plant that typically bears clusters of small pink or white flowers. Native to Eurasia, several species have been introduced to North America.

Varicella congenital syndrome: A rare birth defect that occurs when a pregnant woman contracts varicella (chickenpox) during the early stages of pregnancy.

Varicella: A highly contagious disease caused by the varicella-zoster virus.

Varicella: A highly contagious disease marked by an itchy, blister-like rash caused by the varicella-zoster virus (a type of herpesvirus). The virus is easily spread from person to person through contact with mucus, saliva, or fluid from the blisters of an infected person or through droplets released when an infected person coughs or sneezes. Also called chickenpox.

Vasoconstrictor: A drug or nerve that causes blood vessels to narrow or constrict.

Vasodilators: Medications that open or dilate blood vessels, allowing blood to flow more easily.

Vasopressin: A nonapeptide synthesized in the hypothalamus. Science has recognized its essential roles in controlling the body's osmotic balance, regulating blood pressure, maintaining sodium homeostasis, and kidney function. Also called antidiuretic hormone (ADH) or arginine vasopressin (AVP).

Ventricular fibrillation: A life-threatening arrhythmia occurs when the ventricles twitch instead of beating normally.

Ventricular flutter: A serious heart arrhythmia that starts in the ventricles, marked by a very fast heart rate (150-300 bpm) and an abnormal electrical pattern on an ECG, often called a "sine wave".

Vertigo: A symptom that causes a false sensation of motion or spinning, often described as dizziness.

Vestibular: A network of structures and neural pathways that helps maintain the balance and sense of orientation.

Vestibulitis: A condition that causes inflammation and pain in the vestibule, the area around the vaginal opening and entrance to the vagina.

Villi: Small, finger-like projections that expand the surface area of a membrane and assist in nutrient absorption.

Virion: A fully infectious virus particle made up of a nucleic acid genome encased in a capsid.

Vivid dreams: Realistic, intense dream experiences, often occurring during REM sleep.

Von Willebrand disease: A bleeding disorder that happens when blood doesn't clot properly. It is the most common inherited bleeding disorder worldwide and affects people of all races and genders.

Vulvovaginal candidiasis: An infection of the vagina caused by *Candida* species, usually *C. albicans*. Symptoms typically include a thick, white vaginal discharge and vulvovaginal itching that can range from moderate to severe.

Walking pneumonia: A mild lung infection. Causes may include bacteria, viruses, or mold.

Warts: Benign skin growths caused by the human papillomavirus.

Wheezing: A high-pitched whistling sound that occurs when breathing in or out and can indicate breathing difficulty.

Withdrawal: Discontinuation of the use of an addictive substance.

Wolf-Parkinson-White syndrome: A rare heart condition that causes an arrhythmia. It occurs between the atria and ventricles.

Xerostomia: Dry mouth caused by insufficient production of saliva.

Yeasts: Single-celled, eukaryotic fungi that reproduce by budding and are commonly used in baking and brewing because they ferment sugars into alcohol and carbon dioxide.

Yellow fever: A viral illness transmitted to humans by infected mosquito bites.

Yohimbe: An indole alkaloid derived from the bark of the Central African yohimbe tree that is commonly used as a treatment for erectile dysfunction. Yohimbine use has been associated with occasional severe adverse events, but it has not been linked to serum enzyme elevations or clinically apparent acute liver injury.

Index

Note: Bold page numbers refer to tables; italic page numbers refer to figures.

For Product Safety Concerns and Information please contact our EU
representative GPSR@taylorandfrancis.com
Taylor & Francis Verlag GmbH, Kaufingerstraße 24, 80331 München, Germany

www.ingramcontent.com/pod-product-compliance
Lightning Source LLC
Chambersburg PA
CBHW080912220326
41598CB00034B/5547